ALFRED RUSSEL WALLACE

*

An Anthology of his Shorter Writings

ALFRED RUSSEL WALLACE

ALFRED RUSSEL WALLACE

An Anthology of His Shorter Writings

Edited by
CHARLES H. SMITH

Oxford New York Tokyo
OXFORD UNIVERSITY PRESS
1991

Oxford University Press, Walton Street, Oxford OX2 6DP
Oxford New York Toronto
Delhi Bombay Calcutta Madras Karachi
Petaling Jaya Singapore Hong Kong Tokyo
Nairobi Dar es Salaam Cape Town
Melbourne Auckland
and associated companies in
Berlin Ibadan

Oxford is a trade mark of Oxford University Press

Published in the United States
by Oxford University Press, New York

British Library Cataloguing in Publication Data
Wallace, Alfred Russel 1823–1913
Alfred Russel Wallace: an anthology of his shorter writings.
1. Science. Social criticism. Wallace, Alfred Russel – Biographies.
I. Title II. Smith, Charles H.
575.0092
ISBN 0–19–857725–7

Library of Congress Cataloguing in Publication Data
Wallace, Alfred Russel, 1823–1913.
[Selections. 1991]
Alfred Russel Wallace: an anthology of his shorter writings/
edited by Charles H. Smith.
"Bibliography of the published works of Alfred Russel Wallace"
Includes bibliographical references and index.
1. Science. I. Smith, Charles H. II. Title.
Q158.5.W3525 1991 500–dc20 90–46745
ISBN 0–19–857725–7

Text compiled by the author on disk
Typeset by Pentacor PLC

Printed and bound in Great Britain
by Biddles Ltd, Guildford and King's Lynn

Preface

THROUGH the present anthology I am attempting to accomplish two things. The major portion of the work is given over to a 'guided tour' of Wallace's ideas, primarily as expressed in those of his writings that appeared as part of the periodical literature. The selections drawn therefrom and reprinted here total about one hundred in number, but it should be understood at the outset that I make no claims of having provided an exhaustive review. In addition to his studies on evolutionary theory and biogeography Wallace wrote extensively on a number of other subjects, most notably physical geography, geology, the theory of land ownership, the occult, social evolution, vaccination, systematics, geographic education, astronomy, social and economic problems, and anthropology. His writings, in fact, appeared in nearly 200 *different* serial publications; it is an additionally revealing fact that in no five-year period of the last fifty–five years of his life did his pen yield fewer than forty works that reached print.

I have chosen for survey purposes to bypass his monographs here for several reasons, the most important being the detail inherent in their own considerable bulk (a total of some 7500 first-edition pages) and my belief that the many facets of Wallace's character and intellect are best exposed through study of his shorter works, written to meet the circumstances of a variety of publication contexts. He was, much to his advantage, a first-class writer, equally adept in the roles of essayist, analyst, reviewer, critic, and popularizer. Many of Wallace's books, moreover, are still in print, and are thus easier to obtain than the majority of his shorter writings, most of which appeared in periodicals now held by a few large libraries only.

My 'guided tour' is intended to provide the reader with a framework through which Wallace's ideas can be understood and placed in broader historical perspective. Little attempt is made, however, to follow his views through to the current state of debate on related subjects; there simply is not the space. This is *not*, then, intended in the main as an analytical study. I have tried none the less to organize the materials presented so as to facilitate appreciation of Wallace's unique world–view. As both the meaning and importance of his work are still under debate, I have provided analysis where I feel this is useful to the purpose of survey. The reader is advised, however, that in some instances my interpretations differ sharply from conclusions that have been drawn by other investigators.

In addition to the 'tour', I also give a complete (but see next paragraph) and subject-referenced Bibliography of Wallace's published writings, some 753 items in length. This is designed, in large part, as a point of departure for scholars. Note in this connection that the selections I have reprinted should *not*, purely by themselves, be used as the basis for scholarly analysis. Though some works are reproduced *in toto*, most are to one degree or another edited. I regret the necessity of this compromise, but the spatial constraints imposed by the combination of Wallace's range of interests and publication considerations permit no other resolution. Edits are fully indicated in the text through use of the '. . .' device. Brackets are employed to set aside the occasional editorial interjection. A good deal of what I have edited out consists of auxiliary factual information; I have made every effort to reproduce argumentation in full. The selections have been fairly thoroughly proof-read for transcription errors, but no effort has been made to 'clean up' apparent mistakes in the original printings. In many instances, in fact, it is difficult to decide whether a problematic passage contains a printer's error, or is merely the product of Wallace's idiosyncratic way of conceptualizing things.

Items in the Bibliography are referred to by number within the text and in subject indices presented at the end of each Section. The more significant of the titles are also listed in the Index. Thus, if one knows the title of a particular work but nothing else about it, the Index should be consulted (for references to either, or both, of the Selections and Bibliography divisions of the work). If the approximate date is known, the Bibliography, which is arranged chronologically, should be examined (see p. 473 for further instructions). For subject references, see the listings found at the end of each Section in the Selections division (or, for names of terms related to the Wallace selections actually reproduced here, the Index). On pp. 539–540 a bibliography of writings *on* Wallace is presented; this is by no means complete, but the reader interested in further leads should consult these works.

The reader should consider for a moment the very notion of a 'complete' bibliography of Wallace's writings. I have put considerable effort into trying to round up everything he wrote that was published, yet have not the slightest doubt that dozens of additional 'letters to the Editor' of his that were printed still lie buried in obscure places. It is known, moreover, that Wallace not infrequently wrote anonymously. Several of his most significant essays, in fact, were unsigned book reviews—a matter for some concern, as his voracious reading habits were probably supported in good part by reviews of courtesy copies. It is anyone's guess as to how many of these efforts might have reached print uncredited. I feel, however, that further survey efforts are unmerited (except, possibly, for attention to quite specific issues/incidents). The Bibliography provided here almost certainly covers every subject of *general* importance on which Wallace wrote (or on which he was directly quoted).

I am indebted to a large number of individuals, institutions, and organizations whose co-operation substantially contributed to the completion of this work. At the group level, I should especially like to mention The University of Illinois, The University of Melbourne Library, The British

Library, The Harry Price Library at The University of London, The Bodleian Library of Oxford University, The State Library of Victoria, Australia, The Australian National and Australian National University Libraries (Canberra), The American Antiquarian Society Library, The Societies (British and American) for Psychical Research, Yale University Library, The British Newspaper Library, Harvard University Library, Wesleyan University Library (Middletown, Connecticut), Trinity College Library (Hartford, Connecticut), The Northwestern Connecticut Community College Computer Center and Library (Winsted, Connecticut), and Oxford University Press.

I reserve special thanks for the interlibrary loan group at The University of Illinois, one of the most consistently dependable organizations I have ever had the pleasure of working with.

Further thanks to M. Raymond Lee, Michele Malinchak, D. J. Mabberley, Anne M. Sing, Russell E. Smith, the late T. W. Freeman, George N. Smith, Richard Burkhardt, Max Gubin, Alan Wesencraft, Hazel M. Smith, A. J. R. and R. R. Wallace (Wallace's grandchildren), several anonymous reviewers, and a 'cast of thousands' army of reference librarians who examined scores of publications for me.

New Hartford, Connecticut C.H.S.
February 23, 1991

Contents

INTRODUCTION

> . . . Truth is born into this world only with pangs and tribulations, and every fresh truth is received unwillingly. To expect the world to receive a new truth, or even an old truth, without challenging it, is to look for one of those miracles which do not occur.
>
> ('Alfred Russel Wallace,' an interview published in 1913: ref 753, p. 622)

ALFRED RUSSEL WALLACE (born 8 January 1823, died 7 November 1913), English evolutionist–naturalist, geographer, anthropologist, and social critic and theorist, was one of the Victorian Era's most colourful figures. Though still best known for his connection with Charles Darwin as 'the other man' in the development of the theory of natural selection, his work stands on its own, and he was, in any case, a person cut from quite a different cloth. Shy by nature and, in his own assessment, 'lazy', he none the less poured out ideas over a span of time extending into eight decades, during which period he managed some 20,000 miles of 'applied' travels, and writings totalling well over 10,000 pages in print. Beyond his independent recognition of the principle of natural selection, he was an important contributor to theory in several different fields, and ranks as both the 'father' of modern biogeography and history's pre-eminent tropical naturalist.

The obvious aside, however, Wallace was such a complex figure that no consensus has been reached regarding the overall level of significance of his work. Certainly, he presents a study in contrasts. A self-described 'scientific sceptic', he none the less maintained a belief in spiritualism for most of his career. Though of aggressively critical disposition professionally, he won the highest respect of allies and adversaries alike by keeping up an enviable standard of logical argumentation and fair representation of opposing viewpoints. As a public figure he eventually achieved the status of 'The Grand Old Man of Science'—despite never considering himself a 'scientist'. A stalwart defender of most of Darwin's views, on occasion he just as strongly promoted dissenting arguments. A seemingly bottomless well of new ideas, he did little to ensure that future generations would connect his name with them. Indeed, he continually risked his scientific reputation by supporting—even leading—various radical social movements. Still, and despite his modest social circumstances, the Crown saw fit to award him a civil pension in 1881, and the Order of Merit in 1908. G. K. Chesterton perhaps summed things up best when he stated his reason for considering Wallace one of the two most important figures of the nineteenth century: he (Chesterton) knew of no other individual who could be considered a leader of both a major revolution in thought (materialistic evolutionary theory) and its own counter-revolution (anti-materialistic social theory and psychical studies)!

It is thus not surprising that Wallace has remained a poorly understood figure. Few individuals have written creatively on so many different subjects, over such a long period of time, from such a unique perspective, and in so consistently self-effacing a manner.

Wallace's early years were relatively unremarkable. His parents were middle-class but not well off, and young Alfred was forced to leave school at the age of fourteen to support himself. In 1838 he was apprenticed to his older brother William, a successful surveyor. He liked the outdoor work, leaving in 1844 to become a private school instructor in Leicester only when a slowdown in jobs occurred. During this period of his early adulthood he developed an interest in natural history collecting. This interest was strongly reinforced when at Leicester he met Henry Walter Bates (later of 'Batesian mimicry' fame), already a full-fledged entomologist, though still a teenager. The Wallace–Bates friendship eventually yielded the idea of organizing a collecting expedition to South America, and the two young men left for Pará (now Belém), Brasil, in 1848.

Wallace remained in South America until 1852 (Bates stayed until 1859, amassing a magnificent collection of insects that would absorb his attention for a good many years thereafter). Wallace had been forwarding his collections to his sales agent in England throughout his stay, but through some misunderstanding a large portion of the later shipments was delayed at Pará. He was forced to arrange passage back to England for both himself and what remained. Unfortunately for Wallace, in the middle of the Atlantic his ship caught fire and sank, and he was able to save only the smallest portion of his possessions. He and the others on board were in fact lucky to escape with their lives, being rescued at sea by a passing vessel after spending ten days in a pair of badly leaking lifeboats. (The drama did not end there, either, as their new carrier also proved barely seaworthy, itself very nearly succumbing to a series of storms encountered over the remainder of the voyage.)

In all, it took Wallace some eighty days to make it back to England. There followed for him a period of varied activities: rest, writing, and working visits to museums. He decided to continue with his collecting efforts, choosing what is now known as Indonesia (and then, as the 'Malay Archipelago') as his next centre of operations. Securing a grant from the Royal Geographical Society to cover his passage, he left for Singapore in March 1854.

Wallace spent eight full years in the East, returning to London in April 1862. By that time, he was something of a celebrity in scientific circles. He had been a highly productive collector—an incredible 126,000 specimens in total—and had secured a considerable reputation as a writer on both theoretical and systematic subjects. And there was, of course, that letter to Darwin . . .

In early 1858, Wallace's long-standing preoccupation with the subject of evolution led him to recognize a principle of great general application: 'natural selection'. The underlying concept was the 'survival of the fittest': that, as a rule, better-adapted individuals produce more surviving offspring, thus differentially passing their particular characteristics on to their progeny. Wallace quickly worked up an essay capturing the essence of this notion and sent it off to Darwin (who he knew through previous exchanges of letters was generally interested in the subject) 'for comment' and possible forwarding to Sir Charles Lyell, the geologist, who at that time was England's foremost naturalist.

The ensuing chain of events has been one of the most studied in the history of science. To cut a very long story short, Darwin was taken aback by what he read, as it neatly outlined the essence of the concept he had been painstakingly working out for some twenty years. A priority conflict seemed imminent. Darwin appealed to Lyell and another prominent naturalist, the botanist Joseph Hooker, to help him decide what to do. Lyell and Hooker arranged to have Wallace's draft and two informal but representative synopses of Darwin's ideas on the same subject read at the next meeting of the Linnean Society, and this took place (without obtaining Wallace's permission first) on 1 July 1858. Darwin then began working in earnest to complete an 'abstract' of his views on natural selection; fifteen months later this was published as *On the origin of species*.

For several years after his second return to England most of Wallace's time was taken up in the study of his collections. By 1869, however, he had been able to complete *The Malay Archipelago*, and, one year later, to compile *Contributions to the theory of natural selection*. The first is among the most successful scientific travel books ever written. It went through fifteen printings during Wallace's lifetime alone, and was translated into virtually every major language. The second was a collection of essays on the subject with which his name had become most associated.

During the 1870s Wallace focused his attention on biogeography, the study of the geographical and ecological distribution of living things. His most important writings on this subject were *The geographical distribution of animals* (1876), *Tropical nature and other essays* (1878), and *Island life* (1880). Between 1880 and 1895, he divided his attention between the study of various natural science matters and several social causes, especially land nationalization. After 1895, his interests further diversified, embracing social evolution, cosmology, new developments in evolutionary theory, and political criticism. His most important post–1880 books were *Land nationalisation* (1882), *Darwinism* (1889), *The wonderful century* (1898), *Man's place in the universe* (1903), *My life*, his autobiography (1905), and *The world of life* (1910). His last two books, *Social environment and moral progress* and *The revolt of democracy*, appeared in 1913, the year of his death (two months short of the age of ninety-one—and, it may be noted, even at that point he had begun arrangements for the production of yet two further monographic studies!).

The preceding historical sketch gives but little feeling for the forces that were operating in Wallace's life after 1862. The interested reader should consult the secondary sources listed at the end of this work for details; but we should take note in passing of several important connections. To begin with, in 1865 Wallace became a spiritualist (see Section Two). The teachings of this belief increasingly flavoured his writings after that date, astonishing many (including Darwin) who had come to regard him as one of the period's most consistently insightful thinkers on natural science subjects. Then there were his involvements in various social reform movements (see Section Three). In 1881 he was instrumental in founding the Land Nationalisation Society, an organization dedicated to the abolishment of private land holdings. He became its first President, continuing in that role until his death. By 1890 he was a proclaimed socialist. Over the years 1880 to 1905 he took an active role in the anti-vaccination movement (see Section Four). Many of his social criticism essays—there were dozens in all—were published in the leading review journals of the day, and an unknown number of his letters on related subjects appeared in various newspapers. Many of his contemporaries considered him a crank.

Wallace's life after his return to England in 1862 was complicated by his inability to secure full-time employment. Money became a constant worry, but he found it possible to earn a decent living through his writings, lecturing opportunities, and several additional incidental sources of income. Though initially situated in London, he soon tired of urban life, and after marrying in 1866 began a series of moves that led him ever further into the English countryside. There he and his wife (née Annie Mitten, the daughter of botanist William Mitten, a friend of Wallace's) raised two children (a third died at age four). The major adventure of his post-collecting period was a ten-month lecture tour to the United States and Canada in 1886–7, during which time he sketched out the material that would form the core of his book *Darwinism*. This work is arguably the definitive late-nineteenth-century exposition of the subject.

It is clear from his own writings and from those who knew him personally that Wallace had few, if any, regrets over the way his life turned out. There is little reason why he should have had. His personal and professional conduct in all matters of which we have knowledge seems to have been above reproach, and he surely accomplished all—and then some—of his life's objectives. His is, simply put, an inspiring story of a life well-lived.

But history has not been altogether kind to him. In his later years he was already considered something of a relic of the past; he became, in fact, the last surviving member of the mid-nineteenth-century circle of 'Darwinian' naturalists. Science was moving in new directions, and, despite Wallace's continuing efforts, 'natural selection' in particular was experiencing a decided downswing in popularity. After his death, his name passed rapidly into relative obscurity.

Interest was re-kindled by the 1958 centenary celebrations of the first public presentation of the natural selection concept. Since that time a considerable literature on Wallace's life and writings has accumulated, and a picture of the man and his accomplishments has slowly been emerging.

If the Wallace–Darwin association certainly worked to Darwin's advantage, it is less clear what its implications were for Wallace. What would the outcome have been had he simply sent the first natural selection essay in for publication, rather than to Darwin? So many 'ifs' confound the question that speculation on the matter, though fascinating, produces little of substance. I am not so certain that as much would have changed as some observers seem to think. Of the two, Darwin was without doubt the more meticulous worker, and the evidence he amassed was clearly crucial to the success of the cause. He was, moreover, in a more advantageous social and financial position—and geographic position, initially—to advance it.

More importantly, natural selection and its many extensions were a dominating theme in Darwin's theoretical explorations (though it must be remembered he was responsible for a large and important body of descriptive biology and geology studies as well). Contrary to what is commonly thought, this was not true in Wallace's case. His 'discovery' of natural selection was something of an accident. Right up to the point of his recognition of the principle he was unaware of what was coming; he had been thinking along quite different lines in his musings over the evolutionary process (see Sections Two and Six). The new concept represented an interim stage in his thinking. While consistent with his earlier positions, and resolving some (important) problems inherent in them, it did not in any sense provide him with the 'complete' evolution model he was looking for.

'Natural selection' thus represents a red herring when it comes to trying to understand Wallace's world-view. Although he was not slow to recognize the significance of the principle and to develop it in as many directions as he could (and indeed there were many that he did), he never considered natural selection more than a logical argument demanded by the relationship between adaptation and species divergence. While study of this particular relationship has come to dominate evolutionary biology, Wallace himself was committed to working out a model of natural organization that extended well beyond it. By about 1865 he felt that certain questions left unanswered by natural selection's relation to the more general model had been solved; on the basis of the resulting synthesis he was able to advance thoughts on a very wide range of subjects.

In the first sentence in this Introduction I described Wallace as 'evolutionist–naturalist, geographer, anthropologist, *and* social critic and theorist.' The critical word in this description is 'and'. Whereas many workers have made important contributions to several distinct fields of knowledge, the vast majority, at least, have been unable to refer their various lines of argument to defendable universal principles. Wallace's post–1862 (the date of his return to

England, it will be remembered) studies were, to his distinction, logical applications of a single, comprehensive, model of natural/social organization and causation. But Wallace's efforts to argue the specific from the general have often conflicted with conventional thinking in the many areas of his attention, leading to problems of interpretation. As part of the difficulty may be traced to the interdisciplinary perspective demanded when one tries to wear four 'hats' (i.e., those mentioned above) at once, a short summary of the connections involved in Wallace's case is useful here by way of introduction to the present collection.

There is, of course, little difficulty in styling Wallace an evolutionist and a naturalist. He was one of history's most productive collector-explorers, investigating not only the animals and plants of the lands he visited, but their people, geology, and physical geography as well. Underlying these activities was a desire to learn how evolution took place—at all levels of organization. The key integrating element in his approach to these subjects was geography: especially biogeography, physical geography, economic geography, and ethnography. Many of his evolutionary studies relied on explicitly spatial or regional models. This is even true of the bulk of his social commentary, which emphasized goal-oriented planning (and, in effect, conservation of natural resources); that is, the analysis of imperfections in the present social order, and how to achieve appropriate reforms. To Wallace, in fact, *social change was a manifestation of evolution no less than was the biological process sustained by natural selection.*

The geographical emphasis in his work notwithstanding, Wallace is only rarely thought of as a geographer, even by geographers. This is understandable, given that: (1) he was not a member of the geographical academic–intellectual establishment, and (2) geography, despite being in many respects the oldest science, has an intellectual core only rarely appreciated by workers outside its own ranks (its history and philosophy, in fact, have only very infrequently been studied by other than geographers). In any case, the geographical element in Wallace's work is extremely pervasive. Consider: his extensive travels (many of which verged on exploration) and travel writings (including dozens of short selections and three books); a planned (but never executed) monograph on the physical geography of the Amazon; his position as history's foremost tropical (regions) naturalist, and the recognized 'father' of modern zoogeography; significant contributions to ethnography; his important work in glaciology; extensive writings on both biogeography and physical geography in general (several monographs and some one hundred shorter works); his strong interest in land planning and associated aspects of economic and agricultural geography (monographs, dozens of shorter writings, and his Land Nationalisation Society association); his service as a part-time Assistant Examiner in physical geography for over twenty years; his interest in and contributions to geographic education (especially his travel writings and suggestions for geographically-arranged museum displays). He was a Fellow of the Royal Geographical Society for over fifty years and was awarded their Founder's Medal in 1892; he was also presented with the Gold Medal of the Société de Géographie of France in 1870.

Considering the above list, one cannot help but wonder whether some of the difficulties people have had in interpreting Wallace's work may be a function of miscategorization. He is, arguably, the most important individual in the history of science who might legitimately be labelled *strictly* a geographer; that is, a geographer who happened to be interested in the subject of evolution.

Wallace's fame as an anthropologist rests primarily on his views on the evolution of man. But he was also a competent and contributing primatologist, ethnologist, and ethnographer, and a significant source of ideas on the role of women in society and the future evolution of mankind. His treatment of human evolution, though in part natural selection-based, differed considerably from Darwin's, and drew (and still draws) no little attention. His attraction to spiritualism is fascinating, but, as it turns out, something of a side-issue; it is important to note in this context, however, that he portrayed spiritualism as 'a new branch of anthropology' in a letter to Thomas Huxley in 1866 (Marchant 1916, p. 418).

Lastly, we have Wallace the social critic and theorist. This aspect of his work at first may seem the least relatable to the others. But many of his philosophical explorations are traceable to an interest in utopian socialist theories dating from his teen years, and in turn to a conviction that the overall theme of existence is progressive change. His evolutionary biology studies (including those with an anthropological emphasis) were thus undoubtedly inspired in part—or even largely—by his prior belief that social change continues a process initiated at the physical and biological levels. But he was in no sense a 'Social Darwinist', protesting vehemently, for example, against 'might is right' arguments. To his mind, evolution was leading to the production of sympathetically co-operating beings, not selfishly competing 'human herds'. His social criticism represented an attempt to expose the things he felt were impeding progress in that direction. He came to believe that the rampant capitalism and materialism of the period were great evils; the corrective measures he advocated included, among other things, land nationalization, abolishing court-supported inheritance and trusts, socialism, and yielding to a 'social' form of natural selection depending on woman's release from economic 'indenture'. Although permeated by a general disregard for what was immediately practicable, many of the goals of the causes he supported have in the long run been reached—usually, it seems, without remembrance of the visionary reasoning he had applied to thinking out such goals many years earlier.

Reading Wallace is a challenge. Though his priorities are both timeless and consistent and his writing style exemplary, he was never afraid to stray from the familiar when it came to conclusions. Unorthodox though they often were, however, his positions had not been superficially reasoned. As notable a source as the American pragmatist philosopher Charles Peirce once opined, in fact, that the arguments Wallace set out in support of his views 'ought to be made the basis of a course of lectures on logic' (Peirce 1906, p. 160). For those who are not in the habit of dismissing the unfamiliar out of hand, therefore, the following readings will provide many challenges to familiar assumptions.

§ 1 · ANTHROPOLOGY

. . . My whole argument is founded on certain facts, and on these facts only. My critic admits the facts, does not refute my arguments, yet maintains that I should give up my conclusion, because the theory of Natural Selection *must* apply equally to man and the rest of Nature, or to neither. But why must it do so? Darwin himself claims no such universality for it . . . I maintain . . . that man is descended from a lower animal form, but I adduce facts which go to prove that some other law or power than Natural Selection has specially modified him. If Darwin is not anti-Darwinian in admitting, as he does, the possibility that animals and plants may not have had a common ancestor, I may surely deny that I am anti-Darwinian when I show that there are certain phenomena in the case of man that cannot be wholly explained by the law of Natural Selection . . .

('Man and natural selection', 1870: ref 173, p. 9)

A GENERAL survey of Wallace's writings must properly begin with his views on the subject of man. Wallace was one of his period's most thoroughly dedicated humanists, a fact that those only superficially familiar with him as a 'natural selectionist' may find surprising. But quite early in life (aged thirteen or fourteen) he had been exposed to the teachings of the social reformer Robert Owen, and in his own words (*My life*, ref 729, Vol. 1, p. 104), Owen 'influenced my character more than I then knew . . . I have always looked upon Owen as my first teacher in the philosophy of human nature and my first guide through the labyrinth of social science.'

That 'human nature' continued to intrigue Wallace as he entered adulthood is evident from a lecture entitled 'The advantages of varied knowledge' (ref 1) he prepared at the age of nineteen or twenty, and also suggested by his first attempt at publishing during the same period, a regional study of the 'habits, manners, and mode of life of the Welsh hill farmer' he called 'The South-Wales farmer' (ref 623). The first work consists of reflections on the value of looking at things from as many sides as possible, a prescription for inquiry that Wallace would later studiously observe. The second, though rather unexceptional in either style or content, is interesting for the way it foreshadows the attention given to detail in Wallace's later accounts of tropical peoples. A short but representative example of this penchant is provided by the following excerpt from 'Notes of a voyage to New Guinea' (ref 51), part of a letter read to members of the Royal Geographical Society at their meeting of 27 June 1859:

. . . The inhabitants of Dorey live always on the coast, or more properly in the sea, as they always build their houses at or below low-water mark, raised on posts, and reached by a rough and tottering causeway from the beach. Of all houses I have yet met with, these are the most wretched. They are very low and long, and the roof is shaped like the bottom of a boat. Old mats, cocoa-nut leaves, broken boats, and bits of board, make a dwelling such as some shipwrecked sailors might hastily set up for a temporary night's shelter, but which it seems hardly credible that any people should contentedly live in. The natives of the interior do not differ perceptibly in physical character, but have a distinct language, and are called 'Arfaki' by the Doreyans. Their houses are very similar, but are raised 12 or 15 feet high, on a perfect forest of thin poles, a few of which are put diagonally, and prevent the whole from falling with the first wind. It is singular that these people know the use of diagonal struts, whereas the comparatively civilized Bugis and Macassar-men are quite ignorant of it, their houses being invariably inclined to one side by the prevalent winds, and only kept from falling by the posts being pretty firmly set in the ground, and the building connected with them framed strongly of bamboos. The Doreyans are fishers and traders, the Arfakis are agriculturists. The former catch turtle and tripang, which they sell for breads, knives, and cloth, and purchase of the Arfakis their rice and yams, plantains and breadfruits, and numbers of tame cockatoos and lories, which they sell again to the Ternate and Tidore traders. All these natives have the characters of the Papuan race very strongly marked;—the flat forehead, heavy brows, and large nose, with the apex bent downwards, are almost universal, as well as the harsh curly hair, which often forms an enormous stiff mop, and is then highly esteemed. It has, in fact, a very grand and imposing effect. The colour of the skin varies greatly. In general it is a dirty black, or sooty colour, but varies to a fine brown, which is often quite as light as that of the pure Malay races. Skin disease is very common, and in the children scrofulous diseases abound, which are seldom seen in adults; it is probable, therefore, that the former die from neglect. The men wear the ordinary strip of bark cloth, the women generally a Bugis sarong, or any piece of cloth or matting they can obtain. Tattooing is generally practised; slightly by the men, but much more extensively by the women, who usually have the whole chest covered with elegant tracery, following the curves of the bosom. The females, however, are, without exception, the least engaging specimens of the fair sex it has yet been my fortune to meet.

In mental and moral characteristics the Papuans differ remarkably from the Malay races. They are much more impulsive, and do not conceal their emotions and passions. They are inquisitive, talk much and loudly, and laugh boisterously; reminding one of the negro character, as much as of the negro form and aspect. The natives of Dorey are not to be trusted in anything where payment is concerned. If they do not actually steal, it is, I am inclined to think, only from fear of consequences. They are, however, not a fair sample of the New Guinea tribes, having been too much in contact with the lowest class of Mahommedan traders, with whom they find it necessary to take every advantage in self-defence. They possess the rude artistic genius of so many of the Oceanic tribes, decorating their household utensils and the prows of their canoes with elaborate carving, and the posts of their council-house with obscene caryatides.

The language of the Doreyans resembles that of the Aru and Ké Islands in containing a large number of monosyllabic words, as well as others excessively polysyllabic, offering a remarkable contrast to the striking dissyllabic character of the whole Malayan group of languages. It exhibits also the Polynesian characteristic of several distinct terms for certain objects according to the prefixed pronoun; thus, "my head," "your head," "his head," are expressed by three distinct words. This language, or mutually intelligible forms of it, is spoken by the coast-dwellers over an extensive area—at Amberbaki, 100 miles west, in the islands of Wagiou, Myfor, Jobie, and Mysory—and at Amberpoea and some other islands in the great bay, the natives can converse with those of Dorey, and seem very similar to them in appearance and habits. They are evidently a wandering race, answering to the Bajees, or sea gipsies, of the Indian Archipelago . . .

* * * * *

In 1866 Wallace 'abstracted' his views on the range of coverage of the field of anthropology in a concise address presented as President of the Department of Anthropology, Section of Biology, at the annual meeting of the British Association for the Advancement of Science:

Anthropology (ref 119)

Anthropology is the science which contemplates man under all his varied aspects (as an animal, and as a moral and intellectual being) in his relations to lower organisms, to his fellow men, and to the universe. The anthropologist seeks to collect together and systematize the facts and the laws which have been brought to light by all those branches of study which, directly or indirectly, have man for their object. These are very various. The physiologist, for example, studies man as a wondrous and most complicated machine, whose parts and motions, actions and reactions he seeks thoroughly to understand. The comparative anatomist and the zoologist compare his structure with that of other animals, take note of their likenesses and differences, determine their degrees of affinity, and seek after the common plan of their organization and the law of their development. The psychologist studies the mind of man, its mode of action, and its development, compares it with the instincts and the reasoning faculties of the lower animals, and ever aims at the solution of the greatest of problems—whence and what is mind. The historian collects and arranges the facts of man's progress in recent times; the geographer determines the localities of the various races that inhabit the earth, their manners, customs, and physical characteristics; the archaeologist seeks, by studying the remains of man and his works, to supplement written history and to carry back our knowledge of man's physical, mental and moral condition into prehistoric times; the geologist extends this kind of knowledge to a still earlier epoch by proving that man coexisted with numerous animals now extinct, and inhabited Europe at so remote a period that the very contour of its surface, the form of its hills and valleys, no less than its climate, vegetation, and geology, were materially different from what they now are, or ever have been during the epoch of authentic history; the philologist devotes himself to the study of human speech, and through it seeks to trace out the chief migrations of nations, and the common origin of many of the races of mankind; and, lastly, the phrenologist and the craniologist have created special sciences out of the study of the human brain and skull. Considering the brain as the organ of the mind, the phrenologist seeks to discover in what way they correspond to each other, and to connect mental peculiarities with the form and dimensions of the brain as indicated by the corresponding form of its bony covering. The craniologist, confining his attention to the skull as an indication of race, endeavours to trace out the affinities of modern and ancient races of man, by the forms and dimensions of their crania.

These various studies have hitherto been pursued separately. There has been great division of labour, but no combination of results. Now it is our object as anthropologists to accept the well-ascertained conclusions which have been arrived at by the students of all these various sciences, to search after every new fact which may throw additional light upon any of them, and, as far as we are able, to combine and generalize the whole of the information thus obtained. We cannot, therefore, afford to neglect any facts relating to man, however trivial, unmeaning, or distasteful some of them may appear to us. Each custom, superstition, or belief of savage or of civilized man may guide us towards an explanation of their origin in common tendencies of the human mind. Each peculiarity of form, colour, or constitution may give us a clue to the affinities of an obscure race. The anthropologist must ever bear in mind that, as the object of his study is man, nothing pertaining to or characteristic of man can be unworthy of his attention. It will be only after we have brought together and arranged all the facts and principles which have been established by the various special studies to which I have alluded, that we shall be in a condition to determine the particular lines of investigation most needed to complete our knowledge of man, and may hope ultimately to arrive at some definite conclusions on the great problems which must interest us all—the questions of the origin, the nature, and the destiny of the human race. I would beg you to recollect also that *here* we must treat all of these problems as purely questions of science, to be decided solely by facts and by legitimate deductions from facts. We can accept no conclusions as authoritative that have not been thus established. Our sole object is to find out for ourselves what is our true nature, to feel our way cautiously, step by step, into the dark and mysterious past of human history, to study man under every phase and aspect of his present condition, and from the knowledge thus gained to derive (as we cannot fail to do) some assistance in our attempts to govern and improve uncivilized tribes, some guidance in our own national and individual progress.

* * * * *

A debt to views expressed in the earlier 'The advantages of varied knowledge' is apparent in the preceding address, as is Wallace's leading suggestion that anthropology must treat its subject in an unbiased, scientific fashion. This pronouncement must at least in part have been occasioned by the then-developing furore over the application of the theory of natural selection to the origins of the human species.

In this debate Wallace played a prominent role. It is a mistake, however, to suppose that his thoughts regarding human evolution were more than secondarily shaped by his discovery of the principle of natural selection. Well before 1858 he had been thinking about the origins of man in evolutionary terms; this is evident both from early letters he exchanged with his confidant Bates, and portions of at least one of his pre–1858 publications: 'On the habits of the Orang-utan of Borneo' (ref 26), which appeared in the Annals and Magazine of Natural History *in 1856. The critical passages, from pages 29–32 of that work, are:*

. . . Here then we have an animal which lives solely and exclusively on fruits or other soft vegetable food, and yet has huge canine teeth. It never attacks other animals, and is rarely attacked itself; but when it is, it uses, not these powerful teeth, but its arms and legs to defend itself. And, lastly, the female, which is weaker, which is encumbered by its young, and which would therefore afford a much easier prey, and a more tempting object of attack, is quite unprovided with these supposed means of defence. Do you mean to assert, then, some of my readers will indignantly ask, that this animal, or any animal, is provided with organs which are of no use to it? Yes, we reply, we do mean to assert that many animals are provided with organs and appendages which serve no material or physical purpose. The extraordinary excrescences of many insects, the fantastic and many-coloured plumes which adorn certain birds, the excessively developed horns in some of the antelopes, the colours and infinitely modified forms of many flower-petals, are all cases, for an explanation of which we must look to some general principle far more recondite than a simple relation to the necessities of the individual. We conceive it to be a most erroneous, a most contracted view of the organic world, to believe that every part of an animal or plant exists solely for some material and physical use to the individual,—to believe that all the beauty, all the infinite combinations and changes of form and structure should have the sole purpose and end of enabling each animal to support its existence,—to believe, in fact, that we know the one sole end and purpose of every modification that exists in organic beings, and to refuse to recognize the possibility of there being any other. Naturalists are too apt to *imagine*, when they cannot *discover*, a use for everything in nature: they are not even content to let "beauty" be a sufficient use, but hunt after some purpose to which even *that* can be applied by the animal itself, as if one of the noblest and most refining parts of man's nature, the love of beauty for its own sake, would not be perceptible also in the works of a Supreme Creator . . .

The separate species of which the organic world consists being parts of a whole, we must suppose some dependence of each upon all; some general design which has determined the details, quite independently of individual necessities. We look upon the anomalies, the eccentricities, the exaggerated or diminished development of certain parts, as indications of a general system of nature, by a careful study of which we may learn much that is at present hidden from us; and we believe that the constant practice of imputing, right or wrong, some use to the individual, of every part of its structure, and even of inculcating the doctrine that every modification exists solely for some such use, is an error fatal to our complete appreciation of all the variety, the beauty, and the harmony of the organic world.

It is a remarkable circumstance, that an animal so large, so peculiar, and of such a high type of form as the Orang-Utan, should yet be confined to such a limited district,—to two islands, and those almost at the limits of the range of the higher mammalia; for, eastward of Borneo and Celebes, the Quadrumana and most of the higher mammalia almost disappear. One cannot help speculating on a former condition of this part of the world which should give a wider range to these strange creatures, which at once resemble and mock the "human form divine,"—which so closely approach us in structure, and yet differ so widely from us in many points of their external form. And when we consider that almost all other animals have in previous ages been represented by allied, yet distinct forms,—that the bears and tigers, the deer, the horses, and the cattle of the tertiary period were distinct from those which now exist, with what intense interest, with what anxious expectation must we look forward to the time when the progress of civilization in those hitherto wild countries may lay open the monuments of a former world, and enable us to ascertain approximately the period when the present species of Orangs first made their appearance, and perhaps prove the former existence of allied species still more gigantic in their dimensions, and more or less human in their form and structure! Some such discoveries we may not unreasonably anticipate, after the wonders that geology has already made known to us. Animals the most isolated in existing nature have been shown to be but the last of a series of allied species which have lived and died upon the earth. Every class and every order has furnished some examples, from which we may conclude, that all isolations in nature are apparent only, and that whether we discover their remains or no, every animal now existing has had its representatives in past geological epochs.

* * * * *

By early 1864, Wallace's thoughts on how mankind had evolved had solidified to a degree permitting a formal statement of position. The result was a paper presented to the members of The Anthropological Society of London:

The origin of human races and the antiquity of man deduced from the theory of "Natural Selection" (ref 93)

Among the most advanced students of man, there exists a wide difference of opinion on some of the most vital questions respecting his nature and origin. Anthropologists are now, indeed, pretty well agreed that man is not a recent introduction into the earth. All who have studied the question now admit that his antiquity is very great; and that, though we have to some extent ascertained the minimum of time during which he *must* have existed, we have made no approximation towards determining that far greater period during which he *may* have, and probably *has*, existed. We can with tolerable certainty affirm that man must have inhabited the earth a thousand centuries ago, but we cannot assert that he positively did not exist, or that there is any good evidence against his having existed, for a period of a hundred thousand centuries. We know positively that he was contemporaneous with many now extinct animals, and has survived changes of the earth's surface fifty or a hundred times greater than any that have occurred during the historical period; but we cannot place any definite limit to the number of species he may have outlived, or to the amount of terrestrial change he may have witnessed.

But while on this question of man's antiquity there is a very general agreement,—and all are waiting eagerly for fresh evidence to clear up those points which all admit to be full of doubt,—on other and not less obscure and difficult questions a considerable amount of dogmatism is exhibited; doctrines are put forward as established truth, no doubt or hesitation is admitted, and it seems to be supposed that no further evidence is required, or that any new facts can modify our convictions. This is especially the case when we inquire, *Are the various forms under which man now exists primitive, or derived from preexisting forms; in other words, is man of one or many species*? To this question we immediately obtain distinct answers diametrically opposed to each other: the one party positively maintaining that man is a *species* and is essentially *one*—that all differences are but local and temporary variations, produced by the different physical and moral conditions by which he is surrounded; the other party maintaining with equal confidence that man is a genus of *many species*, each of which is practically unchangeable, and has ever been as distinct, or even more distinct, than we now behold them. This difference of opinion is somewhat remarkable, when we consider that both parties are well acquainted with the subject; both use the same vast accumulation of facts; both reject those early traditions of mankind which profess to give an account of his origin; and both declare that they are seeking fearlessly after truth alone. I believe, however, it will be found to be the old story over again of the shield—gold on one side and silver on the other—about which the knights disputed; each party will persist in looking only at the portion of truth on his own side of the question, and at the error which is mingled with his opponent's doctrine. It is my wish to show how the two opposing views can be combined so as to eliminate the error and retain the truth in each, and it is by means of Mr. Darwin's celebrated theory of "Natural Selection" that I hope to do this, and thus to harmonise the conflicting theories of modern anthropologists.

Let us first see what each party has to say for itself. In favour of the unity of mankind it is argued that there are no races without transitions to others; that every race exhibits within itself variations of colour, of hair, of feature, and of form, to such a degree as to bridge over to a large extent the gap that separates it from other races. It is asserted that no race is homogeneous; that there is a tendency to vary; that climate, food, and habits produce and render permanent physical peculiarities, which, though slight in the limited periods allowed to our observation, would, in the long ages during which the human race has existed, have sufficed to produce all the differences that now appear. It is further asserted that the advocates of the opposite theory do not agree among themselves; that some would make three, some five, some fifty or a hundred and fifty species of man; some would have had each species created in pairs, while others require nations to have at once sprung into existence, and that there is no stability or consistency in any doctrine but that of one primitive stock.

The advocates of the original diversity of man, on the other hand, have much to say for themselves. They argue that proofs of change in man have never been brought forward except to the most trifling amount, while evidence of his permanence meets us everywhere. The Portuguese and Spaniards, settled for two or three centuries in South America, retain their chief physical, mental, and moral characteristics; the Dutch boers at the Cape, and the descendants of the early Dutch settlers in the Moluccas, have not lost the features or the colour of the Germanic races; the Jews, scattered over the world in the most diverse climates, retain the same characteristic lineaments everywhere; the Egyptian sculptures and paintings show us that, for at least 4000 or 5000 years, the strongly contrasted features of the Negro and the Semitic races have remained altogether unchanged; while more recent discoveries prove that, in the case at least of the American aborigines, the mound-builders of the Mississippi valley, and the dwellers on Brazilian mountains, had still in the very infancy of the human race the same characteristic type of cranial formation that now distinguishes them.

If we endeavour to decide impartially on the merits of this difficult controversy, judging solely by the evidence that each party has brought forward, it certainly seems that the best of the argument is on the side of those who maintain the primitive diversity of man. Their opponents have not been able to refute the permanence of existing races as far back as we can trace them, and have failed to show, in a single case, that at any former epoch the well marked varieties of mankind approximated more closely than they do at the present day. At the same time this is but negative evidence. A condition of immobility for four or five thousand years, does not preclude an advance at any earlier epoch, and—if we can show that there are causes in nature which would check any further physical change when certain conditions were unfulfilled—does not even render such an advance improbable, if there are any general arguments to be adduced in its favour. Such a cause, I believe, does exist, and I shall now endeavour to point out its nature and its mode of operation.

In order to make my argument intelligible, it is necessary for me to explain very briefly the theory of "Natural Selection" promulgated by Mr. Darwin, and the power which it possesses of modifying the forms of animals and plants. The grand feature in the multiplication of organic life is that of close general resemblance, combined with more or less individual variation. The child resembles its parents or ancestors more or less closely in all its peculiarities, deformities, or beauties; it resembles them in general more than it does any other individuals; yet children of the same parents are not all alike, and it often happens that they differ very considerably from their parents and from each other. This is equally true of man, of all animals, and of all plants. Moreover, it is found that individuals do not differ from their parents in certain particulars only, while in others they are exact duplicates of them. They differ from them and from each other in every particular: in form, in size, in colour, in the structure of internal as well as of external organs; in those subtle peculiarities which produce differences of constitution, as well as in those still more subtle ones which lead to modifications of mind and character. In other words, in every possible way, in every organ and in every function, individuals of the same stock vary.

Now, health, strength, and long life are the results of a harmony between the individual and the universe that surrounds it. Let us suppose that at any given moment this harmony is perfect. A certain animal is exactly fitted to secure its prey, to escape from its enemies, to resist the inclemencies of the seasons, and to rear a numerous and healthy offspring. But a change now takes place. A series of cold winters, for instance, come on, making food scarce, and bringing an immigration of some other animals to compete with the former inhabitants of the district. The new immigrant is swift of foot, and surpasses its rivals in the pursuit of game; the winter nights are colder, and require a thicker fur as a protection, and more nourishing food to keep up the heat of the system. Our supposed perfect animal is no longer in harmony with its universe; it is in danger of dying of cold or of starvation. But the animal varies in its offspring. Some of these are swifter than others—they still manage to catch food enough; some are hardier and more thickly furred—they manage in the cold nights to keep warm enough; the slow, the weak, and the thinly clad soon die off. Again and again, in each succeeding generation, the same thing takes place. By this natural process, which is so inevitable that it cannot be conceived not to act, those best adapted to live, live; those least adapted, die. It is sometimes said that we have no direct evidence of the action of this selecting power in nature. But it seems to me we have better evidence than even direct observation would be, because it is more universal, viz., the evidence of necessity. It must be so; for, as all wild animals increase in a geometrical ratio, while their actual numbers remain on the average stationary, it follows that as many die annually as are born. If therefore, we deny natural selection, it can only be by asserting that in such a case as I have supposed, the strong, the healthy, the swift, the well clad, the well organised animals in every respect, have no advantage over,—do not on the average live longer than the weak, the unhealthy, the slow, the ill-clad, and the imperfectly organised individuals; and this no sane man has yet been found hardy enough to assert. But this is not all; for the offspring on the average resemble their parents, and the selected portion of each succeeding generation will therefore be stronger, swifter, and more thickly furred than the last; and if this process goes on for thousands of generations, our animal will have again become thoroughly in harmony with the new conditions in which he is placed. But he will now be a different creature. He will be not only swifter and stronger, and more furry, he will also probably have changed in colour, in form, or perhaps have acquired a longer tail, or differently shaped ears; for it is an ascertained fact, that when one part of an animal is modified, some other parts almost always change as it were in sympathy with it. Mr. Darwin calls this "correlation of growth," and gives as instances that hairless dogs have imperfect teeth; blue eyed cats are deaf; small feet accompany short beaks in pigeons; and other equally interesting cases.

Grant, therefore, the premises: 1st. That peculiarities of every kind are more or less hereditary. 2nd. That the offspring of every animal vary more or less in all parts of their organisation. 3rd. That the universe in which these animals live, is not absolutely invariable;—none of which propositions can be denied; and then consider that the animals in any country (those at least which are not dying out) must at each successive period be brought into harmony with the surrounding conditions; and we have all the elements for a change of form and structure in the animals, keeping exact pace with changes of whatever nature in the surrounding universe. Such changes must be slow, for the changes in the universe are very slow; but just as these slow changes become important, when we look at results after long periods of action, as we do when we perceive the alterations of the earth's surface during geological epochs; so the parallel changes in animal form become more and more striking according as the time they have been going on is great, as we see when we compare our living animals with those which we disentomb from each successively older geological formation.

This is briefly the theory of "natural selection," which explains the changes in the organic world as being parallel with, and in part dependent on those in the inorganic. What we now have to inquire is,—Can this theory be applied in any way to the question of the origin of the races of man? or is there anything in human nature that takes him out of the category of those organic existences, over whose successive mutations it has had such powerful sway?

In order to answer these questions, we must consider why it is that "natural selection" acts so powerfully upon animals, and we shall, I believe, find that its effect depends mainly upon their self-dependence and individual isolation. A slight injury, a temporary illness, will often end in death, because it leaves the individual powerless against its enemies. If a herbivorous animal is a little sick and has not fed well for a day or two, and the herd is then pursued by a beast of prey, our poor invalid inevitably falls a victim. So in a carnivorous animal the least deficiency of vigour prevents its capturing food, and it soon dies of starvation. There is, as a general rule, no mutual assistance between adults, which enables them to tide over a period of sickness. Neither is there any division of labour; each must fulfill *all* the conditions of its existence, and, therefore, "natural selection" keeps all up to a pretty uniform standard.

But in man, as we now behold him, this is different. He is social and sympathetic. In the rudest tribes the sick are assisted at least with food; less robust health and vigour than the average does not entail death. Neither does the want of perfect limbs or other organs produce the same effects as among animals. Some division of labour takes place; the swiftest hunt, the less active fish, or gather fruits; food is to some extent exchanged or divided. The action of natural selection is therefore checked; the weaker, the dwarfish, those of less active limbs, or less piercing eyesight, do not suffer the extreme penalty which falls upon animals so defective.

In proportion as these physical characteristics become of less importance, mental and moral qualities will have increasing influence on the well-being of the race. Capacity for acting in concert, for protection and for the acquisition of food and shelter; sympathy, which leads all in turn to assist each other; the sense of right, which checks depredations upon our fellows; the decrease of the combative and destructive propensities; self-restraint in present appetites; and that intelligent foresight which prepares for the future, are all qualities that from their earliest appearance must have been for the benefit of each community, and would, therefore, have become the subjects of "natural selection." For it is evident that such qualities would be for the well-being of man; would guard him against external enemies, and impending famine, more surely than could any merely physical modification. Tribes in which such mental and moral qualities were predominant, would therefore have an advantage in the struggle for existence over other tribes in which they were less developed, would live and maintain their numbers, while the others would decrease and finally succumb.

Again, when any slow changes of physical geography, or of climate, make it necessary for an animal to alter its food, its clothing, or its weapons, it can only do so by a corresponding change in its own bodily structure and internal organisation. If a larger or more powerful beast is to be captured and devoured, as when a carnivorous animal which has hitherto preyed on sheep is obliged from their decreasing numbers to attack buffaloes, it is only the strongest who can hold,—those with most powerful claws, and formidable canine teeth, that can struggle with and overcome such an animal. Natural selection immediately comes into play, and by its action these organs gradually become adapted to their new requirements. But man, under similar circumstances, does not require longer nails or teeth, greater bodily strength or swiftness. He makes sharper spears, or a better bow, or he constructs a cunning pitfall, or combines in a hunting party to circumvent his new prey. The capacities which enable him to do this are what he requires to be strengthened, and these will, therefore, be gradually modified by "natural selection," while the form and structure of his body will remain unchanged. So when a glacial epoch comes on, some animals must acquire warmer fur, or a covering of fat, or else die of cold. Those best clothed by nature are, therefore, preserved by natural selection. Man, under the same circumstances, will make himself warmer clothing, and build better houses; and the necessity of doing this will react upon his mental organisation and social condition—will advance them while his natural body remains naked as before.

When the accustomed food of some animal becomes scarce or totally fails, it can only exist by becoming adapted to a new kind of food, a food perhaps less nourishing and less digestible. "Natural selection" will now act upon the stomach and intestines, and all their individual variations will be taken advantage of to modify the race into harmony with its new food. In many cases, however, it is probable that this cannot be done. The internal organs may not vary quick enough, and then the animal will decrease in numbers, and finally become extinct. But man guards himself from such accidents by superintending and guiding the operations of nature. He plants the seed of his most agreeable food, and thus procures a supply independent of the accidents of varying seasons or natural extinction. He domesticates animals which serve him either to capture food or for food itself, and thus changes of any great extent in his teeth or digestive organs are rendered unnecessary. Man, too, has everywhere the use of fire, and by its means can render palatable a variety of animal and vegetable substances, which he could hardly otherwise make use of, and thus obtains for himself a supply of food far more varied and abundant than that which any animal can command.

Thus man, by the mere capacity of clothing himself, and making weapons and tools, has taken away from nature that power of changing the external form and structure which she exercises over all other animals. As the competing races by which they are surrounded, the climate, the vegetation, or the animals which serve them for food, are slowly changing, they must undergo a corresponding change in their structure, habits, and constitution, to keep them in harmony with the new conditions—to enable them to live and maintain their numbers. But man does this by means of his intellect alone; which enables him with an unchanged body still to keep in harmony with the changing universe.

From the time, therefore, when the social and sympathetic feelings came into active operation, and the intellectual and moral faculties became fairly developed, man would cease to be influenced by "natural selection" in his physical form and structure; as an animal he would remain almost stationary; the changes of the surrounding universe would cease to have upon him that powerful modifying effect which it exercises over other parts of the organic world. But from the moment that his body became stationary, his mind would become subject to those very influences from which his body had escaped; every slight variation in his mental and moral nature which should enable him better to guard against adverse circumstances, and combine for mutual comfort and protection, would be preserved and accumulated; the better and higher specimens of our race would therefore increase and spread, the lower and more brutal would give way and successively die out, and that rapid advancement of mental organisation would occur, which has raised the very lowest races of man so far above the brutes (although differing so little from some of them in physical structure), and, in conjunction with scarcely perceptible modifications of form, has developed the wonderful intellect of the Germanic races.

But from the time when this mental and moral advance commenced, and man's physical character became fixed and immutable, a new series of causes would come into action, and take part in his mental growth. The diverse aspects of nature would now make themselves felt, and profoundly influence the character of the primitive man.*

* M. Guizot says: "If we regard the immediate influence of climate upon men, perhaps it is not so extensive as has been supposed. But the indirect influence of climate, that which, for

When the power that had hitherto modified the body, transferred its action to the mind, then races would advance and become improved merely by the harsh discipline of a sterile soil and inclement seasons. Under their influence, a hardier, a more provident, and a more social race would be developed, than in those regions where the earth produces a perennial supply of vegetable food, and where neither foresight nor ingenuity are required to prepare for the rigours of winter. And is it not the fact that in all ages, and in every quarter of the globe, the inhabitants of temperate have been superior to those of tropical countries? All the great invasions and displacements of races have been from North to South, rather than the reverse; and we have no record of there ever having existed, any more than there exists to-day, a solitary instance of an indigenous inter-tropical civilisation. The Mexican civilisation and government came from the North, and, as well as the Peruvian, was established, not in the rich tropical plains, but on the lofty and sterile plateaux of the Andes. The religion and civilisation of Ceylon were introduced from North India; the successive conquerors of the Indian peninsula came from the North-west, and it was the bold and adventurous tribes of the North that overran and infused new life into Southern Europe.

It is the same great law of "*the preservation of favoured races in the struggle for life*," which leads to the inevitable extinction of all those low and mentally undeveloped populations with which Europeans come in contact. The red Indian in North America, and in Brazil; the Tasmanian, Australian and New Zealander in the southern hemisphere, die out, not from any one special cause, but from the inevitable effects of an unequal mental and physical struggle. The intellectual and moral, as well as the physical qualities of the European are superior; the same powers and capacities which have made him rise in a few centuries from the condition of the wandering savage* with a scanty and stationary population to his present state of culture and advancement, with a greater average longevity, a greater average strength, and a capacity of more rapid increase,—enable him when in contact with the savage man, to conquer in the struggle for existence, and to increase at his expense, just as the more favourable increase at the expense of the less favourable varieties in the animal and vegetable kingdoms, just as the weeds of Europe overrun North America and Australia, extinguishing native productions by the inherent vigour of their organisation, and by their greater capacity for existence and multiplication.

If these views are correct; if in proportion as man's social, moral and intellectual faculties became developed, his physical structure would cease to

example, results from the fact that, in a warm country, men live in the open air, while in a cold country they shut themselves up in their houses; that in the one case they nourish themselves in one manner, in the other in another;—these are facts of great importance, facts which, by the simple difference of material life, act powerfully upon civilisation."——*History of Civilisation in Europe*

* "It is probable that the present state and condition of New Zealand exhibit more nearly than any other the condition of Britain when the Romans entered it."——Turner, *History of the Anglo-Saxons*, i, p. 69. "When the Romans first became acquainted with Germany, the natives had advanced but a few steps beyond the savage state."——*Encyclopaedia Britannica*, article "Germany".

be affected by the operation of "natural selection," we have a most important clue to the origin of races. For it will follow, that those striking and constant peculiarities which mark the great divisions of mankind, could not have been produced and rendered permanent after the action of this power had become transferred from physical to mental variations. They must, therefore, have existed since the very infancy of the race; they must have originated at a period when man was gregarious, but scarcely social, with a mind perceptive but not reflective, ere any sense of *right* or feelings of *sympathy* had been developed in him.

By a powerful effort of the imagination, it is just possible to perceive him at that early epoch existing as a single homogeneous race without the faculty of speech, and probably inhabiting some tropical region. He would be still subject, like the rest of the organic world, to the action of "natural selection," which would retain his physical form and constitution in harmony with the surrounding universe. He must have been even then a dominant race, spreading widely over the warmer regions of the earth as it then existed, and, in agreement with what we see in the case of other dominant species, gradually becoming modified in accordance with local conditions. As he ranged farther from his original home, and became exposed to greater extremes of climate, to greater changes of food, and had to contend with new enemies, organic and inorganic, useful variations in his constitution would be selected and rendered permanent, and would, on the principle of "correlation of growth", be accompanied by corresponding external physical changes. Thus arose those striking characteristics and special modifications which still distinguish the chief races of mankind. The red, black, yellow, or blushing white skin; the straight, the curly, the woolly hair; the scanty or abundant beard; the straight or oblique eyes; the various forms of the pelvis, the cranium, and other parts of the skeleton.

But while these changes had been going on, his mental development had correspondingly advanced, and had now reached that condition in which it began powerfully to influence his whole existence, and would, therefore, become subject to the irresistible action of "natural selection." This action would rapidly give the ascendancy to mind: speech would probably now be first developed, leading to a still further advance of the mental faculties, and from that moment man as regards his physical form would remain almost stationary. The art of making weapons, division of labour, anticipation of the future, restraint of the appetites, moral, social and sympathetic feelings, would now have a preponderating influence on his well being, and would therefore be that part of his nature on which "natural selection" would most powerfully act; and we should thus have explained that wonderful persistence of mere physical characteristics, which is the stumbling-block of those who advocate the unity of mankind.

We are now, therefore, enabled to harmonise the conflicting views of anthropologists on this subject. Man may have been, indeed I believe must have been, once a homogeneous race; but it was at a period of which we have as yet discovered no remains, at a period so remote in his history, that he had not yet acquired that wonderfully developed brain, the organ of the mind, which now, even in his lowest examples, raises him far above the highest brutes;—at a period when he had the form but hardly the nature of man, when he neither possessed human speech, nor those sympathetic and moral feelings which in a greater or less degree everywhere now distinguish the race. Just in proportion as these truly human faculties became developed in him would his physical features become fixed and permanent, because the latter would be of less importance to his well being; he would be kept in harmony with the slowly changing universe around him, by an advance in mind, rather than by a change in body. If, therefore, we are of opinion that he was not really man till these higher faculties were developed, we may fairly assert that there were many originally distinct races of men; while, if we think that a being like us in form and structure, but with mental faculties scarcely raised above the brute, must still be considered to have been human, we are fully entitled to maintain the common origin of all mankind.

These considerations, it will be seen, enable us to place the origin of man at a much more remote geological epoch than has yet been thought possible. He may even have lived in the Eocene or Miocene period, when not a single mammal possessed the same form as any existing species. For, in the long series of ages during which the forms of these primeval mammals were being slowly specialised into those now inhabiting the earth, the power which acted to modify them would only affect the mental organisation of man. His brain alone would have increased in size and complexity and his cranium have undergone corresponding changes of form, while the whole structure of lower animals was being changed. This will enable us to understand how the fossil crania of Denise and Engis agree so closely with existing forms, although they undoubtedly existed in company with large mammalia now extinct. The Neanderthal skull may be a specimen of one of the lowest races then existing, just as the Australians are the lowest of our modern epoch. We have no reason to suppose that mind and brain and skull-modification, could go on quicker than that of the other parts of the organisation, and we must, therefore, look back very far in the past to find man in that early condition in which his mind was not sufficiently developed to remove his body from the modifying influence of external conditions, and the cumulative action of "natural selection." I believe, therefore, that there is no *à priori* reason against our finding the remains of man or his works, in the middle or later tertiary deposits. The absence of all such remains in the European beds of this age has little weight, because as we go further back in time, it is natural to suppose that man's distribution over the surface of the earth was less universal than at present. Besides, Europe was in a great measure submerged during the tertiary epoch, and though its scattered islands may have been uninhabited by man, it by no means follows that he did not at the same time exist in warm or tropical continents. If geologists can point out to us the most extensive land in the warmer regions of the earth, which has not been submerged since eocene or miocene times, it is there that we may expect to find some traces of the very early progenitors of man. It is there that we may trace back the gradually decreasing brain of former races, till we come to a time when the body also begins materially to differ. Then we shall have reached the starting point of the human family. Before that period, he had not mind enough to preserve his body from change, and would, therefore, have been subject to the same comparatively rapid modifications of form as the other mammals.

If the views I have here endeavoured to sustain have any foundation, they give us a new argument for placing man apart, as not only the head and culminating point of the grand series of organic nature, but as in some degree a new and distinct order of being. From those infinitely remote ages, when the first rudiments of organic life appeared upon the earth, every plant, and every animal has been subject to one great law of physical change. As the earth has gone through its grand cycles of geological, climatal, and organic progress, every form of life has been subject to its irresistible action, and has been continually, but imperceptibly moulded into such new shapes as would preserve their harmony with the ever changing universe. No living thing could escape this law of its being; none could remain unchanged and live, amid the universal change around it.

At length, however, there came into existence a being in whom that subtle force we term *mind*, became of greater importance than his mere bodily structure. Though with a naked and unprotected body, *this* gave him clothing against the varying inclemencies of the seasons. Though unable to compete with the deer in swiftness, or with the wild bull in strength, *this* gave him weapons with which to capture or overcome both. Though less capable than most other animals of living on the herbs and the fruits that unaided nature supplies, this wonderful faculty taught him to govern and direct nature to his own benefit, and make her produce food for him when and where he pleased. From the moment when the first skin was used as a covering, when the first rude spear was formed to assist in the chase, the first seed sown or shoot planted, a grand revolution was effected in nature, a revolution which in all the previous ages of the earth's history had had no parallel, for a being had arisen who was no longer necessarily subject to change with the changing universe—a being who was in some degree superior to nature, inasmuch, as he knew how to control and regulate her action, and could keep himself in harmony with her, not by a change in body, but by an advance of mind.

Here, then, we see the true grandeur and dignity of man. On this view of his special attributes, we may admit that even those who claim for him a position as an order, a class, or a sub-kingdom by himself, have some reason on their side. He is, indeed, a being apart, since he is not influenced by the great laws which irresistibly modify all other organic beings. Nay more; this victory which he has gained for himself gives him a directing influence over other existences. Man has not only escaped "natural selection" himself, but he actually is able to take away some of that power from nature which, before his appearance, she universally exercised. We can anticipate the time when the earth will produce only cultivated plants and domestic animals; when man's selection shall have supplanted "natural selection"; and when the ocean will be the only domain in which that power can be exerted, which for countless cycles of ages ruled supreme over all the earth.

Briefly to recapitulate the argument;—in two distinct ways has man escaped the influence of those laws which have produced unceasing change in the animal world. By his superior intellect he is enabled to provide himself with clothing and weapons, and by cultivating the soil to obtain a constant supply of congenial food. This renders it unnecessary for his body, like those of the lower animals, to be modified in accordance with changing conditions—to gain a warmer natural covering, to acquire more powerful teeth or claws, or to become adapted to obtain and digest new kinds of food, as circumstances may require. By his superior sympathetic and moral feelings, he becomes fitted for the social state; he ceases to plunder the weak and helpless of his tribe; he shares the game which he has caught with less active or less fortunate hunters, or exchanges it for weapons which even the sick or the deformed can fashion; he saves the sick and wounded from death; and thus the power which leads to the rigid destruction of all animals who cannot in every respect help themselves, is prevented from acting on him.

This power is "natural selection"; and, as by no other means can it be shewn that individual variations can ever become accumulated and rendered permanent so as to form well-marked races, it follows that the differences we now behold in mankind must have been produced before he became possessed of a human intellect or human sympathies. This view also renders possible, or even requires, the existence of man at a comparatively remote geological epoch. For, during the long periods in which other animals have been undergoing modification in their whole structure to such an amount as to constitute distinct genera and families, man's *body* will have remained generically, or even specifically, the same, while his *head* and *brain* alone will have undergone modification equal to theirs. We can thus understand how it is that, judging from the head and brain, Professor Owen places man in a distinct sub-class of mammalia, while, as regards the rest of his body, there is the closest anatomical resemblance to that of the anthropoid apes, "every tooth, every bone, strictly homologous—which makes the determination of the difference between *Homo* and *Pithecus* the anatomist's difficulty." The present theory fully recognises and accounts for these facts; and we may perhaps claim as corroborative of its truth, that it neither requires us to depreciate the intellectual chasm which separates man from the apes, nor refuses full recognition of the striking resemblances to them which exist in other parts of its structure.

In concluding this brief sketch of a great subject, I would point out its bearing upon the future of the human race. If my conclusions are just, it must inevitably follow that the higher—the more intellectual and moral—must displace the lower and more degraded races; and the power of "natural selection", still acting on his mental organisation, must ever lead to the more perfect adaptation of man's higher faculties to the conditions of surrounding nature, and to the exigencies of the social state.* While his external form will probably ever remain unchanged, except in the development of that perfect beauty which results from a healthy and well organised body, refined and

* M. Guizot says: "For myself, I am convinced that there is a destiny of humanity, a transmission of the aggregate of civilisation."—*Civilisation in Europe*

ennobled by the highest intellectual faculties and sympathetic emotions, his mental constitution may continue to advance and improve till the world is again inhabited by a single homogeneous race, no individual of which will be inferior to the noblest specimens of existing humanity. Each one will then work out his own happiness in relation to that of his fellows; perfect freedom of action will be maintained, since the well balanced moral faculties will never permit any one to transgress on the equal freedom of others; restrictive laws will not be wanted, for each man will be guided by the best of laws; a thorough appreciation of the rights, and a perfect sympathy with the feelings, of all about him; compulsory government will have died away as unnecessary (for every man will know how to govern himself), and will be replaced by voluntary associations for all beneficial public purposes; the passions and animal propensities will be restrained within those limits which most conduce to happiness; and mankind will have at length discovered that it was only required of them to develop the capacities of their higher nature, in order to convert this earth, which had so long been the theatre of their unbridled passions, and the scene of unimaginable misery, into as bright a paradise as ever haunted the dreams of seer or poet.

(The general idea and argument of this paper I believe to be new. It was, however, the perusal of Mr. Herbert Spencer's works, especially *Social Statics*, that suggested it to me, and at the same time furnished me with some of the applications.)

* * * * *

As the first effort to connect natural selection to the touchy problem of the evolution of human races, these ideas were immediately attacked on all fronts. At the meeting itself, Wallace responded to the various objections that were raised to his theory as follows (from pp. clxxxi–clxxxvii of ref 93):

. . . I will now pass on to notice the special objections that have been brought forward to my theory . . . two . . . gentlemen who spoke agreed with me generally. Mr. Bendyshe objected to my statement, that man, to some extent, triumphs over nature; and he argued that man does not triumph over climate, because Europeans cannot live in the tropics, and the natives of the tropics cannot live in Europe. First, I say that there are facts to show that this is not absolutely the case. There are cases in which Europeans have gone and resided in the tropics, and, as far as we can see, live there to this day perfectly well . . .

. . . admitting that man may not be able to stand a sudden change in climate, yet, supposing that the change were a slow one—supposing that Europe were gradually sunk beneath the sea from the north, so that we were gradually shoved, as it were, into a tropical climate at the rate of a few miles in a century,—do you not think that natural selection would act so that the race would stand the climate? I do not think we should all die out. All the facts of nature seem to be opposed to such a supposition. The dog has stood all over the world with us notwithstanding the climate . . .

. . . I have asserted that man, in his progress from a low to a high state, would be assisted by the necessary discipline of a harsh climate, which would make him exert his mental faculties much more than a tropical climate. Now, I think that is almost self-evident, and is not at all affected [for example] by the fact that the Esquimaux are less intelligent than the English. The question is, "Do they build houses at all?" Yes; and very good ones. Travellers describe how ingeniously they build their snow houses; and the manner in which they make their clothing and sledges shows that they are not so low intellectually as most of the inhabitants of tropical countries.

Mr. [James] Reddie . . . wants to know how the intellect came at first. I don't pretend to answer that question, because we must go so long back. If Mr. Reddie denies that any animal has intellect, it is a difficult question to answer; but if animals have intellect in different proportions, and if the human infant, the moment it is born, has not so much intellect as an animal, and if, as the infant grows, the intellect grows with it, I do not see the immense difficulty if you grant the universal process of selection from lower to higher animals. If you throw aside altogether this process of selection, you need not make the objection about the intellect.

Mr. [C. Carter] Blake made a few objections, which may have some little weight. The principal was that we have no evidence to show that when one race, or nation, or people are exterminated, or driven out by another, the one that is so exterminated is necessarily inferior; and he wanted to show either by historic evidence or by remains of bodies that it is impossible to say that the Celtic was inferior to the Teutonic, or the Basque inferior to the race which drove them out. Now, it appears to me that the mere fact of one race supplanting another proves their superiority. It is not a question of intellect only, nor of bodily strength only. We cannot tell what causes may produce it. A hundred peculiarities, that we can hardly appreciate, may cause the one race to melt away, as it were, before the other. But still there is the plain fact that two races came into contact, and that one drives out the other. This is a proof that the one race is better fitted to live upon the world than the other. Mr. Blake says that there is no necessary correlation between man and his habitat; and he endeavoured to show that by proving that the thickness of the crania does not vary in accordance with the heat of the sun. No doubt such an objection is very easy to make; but we must consider, is it at all likely that we shall be able, by our examination, to appreciate this correlation, whatever it may be. For instance, you take two animals; one lives in a northern hemisphere, the other in a southern,—one in a wet country, the other in a dry one. Can you tell me why these two animals are fitted to live in their respective climates? They may be so closely allied that you can hardly find out their differences; and if you cannot find out the difference in animals which serves to adapt them to the climate, is it likely you can find out the difference in man? But there are facts which show that there is a correlation between man and his habitat. For instance, take the case of the inhabitants of West Africa, who stand the fever and malaria of that country; and it is the same in New Orleans. It is asserted in America, I believe, that one-fourth of black blood is enough to save the individual from the yellow fever in New Orleans. This is a striking case, I think, of correlation between man and his habitat. Then again, as to the prevalence of black-skinned races in the tropical regions, I do not believe that there is any special production of the black skin by the heat of the sun; but I believe that because the black skin is correlative to the hot sun, the black-skinned constitution is best adapted to stand the diseases of the climate, and the process of natural selection has preserved them. If we find a people who are apparently not well adapted to stand the climate, we have some reason to believe that they are a comparatively recent immigration into the country. My friend, Mr. Bates, who is not here, has supported this theory from his observations on the Amazon, asserting that the inhabitants of tropical America are a recent introduction. He comes to that conclusion from a great many peculiarities of manners and customs, and if so, it is a corroboration of the argument that races do become correlated to the climate in which they live.

Mr. Blake [also] objected to my statement, that man can to a certain extent control nature. He asserted that man could not control disease; but that was not the point I went upon. I especially mentioned the point on which man can control nature,—raising himself by his intellect above the action of natural selection, which changes the forms of other animals, because they could only be kept in harmony with the universe except by being changed; whereas man is kept in harmony by his mind. Again, no weak animal—no animal born with a sickly constitution—lives to propagate its kind: but man does. Hundreds of weak individuals live to a comparatively healthy and comfortable old age, and have large families. This is a special case, in which man controls nature differently to the animals. He controls nature so much that he is an exception to all the rest of animated beings.

Dr. [James] Hunt made a great many special objections. He says I disappointed him, because I promised to explain everything. I must say, I did not. I simply proposed to myself to explain, or rather to suggest, a theory which should do away with this difficulty of the absolute contradiction between two classes of ethnologists, commonly called the monogenists and polygenists, by showing that both were right. I think that is a most satisfactory way of harmonising people that differ. Again, he objects to my using the expression "must have been". Well, I put in the words "I believe," and "according to the Darwinian theory," because, according to that theory, every group of species arises from one, every group of varieties from one, every group of individuals from a pair; therefore, if you do but go far back enough, you must come to a unity of origin. If that theory is utterly wrong, then my argument goes for nothing . . . Dr. Hunt [then] asserts that archaeology shows that the ancient races were the same as modern. Well, that is a fact I quoted on my own side, and his quoting it against me only shows that you can twist a fact as you like. I quoted it as a proof that you must go to an enormous distance of time to bridge over the difference between the crania of the lower animals and of man. I said, perhaps a million, or even ten millions, of years were necessary. If my argument is correct, it is a logical conclusion. Dr. Hunt objects to my using an expression to the effect that students are rather dogmatic in assertions of this kind. Well, I think I could bring forward facts to prove this; and I should think that anybody who knows anything of the literature of the subject would agree with me that there is the strongest feeling on both sides that they are right, and that they express their feelings in the strongest manner, and that each party is inclined to look down on what it believes to be the absurd ideas of the other. Still, I do not deny that there are some who do not manifest this dogmatic feeling. With respect to the fact about the Portuguese and Spaniards in South America, I can assert it on my own authority, because I have lived among them, and have seen European families in tropical countries who have been there for many generations . . .

. . . You cannot [of course] isolate men [entirely]. They will mix; and there is no possible fact you can bring forward but is liable to the same objection. It was thought at one time by Prichard and the older ethnologists that it was a strong argument for the unity of the race that the Jews were white, black, and brown. Now, it is known that in every case in which the Jews have changed colour apparently, it has been the Jewish converts who have been treated as Jews, simply because they have embraced that religion. But a better proof than colour is physiognomy, which you see maintained in the Jews all over the world. Physiognomy maintains itself much longer than colour; and it seems as if the physiognomy of the superior race maintained itself much longer than the inferior; whereas the colour of the inferior race is often most lasting. For example, I may mention the descendants of the Portuguese in the Malay archipelago. In a great many towns there are thousands of Portuguese; some of them keep the Portuguese language; others have lost it; but still Portuguese words crop up all over the land, and there are Portuguese customs and manners and European features; but still they are generally the same colour as the people of the country in which they live. With respect to Europeans not living in India, that is nothing when we remember what a vile climate it is. We live in it as an exceptional race; and if we could bring instances of the third generation, you would say there was mixed blood in them.

Then again, Dr. Hunt wanted me to explain how I could use such a word as "provident". Why, is it not perfectly clear that if people live in a country where there is a severe winter, in which little or no food is to be had, that they must provide against the scarcity, and that gradually the race would become a provident race? Therefore, I think I am justified in saying that, given two races of the same capacity, and put one in a tropical and the other in a temperate climate, the one in the temperate climate will become the more provident race of the two . . .

. . . Well, then, as to the term "inherent," I do not mean to withdraw it. I mean to maintain it as a very proper expression; and the answer I gave to that last question about a provident race will almost answer for this,—that peculiarities produced gradually by natural selection, or any other cause, become inherent. The very fact of the race being gradually brought into harmony with the climate of the country in which it is, gives it a superior power, and an inherent capacity to maintain it. I do not know whether the words are the same, but the sense is exactly the same as will be found in

Darwin's own book, where he points out this extraordinary fact, the bearing of which had never been noticed before, that in Australia, in the Cape of Good Hope, and to a considerable degree in North America—in fact, to a great extent in all the comparatively limited areas to which Europeans go—the weeds of Europe that are carried accidentally thrive and flourish there. They spread over the country, and maintain themselves in competition with the native weeds, showing that they are better adapted for the country than the plants which were apparently specially created for the country. Mr. Darwin explains it on his theory in this manner,—that Europe and Asia, which now to a great extent are dry land, have been long in existence as dry land; and that in the immense series of ages during which the changes of the northern continent have been going on, becoming modified from one form to another, sometimes to an inland climate, sometimes to a continental climate, sometimes a mountainous region, sometimes a flat region; owing to that great amount of change, its plants have acquired an immense variety of specialities: because, when a speciality is once acquired, it is not lost. It is handed down and kept in store, as it were, so that the immense mutations which the northern hemisphere has undergone, have given these plants a capacity of adapting themselves to a great variety of conditions. The result is that directly they are carried into Australia these properties come into play. They have been adapted, in some previous state of the northern hemisphere, to similar conditions, and they have inherited this peculiarity by transmission, and therefore they are capable of driving out the plants of Australia merely by the inherent vigour they have gained. I applied this in illustration of the way in which civilised man has been developed by a great variety of circumstances. The intermixture of races has been very great. We are a mixed race to a very great extent, and therefore we have the capacities and powers of a great many; therefore, when we come into contact with the lower races, we are enabled in the same manner to drive them out.

Then, it is said, that man without speech is not man. That is one of my points. I said, if you choose to consider he is not man, then so and so follows; but if you consider he is man, then so and so. And as to the argument that if man could take the effects of natural selection away, it must be powerless,—that has not much to do with the subject. We might as well say, how powerless life is, because we can take it away,—when such a slight thing as stopping the mouth with pitch-plaster can destroy it. This only shows how easily it can be changed or destroyed; it does not prove its weakness. And so it does not show the weakness of natural selection, because man is able to modify it by putting himself into certain conditions, instead of leaving nature to select those conditions for him . . .

. . . With regard to the poetical conclusion, I would merely say that I began it by stating that I would point out what I considered to be the bearings of this theory, if it is true. If it is not true, of course my remarks go for nothing; but I do not think myself that the concluding part of the paper is more poetical than true.

* * * * *

*Wallace's application of the theory of natural selection to the special problem of human evolution helped prepare thinkers for Darwin's later, exhaustive, treatment of the subject (*The descent of man, *1871). It can be surmised, however, that Wallace himself never considered natural selection the only force supporting human evolution. Most writers have interpreted the chain of events that followed his writing of the 1858 Ternate essay on natural selection (ref 43) as indicating that Wallace at first believed natural selection to be an all-sufficient evolutionary mechanism, but later changed his mind (perhaps in conjunction with his adoption of spiritualism). The evidence suggests to me, however, that no such 'change of mind' ever took place; rather, he simply never considered natural selection to be an 'all-sufficient' force. (We will pursue this point further in Sections Two and Six.) In the 1864 paper reproduced above there are some hints as to Wallace's position on the matter—for example, the 'above biology' portrayal of man—but for the time being he avoided extending the argument (commentary of similar type, however, appeared in an 1867 letter to the* Anthropological Review *(ref 125) and the same year's 'Creation by law' (ref 140)). The opportunity to speak his mind fully came in 1869, when he was asked by* The Quarterly Review *to notice new editions of Sir Charles Lyell's* Principles of geology *and* Elements of geology. *On pages 391–4 of this review (ref 146) appears the following analysis:*

. . . In adopting the views of Mr. Darwin, Sir Charles Lyell carries them out to their legitimate results, and does not shrink from the logical necessity, of the derivation of man from the lower animals; and he has written a very interesting chapter on the 'Origin and Distribution of Man.' Into this subject, however, we cannot now enter, except to remark briefly on some aspects of the question which all who have hitherto written upon it seem to have neglected.

It would certainly appear in the highest degree improbable that the whole animal kingdom from the lowest zoophytes up to the horse, the dog, and the ape, should have been developed by the simple action of natural laws, and that the animal man, so absolutely identical with them in all the main features and many of the details of his organization, should have been formed in some quite other unknown way. But if the researches of geologists and the investigations of anatomists should ever demonstrate that he was derived from the lower animals in the same way that they have been derived from each other, we shall not be thereby debarred from believing, or from proving, that his intellectual capacities and his moral nature were not wholly developed by the same process. Neither natural selection nor the more general theory of evolution can give any account whatever of the origin of sensational or conscious life. They may teach us how, by chemical, electrical, or higher natural laws, the organized body can be built up, can grow, can reproduce its like; but those laws and that growth cannot even be conceived as endowing the newly-arranged atoms with consciousness. But the moral and higher intellectual nature of man is as unique a phenomenon as was conscious life on its first appearance in the world, and the one is almost as difficult to conceive as originating by any law of evolution as the other. We may even go further, and maintain that there are certain purely physical characteristics of the human race which are not explicable on the theory of variation and survival of the fittest. The brain, the organs of speech, the hand, and the external form of man, offer some special difficulties in this respect, to which we will briefly direct attention.

In the brain of the lowest savages, and, as far as we yet know, of the prehistoric races, we have an organ so little inferior in size and complexity to that of the highest types (such as the average European), that we must believe it capable, under a similar process of gradual development during the space of two or three thousand years, of producing equal average results. But the mental requirements of the lowest savages, such as the Australians or the Andaman islanders, are very little above those of many animals. The higher moral faculties and those of pure intellect and refined emotion are useless to them, are rarely if ever manifested, and have no relation to their wants, desires, or well-being. How, then, was an organ developed so far beyond the needs of its possessor? Natural selection could only have endowed the savage with a brain a little superior to that of an ape, whereas he actually possesses one but very little inferior to that of the average members of our learned societies.

Again, what a wonderful organ is the hand of man; of what marvels of delicacy is it capable, and how greatly it assists in his education and mental development! The whole circle of the arts and sciences are ultimately dependent on our possession of this organ, without which we could hardly have become truly human. This hand is equally perfect in the lowest savage, but he has no need for so fine an instrument, and can no more fully utilise it than he could use without instruction a complete set of joiner's tools. But, stranger still, this marvellous instrument was foreshadowed and prepared in the Quadrumana; and any person who will watch how one of these animals uses its hands, will at once perceive that it possesses an organ far beyond its needs. The separate fingers and the thumb are never fully utilised, and objects are grasped so clumsily, as to show that a much less specialised organ of prehension would have served its purpose quite as well; and if this be so, it could never have been produced through the agency of natural selection alone.

We have further to ask—How did man acquire his erect posture, his delicate yet expressive features, the marvellous beauty and symmetry of his whole external form;—a form which stands alone, in many respects more distinct from that of all the higher animals than they are from each other? Those who have lived much among savages know that even the lowest races of mankind, if healthy and well fed, exhibit the human form in its complete symmetry and perfection. They all have the soft smooth skin absolutely free from any hairy covering on the dorsal line, where all other mammalia from the Marsupials up to the Anthropoid apes have it most densely and strongly developed. What use can we conceive to have been derived from this exquisite beauty and symmetry and this smooth bare skin, both so very widely removed from his nearest allies? And if these modifications were of no physical use to him—or if, as appears almost certain in the case of the naked skin, they were at first a positive disadvantage—we know that they could not have been produced by natural selection. Yet we can well understand that both these characters were essential to the proper development of the perfect human being. The supreme beauty of our form and countenance has probably been the source of all our aesthetic ideas and emotions, which could hardly have arisen had we retained the shape and features of an erect gorilla; and our naked skin, necessitating the use of clothing, has at once stimulated our intellect, and by developing the feeling of personal modesty may have profoundly affected our moral nature.

The same line of argument may be used in connexion with the structural and mental organs of human speech, since that faculty can hardly have been physically useful to the lowest class of savages; and if not, the delicate arrangements of nerves for its production could not have been developed and co-ordinated by natural selection. This view is supported by the fact that, among the lowest savages with the least copious vocabularies, the capacity of uttering a variety of distinct articulate sounds, and of applying to them an almost infinite amount of modulation and inflection, is not in any way inferior to that of the higher races. An instrument has been developed in advance of the needs of its possessor.

. . . enough, we think, has now been said to indicate the possibility of a new stand-point for those who cannot accept the theory of evolution as expressing the whole truth in regard to the origin of man. While admitting to the full extent the agency of the same great laws of organic development in the origin of the human race as in the origin of all organized beings, there yet seems to be evidence of a Power which has guided the action of those laws in definite directions and for special ends. And so far from this view being out of harmony with the teachings of science, it has a striking analogy with what is now taking place in the world, and is thus strictly uniformitarian in character. Man himself guides and modifies nature for special ends. The laws of evolution alone would perhaps never have produced a grain so well adapted to his uses as wheat; such fruits as the seedless banana and the bread-fruit; such animals as the Guernsey milch-cow, or the London dray-horse. Yet these so closely resemble the unaided productions of nature, that we may well imagine a being who had mastered the laws of development of organic forms through past ages, refusing to believe that any new power had been concerned in their production, and scornfully rejecting the theory that in these few cases a distinct intelligence had directed the action of the laws of variation, multiplication, and survival, for his own purposes. We know, however, that this has been done; and we must therefore admit the possibility that in the development of the human race, a Higher Intelligence has guided the same laws for nobler ends.

Such, we believe, is the direction in which we shall find the true reconciliation of Science with Theology on this most momentous problem. Let us fearlessly admit that the mind of man (itself the living proof of a supreme mind) is able to trace, and to a considerable extent has traced, the laws by means of which the organic no less than the inorganic world has been developed. But let us not shut our eyes to the evidence that an Overruling

Intelligence has watched over the action of those laws, so directing variations and so determining their accumulation, as finally to produce an organization sufficiently perfect to admit of, and even to aid in, the indefinite advancement of our mental and moral nature.

* * * * *

In 1870 Wallace expanded this argument into a rather long essay (ref 165) that completed the collection Contributions to the theory of natural selection. *These newly-expressed views troubled Darwin, of course—he feared in them an erosion of the natural selection concept. Some, however (including Lyell), felt that they returned some 'dignity' to the story of our species and awaited further developments. Others attacked Wallace's restrictive treatment of one of natural selection's central tenets, the 'utility hypothesis'. In the following letter, printed in* Scientific Opinion *in 1869 (and later incorporated into ref 165), he spoke out on a related issue:*

The origin of moral intuitions (ref 153)

Sir,—I do not think your correspondent "S. L." is justified in speaking so contemptuously of Mr. Hutton's original and well-reasoned article 'A Questionable Parentage for Morals;' and I am inclined to think that Mr. Herbert Spencer himself would not claim for his views on this subject that demonstrable certainty which "S. L." seems to think attaches to them. The question really depends upon the more fundamental one, of whether man's entire mental and moral nature is the product and outcome of that material organization whose laws of growth and development Mr. Spencer has so well elucidated. If mind with all its powers is simply a function of organized matter, then Mr. Spencer's theory of the origin of morals is the only one which can be held by a student of science. If, however, there is any thing in man more than his physical organization, then it becomes a subject of strict scientific and philosophical inquiry to determine from a study of the phenomena of his mind in various stages of growth and under various conditions, what is the mental substratum required to account for the development of the faculties we actually find in him. Mr. Spencer maintains that an appreciation of utilities is all that is required to develop the moral sense; Mr. Hutton argues that this is insufficient, and that the moral sense itself, the appreciation of right and wrong, with a mental impulse towards the first and away from the second, is an essential part of the mental substratum of our nature.

To go into the question at all fully would be quite out of place here; I will therefore only adduce one group of facts which seem to me inexplicable on the utilitarian hypothesis. The utilitarian sanction for truthfulness is by no means very powerful or universal. Few laws enforce it. No very severe reprobation follows untruthfulness. In all ages and countries falsehood has been held permissible in love, and laudable in war; while, even to the present day, it is held venial by the majority of mankind in trade and commerce. A certain amount of untruthfulness is a necessary part of politeness in the east and west alike, while even severe moralists have held a lie justifiable to elude an enemy or prevent a crime. Such being the difficulties with which this virtue has had to struggle, with so many exceptions to its practice, with so many instances in which it brought ruin or death to its too ardent devotee, how can we believe that considerations of utility could ever invest it with the mysterious sanctity of the highest virtue,—could ever induce men to value truth for its own sake and practice it regardless of consequences?

Yet it is a fact that such a mystical sense of wrong does attach to untruthfulness, not only among the higher classes of civilized people, but among whole tribes of utter savages. Sir Walter Elliott tells us (in his paper 'On the Characteristics of the Population of Central and Southern India,' published in the *Journal* of the Ethnological Society of London, vol. i., p. 107) that the Kurubars and Santals, barbarous hill-tribes of Central India, are noted for veracity. It is a common saying that "a Kurubar *always* speaks the truth;" and Major Jervis says, "the Santals are the most truthful men I ever met with." As a remarkable instance of this quality the following fact is given. A number of prisoners, taken during the insurrection, were allowed to go free on parole, to work at a certain spot for wages. After some time cholera attacked them and they were obliged to leave, but every man of them returned and gave up his earnings to the guard. Two hundred savages with money in their girdles, walked thirty miles back to prison rather than break their word! My own experience among savages has furnished me with similar, although less severely tested, instances, and we cannot avoid asking, how is it that in these few cases "experiences of utility" have left such an overwhelming impression, while in so many others they have left none? The experiences of savage men as regards the utility of truth must, in the long run, be pretty nearly equal. How is it then that in some cases the result is a sanctity which overrides all considerations of personal advantage, while in others there is hardly a rudiment of such a feeling?

The intuitional theory explains this by the supposition that there is a feeling—a sense of right and wrong—in our nature antecedent to and independent of experience of utility. Where free play is allowed to the relations between man and man, this feeling attaches itself to those acts of universal utility or self-sacrifice which are the products of our affections and sympathies, and which we term moral, while it may be, and often is, perverted to give the same sanction to acts of narrow and conventional utility which are really immoral,—as when the Hindoo will tell a lie but will sooner starve than eat unclean food, and looks upon the marriage of adult females as gross immorality.

The strength of the moral feeling will depend upon individual or racial constitution,—the acts to which its sanctions are applied will depend upon how far the simple feelings and affections of our nature have been modified by custom, by law, or by religion.

The question to be considered is, first, whether such an intense and mystical feeling of right and wrong (so intense as to overcome all ideas of personal advantage or utility) could have been developed out of accumulated ancestral experiences of utility; and, in the second place, whether feelings so developed by one set of utilities could be transferred to acts of which the utility was partial, imaginary, or altogether absent.

Although myself an enthusiastic admirer of Mr. Spencer's writings, and a follower of his philosophy, I am decidedly of opinion that there is a limit to the sphere which that philosophy embraces, and that the limit is to be found in the doctrine of the origin of morals.

* * * * *

Wallace's attack on Darwin's application of natural selection to the development of the 'higher faculties' was continued on pages 182–3 of a review of The descent of man *that appeared in print in 1871 (ref 186):*

... Having in the first part of this article made some objections to the theory of sexual selection in its application to man, I will now briefly notice Mr. Darwin's account of the probable mode in which man became developed from his brute ancestor. All the evidence goes to show, that the change from brute to man took place in some limited area, probably tropical. Here he lost his hairy covering, acquired his erect form and his wonderful brain, and became so far advanced in the arts and in morals that natural selection ceased to act upon his mere bodily organization. It is also probable that he learnt to speak language, discovered the use of fire, and perhaps even of canoes, before he spread over the earth, and before the several races of man were differentiated. The agency through which this vast transformation occurred was the struggle for existence and natural selection—a struggle first with other animals, and when they were surpassed between tribe and tribe; and this alone Mr. Darwin thinks would, under favourable conditions, raise man to his present high position in the organic scale.

In this view there are many difficulties. How is it possible to conceive, that during the enormous interval required to change a quadrumanous, hairy, speechless animal, into erect, smooth-skinned, large-brained, fire-using man, while the struggle for existence was most severe (for by the severity of the struggle alone he was raised), he yet never spread over the earth but remained concentrated in a limited area. Had he spread widely during the process of modification, divergence of character would inevitably have occurred, and we should have had several distinct species of man. Mr. Darwin argues that the fact of man, even at his lowest stage of civilisation and intellect, being able to maintain himself, surrounded by the most powerful and ferocious animals, is due to his large brain, which is thus of the most essential use to him. But almost all herbivorous animals also maintain themselves under similar conditions, with no special endowment of brains; and in South America the apparently helpless and almost idiotic sloth is not exterminated, though exposed to the attacks of pumas, tiger-cats, and harpy eagles. Man could have acquired very little of his superiority to animals by a struggle with animals. "Natural selection does not produce absolute perfection but only relative perfection." We have to fall back therefore on the struggle with his fellows—family with family, tribe with tribe. But for this to be at all effectual, one of the most essential conditions is a large population inhabiting an extensive area, and this the conditions of the problem deny to us. The vast amount of the superiority of man to his nearest allies is what is so difficult to account for. His absolute erectness of posture, the completeness of his nudity, the harmonious perfection of his hands, the almost infinite capacities of his brain, constitute a series of correlated advances too great to be accounted for by the struggle for existence of an isolated group of apes in a limited area. And Mr. Darwin himself gives hints of unknown causes which may have aided in the work. He says: "An unexplained residuum of change, perhaps a large one, must be left to the assumed uniform action of those unknown agencies which occasionally induce strongly marked and abrupt deviations of structure in our domestic productions." And again: "If these causes, whatever they may be, were to act more uniformly and energetically during a lengthened period (and no reason can be assigned why this should not sometimes occur), the result would probably be, not mere slight differences, but well-marked, constant modifications."

In concluding this very imperfect account of one of the most remarkable works in the English language, it may be affirmed, that Mr. Darwin has all but demonstrated the origin of man by descent from some inferior animal form—that he has proved the vast importance of sexual influences in modifying the colours and the structure of the more highly organized animals—and that he has thrown fresh light upon the intricate question of the mode of development of the moral and intellectual nature of man. Yet it must be admitted that there are many difficulties in the detailed application of his views; and it seems probable that these can only be overcome by giving more weight to those unknown laws whose existence he admits, but to which he assigns an altogether subordinate part in determining the development of organic forms.

* * * * *

Wallace's study of human evolution extended logically to an interest in the antiquity of the civilized condition. This is the featured subject of 'Rise and progress of modern views as to the antiquity and origin of man' (ref 257), the second half of an address delivered as President of the Section of Biology during the 1876 meeting of the British Association for the Advancement of Science:

. . . It is a curious circumstance that notwithstanding the attention that has been directed to the subject in every part of the world, and the numerous excavations connected with railways and mines which have offered such facilities for geological discovery, no advance whatever has been made for a considerable number of years in detecting the time or mode of man's origin. The Palaeolithic flint weapons first discovered in the North of France more than thirty years ago are still the oldest undisputed proofs of man's existence; and amid the countless relics of a former world that have been brought to light, no evidence of any one of the links that must have connected man with the lower animals has yet appeared.

It is, indeed, well known that negative evidence in geology is of very slender value; and this is, no doubt, generally the case. The circumstances here are, however, peculiar, for many converging lines of evidence show that, on the theory of development by the same laws which have determined the development of the lower animals, man must be immensely older than any traces of him yet discovered. As this is a point of great interest we must devote a few moments to its consideration.

1. The most important difference between man and such of the lower animals as most nearly approach him is undoubtedly the bulk and development of his brain, as indicated by the form and capacity of the cranium. We should therefore anticipate that these earliest races, who were contemporary with the extinct animals and used rude stone weapons, would show a marked deficiency in this respect. Yet the oldest known crania (those of the Engis and Cro-Magnon caves) show no marks of degradation. The former does not present so low a type as that of most existing savages, but is (to use the words of Prof. Huxley) "a fair average human skull, which might have belonged to a philosopher, or might have contained the thoughtless brains of a savage." The latter are still more remarkable, being unusually large and well formed. Dr. Pruner-Bey states that they surpass the average of modern European skulls in capacity, while their symmetrical form without any trace of prognathism, compares favourably not only with those of the foremost savage races, but with many civilized nations of modern times.

One or two other crania of much lower type, but of less antiquity than this, have been discovered; but they in no way invalidate the conclusion which so highly developed a form at so early a period implies, viz. that we have as yet made a hardly perceptible step towards the discovery of any earlier stage in the development of man.

2. This conclusion is supported and enforced by the nature of many of the works of art found even in the oldest cave-dwellings. The flints are of the old chipped type, but they are formed into a large variety of tools and weapons—such as scrapers, awls, hammers, saws, lances, etc., implying a variety of purposes for which these were used, and a corresponding degree of mental activity and civilization. Numerous articles of bone have also been found, including well-formed needles, implying that skins were sewn together, and perhaps even textile materials woven into cloth. Still more important are the numerous carvings and drawings representing a variety of animals, including horses, reindeer, and even a mammoth, executed with considerable skill on bone, reindeer-horns, and mammoth-tusks. These, taken together, indicate a state of civilization much higher than that of the lowest of our modern savages, while it is quite compatible with a considerable degree of mental advancement, and leads us to believe that the crania of Engis and Cro-Magnon are not exceptional, but fairly represent the characters of the race. If we further remember that these people lived in Europe under the unfavourable conditions of a sub-Arctic climate, we shall be inclined to agree with Dr. Daniel Wilson, that it is far easier to produce evidences of deterioration than of progress in instituting a comparison between the contemporaries of the mammoth and later prehistoric races of Europe or savage nations of modern times (*Prehistoric Man*, 3rd ed., vol. i., p. 117).

3. Yet another important line of evidence as to the extreme antiquity of the human type has been brought prominently forward by Prof. [St. George] Mivart (*Man and Apes*, pp. 171–193). He shows, by a careful comparison of all parts of the structure of the body, that man is related not to any one, but almost equally to any of the existing apes—to the orang, the chimpanzee, the gorilla, and even to the gibbons—in a variety of ways; and these relations and differences are so numerous and so diverse that, on the theory of evolution, the ancestral form which ultimately developed into man must have diverged from the common stock whence all these various forms and their extinct allies originated. But so far back as the Miocene deposits of Europe we find the remains of apes allied to these various forms, and especially to the gibbons; so that in all probability the special line of variation which led up to man branched off at a still earlier period. And these early forms, being the initiation of a far higher type, and having to develop by natural selection into so specialized and altogether distinct a creature as man, must have risen at a very early period into the position of a dominant race, and spread in dense waves of population over all suitable portions of the great continent—for this, on Mr. Darwin's hypothesis, is essential to rapid developmental progress through the agency of natural selection.

Under these circumstances we might certainly expect to find some relics of these earlier forms of man along with those of animals, which were presumably less abundant. Negative evidence of this kind is not very weighty, but still it has *some* value. It has been suggested that as apes are mostly tropical, and anthropoid apes are now confined almost exclusively to the vicinity of the equator, we should expect the ancestral forms also to have inhabited these same localities—West Africa and the Malay Islands. But this objection is hardly valid, because existing anthropoid apes are wholly dependent on a perennial supply of easily accessible fruits, which is only found near the equator, while not only had the south of Europe an almost tropical climate in Miocene times, but we must suppose even the earliest ancestors of man to have been terrestrial and omnivorous, since it must have taken ages of slow modification to have produced the perfectly erect form, the short arms, and the wholly non-prehensile foot, which so strongly differentiate man from the arboreal apes.

The conclusion which I think we must arrive at is, that if man has been developed from a common ancestor, with all existing apes, *and by no other agencies than such as have affected their development*, then he must have existed, in something approaching his present form, during the tertiary period—and not merely existed, but predominated in numbers, wherever suitable conditions prevailed. If, then, continued researches in all parts of Europe and Asia fail to bring to light any proofs of his presence, it will be at least a presumption that he came into existence at a much later date, and by a much more rapid process of development. In that case it will be a fair argument that, just as he is in his mental and moral nature, his capacities and aspirations, so infinitely raised above the brutes, so his origin is due, in part, to distinct and higher agencies than such as have affected their development.

There is yet another line of inquiry bearing upon this subject to which I wish to call your attention. It is a somewhat curious fact that, while all modern writers admit the great antiquity of man, most of them maintain the very recent development of his intellect, and will hardly contemplate the possibility of men equal in mental capacity to ourselves having existed in prehistoric times. This question is generally assumed to be settled by such relics as have been preserved of the manufactures of the older races showing a lower and lower state of the arts, by the successive disappearance in early times of iron, bronze, and pottery, and by the ruder forms of the older flint implements. The weakness of this argument has been well shown by Mr. Albert Mott in his very original but little-known presidential address to the Literary and Philosophical Society of Liverpool in 1873. He maintains that "our most distant glimpses of the past are still of a world peopled as now with men both civilized and savage," and "that we have often entirely misread the past by supposing that the outward signs of civilization must always be the same, and must be such as are found among ourselves." In support of this view he adduces a variety of striking facts and ingenious arguments, a few of which I will briefly summarize.

On one of the most remote islands of the Pacific—Easter Island—2000 miles from South America, 2000 from the Marquesas, and more than 1000 from the Gambier Islands, are found hundreds of gigantic stone images, now mostly in ruins, often thirty or forty feet high, while some seem to have been much larger, the crowns on their heads cut out of a red stone being sometimes ten feet in diameter, while even the head and neck of one is said to have been twenty feet high (*J. Roy. Geog. Soc.* 1870, pp. 177, 178). These once stood erect on extensive stone platforms; yet the island has only an area of about thirty square miles, or considerably less than Jersey. Now as one of the smallest images eight feet high weighs four tons, the largest must weigh over a hundred tons, if not much more; and the existence of such vast works implies a large population, abundance of food, and an established goverment. Yet how could these coexist in a mere speck of land wholly cut off from the rest of the world? Mr. Mott maintains that this necessarily implies the power of regular communication with larger islands or a continent, the arts of navigation, and a civilization much higher than now exists in any part of the Pacific. Very similar remains in other islands scattered widely over the Pacific add weight to this argument.

The next example is that of the ancient mounds and earthworks of the North American continent, the bearing of which is even more significant. Over the greater part of the extensive Mississippi valley four well-marked classes of these earthworks occur. Some are camps, or works of defence, situated on bluffs, promontories, or isolated hills; others are vast enclosures in the plains and lowlands, often of geometric forms, and having attached to them roadways or avenues often miles in length; a third are mounds corresponding to our tumuli, often seventy to ninety feet high, and some of them covering acres of ground; while a fourth group consist of representations of various animals modelled in relief on a gigantic scale, and occurring chiefly in an area somewhat to the north-west of the other classes, in the plains of Wisconsin . . .

The antiquity of [the] . . . remarkable race [responsible for these earthworks] may perhaps not be very great as compared with the prehistoric man of Europe, although the opinions of some writers on the subject seemed affected by that "parsimony of time" on which the late Sir Charles Lyell so often dilated. The mounds are all overgrown with dense forest, and one of the large trees was estimated to be eight hundred years old, while other observers consider the forest growth to indicate an age of at least 1000 years. But it is well known that it requires several generations of trees to pass away before the growth on a deserted clearing comes to correspond with that of the surrounding virgin forest, while this forest, once established, may go on growing for an unknown number of thousands of years. The 800 or 1000 years estimate from the growth of existing vegetation is a minimum which has no bearing whatever on the actual age of these mounds . . .

The important thing for us, however, is that when North America was first settled by Europeans, the Indian tribes inhabiting it had no knowledge or tradition of any preceding race of higher civilization than themselves. Yet we find that such a race existed; that they must have been populous and have lived under some established government; while there are signs that they practised agriculture largely, as, indeed, they must have done to have supported a population capable of executing such gigantic works in such vast profusion; for it is stated that the mounds and earthworks of various kinds in the state of Ohio alone amount to between eleven and twelve thousand. In their habits, customs, religion, and arts they differed strikingly from all the Indian tribes: while their love of art and of geometric forms, and their capacity for executing the latter upon so gigantic a scale, render it probable that they were a really civilized people, although the form their civilization took may have been very different from that of later people subject to very different influences, and the inheritors of a longer series of ancestral civilizations. We have here, at all events, a striking example of the transition, over an extensive country, from comparative civilization to comparative barbarism, the former left to tradition and having hardly any trace of influence on the latter.

As Mr. Mott well remarks:—Nothing can be more striking than the fact that Easter Island and North America both gave the same testimony as to the origin of the savage life found in them, although in all circumstances and surroundings the two cases are so different. If no stone monuments had been constructed in Easter Island, or mounds, containing a few relics saved from fire, in the United States, we might never have suspected the existence of these ancient peoples. He argues, therefore, that it is very easy for the records of an ancient nation's life entirely to perish or to be hidden from observation. Even the arts of Nineveh and Babylon were unknown only a generation ago, and we have only just discovered the facts about the mound-builders of North America . . .

There is one other striking example of a higher being succeeded by a lower degree of knowledge, which is in danger of being forgotten because it has been made the foundation of theories which seem wild and fantastic, and are probably in great part erroneous. I allude to the Great Pyramid of Egypt, whose form, dimensions, structure, and uses have recently been the subject of elaborate works by Prof. Piazzi Smyth. Now the admitted facts about this pyramid are so interesting and apposite to the subject we are considering, that I beg to recall them to your attention. Most of you are aware that this pyramid has been carefully explored and measured by successive Egyptologists, and that the dimensions have lately become capable of more accurate determination, owing to the discovery of some of the original casing-stones and the clearing away of the earth from the corners of the foundation showing the sockets in which the corner-stones fitted. Prof. Smyth devoted many months of work with the best instruments in order to fix the dimensions and angles of all accessible parts of the structure; and he has carefully determined these by a comparison of his own and all previous measures, the best of which agree pretty closely with each other. The results arrived at are:—

1. That the pyramid is truly square, the sides being equal and the angles right angles.
2. That the four sockets on which the four first stones of the corners rested are truly on the same level.
3. That the direction of the sides are accurately to the four cardinal points.
4. That the vertical height of the pyramid bears the same proportion to its circumference at the base, as the radius of a circle does to its circumference.

Now all these measures, angles, and levels are accurate, not as an ordinary surveyor or builder could make them, but to such a degree as requires the very best modern instruments and all the refinements of geodetical science to discover any error at all. In addition to this we have the wonderful perfection of the workmanship in the interior of the pyramid, the passages and chambers being lined with huge blocks of stones fitted with the utmost accuracy, while every part of the building exhibits the highest structural science.

In all these respects this largest pyramid surpasses every other in Egypt. Yet it is universally admitted to be the oldest, and also the oldest historical building in the world.

Now these admitted facts about the Great Pyramid are surely remarkable, and worthy of the deepest consideration. They are facts which, in the pregnant words of the late Sir John Herschel, "according to received theories ought not to happen," and which, he tells us, should therefore be kept ever present to our minds, since "they belong to the class of fact which serve as the clue to new discoveries." According to modern theories, the higher civilization is ever a growth and an outcome from a preceding lower state; and it is inferred that this progress is visible to us throughout all history and in all the material records of human intellect. But here we have a building which marks the very dawn of history, which is the oldest authentic monument of man's genius and skill, and which, instead of being far inferior, is very much superior to all which followed it. Great men are the products of their age and country, and the designer and constructors of this wonderful monument could never have arisen among an unintellectual and half-barbarous people. So perfect a work implies many preceding less perfect works which have disappeared. It marks the culminating point of an ancient civilization, of the early stages of which we have no record whatever.

The three cases to which I have now adverted (and there are many others) seem to require for their satisfactory interpretation a somewhat different view of human progress from that which is now generally accepted. Taken in connexion with the great intellectual power of the ancient Greeks—which Mr. Galton believes to have been far above that of the average of any modern nation—and the elevation, at once intellectual and moral, displayed in the writings of Confucius, Zoroaster, and in the Vedas, they point to the conclusion that, while in material progress there has been a tolerably steady advance, man's intellectual and moral development reached almost its highest level in a very remote past. The low, the more animal, but often the more energetic types have, however, always been far the more numerous; hence such established societies as have here and there arisen under the guidance of higher minds have always been liable to be swept away by the incursions of barbarians. Thus in almost every part of the globe there may have been a long succession of partial civilizations, each in turn succeeded by a period of barbarism; and this view seems supported by the occurrence of degraded types of skull along with such "as might have belonged to a philosopher," at a time when the mammoth and the reindeer inhabited southern France.

Nor need we fear that there is not time enough for the rise and decay of so many successive civilizations as this view would imply; for the opinion is now gaining ground among geologists that palaeolithic man was really preglacial, and that the great gap (marked alike by a change of physical conditions and of animal life) which in Europe always separates him from his neolithic successor, was caused by the coming on and passing away of the great ice age.

If the views now advanced are correct, many, perhaps most, of our existing savages are the successors of higher races; and their arts, often showing a wonderful similarity in distant continents, may have been derived from a common source among more civilized peoples.

I must now conclude this very imperfect sketch of a few of the offshoots from the great tree of Biological study. It will, perhaps, be thought by some that my remarks have tended to the depreciation of our science, by hinting at imperfections in our knowledge and errors in our theories where more enthusiastic students see nothing but established truths. But I trust that I may have conveyed to many of my hearers a different impression . . . as regards man, I have endeavoured to fix your attention on a class of facts which indicate that the course of his development has been far less direct and simple than has hitherto been supposed; and that, instead of resembling a single tide with its advancing and receding ripples, it must rather be compared to the progress from neap to spring tides, both the rise and the depression being comparatively greater as the waters of true civilization slowly advance towards the highest level they can reach.

And if we are thus led to believe that our present knowledge of nature is somewhat less complete than we have been accustomed to consider it, this is only what we might expect; for however great may have been the intellectual triumphs of the nineteenth century, we can hardly think so highly of its achievements as to imagine that, in somewhat less than twenty years, we have passed from complete ignorance to almost perfect knowledge on two such vast and complex subjects as the origin of species and the antiquity of man.

* * * * *

Wallace's preoccupation with human origins led him to studies of the geographical affinities of individual races and peoples. An important early work of his on this subject appeared in The Quarterly Journal of Science *in 1867:*

The Polynesians and their migrations (ref 131)

The origin of the various races of the islands of the Pacific has always been one of the most difficult problems for the believers in the unity and the recent origin of man. Their diversity of physical features, of civilization, and of language, the absence of any continental races to which they could be affiliated, and the wide spaces of ocean over which they are distributed, have hitherto seemed to indicate that their origin dates from a period so remote that we cannot hope to determine it with any approach to certainty.

M. [Armand de] Quatrefages, however, an eminent anthropologist, has courageously attempted to solve the enigma of the origin of the Polynesians, the most important of the Pacific races. He very properly limits this term to the brown races spread over a wide area from the Sandwich Islands in the north to New Zealand on the south, and from Easter Island on the east to the Tonga and Samoan groups on the west, but all speaking dialects of one well-marked language. Now what M. Quatrefages attempts to prove is, that these people are simply Malays, who migrated from some islands of the Malayan Archipelago (probably Bouru in the Moluccas), and have more or less intermingled with the races of Melanesia and Micronesia. His evidence to prove this is of two kinds:—first, he endeavours to show that a migration has taken place; secondly, that the Polynesians are in their physical, mental, and moral characteristics, a true Malayan race.

1. Migrations.—We find in M. Quatrefages' volume a very careful summary of all the native accounts of their migrations, and also of the involuntary migrations that have recently occurred. These, no doubt, prove that the Sandwich Islands and New Zealand have been peopled by emigrants from the Marquesas and Tahiti, and the fact of this emigration is confirmed by the independent evidence of language. It is proved, therefore, that the Polynesians have passed over immense spaces of ocean, in directions not especially favoured by winds or currents, and thus the difficulty of any migration, merely from its distance, is quite overcome. It is further shown that all the traditions point to the Samoan group and the Fiji Islands as the central points to which almost all Polynesians trace their origin. It is to be observed here that these are the largest of all the islands in the central Pacific inhabited by the Polynesian race, and it is these, therefore, that we should naturally expect to have sent out colonies to the smaller islands. So far we have the strongest corroboration of there having actually been a migration in the fact of

the community of language, and all the legends of these migrations speak of them as having been made by simple men, the natural ancestors of the existing Polynesians. But in the legend which refers the origin of the Samoans themselves to a migration from a large country "further west," we get into pure legend,—for the mythic Boulotou, whence the first inhabitants are said to have come, is a spiritual and not a real country, and these inhabitants are believed to have been not men, but inferior gods. And even the direct evidence of migration having been generally from the west, is by no means so clear as M. Quatrefages appears to believe; for one of the latest authorities on the subject, Mr. W. T. Pritchard, who . . . is well qualified to give an opinion on this matter, says it is just the contrary. In his *Polynesian Reminiscences*, p. 402, he observes: "It is, however, remarkable that in all these many instances of authenticated driftings, the course of the drifted canoes has been *from east to west*, before the prevailing trade winds, and *not from west to east* before the westerly winds;" during the prevalence of which he tells us the natives do not usually venture out on fishing or travelling expeditions. In this case, too, the corroborative proof by language completely fails, for though there is an undoubted Malay element in the Polynesian language, it is an element derived from the civilized Malay and Javanese tongues, not from those of the Moluccas, which are totally distinct.

It is to be noted also that this Malay element in the language has all the character of a recent introduction, since the Malay words are hardly changed, except by the phonetic character of the language which has received them.

2. Physical Characters.—The Malayan origin of the Polynesians at a comparatively recent date implies much physical similarity; for even if the Malay formed a still larger portion of the Polynesian language than it does, this would not prove a community of race, unless the physical characters also in some degree corresponded. It is here that we find an absolute defect of all evidence bearing upon the point in question—the similarity of the Polynesians to any race speaking the Malay language. Almost the only evidence adduced by M. Quatrefages goes to show the similarity of the brown race of Timor to those of Polynesia. But the Timorese are not Malays at all; they belong to that curious race which has close affinities to the Papuan in all moral and physical characteristics except colour, and their languages are much further removed from the Malay than even the Polynesian itself. The resemblance physically of this race with the Polynesian proves absolutely nothing with regard to the Malay question.

Now let us compare the most important and thoroughly well-established physical and mental characteristics of the two races:

Polynesians: tall, averaging 5 ft. 10 in., per Wilkes (Samoans), or 6 ft., per Dupurey (Tahitians); hair wavy, curly, or frizzly; beard often full; face handsome, European type; nose often aquiline; disposition active and joyous; character open and frank; often erected stone edifaces; use double canoes.

Malays: short, 5 ft. 4 in. or 5 ft. 6 in.; hair always straight; beard scanty or none; face never of European type; nose never aquiline; disposition slow and morose; character eminently secretive; never used stone for building; use single canoes.

In the following passages from Consul Hopkins' recent work on Hawai, the italicized passages show points in which the Polynesian is the exact opposite of the Malay:—

"The hair of the Hawaians is black or brown, strong, and frequently *curly.*"

"The Hawaians are strong, well-made, and active, in height rather *above the average of Englishmen* . . . The Hawaians possess the virtue of *courage in an unquestionably high degree* . . . They are now as peaceful a people as any upon earth; they are more *free from crimes of violence* than any nation that can be named . . . The natural disposition of the Hawaians is everything that is *opposite to the gloomy and morose.* The pleasant universal '*aloha*' or salutation, the *merry ringing laughter* of the women wherever found, proclaim the people to be a *light-hearted race.*"

Taking the whole of these differences, they appear to indicate a radical diversity of race, not to be overcome by any mere similarity of colour and some common words in language, which is all that really exists to prove identity of race. The one single fact of stature is conclusive against any such comparatively recent common origin as M. Quatrefages argues for. A race which averages 5 ft. 10 in., and has many men 6 ft. 2 in. or more, can hardly have been derived, at such a recent period as to have retained community of language, from a race averaging 5 ft. 5 in. or 5 ft. 6 in., and among whom a single individual of 5 ft. 10 in. is rarely, perhaps never, found. Again the hair of the Malay is of the true Mongol type—black, coarse, and perfectly straight. The least approach to wavy or curly hair is never found among the unmixed Malay. I cannot find evidence that the Polynesians *ever* have this character of hair, while it is undoubtedly often as frizzly as the most decided Papuan. Again, the mental character of two races in a parallel state of civilization and inhabiting very similar countries, is surely of great importance; yet what contrast can be greater than between the phlegmatic, suspicious, undemonstrative Malay, and active, frank, and joyous Tahitian? Are we to throw down all these barriers of diversity for the sake of solving by main force a problem that is probably insoluble?

3. Geological and Zoological Evidence.—M. Quatrefages dismisses with a very brief notice the proofs of a former much greater extent of land in the Polynesian area than now exists. These proofs are of two kinds: first, the existence of numerous groups of coral islands, which are admitted to indicate sunken land; and secondly, the distribution of animals in the existing lands. That coral reefs and atolls are proofs of subsidence of the land, has never been seriously denied since Darwin's work on Coral Reefs was published; and as immense areas of the Pacific are occupied by such coral islands alternating with volcanic groups and such as show signs of elevation, it is only a question of time as to whether man could have inhabited these sunken lands.

Can we form any notion how long it is since the Pacific lands have disappeared? This can perhaps be approximated by the existing distribution of animals in these islands. The only group we yet know with any approach to completeness are the Birds; and though these may not be supposed to be the best adapted to test a question of this kind, yet ornithologists know that a very moderate extent of ocean practically limits the range of most land birds. The total number of species found in any of these islands is very small. For example, if we exclude the waders, swimmers, and birds of prey as having roving habits and great powers of flight, we find that according to the best information only twenty-two species of land birds inhabit the group of the Society Islands, and of these seven are found in other groups ranging as far as the Marquesas, the Fiji Islands, and even to the Sandwich Islands. These are almost all birds of moderate powers of flight and such as inhabit the forests and mountains, and do not generally range far. In Mr. G. R. Gray's list of the birds of the Pacific Islands I find eleven species of the genera *Myzomela*, *Meliphaga*, *Tatare*, *Monarcha*, *Coriphilus*, *Eudynamis*, and *Ptilonopus*, which are known from two or more of the distinct and well separated groups of islands in the central Pacific, and some of them have a very wide range. Among these are two very distinct genera, *Tatare* and *Coriphilus*, which are entirely confined to the Polynesian areas. Now these facts would certainly indicate a more intimate connection of the various groups of islands within the period of living species, and therefore within the human period, than now exists. The phenomena presented by the distribution of man are thus to some extent reproduced by the distribution of land-birds in the same area, and entitle us to believe that the subsidence of land indicated by coral reefs took place since man inhabited the earth. This subsidence was probably coincident with, perhaps caused by, the elevation of the existing volcanic islands; and while man and birds were able to migrate to these, the mammalia dwindled away and finally perished, when the last mountain-top of the old Pacific sank beneath the Ocean.

This hypothesis is one which does not outrage nature, as does that of the direct and recent derivation of the Polynesians from the Malays. It harmonizes at once with the Geological, the Zoological, and the Anthropological phenomena; and if it is held that the facts are not sufficient to prove it, or that even if proved it only removes the origin of the race in question one step further back into the obscurity of the past, it may be suggested that in a case of such admitted difficulty we can hardly do more. We ought not to expect that the beginnings of every race are to be discovered within the short epoch of human history or tradition, and we have every reason to be suspicious of the theory that professes such a discovery. In the present case, the very erroneous views prevalent on the subject arise from two causes. One is the occurrence of a number of Malay words in the Polynesian language; the other, the similarity of the brown tint of the Malays and Polynesians, while they are separated by a group of people of a much darker colour. The similarity of tint has led many travellers in the one area to jump to the conclusion that the people of the other area, of which they have little knowledge, are the same race. It unfortunately happens that not a single traveller appears to be well acquainted with both races, and for that reason their opinions as to the similarity of the two should be received with great doubt. If, on the contrary, my account of the physical and mental characteristics of the Malays be taken as correct (and I resided among them for eight years), and if it be compared with that of the Polynesians given by Cook, and by recent travellers and missionaries, the differences will be seen to be so striking and radical, that all idea of their being the same race must be given up. In the case of the Malays in particular, much confusion has arisen from travellers having confounded with them the many peoples of distinct race which inhabit the eastern parts of the Malayan Archipelago, such as the Timorese, the mountaineers of Ceram and Gilolo, and of the small islands near New Guinea; and this mistake has been rendered excusable by the number of half-breeds between the two races to be found everywhere. Many of these people are, perhaps, allied to the Polynesians (the mountaineers of Gilolo and Ceram are perhaps true outliers of the Polynesians, and may represent the effect of that *westerly* migration from Samoa, of which Mr. Pritchard speaks), but they are certainly not Malays, who are essentially a Mongol race, with many of the Mongol characteristics very strongly marked. The Papuans of New Guinea form the extreme type of another and a widely different race, and all the evidence goes to show that in every characteristic except colour, the Polynesians are nearer to the Papuans than they are to the Malays, although it is not improbable that they are equally distinct from both.

* * * * *

Wallace's best-known contribution to ethnography is probably his recognition that the aborigines of Australia should be classed as a Caucasian people. The reasoning behind this judgement was set out at the end of Chapter 5 of Volume 1 of the 1893 edition of Australasia *(ref 720), and subsequently expanded into a much longer work (ref 583) for inclusion in* Studies scientific and social *(ref 727) in 1900. Most of the earlier passage is reproduced here:*

Probable origin of the Australian races

The aborigines of Australia differ remarkably from those of all the surrounding countries, while they agree so closely among themselves in every part of the continent that they evidently form a single race. To recapitulate their main characteristics: they may be described as men of medium stature, muscular, but with slender arms and legs, rather large heads with broad foreheads and overhanging brows, the nose thick and very broad at the nostrils, as with many of the lower races, the mouth large and lips thick, but far less so than in the negro races. In colour they are a deep copper or chocolate, never sooty black as in the negro; the hair is long, glossy black or very deep auburn, usually wavy or curly and very abundant, and the face is adorned with a luxuriant growth of moustache, beard, and whiskers, usually with an auburn tinge. These characters, in their combination, give to the face as a whole a familiar appearance resembling that of the coarser and more sensual types of western Europeans; while they are as totally removed from any of the beardless

Malayan and Polynesian tribes, or the woolly or frizzly haired Papuans. They have sometimes a considerable growth of hair on the body, especially on the chest and back, while the infants are much lighter in colour than the adults, and have a considerable amount of soft fur on the neck and back.

If we turn to habits and customs for some light as to their probable derivation, we find equally clear proofs that we must go far beyond the limits of all immediately surrounding peoples. While Malays, Papuans, and Polynesians all cultivate the ground, and all build good permanent houses, the Australians never do one or the other. The pottery of the Malays and Papuans, the bows and arrows of the Papuans and other Melanesians, and the elaborate canoes of all these races are equally unknown to the Australians, who thus remain among the savage tribes who have made the very smallest advance on the road to material civilisation.

In Mr. Curr's elaborate work on *The Australian Race* an attempt is made to show that there is a great resemblance in language and in many customs between the native races of Australia and Africa. By comparing some words of the numerous languages of Australia with the words of the same meaning in some of the still more numerous African languages and dialects, a considerable number of striking resemblances are found. But it is probable that an equal number of resemblances could be found if the languages of any other part of the world were taken for comparison, because many of the words used are either onomatopoeic or what may be described as functional. Thus words for the "lips," "mouth," or "head" are expressed by labial sounds, for the teeth by dental sounds, for the nose by nasals, such as *n*, *ng*, or *ny*; and thus the names for these parts of the body often resemble each other in the most remote parts of the world. So in the names for father and mother the infantile *ba*, *pa*, and *ma* occur in every part of the world, though sometimes transposed in their meaning. Similar resemblances have been found by other writers among the languages of the hill tribes of India, and with these races there is rather more accordance in physical characters. Mr. Curr also adduces a considerable number of curious customs which are common to Australia and Africa: such as the great importance attached to sorcery, the forming raised scars on the body, the habit of knocking out the front teeth, of circumcision, of prohibition of marriage within certain class divisions, the forms of burials, and some others. But some of these customs are very widespread among savages; and unless it can be ascertained that a considerable number of them are strictly limited to Africa and Australia, they can afford no proof of a common origin to the two races. More interesting is the fact that the peculiar Australian weapon, the boomerang, finds its nearest representative in Abyssinia and among the ancient Egyptians. This may indicate that the weapon had a wider range in early times, but was gradually supplanted by the bow and other superior weapons, but can hardly be held to prove identity of race in opposition to so many characteristic differences.

Looking broadly, and without prejudice, at the physical features of the Australians, they evidently belong neither to the Negroid nor to the Mongoloid types of man, while in all essential characters they must be classed as

Caucasians. If we look abroad for other isolated fragments of the same type, we find one in the Ainos of Japan. These singular people agree wonderfully with the Australian type, but are somewhat more hairy and of a lighter colour. They are also in a more advanced stage of material civilisation, and are probably on a somewhat higher intellectual and moral plane. Other fragments of the same great primitive race exist in the Khmers and Chams of Cambodia, who are said to be decidely Caucasian in type; while their language has affinities with those of Polynesia, where also Caucasian affinities are shown, especially in some of the inhabitants of Micronesia. Of all these widely scattered Caucasian fragments we must look upon the Australians as the lowest and probably most primitive. Their antiquity is, in all probability, very great, since they must have entered their present country at a time when their ancestors had not acquired the arts of making pottery, of cultivating the soil, of domesticating animals, of constructing houses, or of fabricating the bow and arrow. They thus afford us an example of one of the most primitive types of humanity yet discovered.

What renders their uniform low condition more remarkable is, that they must have been many times brought into contact with more advanced races. There are some signs of intermixture in the north with both Malays and Papuans, but this has had little or no effect on the customs of the people. Some still higher race has evidently at one time formed a settlement on the north-west coast, as indicated by the very remarkable cave-paintings and sculptures discovered by Sir George Grey . . . in the valley of the Glenelg River in North-West Australia . . .

The conclusion here reached that the Australians, usually classed as one of the lowest of existing races, are really of Caucasian type, and are more nearly allied to ourselves than the civilised Japanese or the brave and intelligent Zulus, may appear to some to be improbable or even absurd. But I venture to think that it nevertheless most nearly accords with all the facts of the case; and since it has been admitted that even some of the darkest Hindoos are nearly allied to Europeans, there is less improbability in the existence of some more archaic and less developed examples of the same type. It also accords with all we are now learning of the vast antiquity of the human race; since, if all the tribes now living can be classed in one or other of the three great divisions of mankind, Negroid, Mongolian, and Caucasian, or as probable mixtures of them, we are impressed with the conviction that we must go back to periods to which the earliest historical dates are but yesterday in order to arrive at an epoch when the common ancestors of these three well-marked types alone inhabited the earth. Even then we shall have made no perceptible approach to the "missing link"—to the common ancestor of man and the higher quadrumana . . .

* * * * *

In his later years, Wallace increasingly turned his attention toward what he felt were moral and sociological issues affecting the future evolution of mankind. Four of his works in particular stand out in this regard: 'Human selection' (ref 427), 'Human progress: past and future' (ref 445), 'Evolution and character' (ref 649), and the monograph Social environment and moral progress *(ref 733). Wallace considered the first of these, written in 1890 and published in* Fortnightly Review, *his single most important contribution to the study of man.*

Human selection (ref 427)

In one of my latest conversations with Darwin he expressed himself very gloomily on the future of humanity, on the ground that in our modern civilisation natural selection had no play, and the fittest did not survive. Those who succeed in the race for wealth are by no means the best or the most intelligent, and it is notorious that our population is more largely renewed in each generation from the lower than from the middle and upper classes. As a recent American writer well puts it,

> We behold the melancholy spectacle of the renewal of the great mass of society from the lowest classes, the highest classes to a great extent either not marrying or not having children. The floating population is always the scum, and yet the stream of life is largely renewed from this source. Such a state of affairs, sufficiently dangerous in any society, is simply suicidal in the democratic civilisation of our day (Hiram M. Stanley in the *Arena* for June, 1890).

That the check to progress here indicated is a real one few will deny, and the problem is evidently felt to be one of vital importance, since it has attracted the attention of some of our most thoughtful writers, and has quite recently furnished the theme for a perfect flood of articles in our best periodicals. I propose here to consider very briefly the various suggestions made by these writers; and afterwards shall endeavour to show that when the course of social evolution shall have led to a more rational organization of society, the problem will receive its final solution by the action of physiological and social agencies, and in perfect harmony with the highest interests of humanity.

Before discussing the question itself it will be well to consider whether there are in fact any other agencies than some form of selection to be relied on. It has been generally accepted hitherto that such beneficial influences as education, hygiene, and social refinement had a cumulative action, and would of themselves lead to a steady improvement of all civilised races. This view rested on the belief that whatever improvement was effected in individuals was transmitted to their progeny, and that it would be thus possible to effect a continuous advance in physical, moral, and intellectual qualities without any selection of the better or elimination of the inferior types. But of late years grave doubts have been thrown on this view, owing chiefly to the researches of Galton and Weismann as to the fundamental causes to which heredity is due. The balance of opinion amongst physiologists now seems to be against the heredity of any qualities acquired by the individual after birth, in which case the question we are discussing will be much simplified, since we shall be limited to some form of selection as the only possible means of improving the race.

In order to make the difference between the two theories clear to those who may not have followed the recent discussions on the subject an illustration may be useful. Let us suppose two persons, each striving to produce two distinct types of horse—the cart-horse and the racer—from the wild prairie horses of America, and that one of them believes in the influence of food and training, the other in selection. Each has a lot of a hundred horses to begin with, as nearly as possible alike in quality. The one who trusts to selection at once divides his horses into two lots, the one stronger and heavier, the other lighter and more active, and, breeding from these, continually selects, for the parents of the succeeding generation, those which most nearly approach the two types required. In this way it is perfectly certain that in a comparatively short period—thirty or forty years perhaps—he would be able to produce two very distinct forms, the one a very fair racehorse, the other an equally good specimen of cart-horse; and he could do this without subjecting the two strains to any difference of food or training, since it *is* by selection alone that our various breeds of domestic animals have in most cases been produced.

On the other hand, the person who undertook to produce similar results by food and training alone, without allowing selection to have any part in the process, would have to act in a very different manner. He would first divide his horses into two lots as nearly as possible identical in all points, and thereafter subject the one lot to daily exercise in drawing loads at a slow pace, the other lot to equally constant exercise in running, and he might also supply them with different kinds of food if he thought it calculated to aid in producing the required effect. In each successive generation he must make no selection of the swiftest or the strongest, but must either keep the whole progeny of each lot, or carefully choose an average sample of each to be again subjected to the same discipline. It is quite certain that the very different kinds of exercise would have some effect on the individuals so trained, enlarging and strengthening a different set of muscles in each, and if this effect were transmitted to the offspring then there ought to be in this case also a steady advance towards the racer and the cart-horse type. Such an experiment, however, has never been tried, and we cannot therefore say positively what would be the result; but those who accept the theory of the non-heredity of acquired characters would predict with confidence that after thirty or forty generations of training without selection, the last two lots of colts would have made little or no advance towards the two types required but would be practically indistinguishable.

It is exceedingly difficult to find any actual cases to illustrate this point, since either natural or artificial selection has almost always been present. The apparent effects of disuse causing the diminution of certain organs, such as the reduced wings of some birds in oceanic islands and the very small or aborted eyes of some of the animals inhabiting extensive caverns, can be as well explained by the withdrawal of the cumulative agency of natural selection and by economy of growth, as by the direct effects of disuse. The following facts, however, seem to show that special skill derived from practice, when continued for several generations, is not inherited, and does not therefore tend to increase. The wonderful skill of most of the North American Indians in following a trail by indications quite imperceptible to the ordinary European has been dwelt upon by many writers, but it is now admitted that the white trappers equal and often excel them, though these trappers have in almost every case acquired their skill in a comparatively short period, without any of the inherited experience which might belong to the Indian. Again, for many generations a considerable portion of the male population of Switzerland have practised rifle-shooting as a national sport, yet in international contests they show no marked superiority over our riflemen, who are, in a large proportion, the sons of men who never handled a gun. Another case is afforded by the upper classes of this country who for many generations have been educated at the universities, and have had their classical and mathematical abilities developed to the fullest extent by rivalry for honours. Yet now, that for some years these institutions have been opened to dissenters whose parents usually for many generations have had no such training, it is found that these dissenters carry off their full share or even more than their share of honours. We thus see that the theory of the non-heredity of acquired characters, whether physical or mental, is supported by a considerable number of facts, while few if any are directly opposed to it. We therefore propose to neglect the influence of education and habit as possible factors in the improvement of our race, and to confine our argument entirely to the possibility of improvement by some form of selection. (Those who desire more information on this subject should read August Weismann's *Essays on Heredity*.)

Among the modern writers who have dealt with this question the opinions of Mr. Francis Galton are entitled to be first considered, because he has studied the whole subject of human faculty in the most thorough manner, and has perhaps thrown more light upon it than any other writer. The method of selection by which he has suggested that our race may be improved is to be brought into action by means of a system of marks for family merit, both as to health, intellect, and morals, those individuals who stand high in these respects being encouraged to marry early by state endowments sufficient to enable the young couples to make a start in life. Of all the proposals that have been made tending to the systematic improvement of our race, this is one of the least objectionable, but it is also I fear among the least effective. Its tendency would undoubtedly be to increase the number and to raise the standard of our highest and best men, but it would at the same time leave the bulk of the population unaffected, and would but slightly diminish the rate at which the lower types tend to supplant or to take the place of the higher. What we want is, not a higher standard of perfection in the few but a higher average, and this can best be produced by elimination of the lowest of all and a free intermingling of the rest.

Something of this kind is proposed by Mr. Hiram M. Stanley in his article on 'Our Civilisation and the Marriage Problem,' already referred to. This writer believes that civilisations perish because, as wealth and art increase, corruption creeps in, and the new generations fail in the work of progress because the renewal of individuals is left chiefly to the unfit. The two great factors which secure perfection in each animal race—sexual selection by which the fit are born, and natural selection by which the fittest survive—both fail in the case of mankind, among whom are hosts of individuals which in any other class of beings would never have been born, or, if born would never survive. He argues that, unless some effective measures are soon adopted and strictly enforced, our case will be irremediable; and, since natural selection fails so largely, recourse must be had to artificial selection. "The drunkard, the criminal, the diseased, the morally weak should never come into society. Not reform but prevention should be the cry." The method by which this is proposed to be done is hinted at in the following passages: "In the true golden age, which lies not behind but before us, the privilege of parentage will be esteemed an honour for the comparatively few, and no child will be born who is not only sound in body and mind, but also above the average as to natural ability and moral force,"—and again—"The most important matter in society, the inherent quality of the members which compose it, should be regulated by trained specialists."

Of this proposal and all of the same character we may say, that nothing can possibly be more objectionable, even if we admit that they might be effectual in securing the object aimed at. But even this is more than doubtful; and it is quite certain that any such interference with personal freedom in matters so deeply affecting individual happiness will never be adopted by the majority of any nation, or if adopted would never be submitted to by the minority without a life-and-death struggle.

Another popular writer of the greatest ability and originality, who has recently given us his solution of the problem, is Mr. Grant Allen. His suggestion is in some respects the very reverse of the last, yet it is, if possible, even more objectionable. Instead of any interference with personal freedom he proposes the entire abolition of legal restrictions as to marriage, which is to be a free contract to last only so long as either party desires. This alone, however, would have no effect on race-improvement, except probably a prejudicial one. The essential part of his method is, that girls should be taught both by direct education and by the influence of public opinion, that the duty of all healthy and intellectual women is to be the mothers of as many and as perfect children as possible. For this purpose they are recommended to choose as temporary husbands the finest, healthiest, and most intellectual men, thus ensuring a variety of combinations of parental qualities which would lead to the production of offspring of the highest possible character and to the continual advancement of the race. (See 'The Girl of the Future' in *The Universal Review*, May, 1890, and a previous article entitled 'Plain Words on the Woman Question,' in *The Fortnightly Review*, October, 1889.)

I think I have fairly summarized the essence of Mr. Grant Allen's proposal, which, though enforced with all his literary skill and piquancy of illustration, can, in my opinion, only be fitly described by the term already applied to it by one of his reviewers, "detestable." It purports to be advanced in the interests of the children and of the race; but it would necessarily impair that family life and parental affection which are the prime essentials to the well-being of children; while, though it need not necessarily produce, it would certainly favour, the increase of pure sensualism, the most degrading and most fatal of all the qualities that tend to the deterioration of races and the downfall of nations. One of the modern American advocates of greater liberty of divorce, in the interest of marriage itself, thus admirably summarises the essential characteristics and purport of true marriage.

> In a true relation, the chief object is the loving companionship of man and woman, their capacity for mutual help and happiness, and for the development of all that is noblest in each other. The second object is the building up a home and family, a place of rest, peace, security, in which child-life can bud and blossom like flowers in the sunshine (Elizabeth Cady Stanton in the *Arena*, April, 1890).

For such rest, peace, and security, permanence is essential. This permanence need not be attained by rigid law, but by the influence of public opinion, and, more surely still, by those deep-seated feelings and emotions which, under favourable conditions, render the marriage tie stronger and its influence more beneficial the longer it endures. To me it appears that no system of the relations of men and women could be more fatal to the happiness of individuals, the well-being of children, or the advancement of the race, than that proposed by Mr. Grant Allen.

Before proceeding further with the main question it is necessary to point out that, besides the special objections to each of the proposals here noticed, there is a general and fundamental objection. They all attempt to deal at once, and by direct legislative enactment, with the most important and most vital of all human relations, regardless of the fact that our present phase of social development is not only extremely imperfect but vicious and rotten at the core. How can it be possible to determine and settle the relations of women to men which shall be best alike for individuals and for the race, in a society in which a very large proportion of women are obliged to work long hours daily for the barest subsistence, while another large proportion are forced into more or less uncongenial marriages as the only means of securing some amount of personal independence or physical well-being. Let any one consider, on the one hand, the lives of the wealthy as portrayed in the society newspapers and even in the advertisements of such papers as *The Field* and *The Queen*, with their endless round of pleasure and luxury, their almost inconceivable wastefulness and extravagance, indicated by the cost of female dress and such facts as the expenditure of a thousand pounds on the flowers for a single entertainment; and, on the other hand, the terrible condition of millions of workers—men, women, and children—as detailed in the *Report of the Lords Commission on Sweating*, on absolutely incontestable evidence, and the still more awful condition of those who seek work of any kind in vain and, seeing their children slowly dying of starvation, are driven in utter helplessness and despair to murder and suicide. Can any thoughtful person admit for a moment that, in a society so constituted that these overwhelming contrasts of luxury and privation are looked upon as necessities, and are treated by the Legislature as matters with which it has practically nothing to do, there is the smallest probability that we can deal successfully with such tremendous social problems as those which involve the marriage tie and the family relation as a means of promoting the physical and moral advancement of the race? What a mockery to still further whiten the sepulchre of modern society, in which is hidden "all manner of corruption," with schemes for the moral and physical advancement of the race!

It is my firm conviction, for reasons which I shall state presently, that, when we have cleansed the Augean stable of our existing organization, and have made such arrangements that *all* shall contribute their share of either physical or mental labour, and that all workers shall reap the *full* reward of their work, the future of the race will be ensured by those laws of human development that have led to the slow but continuous advance in the higher qualities of human nature. When men and women are alike free to follow their best impulses; when idleness and vicious or useless luxury on the one hand, oppressive labour and starvation on the other, are alike unknown; when all receive the best and most thorough education that the state of civilisation and knowledge at the time will admit; when the standard of public opinion is set by the wisest and the best, and that standard is systematically inculcated on the young; then we shall find that a system of selection will come spontaneously into action which will steadily tend to eliminate the lower and more degraded types of man, and thus continuously raise the average standard of the race. I therefore strongly protest against any attempt to deal with this great question by legal enactments, or by endeavouring to modify public opinion as to the beneficial character of monogamy and permanence in marriage. That the existing popular opinion is the true one is well and briefly shown by Miss Chapman in a recent number of *Lippincott's Magazine*; and as her statement of the case expresses my own views, and will, I think, be approved by most thinkers on the subject, I give it:

> 1. *Nature* indicates permanent marriage as the true human relation. The young of the human pair need parental care and supervision for a great number of years.
> 2. *Instinct* is strongly on the side of indissoluble marriage. In proportion as men leave brutedom behind and enter into the fulness of their human heritage, they will cease to tolerate the idea of two or more living partners.
> 3. *History* shows conclusively that where divorce has been easy, licentiousness, disorder, and often complete anarchy have prevailed. The history of civilisation is the history of advance in monogamy, of the fidelity of one man to one woman, and one woman to one man.
> 4. *Science* tells the same tale. Physiology and Hygiene point to temperance, not riot. Sociology shows how man, in spite of himself, is ever striving, through lower forms, upward, to the monogamic relation.
> 5. *Experience* demonstrates to every one of us, individually, the superiority of the indissoluble marriage. We know that, speaking broadly, marriages turn out well or ill in proportion as husband and wife are—let me not say loving—but loyal, sinking differences and even grievances for the sake of children and for the sake of example.

We have now to consider what would be the probable effect of a condition of social advancement, the essential characteristics of which have been already hinted at, on the two great problems—the increase of population, and the continuous improvement of the race by some form of selection which we have reason to believe is the only method available. In order to make this clear, however, and in order that we may fully realise the forces that would come into play in a just and rational state of society, such as may certainly be realised in the not distant future, it will be necessary to have a clear conception of its main characteristics. For this purpose, and without committing myself in any way to an approval of all the details of his scheme, I shall make use of Mr. [Edward] Bellamy's clear and forcible picture of the society of the future, as he supposes it may exist in America in little more than a century hence (*Looking Backward*. See especially chapters vii, ix, xii, and xxv.)

The essential principle on which society is supposed to be founded is that of a great family. As in a well-regulated modern family the elders, those who have experience of the labours, the duties, and the responsibilities of life, determine the general mode of living and working, with the fullest consideration for the convenience and real well-being of the younger members, and with a recognition of their essential independence. As in a family, the same comforts and enjoyments are secured to all, and the very idea of making any difference in this respect to those who from mental or physical disability are unable to do so much as others, never occurs to anyone, since it is opposed to the essential principles on which a true society is held to rest. As regards education all have the same advantages, and all receive the fullest and best training, both intellectual and physical; every one is encouraged to follow out those studies or pursuits for which they are best fitted, or for which they exhibit the strongest inclination. This education, the complete and thorough training for a life of usefulness and enjoyment, continues in both sexes till the age of twenty-one (or thereabouts) when all alike, men and women, take their place in the ranks of the industrial army in which they serve for three years. During the latter years of their education, and during the succeeding three years of industrial service, every opportunity is given to them to see and understand every kind of work that is carried on by the community, so that at the end of the term of probation they can choose what department of the public service they prefer to enter. As every one—men, women, and children alike—receive the same amount of public credit—their equal share of the products of the labour of the community, the attractiveness of various pursuits is equalized by differences in the hours of labour, in holidays, or in special privileges attached to the more disagreeable kinds of necessary work, and these are so modified from time to time that the volunteers for every occupation are always about equal to its requirements. The only other essential feature that it is necessary to notice for our present purpose is the system of grades, by which good conduct, industry, and intelligence in every department of industry and occupation are fully recognised, and lead to appointments as overseers, superintendents, or general managers, and ultimately to the highest offices of the state. Every one of these grades and appointments is made public; and as they constitute the only honours and the only differences of rank, with corresponding insignia and privileges, in an otherwise equal body of citizens, they are highly esteemed and serve as ample inducements to industry and zeal in the public service.

At first sight it may appear that in any state of society whose essential features were at all like those here briefly outlined, all the usual restraints to early marriage as they now exist would be removed, and that a rate of increase of the population unexampled in any previous era would be the result, leading in a few generations to a difficulty in obtaining subsistence, which Malthus has shown to be the inevitable result of the normal rate of increase of mankind when all the positive as well as the preventive checks are removed. As the positive checks—which may be briefly summarised as war, pestilence, and famine—are supposed to be non-existent, what, it may be asked, are the preventive checks which are suggested as being capable of reducing the rate of increase within manageable limits? This very reasonable question I will now endeavour to answer.

The first and most important of the checks upon a too rapid increase of population will be the comparatively late average period of marriage, which will be the natural result of the very conditions of society, and will besides be inculcated during the period of education, and still further enforced by public opinion. As the period of systematic education is supposed to extend to the age of twenty-one, up to which time both the mental and physical powers will be trained and exercised to their fullest capacity, the idea of marriage during this period will rarely be entertained. During the last year of education, however, the subject of marriage will be dwelt upon, in its bearing on individual happiness and on social well-being, in relation to the welfare of the next generation and to the continuous development of the race. The most careful and deliberate choice of partners for life will be inculcated as the highest social duty; while the young women will be so trained as to look with scorn and loathing on all men who in any way wilfully fail in their duty to society—on idlers and malingerers, on drunkards and liars, on the selfish, the cruel, or the vicious. They will be taught that the happiness of their whole lives will depend on the care and deliberation with which they choose their husbands, and they will be urged to accept no suitor till he has proved himself to be worthy of respect by the place he holds and character he bears among his fellow-labourers in the public service.

Under social conditions which render every woman absolutely independent, so far as the necessaries and comforts of existence are concerned, surrounded by the charms of family life and the pleasures of society, which will be far greater than anything we now realise when all possess the refinements derived from the best possible education, and all are relieved from sordid cares and the struggle for mere existence, is it not in the highest degree probable that marriage will rarely take place till the woman has had three or four years' experience of the world after leaving college—that is, till the age of 25, while it will very frequently be delayed till 30 or upwards? Now Mr. Galton has shown, from the best statistics available, that if we compare women married at 20 with those married at 29, the proportionate fertility is about as 8 to 5. But this difference, large as it is, only represents a portion of the effect on the rate of increase of population caused by a delay in the average period of marriage. For when the age of marriage is delayed the time between successive generations is correspondingly lengthened; while a still further effect is produced by the fact that the greater the average age of marriage the fewer generations are alive at the same time, and it is the combined effect of these three factors that determines the actual rate of increase of the population. (See *Inquiries into Human Faculty and Its Development*, p. 321; and *Hereditary Genius*, p. 353.)

But there is yet another factor tending to check the increase of population that would come into play in a society such as we have been considering. In a remarkable essay on the *Theory of Population* Herbert Spencer has shown, by an elaborate discussion of the phenomena presented by the whole animal kingdom, that the maintenance of the individual and the propagation of the race vary inversely, those species and groups which have the shortest and most uncertain lives producing the greatest number of offspring; in other words, individuation and reproduction are antagonistic. But individuation depends almost entirely on the development and specialisation of the nervous system, through which, not only are the several activities and co-ordinations of the various organs carried on, but all advance in instinct, emotion, and intellect is rendered possible. The actual rate of increase in man has been determined by the necessities of the savage state, in which, as in most animal species, it has usually been only just sufficient to maintain a limited average population. But with civilisation the average duration of life increases, and the possible increase of population under favourable conditions becomes very great, because fertility is greater than is needed under the new conditions. The advance in civilisation as regards the preservation of life has in recent times become so rapid, and the increased development of the nervous system has been limited to so small a portion of the whole population, that no general diminution in fertility has yet occurred. That the facts do, however, accord with the theory is indicated by the common observation that highly intellectual parents do not as a rule have large families, while the most rapid increase occurs in those classes which are engaged in the simpler kinds of manual labour. But in a state of society in which all have their higher faculties fully cultivated and fully exercised throughout life, a slight general diminution of fertility would at once arise, and this diminution, added to that caused by the later average period of marriage, would at once bring the rate of increase of population within manageable limits. The same general principle enables us to look forward to that distant future when the world will be fully peopled, in perfect confidence that an equilibrium between birth and death rates will then be brought about by a combination of physical and social agencies, and the bugbear of over-population become finally extinct. (*A Theory of Population Deduced from the General Law of Animal Fertility*. Republished from the *Westminster Review* for April, 1852.)

There now only remains for consideration the means by which, in such a society, a continuous improvement of the race could be brought about, on the assumption that for this purpose education is powerless as a direct agency, since its effects are not hereditary, and that some form of selection is an absolute necessity. This improvement I believe will certainly be effected through the agency of female choice in marriage. Let us, therefore, consider how this would probably act.

It will be generally admitted that, although many women now remain unmarried from necessity rather than from choice, there are always a considerable number who feel no strong inclination to marriage, and who accept husbands to secure a subsistence or a home of their own rather than from personal affection or sexual emotion. In a society in which women were all pecuniarily independent, were all fully occupied with public duties and intellectual or social enjoyments, and had nothing to gain by marriage as regards material well-being, we may be sure that the number of the unmarried from choice would largely increase. It would probably come to be considered a degradation for any woman to marry a man she could not both love and esteem, and this feeling would supply ample reasons for either abstaining from marriage altogether or delaying it till a worthy and sympathetic husband was encountered. In man, on the other hand, the passion of love is more general, and usually stronger; and as in such a society as is here postulated there would be no way of gratifying this passion but by marriage, almost every woman would receive offers, and thus a powerful selective agency would rest with the female sex. Under the system of education and of public opinion here suggested there can be no doubt how this selection would be exercised. The idle and the selfish would be almost universally rejected. The diseased or the weak in intellect would also usually remain unmarried; while those who exhibited any tendency to insanity or to hereditary disease, or who possessed any congenital deformity would in hardly any case find partners, because it would be considered an offence against society to be the means of perpetuating such diseases or imperfections.

We must also take into account a special factor hitherto, I believe, unnoticed in this connection, that would in all probability intensify the selection thus exercised. It is well known that females are largely in excess of males in our existing population, and this fact, if it were a necessary and permanent one, would tend to weaken the selective agency of women, as it undoubtedly does now. But there is good reason to believe that it will not be a permanent feature of our population. The births always give a larger proportion of males than females, varying from 3½ to 4 per cent. But boys die so much more rapidly than girls that when we include all under the age of five the numbers are nearly equal. For the next five years the mortality is nearly the same in both sexes; then that of females preponderates up to 30 years of age, then up to 60 that of men is the larger, while for the rest of life female mortality is again greatest. The general result is that at the ages of most frequent marriage—from 20 to 35—females are between 8 and 9 per cent in excess of males. But during the ages 5 to 35 we find a wonderful excess of male deaths from two preventible causes—"accident" and "violence." For the year 1888 the deaths from these causes in England and Wales was as follows: males (5 to 35 years), 4,158; females (5 to 35 years), 1,100 (*Annual Report of the Registrar General*, 1888, pp. 106–7).

Here we have an excess of male over female deaths in one year of 3,058, all between the ages of 5 and 35, a very large portion of which is no doubt due to the greater risks run by men and boys in various industrial occupations. In a state of society in which the bulk of the population were engaged in industrial work it is quite certain that almost all these deaths would be prevented, and thus bring the male population more nearly to an equality with the female. But there are also many unhealthy employments in which men are exclusively engaged, such as the grinders of Sheffield, the white-lead manufacturers, and many others; and many more men have their lives shortened by labour in unventilated workshops, to say nothing of the loss of life in war. When the lives of all its citizens are accounted of equal value to the community, no one will be allowed to suffer from such preventible causes as these; and this will still further reduce the mortality of men as compared with that of women. On the whole, then, it seems highly probable that in the society of the future the superior numbers of males at birth will be maintained throughout life, or, at all events, during what may be termed the marriageable period. This will greatly increase the influence of women in the improvement of the race. Being a minority they will be more sought after, and will have a real choice in marriage, which is rarely the case now. This actual minority being further increased by those who, from the various causes already referred to, abstain from marriage, will cause considerable numbers of men to remain permanently unmarried, and as these will consist very largely, if not almost wholly, of those who are the least perfectly developed either mentally or physically, the constant advance of the race in every good quality will be ensured.

This method of improvement by elimination of the worst has many advantages over that of securing the early marriages of the best. In the first place it is the direct instead of the indirect way, for it is more important and more beneficial to society to improve the average of its members by getting rid of the lowest types than by raising the highest a little higher. Exceptionally great and good men are always produced in sufficient numbers, and have always been so produced in every phase of civilisation. We do not need more of these so much as we need less of the weak and the bad. This weeding-out system has been the method of natural selection, by which the animal and vegetable worlds have been improved and developed. The survival of the fittest is really the extinction of the unfit. In nature this occurs perpetually on an enormous scale, because, owing to the rapid increase in most organisms, the unfit which are yearly destroyed form a large proportion of those that are born. Under our hitherto imperfect civilisation this wholesome process has been checked as regards mankind; but the check has been the result of the development of the higher attributes of our nature. Humanity—the essentially *human* emotion—has caused us to save the lives of the weak and suffering, of the maimed or imperfect in mind or body. This has to some extent been antagonistic to physical and even intellectual race-improvement; but it has improved us morally by the continuous development of the characteristic and crowning grace of our human, as distinguished from our animal, nature.

In the society of the future this defect will be remedied, not by any diminution of our humanity, but by encouraging the activity of a still higher human characteristic—admiration of all that is beautiful and kindly and self-sacrificing, repugnance to all that is selfish, base, or cruel. When we allow ourselves to be guided by reason, justice, and public spirit in our dealings with our fellow-men, and determine to abolish poverty by recognising the equal rights of all the citizens of our common land to an equal share of the wealth which all combine to produce,—when we have thus solved the lesser problem of a rational social organisation adapted to secure the equal well-being of all, then we may safely leave the far greater and deeper problem of the improvement of the race to the cultivated minds and pure instincts of the Women of the Future.

* * * * *

In concluding this first Section, mention should be made of Wallace's popularization of the mouth-gesture theory of the evolution of language. Though he claimed no special talent for learning languages or for linguistics in general, he studiously recorded observations on the languages of the peoples among whom he lived (along the way collecting dozens of dialectal vocabularies), and was able to apply this experience to an 1881 Nature *review (ref 337) of Edward Tylor's* Anthropology, *from which the following is excerpted:*

. . . In treating of the origin of language Mr. Tylor doubts the sufficiency of the theory that emotional, imitative, and suggestive sounds were the basis on which all languages were founded, though he gives tolerably full illustrations of how roots thus obtained became modified in an infinite variety of ways to serve the growing needs of mankind in expressing their wants or their feelings. He impresses on his readers the important fact that language is always growing and that new words are continually made "by choosing fit and proper sounds." He shows how words once imitative or emotional have been often so changed and modified as to have their original character totally concealed; yet he concludes, that—"it would be unscientific to accept all this as a complete explanation of the origin of language"—because "other causes may have helped." It seems, however, to the present writer, that the imitative and emotional origin of language is demonstrated by a body of facts almost as extensive and complete as that which demonstrates the origin of species by natural selection; and that the "other causes" are in both cases exceptional and subordinate. As the examples of imitative words given by Mr. Tylor are comparatively trivial and altogether inadequate, it may be well to call attention to the wide and far-reaching character of such words, and to show how much of the force, expressiveness, and beauty of our language (as of most others) depends upon them.

Putting aside all mere representations of animal sounds—as the *whinny* of the colt, the *mew* of the cat, or the *bleat* of the sheep—let us consider what an immense number of natural sounds are named by words which we at once see to be appropriate representations or imitations of them. Such are—*crash*, *whizz*, *fizz*, *hiss*, *creak*, *whistle*, *rattle*, *bang*, *clang*, *flop*, *thud*, *clap*, *roar*, *snore*, *groan*, *moan*, *wail*, *thunder*. In other cases sights, sounds, or feelings, are represented by their accompanying or appropriate sounds. We see a *splash*, or a *slop*; we feel a thing to be *smooth* or *rough*, or to *vibrate*; and we *shiver* with cold or *terror*. Again, how many actions and qualities are represented by words expressing the sounds which sometimes accompany them—as *knock*, *shock*, *crack*, *snap*, *ring*, *whisper*, *hush*, *sigh*, *sob*, *wash*, *squash*, *crush*, *crunch*, *rip*, *rend*, *grind*, *scratch*, *split*, *spit*, *cough*, *sneeze*, *wheeze*. How characteristic are such words as *sticky*, *flicker*, *flutter*, *hurry*, *flurry*, *stumble*, *hobble*, *wobble*. Here we have not only sound, but motion and quality, represented by the arrangement of letters and syllables. How clearly do such words as *slide*, *glide*, and *wave* imply slow and continuous motion, the movement of the lips while pronouncing the latter word being a perfect double undulation. How curiously do the tongue and palate seem to be pulled apart from each other while pronouncing the words *glue* or *sticky*. How marked is the contrast between the harsh consonants used to express *rough*, *rugged*, and *gritty*, as compared with the soft flow of sounds in *smooth*, *oily*, *even*, *polished*. Look again at the sense of effort and feeling of grandeur in pronouncing the words *strong*, *strength*, *power*, *might*, as compared with the opposites, *weak*, *faint*; or the open-mouthed sounds of *grand*, *huge*, *monstrous*, *vast*, *immense*, *giant*, *gigantic*, as contrasted with the almost closed lips with which we say *small*, *little*, *tiny*, *minute*, *pigmy*, *midget*. So *crawl* and *drag* are pronounced slowly as compared with *run*, *fly*, or *swim*; while *difficult* and *easy* express their own meaning while we pronounce them. Many objects and substances have names curiously corresponding with their qualities. We have already noticed *glue* as indicating stickiness, but no less clearly is *oil* smooth; while *brass* and *glass* indicate resonance; *tin* a tinkling sound; *lead* and *wood* a dull sound or *thud*; in *bell* we imitate its sound, while the word *jelly* indicates the shaking of the substance. In *ice* we hear the interjectional *sh* of shivering with cold; in *fire* the flicker of the ascending flame. In other cases the motion of the breath gives an indication of meaning; *in* and *out*, *up* and *down*, *elevate* and *depress*, are pronounced with an inspiration and expiration respectively, the former being necessarily accompanied with a raising, the latter with a depression, of the head. When we name the *mouth* or *lips* we use labials; for *tooth* and *tongue*, dentals; for the *nose* and things relating to it, nasal sounds; and this peculiarity is remarkably constant in most languages, civilised and savage. Among the Malay races, for instance, we find such words as *mulut*, *bawa*, *mohon*, and *moda* for mouth; *gigit*, *nisinen*, *nigni*, and *niki* for teeth, and *idong*, *ugerun* and *usnut*, for nose. So in words for *large* we find a prevalence of broad sounds involving a wide opening of the mouth, as *busar*, *bākè*, *bāgut*, *lāmu*, *elāmo*, *ilāhé*, *eräāmei*, *aiyuk*, *mäina*—and for *small*, words that are pronounced quickly and with slight opening of the lips, as *kichil*, *chili*, *kidi*, *köi*, *roit*, *kemi*, *anan*, *kiiti*, *fek*, *didiki*, all taken from languages of the Malay Archipelago.

These few examples, which might be greatly increased, indicate a variety of ways in which, even now, after all the modifications and development which language has undergone, sound still corresponds to sense; and if the reader will turn to Dr. Farrar's suggestive little work on the *Origin of Language*, he will find how wonderfully, by the help of analogy and metaphor, the uses and meanings of simple words and sounds have been indefinitely increased, so as to subserve the growing need of mankind to express more and more complex ideas. Mr. Tylor is rather unfortunate in his illustration of words for the form of which no cause can be assigned, when he says: "There is no apparent reason why the word *go* should not have signified the idea of coming, and the word *come* the idea of going." But, in accordance with the examples already given, there *is* a very good and sufficient reason. We pronounce *come* with a closure and contraction of the lips and usually during inspiration, *go* with open and protruding lips and usually during expiration. Now many savages point with the lips as we do with the finger, signifying *there*, by protruding the lips in the direction to be indicated; and any one who has seen this curious gesture must be struck with its close similarity to the protrusion of the lips in pronouncing the word *go*. The same difference of the nearly closed or open lips characterises the words for these two ideas in many other languages. In French we have *viens* and *va*, in the German *komm* and *geh*, in Italian *vieni* and *vai*, showing that words in distinct languages differing greatly in spelling and pronunciation may yet have a common character in the mode of speaking which indicates their common meaning . . .

* * * * *

In 1895 Wallace enlarged upon this treatment (which itself was anticipated in a review of another Tylor book in 1870—see ref 169) in an essay entitled 'The expressiveness of speech, or mouth-gesture as a factor in the origin of language' (ref 518).

In Section One I have provided a review of Wallace's most significant contributions to anthropology. It is not difficult to recognize in these writings how easily his train of thought integrated different subjects, and in turn the inter-dependent nature of his ideas. This interdisciplinary focus will be evident in materials presented in the remaining Sections as well; whatever their particular emphases, however, they are with but relatively few exceptions never far removed from Wallace's views on man. The condition of the human race was the central concern of his life.

Related writings

on non-human primates: refs 8, 23, 24, 26, 30, 237, 350, 433, 712, 714, 715, 724, 727

descriptions of the characteristics and lives of tropical peoples and societies: refs 10, 11, 18, 19, 22, 27, 29, 31, 32, 34, 37, 38, 41, 51, 65, 67, 76, 82, 91, 104, 113, 235, 301, 713, 714, 715, 720, 727, 729

on ethnology and ethnography: refs 29, 34, 41, 51, 82, 86, 87, 92, 95, 97, 99, 100, 104, 106, 113, 120, 131, 167, 169, 207, 234, 235, 266, 296, 297, 301, 317, 323, 518, 583, 615, 645, 712, 714, 715, 720, 727

on the (past) evolution of the mind and body of man: refs 82, 90, 93, 100, 120, 125, 146, 150, 165, 173, 175, 186, 221, 237, 303, 305, 337, 445, 461, 524, 574, 600, 645, 649, 716, 724, 725, 726, 727, 742, 750, 753

on the antiquity/progress of civilized man (including the "degeneration hypothesis"): refs 87, 93, 150, 152, 161, 207, 233, 234, 237, 257, 314, 324, 337, 406, 433, 445, 496, 524, 600, 645, 649, 716, 725, 726, 727, 733, 752 on archaeological subjects: refs 122, 233, 237, 257, 266, 404, 406, 433, 719, 725, 727

on the future of mankind: refs 93, 125, 221, 427, 445, 450, 496, 545, 549, 690, 716, 725, 727, 733, 736, 747

on individual and social morality and ethics: refs 1, 95, 104, 111, 113, 150, 152, 153, 165, 186, 217, 243, 379, 398, 427, 445, 451, 461, 505, 506, 540, 545, 549, 580, 623, 630, 649, 715, 716, 717, 724, 725, 726, 727, 729, 733, 736, 747

on women's rights and their role in social evolution: refs 427, 445, 497, 549, 647, 649, 650, 671, 727, 733, 736, 737, 738

on language and linguistics: refs 29, 51, 82, 131, 169, 199, 337, 518, 623, 714, 715, 720, 727, 729

on miscellaneous topics: refs 43, 92, 119, 120, 133, 151, 161, 164, 165, 169, 207, 208, 217, 220, 221, 232, 242, 248, 253, 303, 432, 439, 562, 602, 623, 629, 630, 715, 716, 725, 726, 727, 729

book reviews and notices: refs 131, 146, 161, 169, 186, 207, 217, 220, 221, 232, 233, 234, 237, 253, 266, 303, 305, 314, 324, 337, 342, 355, 421, 433, 450, 461, 496, 523, 524, 727

secondary literature: Cooter (1984), Eiseley (1958), Kottler (1974), McKinney (1972), Malinchak (1987), Schwartz (1984*a*), R. Smith (1972), Turner (1974)

§ 2 · SPIRITUALISM

... we are, all of us, in every act and thought of our lives, helping to build up a mental fabric which will be and constitute ourselves in the future life, even more completely than now. Just in proportion as we have developed our higher intellectual and moral nature, or starved it by disuse, shall we be well or ill fitted for the new life we shall enter on. The Spiritualist who, by repeated experiences, becomes convinced of the absolute reality and the complete reasonableness of these facts regarding the future state—who knows that, just in proportion as he indulges in passion, or selfishness, or the reckless pursuit of wealth, and neglects to cultivate his moral and intellectual nature, so does he inevitably prepare for himself misery in a world in which there are no physical wants to be provided for, no struggle to maintain mere existence, no sensual enjoyments except those directly associated with sympathy and affection, no occupations but those having for their object social, moral, and intellectual progress—is impelled towards a pure and moral life by motives far stronger than any which either philosophy or religion can supply . . .

('Why live a moral life?', 1894: ref 506, p. 9)

WALLACE'S attraction to spiritualism is not likely to be fully appreciated until such time as the belief itself is demonstrated to have—or *not* to have—connection to phenomena we can comfortably incorporate into our overall understanding of 'nature'. That Wallace 'believed in' spiritualism is certain enough; it is quite another matter, however, to come up with an interpretation of his belief that also accounts for the way he dealt with other subjects.

On pages 645–6 of the entry 'Spiritualism' which Wallace contributed to the 1892 edition of Volume 9 of *Chambers's encyclopaedia* (ref 451), he defined the term as:

. . . the name applied to a great and varied series of abnormal or preter-normal phenomena purporting to be for the most part caused by spiritual beings, together with the belief thence arising of the intercommunion of the living and the so-called dead.

One cannot help but conjure up some rather unattractive images of the subject when it is reduced to such terms, and indeed it seems to be on the basis of this kind of reaction that Wallace's association with spiritualism has generally been regarded as his greatest intellectual inconsistency. A little probing, however, reveals that there is considerably more intellectual content to spiritualism than a simple belief in disembodied spirits. On page 648 of the same *Chambers's encyclopaedia* entry cited above, Wallace encapsulates the *object* of spiritualist belief rather differently:

The universal teaching of modern spiritualism is that the world and the whole material universe exist for the purpose of developing spiritual beings—that death is simply a transition from material existence to the first grade of spirit-life—and that our happiness and the degree of our progress will be wholly dependent upon the use we have made of our faculties and opportunities here. It is urged that the present life will assume a new value and interest when men are brought up not merely in the vascillating and questionable *belief*, but in the settled, indubitable *conviction*, that our existence in this world is really but one of the stages in an endless career, and that the thoughts we think and the deeds we do here will certainly affect our condition and the very form and organic expression of our personality hereafter.

It has sometimes been suggested (for example, by Kottler 1974; McKinney 1976; and Schwartz 1984*a*) that spiritualism was in effect Wallace's religion, a means through which he could reconcile his 'incompatible' views on 'natural' and 'human' evolution. I regard this as a serious oversimplification. A much better starting point can be gleaned from the excerpts presented above: Wallace regarded spiritualization as a *process continuing the evolutionary progression in a manner rather analogous to the way organic evolution depended on—and itself complemented—inorganic evolution*. It is inaccurate to regard his views on the way each of the primary realms of nature experience this progression as being 'incompatible'; indeed, whatever degree of *accuracy* they may turn out to have, they are the very *model* of internal consistency. For Wallace, the 'spiritual landscape' was linked to the biological landscape in a fashion not unlike the way in which the biological landscape interacts with the inorganic world. It did not particularly upset him that evidence regarding the reality of the 'spirit realm' was slim, because: (1) there appeared to be *some*, at least, that was convincing; (2) he was fully satisfied that natural selection—whatever its vast range of applications to other subjects—could not account for the emergence and domination of mankind's higher faculties; and (3) the teachings of spiritualism appeared to reconcile a variety of otherwise inexplicable events in human history. His position appears to have been that while natural selection could operate to produce the fundamental biological structure of human beings, truly 'human' (or, better yet, 'humanizing') evolution described a process of ascendance culminating in a higher, entirely spiritual form of existence that lay beyond the individual's biological life span. (Further analysis of Wallace's cosmology will be presented in Section Six.) 'Spiritualism' *per se* was the rubric he (unfortunately) attached to both the overall process of 'spiritualization' itself, and its formal study.

Three connections in particular make it easier to understand Wallace's adoption of spiritualism. First, he was an accomplished mesmerist, having learned the technique in 1844 after attending—at that point, as a non-believer—a lecture given by Spencer Hall, an early proponent. At that time most people considered mesmerism a hoax; but on discovering his own latent talent, Wallace learned, as he later put it (ref 478, p. 440):

my first great lesson in the inquiry into these obscure fields of knowledge, never to accept the disbelief of great men, or their accusations of imposture or of imbecility, as of any weight when opposed to the repeated observation of facts by other men admittedly sane and honest.

From that point onward, Wallace would trust to his own appraisals of the 'facts', not to what he interpreted as ignorant hearsay. Twenty years later he would perceive in the general response to spiritualism a case of history repeating itself. Further, he soon discovered that there were many connections between mediumistic abilities and the characteristics of mesmeristic trance.

An equally significant influence was Wallace's conclusion—in part arising from his years of living among tropical peoples—that many strictly human faculties apparently had no survival utility. With his recognition of the principle of natural selection in 1858, Wallace probably tentatively accepted that they had a utilitarian function after all—though not one that could be understood in terms of the 'survival of the fittest'. His adoption of spiritualism about 1865 signalled his decision that this use must be integrally connected to mankind's spiritualization (i.e., intellectual/moral evolution) in a 'world [that] is really but one of the stages in an endless career . . .'.

A third influence pre-disposing Wallace toward spiritualism is alluded to at the end of this Section.

Wallace's writings on spiritualism mirror both his interest in the condition and evolution of man and his concern that 'occult' subjects be regarded as deserving of serious scientific attention. His assessment of the evidence available—including many first-hand experiences at seances—had convinced him that legitimate phenomena referable to a 'spirit realm' existed, and might be critically examined by unprejudiced minds. His first major statement to that effect, published in instalments in the obscure secularist journal The English Leader *in 1866, was 'The scientific aspect of the supernatural' (ref 118). That work is too long to reproduce here, but his next extended treatment of the subject, composed in 1870, serves to illustrate his style of argument:*

No antecedent impossibility in miracles. A reply to modern objectors (ref 174)

The Supernatural and Modern Thought. It is now generally admitted, that those opinions and beliefs in which men have been educated generation after generation, and which have thus come to form part of their mental nature, are especially liable to be erroneous, because they keep alive and perpetuate the ideas and prejudices of a bygone and less enlightened age. It is therefore in the interest of truth, that every doctrine or belief, however well established or sacred they may appear to be, should at certain intervals be challenged to arm themselves with such facts and reasonings as they possess, to meet their opponents in the open field of controversy, and do battle for their right to live. Nor can any exemption be claimed in favour of those beliefs which are the product of modern civilisation, and which have for several generations been unquestioned by the great mass of the educated community; for the prejudice in their favour will be proportionately great, and, as was the case with the doctrines of Aristotle, and the dogmas of the schoolmen, they may live on by mere weight of authority and force of habit, long after they have been shown to be opposed alike to fact and to reason. There have been times when popular beliefs were defended by the terrors of the law, and when the sceptic could only attack them at the peril of his life. Now we all admit that truth can take care of itself, and that only error needs protection. But there is another mode of defence which equally implies a claim to certain and absolute truth, and which is therefore equally unworthy and unphilosophical—that of ridicule, misrepresentation, or a contemptuous refusal to discuss the question at all. This method is used among us even now, for there is one belief, or rather disbelief, whose advocates claim more than papal infallibility, by refusing to examine the evidence brought against it, and by alleging general arguments which have been in use for two centuries to prove that it cannot be erroneous. The belief to which I allude is, that all alleged miracles are false; that what is commonly understood by the term *supernatural* does not exist, or if it does, is incapable of proof by any amount of human testimony; that all the phenomena we can have cognizance of depend on ascertainable physical laws, and that no other intelligent beings than man and the inferior animals can or do act upon our material world. These views have been now held almost unquestioned for many generations; they are inculcated as an essential part of a liberal education; they are popular, and are held to be one of the indications of our intellectual advancement; and they have become so much a part of our mental nature, that all facts and arguments brought against them are either ignored as unworthy of serious consideration, or listened to with undisguised contempt. Now this frame of mind is certainly not one favourable to the discovery of truth, and strikingly resembles that by which, in former ages, systems of error have been fostered and maintained. The time has therefore come when it must be called upon to justify itself.

This is the more necessary, because the doctrine, whether true or false, actually rests upon a most unsafe and rotten foundation. I propose to show you that the best arguments hitherto relied upon to prove it are, one and all, fallacious—and prove nothing of the kind. But a theory or belief may be supported by very bad arguments, and yet be true; while it may be supported by some good arguments, and yet be false. But there was never a true theory which had no good arguments to support it. If therefore all the arguments hitherto used against miracles in general can be shown to be bad, it will behove sceptics to discover good ones; and if they cannot do so, the evidence in favour of miracles must be fairly met and judged on its own merits, not ruled out of court as it is now.

It will be perceived therefore, that my present purpose is to clear the ground for the discussion of the great question of the so-called supernatural. I shall not attempt to bring arguments either for or against the main proposition, but shall confine myself to an examination of the allegations and the reasonings which have been supposed to settle the whole question on general grounds.

David Hume, and His False Definition of a Miracle. One of the most remarkable works of the great Scotch philosopher, David Hume, is *An Inquiry concerning Human Understanding*, and the tenth chapter of this work is 'On Miracles,' in which occur the arguments which are so often quoted to show that no evidence can prove a miracle. Hume himself had a very high opinion of this part of his work, for he says at the beginning of the chapter—

> I flatter myself that I have discovered an argument which, if just, will with the wise and learned be an everlasting check to all kinds of superstitious delusion, and consequently will be useful as long as the world endures; for so long, I presume, will the accounts of miracles and prodigies be found in all history, sacred and profane.

After a few general observations on the nature of evidence, and the value of human testimony in different cases, he proceeds to define what he means by miracle. And here, at the very beginning of the subject, we find that we have to take objection to Hume's definition of a miracle, which exhibits unfounded assumptions and false premises. He gives two definitions in different parts of his essay. The first is—"A miracle is a violation of the laws of nature." The second is—"A miracle is a transgression of a law of nature, by a particular volition of the Deity, or by the interposition of some invisible agent." Now both these definitions are bad or imperfect. The first assumes that we know all the laws of nature—that the particular effect could not be produced by some unknown law of nature overcoming the law we do know; it assumes also, that if an invisible intelligent being held an apple suspended in the air, the act would violate the law of gravity. The second is not precise; it should be "some invisible *intelligent* agent," otherwise the action of galvanism or electricity, when these agents were first discovered, and before they were ascertained to form part of the order of nature, would answer accurately to this definition of a miracle. The words "violation" and "transgression" are both improperly used, and really beg the question by the definition. How does Hume know that any particular miracle is a violation of a law of nature? He assumes this without a shadow of proof, and on these words, as we shall see, rests his whole argument.

The True Definition of a Miracle. Before proceeding any further, it is necessary for us to consider what is the true definition of a miracle, or what is most commonly meant by that word. A miracle, as distinguished from a new and unheard-of natural phenomenon, supposes an intelligent superhuman agent either visible or invisible;—it is not necessary that what is done should be beyond the power of man to do. The simplest action, if performed independently of human or visible agency, such as a tea-cup lifted in the air at request, as by an invisible hand and without assignable cause, would be universally admitted to be a miracle, as much so as the lifting of a house into the air, the instantaneous healing of a wound, or the instantaneous production of an elaborate drawing. My definition of a miracle therefore is as follows:—"Any act or event implying the existence and agency of superhuman intelligences," considering the human soul or spirit, if manifested out of the body, as one of these superhuman intelligences. This definition is more complete than that of Hume, and defines more accurately the essence of that which is commonly termed a miracle.

Hume's First Argument a Radical Fallacy. We now have to consider Hume's arguments. The first is as follows:—

> A miracle is a *violation of the laws of nature*; and as a firm and *unalterable experience* has established these laws, the proof against a miracle, from the very nature of the fact, is as entire as any argument from experience can possibly be imagined. Why is it more than probable that all men must die; that lead cannot *of itself remain suspended in the air*; that fire consumes wood, and is extinguished by water; unless it be, that these events are found agreeable to the laws of nature, and there is required a *violation of these laws*, or, in other words a *miracle*, to prevent them? Nothing is esteemed a miracle, if it ever happened in the *common* course of nature. It is no miracle that a man seemingly in good health should die on a sudden; because such a kind of death, though more unusual than any other, has yet been frequently observed to happen. But it is a miracle that a dead man should come to life; because *that has never been observed in any age or country*. There must, therefore, be an uniform experience against every miraculous event, otherwise the event would not merit that appellation. And as a *uniform* experience amounts to a *proof*, there is here a direct and full proof, from the nature of the fact, against the existence of any miracle; nor can such a proof be destroyed, or the miracle rendered credible, but by an opposite proof, which is superior.

This argument is radically fallacious, because if it were sound, no perfectly new fact could ever be proved, since the first and each succeeding witness would be assumed to have universal experience against him. Such a simple fact as the existence of flying fish could never be proved, if Hume's argument is a good one; for the first man who saw and described one, would have the universal experience against him that fish do not fly, or make any approach to flying, and his evidence being rejected, the same argument would apply to the second, and to every subsequent witness, and thus no man at the present day who has not seen a flying fish ought to believe that such things exist.

Again—painless operations in a state produced by mere passes of the hand, were, twenty-five years ago, maintained to be contrary to the laws of nature, contrary to all human experience, and therefore incredible. On Hume's principles they were miracles, and no amount of testimony could ever prove them to be real. But miracles do not stand alone, single facts opposed to uniform experience. Reputed miracles abound in all periods of history; every one has a host of others leading up to it; and every one has strictly analogous facts testified to at the present day. The uniform opposing experience therefore on which Hume lays so much stress does not exist. What, for instance, can be a more striking miracle than the levitation or raising of the human body into the air without visible cause, yet this fact has been testified to during a long series of centuries . . . Butler, in his *Lives of the Saints*, says that many such facts are related by persons of undoubted veracity, who testify that they themselves were eye-witnesses of them. So we all know that at least fifty persons of high character may be found in London, who will testify that they have seen the same thing happen to Mr. [Daniel] Home [a famous nineteenth-century medium]. I do not adduce this testimony as proving that the circumstances related really took place; I merely bring it forward now to show

how utterly unfounded is Hume's argument, which rests upon universal testimony on the one side, and no testimony on the other.

Hume Self-Contradictory. I now have to show that in Hume's efforts to prove his point, he contradicts himself in a manner so gross and complete as is perhaps not to be found in the works of any other eminent author. The first passage I will quote is as follows:

> For, first, there is *not to be found*, in *all history*, any miracle attested by a *sufficient number* of men, of such unquestioned *good sense, education*, and *learning*, as to secure us against all delusion in themselves; of such undoubted *integrity*, as to place them beyond all suspicion of any design to deceive others; of such credit and reputation in the eyes of mankind, as to have a great deal to lose in case of their being detected in any falsehood; and at the same time attesting facts performed in such a *public manner*, and in so *celebrated a part of the world*, as to render the detection unavoidable: all which circumstances are requisite to give us a full assurance in the testimony of men.

A few pages further on we find this passage:

> There surely never was a greater number of miracles ascribed to one person, than those which were lately said to have been wrought in France upon the tomb of Abbé Paris, the famous Jansenist, with whose sanctity the people were so long deluded. The curing of the sick, giving hearing to the deaf, and sight to the blind, were everywhere talked of as the usual effects of that holy sepulchre. But what is more extraordinary, many of the miracles were *immediately proved upon the spot*, before *judges of unquestioned integrity*, attested by *witnesses* of *credit and distinction*, in a *learned age*, and on the most *eminent theatre* that is *now in the world*. Nor is this all. A relation of them was published and dispersed everywhere; nor were the Jesuits, though a learned body, supported by the civil magistrates, and determined enemies to those opinions, in whose favour the miracles were said to have been wrought, ever able *distinctly to refute or detect them*. Where shall we find such a number of circumstances, agreeing to the corroboration of one fact? And what have we to oppose to such a cloud of witnesses, but the absolute *impossibility*, or *miraculous nature* of the events which they relate? And this, surely, in the eyes of all reasonable people, will alone be regarded as a sufficient refutation.

In the second passage he affirms the existence of every single fact and quality which in the first passage he declared never existed, and he entirely changes his ground of argument by appealing to the inherent impossibility of the fact, and not at all to the insufficiency of the evidence. He even makes this contradiction still more remarkable by a note which he has himself given to this passage, a portion of which is as follows:

> This book was wrote by Mons. Montgeron, Councillor or Judge of the Parliament of Paris, a man of figure and character, who was also a martyr to the cause, and is now said to be somewhere in a dungeon on account of his book . . .
>
> Many of the miracles of Abbé Paris were proved immediately by witnesses before the officialty or bishop's court at Paris, under the eye of Cardinal Noailles; whose character for integrity and capacity was never contested, even by his enemies.
>
> His successor in the archbishopric was an enemy to the Jansenists, and for that

reason promoted to the see by the court. Yet 22 rectors or curés of Paris, with infinite earnestness, press him to examine those miracles, which they assert to be known to the whole world, and indisputably certain; but he wisely forebore . . .

All who have been in France about that time have heard of the reputation of Mons. Herault, the lieutenant of police, whose vigilance, penetration, activity, and extensive intelligence, have been much talked of. This magistrate who, by the nature of his office, is almost absolute, was invested with full powers, on purpose to suppress or discredit these miracles; and he frequently seized immediately, and examined the witnesses and subjects to them: *but never could reach anything satisfactory against them.*

In the case of Mademoiselle Thibaut he sent the famous De Sylva to examine her; whose evidence is very curious. The physician declares, that it was impossible that she could have been so ill as was proved by witnesses; because it was impossible she could in so short a time have recovered so perfectly as he found her. He reasoned like a man of sense, from natural causes; but the opposite party told him that the whole was a miracle, and that his evidence was the very best proof of it . . .

No less a man than the Duc de Chatillon, a duke and peer of France, of the highest rank and family, gives evidence of a miraculous cure performed upon a servant of his, who had lived for several years in his house with a visible and palpable infirmity.

I shall conclude with observing, that no clergy are more celebrated for strictness of life and manners than the regular clergy of France, particularly the rectors or curés of Paris, who bear testimony to these impostures.

The learning, genius, and probity of the gentlemen, and the austerity of the nuns of Port-Royal have been much celebrated all over Europe. Yet they all give evidence for a miracle, wrought on the niece of the famous Pascal, whose sanctity of life, as well as extraordinary capacity, is well known. The famous Racine gives an account of this miracle in his famous history of Port-Royal, and fortifies it with all the proofs, which a multitude of nuns, priests, physicians and men of the world, all of them of undoubted credit could bestow upon it. Several men of letters, particularly the Bishop of Tournay, thought this miracle so certain, as to employ it in the refutation of Atheists and Free-thinkers. The queen-regent of France, who was *extremely prejudiced against the Port-Royal*, sent *her own physician to examine the miracle*, who returned an *absolute convert*. In short the supernatural cure was so incontestible, that it saved, for a time, that famous monastery from the ruin with which it was threatened by the Jesuits. *Had it been a cheat, it had certainly been detected by such sagacious and powerful antagonists, and must have hastened the ruin of the contrivers.*

It seems almost incredible that this can have been written by the great sceptic David Hume, and written in the same work in which he has already affirmed that in all history no such evidence is to be found . . . [These are only a few of many] cases equally marvellous, and equally well attested, and we therefore cannot be surprised at Hume's being obliged to give up the argument of the insufficiency of the evidence for miracles and of the uniform experience against them; the wonder being that he ever put forth an argument which he was himself able to refute so completely.

Another Argument of Hume's a Palpable Fallacy. We now have another argument which Hume brings forward, but which is, if possible, still weaker than the last. He says:

> I may add, as a fourth reason, which diminishes the authority of prodigies, that there is no testimony for any, even those which have not been expressly detected, that is not opposed by any infinite number of witnesses; so that not only the miracle destroys the credit of the testimony, but the testimony destroys itself. To make this the better understood, let us consider that, in matters of religion, whatever is different is contrary: and that it is impossible the religions of ancient Rome, of Turkey, and Siam, and of China, should, all of them, be established on any solid foundation. Every miracle, therefore, pretended to have been wrought in any of these religions (and all of them abound in miracles), as its direct scope is to establish the particular system to which it is attributed; so has it the same force, though more indirectly, to overthrow every other system. In destroying a rival system it likewise destroys the credit of those miracles on which that system was established; so that all the prodigies of different religions are to be regarded as contrary facts; and the evidences of these prodigies, whether weak or strong, as opposite to each other. According to this method of reasoning, when we believe any miracle of Mahomet or his successors, we have for our warrant the testimony of a few barbarous Arabians. And, on the other hand, we are to regard the authority of Titus, Livius, Plutarch, and in short of all the authors and witnesses, Grecian, Chinese, and Roman Catholic, who have related any miracle in their particular religion; I say, we are to regard their testimony in the same light as if they had mentioned that Mahometan miracle, and had in express terms contradicted it, with the same certainty as they have for the miracle they relate.

Now this argument, if argument it can be called, rests upon the extraordinary assumption that a miracle, if real, can only come from God, and must therefore support only a true religion. It assumes also that religions cannot be true unless given by God. Mr. Hume assumes, therefore, to know that nothing which we term a miracle can possibly be performed by any of the probably infinite number of intelligent beings who may exist in the universe between ourselves and the Deity. He confounds the evidence for the fact with the theories to account for the fact, and most illogically and unphilosophically argues, that if the theories lead to contradictions, the facts themselves do not exist.

I think, therefore, that I have now shown that—1. Hume gives a false definition of miracles, which begs the question of their possibility. 2. He states the fallacy that miracles are isolated facts, to which the entire course of human testimony is opposed. 3. He deliberately and absolutely contradicts himself as to the amount and quality of the testimony in favour of miracles. 4. He propounds the palpable fallacy as to miracles connected with opposing religions destroying each other.

Recent Objections to Miracles. We will now proceed to some of the more modern arguments against miracles. One of the most popular modern objections consists of making a supposition and drawing an inference, which looks like a dilemma, but which is really none at all.

This argument has been put in several forms. One is, "If a man tells me he came from York by the telegraph-wire, I do not believe him. If fifty men tell me they came from York by telegraph wires, I do not believe them. If any number of men tell me the same, I do not believe them. Therefore, Mr. Home did not float in the air, notwithstanding any amount of testimony you may bring to prove it."

Another is, "If a man tells me that he saw the lion on Northumberland House descend into Trafalgar Square and drink water from the fountains, I should not believe him. If fifty men, or any number of men, informed me of the same thing, I should still not believe them."

Hence it is inferred that there are certain things so absurd and so incredible, that no amount of testimony could possibly make a sane man believe them.

Now, these illustrations look like arguments, and at first sight it is not easy to see the proper way to answer them; but the fact is that they are utter fallacies, because their whole force depends upon an assumed proposition which has never been proved, and which I challenge any one to prove. The proposition is, that a large number of independent, honest, sane, and sensible witnesses, *can* testify to a plain matter of fact which never happened at all.

Now, no evidence has ever been adduced to show that this ever has happened or ever could happen. But the assumption is rendered still more monstrous when we consider the circumstances attending such cases as those of the curés at the tomb of the Abbé Paris, and the cases of modern scientific men being converted to a belief in the reality of the phenomena of modern Spiritualism; for we must assume that, being warned that the alleged facts are impossible and are therefore delusions, and having the source of the supposed delusion pointed out, and all the prejudices of the age and the whole tone of educated thought being against the reality of such facts, yet numbers of educated men, including physicians and men of science, are convinced of the reality of such facts after the most searching investigation. Yet the assumption that such an amount and quality of independent converging evidence *can* be all false, must be proved to be a fact if the argument is to have the slightest value, otherwise it is merely begging the question. It must be remembered that we have to consider, not absurd beliefs or false inferences, but plain matters of fact; and it cannot be proved, and never has been proved, that any large amount of cumulative evidence of disinterested and sensible men, was ever obtained for an absolute and entire delusion. To put the matter in a simple form, the asserted fact is either possible or not possible. If possible, such evidence as we have been considering would prove it; if not possible, such evidence could not exist. The argument is, therefore, an absolute fallacy, since its fundamental assumptions cannot be proved. If it is intended merely to enunciate the proposition, that the more strange and unusual a thing is the more and the better evidence we require for it, that we all admit; but I maintain that human testimony increases in value in such an enormous ratio with each additional independent and honest witness, that no fact ought to be rejected when attested by such a body of evidence as exists for many of the events termed miraculous or supernatural, and which occur now daily among us. The burden of proof lies on those who maintain that such evidence can possibly be fallacious. Let them point out one case in which such cumulative evidence existed, and which yet proved to be false; let them give not supposition, but proof.

Another modern argument is used more especially against the reality of the so-called spiritual phenomena. It is said, "These phenomena are so uncertain, you have no control over them, they follow no law; prove to us that they follow definite laws like all other groups of natural phenomena, and we will believe them." This argument appears to have weight with some persons, and yet it is really an absurdity. The essence of the alleged phenomena (whether they be true or not, is of no importance) is, that they seem to be the result of the action of independent intelligences, and are therefore deemed to be spiritual or superhuman. If they had been found to follow strict law and not independent will, no one would have ever supposed them to be spiritual. The argument therefore, is merely the statement of a foregone conclusion, namely, "As long as your facts go to prove the existence of distinct intelligences, we will not believe them; demonstrate that they follow fixed law, and not intelligence, and then we will believe them." This argument appears to me to be childish, and yet it is used by some persons who claim to be philosophical.

Scientific Men in Error when They Deny the Facts of Investigators. Another objection which I have heard stated in public, and received with applause, is that it requires immense scientific knowledge to decide on the reality of any uncommon or incredible facts, and that till scientific men investigate and prove them, they are not worthy of credit. Now I venture to say that a greater fallacy than this was never put forth. The subject is a very important one, and the error is a very common one, but the fact is the exact opposite of what is stated; for I assert that whenever the scientific men of any age have denied the facts of investigators on *à priori* grounds, they have *always been wrong.*

It is not necessary to do more than refer to the world-known names of Galileo, Harvey, and Jenner; the great discoveries they made were, as we know, violently opposed by all their scientific contemporaries, to whom they appeared absurd and incredible; but we have equally striking examples much nearer to our own day. When Benjamin Franklin brought the subject of lightning conductors before the Royal Society, he was laughed at as a dreamer, and his paper was not admitted to the *Philosophical Transactions*. When Young put forth his wonderful proofs of the undulatory theory of light, he was equally hooted at as absurd by the popular scientific writers of the day. The *Edinburgh Review* called upon the public to put Thomas Gray into a straight jacket for maintaining the practicability of railroads. Sir Humphry Davy laughed at the idea of London ever being lighted with gas. When Stephenson proposed to use locomotives on the Liverpool and Manchester Railway, learned men gave evidence that it was impossible that they could go even twelve miles an hour. Another great scientific authority declared it to be equally impossible for ocean steamers ever to cross the Atlantic. The French Academy of Sciences ridiculed the great astronomer Arago, when he wanted even to discuss the subject of the electric telegraph. Medical men ridiculed the stethoscope when it was first discovered. Painless operations during the mesmeric coma were pronounced impossible, and therefore impostures.

But one of the most striking, because one of the most recent cases of this opposition to, or rather disbelief in facts opposed to the current belief of the day, among men who are generally charged with going too far in the other direction, is that of the doctrine of the "Antiquity of Man." Boué, an experienced French geologist, in 1823, discovered a human skeleton eighty feet deep in the lees or hardened mud of the Rhine. It was sent to the great anatomist Cuvier, who so utterly discredited the fact, that he threw aside this invaluable fossil as worthless, and it was lost. Sir. C. Lyell, from personal investigation on the spot, now believes that the statements of the original observer were quite accurate. So early as 1715, flint weapons were found with the skeleton of an elephant in an excavation in Gray's Inn Lane, in the presence of Mr. Conyers, who placed them in the British Museum, where they remained utterly unnoticed till quite recently. In 1800, Mr. Frere found flint weapons along with the remains of extinct animals at Hoxne, in Suffolk. From 1841 to 1846, the celebrated French geologist, Boucher de Perthes, discovered great quantities of flint weapons in the drift gravels of the North of France, but for many years he could convince none of his fellow scientific men that they were works of art, or worthy of the slightest attention. At length, however, in 1853, he began to make converts. In 1859–60, some of our own most eminent geologists visited the spot, and fully affirmed the truth of his observations and deductions.

Another branch of the subject was, if possible, still worse treated. In 1825, Mr. McEnery, of Torquay, discovered worked flints along with the remains of extinct animals in the celebrated Kent's Hole Cavern, but his account of his discoveries was simply laughed at. In 1840, one of our first geologists, Mr. Godwin Austin, brought this matter before the Geological Society, and Mr. Vivian, of Torquay, sent in a paper fully confirming Mr. McEnery's discoveries, but it was thought too improbable to be published. Fourteen years later, the Torquay Natural History Society made further observations, entirely confirming the previous ones, and sent an account of them to the Geological Society of London, but the paper was rejected as too improbable for publication. Now, however, for five years past, the cave has been systematically explored under the superintendence of a Committee of the British Association, and all the previous reports for forty years have been confirmed, and have been shown to be even less wonderful than reality. It may be said that "this was proper scientific caution." Perhaps it was; but at all events, it proves this important fact, that in this, as in every other case, the discoverers have been right, those who rejected their observations have been wrong.

Now, are the modern discoverers of some phenomena usually termed supernatural and incredible, less worthy of attention than those already quoted? Let us take, first, the reality of what is called clairvoyance. The men who have observed this phenomenon, and have carefully tested it through long years or through their whole lives, will rank in scientific knowledge, and in intellectual ability, as quite equal to any observers in any other branch of discovery. We have no less than seven eminent medical men, Drs. Elliotson, Gregory, Ashburner, Lee, Herbert, Mayo, Esdaile, and Haddock, besides persons of such high ability as Miss Martineau, Mr. H. G. Atkinson, Mr. Charles Bray, and Baron Reichenbach. With the history of previous discoverers before us, is it more likely that these eleven educated persons, knowing all the arguments against the facts, and investigating them carefully, should be all wrong, and those who say *à priori* that the thing is impossible should be all right, or the contrary? If we are to learn anything by history and experience, then we may safely prognosticate that, in this case as in so many others, the disbelievers in other men's observations will be found to be in the wrong.

Fallacies of Mr. Lecky. We now come to the modern philosophical objectors, most eminent among whom is Mr. [William] Lecky, author of the *History of Rationalism* and the *History of Morals*. In the latter work he has devoted some space to this question, and his clear and well expressed views may be taken to represent the general opinions and feelings of the educated portion of modern society.

He says:

> The attitude of ordinary educated people towards miracles is not that of doubt, of hesitation, of discontent with the existing evidence, but rather of absolute, derisive, and even unexamining incredulity . . . In certain stages of society, and under the action of certain influences, an accretion of miracles is *invariably formed* around every *prominent person* or *institution*. We can analyse the general causes that *have impelled men towards the miraculous*; we can show that these causes have never failed to produce the effect; and we can trace the gradual alteration of mental conditions *invariably accompanying* the decline of the belief.
>
> When men are *destitute of the critical spirit*, when the notion of *uniform law is yet unborn*, and when their imaginations are still incapable of rising to abstract ideas, histories of miracles are always formed and always believed; and they continue to flourish and to multiply until these conditions are altered. Miracles cease when men cease to believe and expect them . . .
>
> We do not say they are impossible, or even that they are not authenticated by as much evidence as many facts we believe. We only say that, *in certain states of society, illusions* of this kind inevitably appear.
>
> Sometimes we can discover the precise natural fact which the superstition has misread, but more frequently we can give only a general explanation, enabling us to assign these legends to their place, as the *normal expression* of a *certain stage* of knowledge or intellectual power; and this explanation is their refutation.

Now, in these statements and arguments of Mr. Lecky, we find some fallacies hardly less striking than those of Hume. His assertion that in certain stages of society an accretion of miracles is invariably formed round every prominent person or institution, appears to me to be absolutely contradicted by certain well-known historical facts.

The Church of Rome has ever been the great theatre of miracles, whether ancient or modern. The most prominent person in the Church of Rome is the Pope; the most prominent institution is the Papacy. We should expect, therefore, if Mr. Lecky's statement be correct, that the Popes would be pre-eminently miracle-workers. But the fact is, that with the exception of one or two very early ones, no miracles whatever are recorded of the great majority of

Popes. On the contrary, it has geen generally among the very humblest members of the Romish Church, whether clergy or laity, that the power of working miracles has appeared, and which has led to their being canonized as saints.

Again, to take another instance, the most prominent person connected with the Reformed Churches is Luther. He himself believed in miracles. The whole world in his day believed in miracles—and miracles, though generally of a demoniac character, continued rife in all Protestant churches for many generations after his death; yet there has been no accretion of miracles round this remarkable man.

Nearer to our own day we have Irving, at the head of a church of miracle-workers; and Joe Smith, the founder of the miracle-working Mormons; yet there is not the slightest sign of any tendency to impute any miracles to either of these men, other than those which the latter individual claimed for himself before his sect was established. These very striking facts seem to me to prove that there must be some basis of truth in nearly every alleged miracle, and that the theory of any growth or accretion round prominent individuals is utterly without evidence to support it. It is one of those convenient general statements which sound very plausible and very philosophical, but for which no proof whatever is offered.

The Decline of Belief in Miracles. Another of Mr. Lecky's statements is, that there is an alteration of mental conditions invariably accompanying the decline of belief. But this "*invariable accompaniment*" certainly cannot be proved, because the decline of the belief has only occurred once in the history of the world; and, what is still more remarkable, while the mental conditions which accompanied that one decline have continued in force or have even increased in energy and are much more widely diffused, belief has now for twenty years been growing up again. In the highest states of ancient civilisation, both among the Greeks and Romans, the belief existed in full force, and has been testified to by the highest and most intellectual men of every age. The decline which in the present century has certainly taken place, cannot, therefore, be imputed to any general law, since it is but an exceptional instance.

Again, Mr. Lecky says that the belief in the supernatural only exists "when men are destitute of the critical spirit, and when the notion of uniform law is yet unborn." Mr. Lecky in this matter contradicts himself almost as much as Hume did. One of the greatest advocates for the belief in the supernatural was [Joseph] Glanvil, and this is what Mr. Lecky says of Glanvil.

He says that Glanvil "has been surpassed in genius by few of his successors."

> The predominating characteristic of Glanvil's mind was an intense scepticism. He has even been termed by a modern critic the first English writer who has thrown scepticism into a definite form; and if we regard this expression as simply implying a profound distrust of human faculties, the judgment can hardly be denied. And certainly it would be difficult to find a work displaying less of credulity and superstition than the treatise on *The Vanity of Dogmatising*, afterwards published as *Scepsis Scientifica*, in which Glanvil expounded his

> philosophical views . . . The *Sadducisimus Triumphatus* is probably the ablest book ever published in defence of the reality of witchcraft. Dr. Henry More, the illustrious Boyle, and the scarcely less eminent Cudworth, warmly supported Glanvil; and no writer comparable to these in ability or influence appeared on the other side; yet the scepticism steadily increased . . . It was between the writings of Bacon and Locke that that latitudinarian school was formed which was irradiated by the genius of Taylor, *Glanvil*, and Hales, and which became the very centre and seedplot of religious liberty.

These are *the men* and these the *mental conditions* which are favourable to *superstition and delusion*!

The critical spirit and the notion of uniform law are certainly powerful enough in the present day, yet in every country in the civilised world there are now hundreds and thousands of intelligent men who believe, on the testimony of their own senses, in phenomena which Mr. Lecky and others would term miraculous, and therefore incredible. Instead of being, as Mr. Lecky says, an indication of "certain states of society"—"the normal expression of a certain stage of knowledge or intellectual power"—this belief has existed in all states of society, and has accompanied every stage of intellectual power. Socrates, Plutarch, and St. Augustine alike, give personal testimony to supernatural facts; this testimony never ceased through the middle ages; the Early Reformers, Luther and Calvin, throng the ranks of witnesses; all the philosophers, and all the judges of England, down to Sir Matthew Hale, admitted that the evidence for such facts was irrefutable. Many cases have been rigidly investigated by the police authorities of various countries, and, as we have already seen, the miracles at the tomb of the Abbé Paris, which occurred in the most sceptical period of French history, in the age of Voltaire and the Encyclopaedists, were proved by such an array of evidence, and were so open to investigation, that one of the noblemen of that court, convinced of their reality after the closest scrutiny, suffered the martyrdom of imprisonment in the Bastile for insisting upon making them public. And in our own day we have, at the lowest estimate, many millions of believers in modern Spiritualism in all classes of society; so that the belief which Mr. Lecky imputes to a certain stage of intellectual culture only, appears on the contrary to have all the attributes of universality.

Mr. Tylor's Objection that Belief in Miracles is an Example of the Survival of Savage Thought. The philosophical argument has been put in another form by Mr. E. B. Tylor, in a lecture at the Royal Institution, and in several passages in his other works. He maintains that all Spiritualistic and other beliefs in the supernatural, are examples of the survival of savage thought among civilised people; but he ignores the facts which compel the beliefs. The thoughts of those educated men who know, from the evidence of their own senses, that things called supernatural are true and real facts, are as totally distinct from those of savages, as are their thoughts respecting the sun, or thunder, or disease, or any other natural phenomenon. As well might he maintain that the modern belief that the sun is a fiery mass is a survival of savage thought, because some savages believe so too; or that our belief that certain diseases are contagious is a similar survival of the savage idea that a man can convey a disease to his enemy. The question is a question of facts, not of theories or thoughts, and I entirely deny the value or relevance of any general arguments, theories, or analogies, when we have to decide on matters of fact.

Thousands of intelligent men now living know from personal observation that some of the strange phenomena which have been pronounced absurd and impossible by scientific men, are nevertheless true. It is no answer to these, and no explanation of the facts, to tell them that such beliefs only occur when men are destitute of the critical spirit, and when the notion of uniform law is yet unborn; that in certain states of society illusions of this kind inevitably appear, that they are only the normal expression of certain stages of knowledge and of intellectual power, and that they clearly prove the survival of savage modes of thought in the midst of modern civilisation.

Concluding Observations. I believe that I have now shown—1. That Hume's arguments against miracles are full of unwarranted assumptions, fallacies, and contradictions; 2. That the modern argument of the telegraph-wire conveyance and drinking stone lion are positively no arguments at all, since they rest on false or assumed premises; 3. That the argument that dependence is to be placed upon men of science and upon them only, is opposed to universal experience and the whole history of science; 4. That the philosophical argument so well put by Mr. Lecky and Mr. Tylor, rests on false or unproved assumption, and is therefore valueless.

In conclusion, I must again emphatically declare that the question I have been discussing is—in no way whether miracles are true or false, or whether modern Spiritualism rests upon a basis of fact or of delusion,—but solely, whether the arguments that have hitherto been supposed conclusive against them have any weight or value. If I have shown, as I flatter myself I have done, that the arguments which have been supposed to settle the general question so completely as to render it quite unnecessary to go into particular cases, are all utterly fallacious, then I shall have cleared the ground for the production of evidence, and no honest man desirous of arriving at truth will be able to evade an enquiry into the nature and amount of that evidence, by moving the previous question—that miracles are unprovable by any amount of human testimony. It is time that the "derisive and unexamining incredulity" which has hitherto existed should give way to a less dogmatic and more philosophical spirit, or history will again have to record the melancholy spectacle of men, who should have known better, assuming to limit the discovery of new powers and agencies in the universe, and deciding, without investigation, whether other men's observations are true or false.

* * * * *

The reference near the end of the preceding analysis to Tylor's studies on animism previewed Wallace's notice of that anthropologist's two-volume work Primitive culture *in 1872 (ref 207), from which the following passages are excerpted:*

. . . More than half the work is occupied with the subject of "Animism" or the doctrine of souls. We are overwhelmed with elaborate details of the endlessly varied ideas and beliefs of men as to the soul, spirits, and gods. We are constantly told that each such belief or idea "finds its place" with the implication that it is thus sufficiently accounted for. But this capacity of being classified necessarily arises from the immense variety of such beliefs and from the fact that they are founded on natural phenomena common to all races, while the faculties by which these phenomena are interpreted are essentially the same in every case. Any great mass of facts or phenomena whatever can be classified, but the classification does not necessarily add anything to our knowledge of the causes which produced the facts or phenomena. We find at times great looseness of statement when Mr. Tylor attempts to account off-hand for superstitions. He tells us, for example, that when the devil with horns and hoofs and a tail had once become a fixed image in the popular mind, "of course men saw him in this conventional shape." Now this general statement is simply not true. In the records of witchcraft-trials it will be found that witches generally described the devil as "a man"—"a dark man"—"a black man"—"a gentleman in black clothes"—"a gentleman richly dressed," and seldom, if ever, as appearing in the full conventional form. The theory of expectant attention determining the form taken by a delusion does not cover these facts, and this is even more strikingly shown by another discordance of a similar nature. It is certainly a popular belief that the devil is hot, and that his touch burns. Yet the witches, whether in Scotland, England, or France, almost invariably describe him as cold to the touch; and this statement, so often made by persons who could have had no knowledge of what others had said, curiously agrees with the phenomenon described by modern spiritualists, of a cold wind passing over the hands during a séance. Such a correspondence of testimony in a direction exactly opposed to popular belief points to some substratum of unrecognised facts even in witchcraft, and it is not satisfactory to find the nature of this testimony misstated to make it fit in with a foregone conclusion. A recognition of the now well-established phenomena of mesmerism would have enabled Mr. Tylor to give a far more rational explanation of were-wolves and analogous beliefs than that which he offers us. Were-wolves were probably men who had exceptional power of acting upon certain sensitive individuals, and could make them, when so acted upon, believe they saw what the mesmeriser pleased; and who used this power for bad purposes. This will explain most of the alleged facts without resorting to the short and easy method of rejecting them as the results of mere morbid imagination and gross credulity. Again, we are told that "the ghost or phantasm seen by the dreamer or visionary is like a shadow, and thus the familiar term of the *shade* comes in to express the soul." But the dreamer sees what appear real substantial bodies, not shadows or images; and it is only the waking seer who, by seeing other objects through the phantasms or by testing their unsubstantiality by means of touch, can arrive at the conclusion that they are of a spiritual or shadowy nature. So, the general belief in the ghost of a man being seen in or near the house where he lived, is not at all accounted for by dreams, which are bound by no limits of locality, and generally show persons in the most incongruous places. Accounts of the other world seen in visions are said to be "just what the seer has been taught to expect"; but at p. 47, vol. ii the seer is *surprised* to find the trees, shrubs, and paths such as she had been used to on earth. It was not therefore what she had been "taught to expect"; and the remark becomes both valueless and misleading, instead of helping us to understand how such visions originate.

Although the details given on these subjects are so numerous and so heterogeneous as to be wearisome in the last degree, they are yet altogether one-sided. They have been amassed with one object and selected, no doubt unconsciously, so as to harmonize with the *à priori* convictions of the writer. All narratives tending to prove that anything which goes under the general term supernatural really exists as fact, are either entirely omitted or just mentioned in such a manner as to imply that they are necessarily impostures or delusions, and therefore unworthy of discussion. There is, however, on record a mass of facts or alleged facts ranging through every period of history down to the present day, and going to prove that the so-called supernatural is not all delusion, and that many of the beliefs of all ages classed as superstitions, have at least a substratum of reality. In the works of Dr. Kerner, Ennemoser, Gorres, and Dale Owen, and in Mr. Howitt's *History of the Supernatural*, are collections of these facts which, although by no means exhaustive, are yet far more extensive than those Mr. Tylor has cited to support his argument; while many of them are so thoroughly well established that they cannot be explained away. It is therefore at least a possible solution of the problem of animism, that the uniformity of *belief* is due in great part to the uniformity of underlying *facts*; and a work on the development of religion and mythology should fairly grapple with the question, "How much of truth is at the bottom of the so-called superstitious beliefs of mankind?" But our author avoids all such inconvenient enquiries by means of his infallible nostrum. A fact or a belief occurring once only might require explanation, but if a second or an analogous fact or belief can be found elsewhere, the whole thing becomes clear. "Second sight," for instance, occurs among savages as well as in Scotland. Nothing more is required, according to Mr. Tylor, to prove that it has no existence at all, except as a mere "belief." Those curious [spiritualistic] phenomena which have been recently investigated by Mr. [William] Crookes and other Fellows of the Royal Society, and which are declared to be realities by members of the French Institute, by American judges and senators, and by many medical and scientific men in this country, are treated in exactly the same way. Something closely related to them is recorded by classical writers, and occurs now among savage tribes. It is therefore clearly a case of "survival of old beliefs," and no further notice need be taken of it. Mr. Tylor even goes so far as to say that for his purpose it really matters little whether they are true or not. In order to arrive at true results as to the origin, nature, and development of men's beliefs, it matters not whether their foundation is fact or imagination! This belief of Mr. Tylor seems to the present writer as completely an hallucination as any to be found recorded in his volumes.

It is pertinent to recall the fact that even matters of pure science, when they have run strongly counter to popular opinion, have been treated just as Mr. Tylor treats superstitions . . . [This] should teach us, that it is unsafe to deny facts which have been vouched for by men of reputation after careful enquiry, merely because they are opposed to our prepossessions. A work like the present, one-sided though it be, furnishes much evidence to support the views of those who maintain that a considerable portion of the so-called superstitions of mankind repose upon facts; that these facts have been almost always misunderstood and misinterpreted in past ages, as they are now by the ignorant and among savages; and that, until they are recognised as possible realities, and studied with thoroughness and devotion and a complete freedom from foregone conclusions, it is hopeless to expect a sound philosophy of religion or any true insight into the mysterious depths of our spiritual nature.

* * * * *

Tylor responded to this assessment with a letter to Nature *that appeared in its 29 February 1872 number. Wallace returned the following comments, printed the next week:*

Ethnology and spiritualism (ref 208)

There is only one point in Mr. Tylor's communication . . . on which it seems desirable that I should say a few words, in order that I may not be supposed to assent to what I conceive to be a most erroneous view. Mr. Tylor suggests that the phenomena that occur in the presence of what are called mediums, are or may be of the same nature as the subjective impressions of persons under the influence of a powerful mesmeriser. Five and twenty years ago I was myself a practised mesmeriser, and was able to produce on my own patients almost the whole range of phenomena which are exhibited in public as illustrative of "mesmerism" or "electro-biology." I carried on numerous experiments in private, and paid especial attention to the conditions under which the phenomena occur. During the last seven years I have had repeated opportunities of examining the phenomena that occur in the presence of so-called "mediums," often under such favourable conditions as to render trick or imposture simply impossible. I believe, therefore, I may lay claim to some qualifications for comparing the mesmeric with the mediumistic phenomena with especial reference to Mr. Tylor's suggestion, and I find that there are two great characteristics that broadly distinguish the one from the other.

1. The mesmerised patient never has *doubts* of the reality of what he sees or hears. He is like a dreamer to whom the most incongruous circumstances suggest no idea of incongruity, and he never inquires if what he thinks he perceives harmonises with his actual surroundings. He has, moreover, lost his memory of what and where he was a few moments before, and can give no account, for instance, of how he has managed to get out of a lecture-room in London to which he came as a spectator half an hour before, on to an Atlantic steamer in a hurricane, or into the recesses of a tropical forest.

The assistants at the séances of Mr. Home or Mrs. Guppy are not in this state, as I can personally testify, and as the almost invariable *suspicion* with which the phenomena are at first regarded clearly demonstrates. They do not lose memory of the immediately preceding events; they criticise, they examine, they take notes; they suggest tests—none of which the mesmerised patient ever does.

2. The mesmeriser has the power of acting on "certain sensitive individuals" (not on "assemblies" of people, as Mr. Tylor suggests), and all experience shows that those who are thus sensitive to any one operator are but a small proportion of the population, and these almost always require previous manipulation with passive submission to the operator. The number who can be acted upon without such previous manipulation is very small, probably much less than one per cent. But there is no such limitation to the number of persons who simultaneously see the mediumistic phenomena. The visitors to Mr. Home or Mrs. Guppy all see whatever occurs of a physical nature, as the records of hundreds of sittings demonstrate.

The two classes of phenomena, therefore, differ fundamentally; and it is a most convincing proof of Mr. Tylor's very slender acquaintance with either of them, that he should even suggest their identity. The real connection between them is quite in an opposite direction. It is the mediums, not the assistants, who are "sensitives." They are almost always subject to the mesmeric influence, and they often exhibit all the characteristic phenomena of coma, trance, rigidity, and abnormal sense-power. Conversely, the most sensitive mesmeric patients are almost invariably mediums. The idea that it is necessary for me to inform "spiritualists" that I believe in the power of mesmerisers to make their patient believe what they please, and that this "information" might "bring about investigations leading to valuable results," is really amusing, considering that such investigations took place twenty years ago, and led to this important result—that almost all the most experienced mesmerists (Prof. Gregory, Dr. Elliotson, Dr. Reichenbach, and many others) became spiritualists! If Mr. Tylor's suggestion had any value, these are the very men who ought to have demonstrated the subjective nature of mediumistic phenomena; but, on the contrary, as soon as they had the opportunity of personally investigating them, they all of them saw and admitted their objective reality.

* * * * *

Wallace's most influential essay on spiritualism was undoubtedly his 'A defence of modern spiritualism' (ref 243), which appeared in two parts in Fortnightly Review *in 1874. This long work was widely reprinted; notably, as a monographic pamphlet and as part of the collection* On miracles and modern spiritualism *in 1875 (ref 717). Reprinted here are its last two sections, which describe a number of themes Wallace viewed as the 'applications' of spiritualism:*

Historical teachings of spiritualism

The lessons which modern Spiritualism teaches may be classed under two heads. In the first place, we find that it gives a rational account of various phenomena in human history which physical science has been unable to explain, and has therefore rejected or ignored; and, in the second, we derive from it some definite information as to man's nature and destiny, and, founded on this, an ethical system of great practical efficacy. The following are some of the more important phenomena of history and of human nature which science cannot deal with, but which Spiritualism explains:

1. It is no small thing that the spiritualist finds himself able to rehabilitate Socrates as a sane man, and his "demon" as an intelligent spiritual being who accompanied him through life,—in other words, a guardian spirit. The non-spiritualist is obliged to look upon one of the greatest men in human history, not only as subject all his life to a mental illusion, but as being so weak, foolish, or superstitious as never to discover that it was an illusion. He is obliged to disbelieve the fact asserted by contemporaries and by Socrates himself, that it forewarned him truly of dangers; and to hold that this noble man, this subtle reasoner, this religious sceptic, who was looked up to with veneration and love by the great men who were his pupils, was imposed upon by his own fancies, and never during a long life found out that they were fancies, and that their supposed monitions were as often wrong as right. It is a positive mental relief not to have to think thus of Socrates.

2. Spiritualism allows us to believe that the oracles of antiquity were not all impostors; that a whole people, perhaps the most intellectually acute who ever existed, were not all dupes. In discussing the question, "Why the Prophetess Pythia giveth no Answers now from the Oracle in Verse," Plutarch tells us that when kings and states consulted the oracle on weighty matters that might do harm if made public, the replies were couched in enigmatical language; but when private persons are asked about their own affairs they get direct answers in the plainest terms, so that some people even complained of their simplicity and directness, as being unworthy of a divine origin. And he adds this positive testimony: "Her answers, though submitted to the severest scrutiny, have never proved false or incorrect. On the contrary, the verification of them has filled the temple with gifts from all parts of Greece and foreign countries." And again, "The answer of Pythoness proceeds to the very truth, without any diversion, circuit, fraud, or ambiguity. It has never yet, in a single instance, been convicted of falsehood." Would such statements be made by such a writer, if these oracles were all the mere guesses of impostors? The fact that they declined and ultimately failed, is wholly in their favour; for why should imposture cease as the world became less enlightened and more superstitious? Neither does the fact that the priests could sometimes be bribed to give out false oracles prove anything, against such statements as that of Plutarch and the belief during many generations, supported by ever-recurring experiences, of the greatest men of antiquity. That belief could only have been formed by demonstrative facts; and modern Spiritualism enables us to understand the nature of those facts.

3. Both the Old and New Testaments are full of Spiritualism, and spiritualists alone can read the record with an enlightened belief. The hand that wrote upon the wall at Belshassar's feast, and the three men unhurt in

Nebuchadnezzar's fiery furnace, are for them actual facts which they need not explain away. St. Paul's language about "spiritual gifts," and "trying the spirits," is to them intelligible language, and the "gift of tongues" a simple fact. When Christ cast out "devils" or evil spirits, he really did so—not merely startle a madman into momentary quiescence; and the water changed into wine, as well as the bread and fishes continually renewed till five thousand men were fed, are credible as extreme manifestations of a power which is still daily at work among us.

4. The miracles of the saints, when well attested, come into the same category. Those of St. Bernard, for instance, were often performed in broad day before thousands of spectators, and were recorded by eye-witnesses. He was himself greatly troubled by them, wondering why this power was bestowed upon him, and fearing lest it should make him less humble. This was not the frame of mind, nor was St. Bernard's the character, of a deluded enthusiast. The spiritualist need not believe that all this never happened; or that St. Francis d'Assisi and St. Theresa were not raised into the air, as eye-witnesses declared they were.

5. Witchcraft and witchcraft trials have a new interest for the spiritualist. He is able to detect hundreds of curious and minute coincidences with phenomena he has himself witnessed; he is able to separate the *facts* from the absurd *inferences*, which people imbued with the frightful superstition of diabolism drew from them, and from which false inferences all the horrors of the witchcraft mania arose. Spiritualism, and Spiritualism alone, gives a rational explanation of witchcraft, and determines how much of it was objective fact, how much subjective illusion.

6. Modern Roman Catholic miracles become intelligible facts. Spirits whose affections and passions are strongly excited in favour of Catholicism, produce those appearances of the Virgin and of saints which they know will tend to increase religious fervour. The appearance itself may be an objective reality; while it is only an inference that it is the Virgin Mary,—an inference which every intelligent spiritualist would repudiate as in the highest degree improbable.

7. Second-sight, and many of the so-called superstitions of savages may be realities. It is well known that mediumistic power is more frequent and more energetic in mountainous countries; and as these are generally inhabited by the less civilised races, the beliefs that are more prevalent there may be due to the facts which are more prevalent, and be wrongly imputed to the coincident ignorance. It is known to spiritualists that the pure dry air of California led to more powerful and more startling manifestations than in any other part of the United States.

8. The recently discussed question of the efficacy of prayer receives a perfect solution by Spiritualism. Prayer may be often answered, though not directly by the Deity. Nor does the answer depend wholly on the morality or the religion of the petitioner; but as men who are both moral and religious, and are firm believers in a divine response to prayer, will pray more frequently, more earnestly, and more disinterestedly, they will attract towards them a

number of spiritual beings who sympathise with them, and who, when the necessary mediumistic power is present, will be able, as they are often willing, to answer the prayer. A striking case is that of George Müller, of Bristol, who has now for forty-four years depended wholly for his own support, and that of his wonderful charities, on answer to prayer. His *Narrative of Some of the Lord's Dealings with George Müller* (6th Ed., 1860), should have been referred to in the late discussion, since it furnishes a better demonstration that prayer is sometimes really answered than the hospital experiment proposed by Sir Henry Thomson could possibly have done. In this work we have a precise yearly statement of his receipts and expenditure for many years. He never asked any one or allowed any one to be asked, directly or indirectly, for a penny. No subscriptions or collections were ever made; yet from 1830 (when he married without any income whatever) he has lived, brought up a family, and established institutions which have steadily increased, till now four thousand orphan children are educated and in part supported. It has happened hundreds of times, that there has been no food in his house and no money to buy any, or no food or milk or sugar for the children. Yet he never took a loaf or any other article on credit even for a day; and during the thirty years over which his narrative extends, neither he nor the hundreds of children dependent upon him for their daily food have ever been without a regular meal! They have lived, literally, from hand to mouth; and his one and only resource has been secret prayer. Here is a case which has been going on in the midst of us for forty years, and is still going on; it has been published to the world for many years, yet a warm discussion is being carried on by eminent men as to the fact whether prayer is or is not answered, and not one of them exhibits the least knowledge of this most pertinent and illustrative phenomenon! The spiritualist explains all this as a personal influence. The perfect simplicity, faith, boundless charity, and goodness of George Müller, have enlisted in his cause beings of a like nature; and his mediumistic powers have enabled them to work for him by influencing others to send him money, food, clothes, etc., all arriving, as we should say, just in the nick of time. The numerous letters, he received with these gifts, describing the sudden and uncontrollable impulse the donors felt to send him a certain definite sum at a certain fixed time, such being the exact sum he was in want of, and had prayed for, strikingly illustrates the nature of the power at work. All this might be explained away, if it were partial and discontinuous; but when it continued to supply the daily wants of a life of unexampled charity, *for which no provision in advance was ever made* (for that Müller considered would show want of trust in God), no such explanation can cover the facts.

9. Spiritualism enables us to comprehend and find a place for, that long series of disturbances and occult phenomena of various kinds, which occurred previous to what are termed the modern Spiritual manifestations. Robert Dale Owen's works give a rather full account of this class of phenomena, which are most accurately recorded and philosophically treated by him. This is not the place to refer to them in detail; but one of them may be mentioned as showing how large an amount of unexplained mystery there was, even in our own

country, before the world heard anything of modern Spiritualism. In 1841, Major Edward Moor, F.R.S., published a little book called *Bealings Bells*, giving an account of mysterious bell-ringing in his house at Great Bealings, Suffolk, and which continued for fifty-three days. Every attempt to discover the cause, by himself, friends, and bell-hangers, were fruitless; and by no efforts, however violent, could the same clamorous and rapid ringing be produced. He wrote an account to the newspapers, requesting information bearing on the subject, when, in addition to certain wise suggestions—of rats or a monkey as efficient causes—he received fourteen communications, all relating cases of mysterious bell-ringing in different parts of England, many of them lasting much longer than Major Moor's, and all remaining quite unexplained. One lasted eighteen months; another was in Greenwich Hospital, where neither clerk-of-the-works, bellhanger, nor men of science could discover the cause. One clergyman wrote of disturbances of a most serious kind continued in his parsonage for *nine years*, and he was able to trace back their existence in the same house for *sixty years*. Another case had lasted *twenty years*, and could be traced back for a *century*. Some of the details of these cases are most instructive. Trick is absolutely the most incredible of all explanations. Spiritualism furnishes the explanation by means of analogous facts occurring every day, and forming part of the great system of phenomena which demonstrates the spiritual theory. Major Moor's book is very rare; but a good abstract of it is given in Owen's *Debatable Land*, pp. 239–258.

Moral teachings of spiritualism

We have now to explain the Theory of Human Nature, which is the outcome of the phenomena taken in their entirety, and is also more or less explicitly taught by the communications which purport to come from spirits. It may be briefly outlined as follows:

1. Man is a duality, consisting of an organised spiritual form, evolved coincidently and permeating the physical body, and having corresponding organs and development.
2. Death is the separation of this duality, and effects no change in the spirit, morally or intellectually.
3. Progressive evolution of the intellectual and moral nature is the destiny of individuals; the knowledge, attainments, and experience of earth-life forming the basis of spirit-life.
4. Spirits can communicate through properly-endowed mediums. They are attracted to those they love or sympathise with, and strive to warn, protect, and influence them for good, by mental impression when they cannot effect any more direct communication; but, as follows from clause (2), their communications will be fallible, and must be judged and tested just as we do those of our fellow-men.

The foregoing outline of propositions will suggest a number of questions and difficulties, for the answers to which readers are referred to the works of R. D. Owen, Hudson Tuttle, Professor Hare, and the records of Spiritualism *passim*. Here I must pass on to explain, with some amount of detail, how the theory leads to a pure system of morality with sanctions far more powerful and effective than any which either religious systems or philosophy have put forth.

This part of the subject cannot, perhaps, be better introduced than by referring to some remarks by Professor [Thomas] Huxley in a letter to the Committee of the Dialectical Society. He says:

> But supposing the phenomena to be genuine—they do not interest me. If anybody would endow me with the faculty of listening to the chatter of old women and curates at the nearest cathedral town, I should decline the privilege, having better things to do. And if the folk in the spiritual world do not talk more wisely and sensibly than their friends report them to do, I put them in the same category.

This passage, written with the caustic satire in which the kind-hearted Professor occasionally indulges, can hardly mean, that if it were proved that men really continued to live after the death of the body, that fact would not interest him, merely because some of them talked twaddle? Many scientific men deny the spiritual source of the manifestations, on the ground that real, genuine spirits might reasonably be expected not to indulge in the commonplace trivialities which do undoubtedly form the staple of ordinary spiritual communications. But surely Professor Huxley, as a naturalist and philosopher, would not admit this to be a reasonable expectation. Does he not hold the doctrine that there can be no effect, mental or physical, without an adequate cause; and that mental states, faculties, and idiosyncracies, that are the result of gradual development and life-long—and even ancestral—habit, cannot be suddenly changed by any known or imaginable cause? And if (as the Professor would probably admit) a very large majority of those who daily depart this life are persons addicted to twaddle, persons who spend much of their time in low or trivial pursuits, persons whose pleasures are sensual rather than intellectual—whence is to come the transforming power which is suddenly, at the mere throwing of the physical body, to change these into beings able to appreciate and delight in high and intellectual pursuits? The thing would be a miracle, the greatest of miracles, and surely Professor Huxley is the last man to contemplate innumerable miracles as part of the order of nature; and all for what? Merely *to save these people from the necessary consequences of their misspent lives*. For the essential teaching of Spiritualism is, that we are, all of us, in every act and thought, helping to build up a "mental fabric" which will be and constitute ourselves more completely after the death of the body than it does now. Just as this fabric is well or ill built, so will our progress and happiness be aided or retarded. Just in proportion as we have developed our higher intellectual and moral nature, or starved it by disuse and by giving undue prominence to those faculties which secure us mere physical or selfish enjoyment, shall we be well or ill fitted for the new life we enter on. The noble teaching of Herbert Spencer, that men are best educated by being left to suffer the natural consequences of their actions, is the teaching of Spiritualism as regards the transition to another phase of life. There will be no imposed rewards or punishments; but every one will suffer the natural and inevitable consequences of a well or ill-spent life. The well-spent life is that in which those faculties which regard our personal physical well-being, are subordinated to those which regard our social and intellectual well-being, and the well-being of others; and that inherent feeling—which is so universal and difficult to account for—that these latter constitute our higher nature, seems also to point to the conclusion that we are intended for a condition in which the former will be almost wholly unnecessary, and will gradually become rudimentary through disuse, while the latter will receive a corresponding development.

Although, therefore, the twaddle and triviality of so many of the communications is not one whit more interesting to sensible spiritualists than it is to Professor Huxley, and is never voluntarily listened to, yet the fact that such poor stuff is talked (supposing it to come from spirits) is both a fact that might have been anticipated and a lesson of deep import. We must remember, too, the character of the séances at which these commonplace communications are received. A miscellaneous assemblance of believers of various grades and tastes, but mostly in search of an evening's amusement, and of sceptics who look upon all the others as either fools or knaves, is not likely to attract to itself the more elevated and refined denizens of the higher spheres, who may well be supposed to feel too much interest in their own new and grand intellectual existence to waste their energies on either class. If the fact is proved, that people continue to talk after they are dead with just as little sense as when alive, but that, being in a state in which sense, both common and uncommon, is of far greater importance to happiness than it is here (where fools pass very comfortable lives), they suffer the penalty of having neglected to cultivate their minds; and being so much out of their element in a world where all pleasures are mental, they endeavour to recall old times by gossiping with their former associates whenever they can find the means—Professor Huxley will not fail to see its vast importance as an incentive to that higher education which he is never weary of advocating. He would assuredly be interested in anything having a really practical bearing on the present and on the future condition of men; and it is evident that even these low and despised phenomena of Spiritualism, "if true," having this bearing, and combined with its higher teachings, constitute a great moral agency which may yet regenerate the world.

. . . the spiritualist . . . is impelled towards a pure, a sympathetic, and an intellectual life by motives far stronger than any which either religion or philosophy can supply. He dreads to give way to passion or to falsehood, to selfishness or to a life of luxurious physical enjoyment, because he knows that the natural and inevitable consequences of such habits are future misery, necessitating a long and arduous struggle in order to develop anew the faculties, whose exercise long disuse has rendered painful to him. He will be deterred from crime by the knowledge that its unforeseen consequences may cause him ages of remorse; while the bad passions which it encourages will be a perpetual torment to himself in a state of being in which mental emotions cannot be laid aside or forgotten amid the fierce struggles and sensual pleasures of a physical existence. It must be remembered that these beliefs (unlike those of theology) will have a living efficacy, because they depend on *facts* occurring again and again in the family circle, constantly reiterating the same truths as the result of personal knowledge, and thus bringing home to the mind of the most obtuse, the absolute reality of that future existence in which our degree of happiness or misery will be directly dependent on the "mental fabric" we construct by our daily thoughts, and words, and actions here.

Contrast this system of natural and inevitable reward and retribution, dependent wholly on the proportionate development of our higher mental and moral nature, with the arbitrary system of rewards and punishments dependent on stated facts and beliefs only, as set forth by all dogmatic religions; and who can fail to see that the former is in harmony with the whole order of nature—the latter opposed to it. Yet it is actually said that Spiritualism is altogether either imposure or delusion, and all its teachings but the product of "expectant attention" and "unconscious cerebration!" If none of the long series of demonstrative facts which [earlier were] . . . sketched out, existed, and its only product were this theory of a future state, that alone would negative such a supposition. And when it is considered that mediums of all grades, whether intelligent or ignorant, and having communications given through them in various direct and indirect ways, are absolutely in accord as to the main features of this theory, what becomes of the gross misstatement that nothing is given through mediums but what they know and believe themselves? The mediums have, almost all, been brought up in some of the usual orthodox beliefs. How is it, then, that the usual orthodox notions of heaven are *never* confirmed through them? In the scores of volumes and pamphlets of spiritual literature I have read, I have found no statement of a spirit describing "winged angels," or "golden harps," or the "throne of God"—to which the humblest orthodox Christian thinks he will be introduced if he goes to heaven at all. There is no more startling and radical opposition to be found between the most diverse religious creeds, than that between the beliefs in which the majority of mediums have been brought up and the doctrines as to a future life that are delivered through them; there is nothing more marvellous in the history of the human mind than the fact that, whether in the back-woods of America or in country towns in England, ignorant men and women having almost all been brought up in the usual sectarian notions of heaven and hell, should, the moment they become seized by the strange power of mediumship, give forth teachings on this subject which are philosophical rather than religious, and which differ wholly from what had been so deeply ingrained into their minds. And this statement is not affected by the fact that communications purport to come from Catholic or Protestant, Mahomedan or Hindoo spirits. Because, while such communications maintain special *dogmas* and *doctrines*, yet they confirm the *very facts* which really constitute the spiritual theory, and which in themselves contradict the theory of the sectarian spirits. The Roman Catholic spirit, for instance, does not describe himself as being in either the orthodox purgatory, heaven, or hell; the Evangelical Dissenter who dies in the firm conviction that he should certainly "go to Jesus," never describes himself as being with Christ, or as ever having seen Him, and so on throughout. Nothing is more common than for religious people at séances to ask questions about God and Christ. In reply they never get more than opinions, or more frequently the statement that they, the spirits, have no more actual knowledge of those subjects than they had while on earth. So that the facts are all harmonious; and the very circumstance of there being sectarian spirits bears witness in two ways to the truth of the spiritual theory—it shows that the mind, with its ingrained beliefs, is not suddenly changed at death; and it shows that the communications are not the reflection of the mind of the medium, who is often of the same religion as the communicating spirit, and, because he does not get his own ideas confirmed, is obliged to call in the aid of "Satanic influence" to account for the anomaly.

The doctrine of a future state and of the proper preparation for it as here developed, is to be found in the works of all spiritualists, in the utterances of all trance-speakers, in the communications through all mediums; and this could be proved, did space permit, by copious quotations. But it varies in form and detail in each; and just as the historian arrives at the opinions or beliefs of any age or nation, by collating the individual opinions of its best and most popular writers, so do spiritualists collate the various statements on this subject. They know well that absolute dependence is to be placed on no individual communications. They know that these are received by a complex physical and mental process, both communicator and recipient influencing the result; and they accept the teachings as to the future state of man only so far as they are repeatedly confirmed in substance (though they may differ in detail) by communications obtained under the most varied circumstances, through mediums of the most different characters and acquirements, at different times, and in different places. Fresh converts are apt to think, that, once satisfied the communications come from their deceased friends, they may implicitly trust to them, and apply them universally; as if the vast spiritual world was all moulded to one pattern, instead of being, as it almost certainly is, a thousand times more varied than human society on the earth is, or ever has been. The fact that the communications do not agree as to the condition, occupations, pleasures, and capacities of individual spirits, so far from being a difficulty, as has been absurdly supposed, is what ought to have been expected; while the agreement on the essential features of what we have stated to be the spiritual theory of a future state of existence, is all the more striking, and tends to establish that theory as a fundamental truth.

The assertion, so often made, that Spiritualism is the survival or revival or old superstitions, is so utterly unfounded as to be hardly worth notice. A science of human nature which is founded on observed facts; which appeals only to facts and experiment; which takes no beliefs on trust; which inculcates investigation and self-reliance as the first duties of intelligent beings; which teaches that happiness in a future life can be secured by cultivating and developing to the utmost the higher faculties of our intellectual and moral nature, *and by no other method*,—is and must be the natural enemy of all superstition. Spiritualism is an experimental science, and affords the only sure foundation for a true philosophy and pure religion. It abolishes the terms "supernatural" and "miracle" by an extension of the sphere of law and the realm of nature; and in doing so it takes up and explains whatever is true in the superstitions and so-called miracles of all ages. It, and it alone, is able to harmonise conflicting creeds; and it must ultimately lead to concord among mankind in the matter of religion, which has for so many ages been the source of inceasing discord and incalculable evil;—and it will be able to do this because it appeals to evidence instead of faith, and substitutes facts for opinions; and is thus able to demonstrate the source of much of the teaching that men have so often held to be divine.

It will thus be seen, that those who can form no higher conception of the uses of Spiritualism, "even if true," than to detect crime or to name in advance the winner of the Derby, not only prove their own ignorance of the whole subject, but exhibit in a marked degree that partial mental paralysis, the result of a century of materialistic thought, which renders so many unable seriously to conceive the possibility of a natural continuation of human life after the death of the body. It will be seen also that Spiritualism is no mere "physiological" curiosity, no mere indication of some hitherto unknown "law of nature;" but that it is a science of vast extent, having the widest, the most important, and the most practical issues, and as such should enlist the sympathies alike of moralists, philosophers, and politicians, and of all who have at heart the improvement of society and the permanent elevation of human nature.

In concluding this necessarily imperfect though somewhat lengthy account of a subject about which so little is probably known to most of the readers of the *Fortnightly Review*, I would earnestly beg them not to satisfy themselves with a minute criticism of single facts, the evidence for which, in my brief survey, may be imperfect; but to weigh carefully the mass of evidence I have adduced, considering its wide range and various bearings. I would ask them to look rather at the results produced by the evidence than at the evidence itself as imperfectly stated by me; to consider the long roll of men of ability who, commencing the inquiry as sceptics left it as believers, and to give these men credit for not having overlooked, during years of patient inquiry, difficulties which at once occur to themselves. I would ask them to ponder well on the fact, that no earnest inquirer has ever come to a conclusion adverse to the reality of the phenomena; and that no spiritualist has ever given them up as false. I would ask them, finally, to dwell upon the long series of facts in human history that Spiritualism explains, and on the noble and satisfying theory of a future life that it unfolds. If they will do this, I feel confident that the result I have alone aimed at will be attained; which is, to remove the prejudices and misconceptions with which the whole subject has been surrounded and to incite to unbiassed and persevering examination of the facts. For the cardinal maxim of Spiritualism is, that every one must find out the truth for himself. It makes no claim to be received on hearsay evidence; but on the other hand, it demands that it be not rejected without patient, honest, and fearless inquiry.

* * * * *

Possible 'applications' of spiritualism notwithstanding, the phenomena alleged to prove its legitimacy were of such extraordinary nature that it appeared nothing short of a strictly 'scientific' approach to their study could ever convince a sceptical majority they were genuine. In consequence, a number of the interested members of the scientific community—including some very prominent figures—set out to prove, so far as was possible, that the effects were not merely the products of deception by unscrupulous hucksters. Wallace himself took part in a considerable number of 'test' seances. Fully controlled experiments were difficult to arrange, however, and for some persons no amount of examination of what Wallace often referred to as 'the facts' of spiritualism could be expected to lead to the conclusion that deception or delusion was not at the root of the matter. The result was a good deal of harsh criticism, descending at times to shallow ridicule. Wallace devoted considerable effort to answering what he felt were irrational criticisms of spiritualism research. The following letter, printed in Athenaeum *in early 1878, represents one of his many responses to one of the movement's most persistent and vocal adversaries, Dr William B. Carpenter:*

The curiosities of credulity (ref 285)

Owing to absence from home I have only just seen Dr. Carpenter's letter in the *Athenaeum* of December 22nd, to which I now beg leave very briefly to reply.

I must first remark on the extreme inconvenience of Dr. Carpenter's erratic mode of carrying on a discussion. As soon as his lectures on *Mesmerism, Spiritualism, etc.*, were published, I wrote a review of them in the *Quarterly Journal of Science* of July last. To this Dr. Carpenter replied in *Fraser's Magazine* of November, promising a fuller reply to certain points in the new edition of his *Lectures*, then in the press. As the article in *Fraser* was of a very personal character, I issued a rejoinder in the same periodical the following month. A discussion has also been carried on in *Nature*, and the scene of the contest is now removed to *Athenaeum*, many of whose readers are probably ignorant of its previous phases.

Dr. Carpenter comes before a fresh audience in order to reply to a specific charge of mis-statement which I made against him in the *Quarterly Journal of Science* (July, 1877, p. 398), which charge, as I will proceed to show, he endeavours to evade by a wordy defence, which really amounts to an admission of it. In his *Lectures* (p. 71) is the following passage:

> It was in France that the pretensions of mesmeric clairvoyance were first advanced; and it was by the French Academy of Medicine, in which the mesmeric trance state had been previously discussed with reference to the performance of surgical operations, that this new and more extraordinary claim was first carefully sifted, in consequence of an offer made in 1837 by M. Burdin (himself a member of that Academy) of a prize of 3,000 francs to any one who should be found capable of reading through opaque substances. The money was deposited in the hands of a notary for a period of two years, afterwards extended to three; the announcement was extensively published; numerous cases were

> offered for examination; every imaginable concession was made to the competitors that was compatible with a thorough testing of the asserted power; and not one was found to stand the trial.

My readers will observe that this is deliberately stated to be the *first* time that clairvoyance was carefully sifted in France; yet it now appears that Dr. Carpenter perfectly well knew of the Commission of the same Academy about ten years earlier, which, after five years of most careful and elaborate experiments, gave a unanimous Report positively in favour of the reality of clairvoyance.

But Dr. Carpenter would have us believe that he studiously avoided all mention of this Report because it had been proved to be wholly founded on imposture or error; and he endeavours to establish this by giving a single hearsay case of a confession of imposture *on another person not even a member of the Commission*! I feel sure that the impression conveyed to the readers of Dr. Carpenter's letters would be that the case of alleged imposture by one of the mesmeric patients of MM. Georget and Rostan occurred to members of the Commission, and that the case had been examined by them and reported on as genuine. But this impression would be entirely erroneous. The members of the Commission, whose names are appended to the Report, are as follows: 1, Bourdois de la Motte (President); 2, Fouquier; 3, Gueneau de Mussy; 4, Guersent; 5, Itard; 6, Leroux; 7, Marc; 8, Thillaye; 9, Husson (Reporter). Against the voluminous and interesting details of this Report, its carefully repeated experiments, its cautious deductions, its amazing facts, not one particle of rebutting evidence is adduced. Yet Dr. Carpenter thought himself justified not only in ignoring its existence, but in giving his readers to understand, by an express form of words, that no such inquiry was ever made! This was the accusation I made against him, and the readers of *Athenaeum* can now judge as to the candour and sufficiency of the reply.

I must add a few words on the way in which Dr. Carpenter treats M. Rostan, "one of the ablest medical psychologists of his day." Dr. Carpenter states, as a fact, that, "when a second edition of the *Dictionnaire de Médecine* came out in 1838, he (M. Rostan) withdrew the article he had contributed to the first"; and then, further on, it is stated that "M. Rostan, by his own confession," had been led away by cunning cheats in the matter of clairvoyance. Now I have always understood that M. Rostan was much annoyed at his article being superseded in the second edition of the Dictionnaire; and, as this is *à priori* probable, I require some direct evidence of Dr. Carpenter's assertion that he voluntarily withdrew it. This is the more necessary because the still more important and damaging statement—that M. Rostan made a "confession" that he had been led away by cunning cheats—is also given as a hearsay report without any reference or authority; and it looks very much as if Dr. Carpenter's logic had deduced the "confession" as an inference from the "withdrawal," no evidence whatever being offered for either of them. If this should really be the case, then the severest things I have said as to Dr. Carpenter's mode of carrying on this discussion will be more than justified.

Throughout my discussion of this subject with Dr. Carpenter I have strictly confined myself to questions of *fact* and *evidence*, and have maintained that these are of more value than *opinions*, however numerous or weighty. My criticisms have, for the most part, been directed to misrepresentations of facts and suppressions of evidence on the part of my opponent. The readers of the *Athenaeum* will now be able to judge, as regards one case, whether that criticism is sound; and for numerous other cases I refer them to my articles in the *Quarterly Journal of Science* and in *Fraser's Magazine*. If they read these, they will, I think, agree with me that the cause of truth will not be advanced by the further continuance of a discussion in which one of the parties perpetually evades or obscures the most important points at issue, and at every step introduces fresh mis-statements to be corrected and fresh insinuations to be rebutted, as I have shown that Dr. Carpenter has done in his numerous writings on this subject.

* * * * *

Several years later, in 1885, Wallace composed a short essay which reaffirmed his position that spiritualist teachings did not conflict with science (ref 379). In the following note, published in Light, *a spiritualist weekly, he responded to a criticism of that work:*

The *Journal of Science* on spiritualism (ref 382)

My article on the 'Harmony of Spiritualism and Science,' written for an American newspaper, and republished in *Light* of May 30th, has been honoured by a notice in the *Journal of Science*, and I have been requested to make a few remarks in reply to the same. I cannot say that I myself think the criticism worth answering, because it is founded on assumptions which will, I am sure, not be granted by men of science in general; still, as they may present difficulties to some readers, it is perhaps as well to show their weakness.

The writer's main and fundamental objection is stated as follows:—

> Science is based upon what we, for want of a better name, term *law*. Spiritualism rests upon *will*. Science, and not merely our present science, but any possible science, so far as I can conceive it takes its stand upon the causal nexus, upon the regular sequence of cause and effect. Iron always sinks in mercury, and always dissolves in hydrochloric acid, etc., etc.

In this passage and in what follows, the term "science" is completely misused. It is taken as synonymous with a limited branch of science, namely—physics. There are, however, whole regions of science in which there is no such regular sequence of cause and effect and no power of prediction. Even within the domain of physics we have the science of meteorology in which there is no precise sequence of effects; and when we come to the more complex phenomena of life we can rarely predict results and are continually face to face with insoluble problems; yet no one maintains that meteorology and biology are not sciences—still less that they are out of harmony with or opposed to science. The absence of uniformity, and the impossibility of predicting what will happen under all circumstances are not, therefore, confined to Spiritualistic phenomena alone. Assuming that they are so, however, the writer thus continues:—

> With the advent of Spiritualism all this beautiful simplicity has been swept away. If Spiritualists are not mistaken there are around us numbers of finite invisible beings, of unknown powers, and of unknown intentions capable of interfering with the order of nature. They can raise bodies in the air against the force of gravitation. They can kindle fires at pleasure, or deprive fire of the power of destroying organised beings or of occasioning pain . . . To me it seems that, if these contentions are true, if there exist beings around us capable of exerting such powers, there are introduced, so to speak, into every equation a number of unknown quantities, rendering it for ever insoluble. We can only say "such results follow under such conditions, if no spirits think proper to interfere." It seems to me that before any harmony can be shown between Spiritualism and science it must be ascertained what are the limits of the powers of these "spirits" and under what conditions can they be exerted. In that manner only can a basis for science be saved.

In these passages there are both misstatements of fact and illogical conclusions. There is little or no proof that the "spirits" around us can of *themselves* do any of the things alleged. They require in almost every case, perhaps in every case, the assistance of human beings, and not only so, but of particular human beings with special organisations—those we term mediums. Here at once is a limitation to their power, and so great a limitation that the cases in which they can interfere with the ordinary effects of natural law are but very rare exceptions. Unless specially sought after, not one person in a thousand ever comes in contact with these phenomena, and even when sought for the general complaint is that they are exceedingly hard to find. To maintain that all science is impossible because once or twice in the lives of one person in a thousand some interference with the ordinary course of nature may occur, is about as sensible as to maintain that agriculture is impossible because phenomenal hailstorms may destroy, or exceptional whirlwinds may carry away, crops, or to give up all quantitative astronomical observation because earthquakes or terrestrial tremors, which cannot be predicted, may alter the level or the orientation of the instruments. And when we come to vital, and mental, and moral phenomena, we are still more subject to "unknown quantities in our equations." The apparently healthy man dies suddenly, while one who has always been weak and ailing lives to a good old age. The sober, moral, and religious citizen suddenly commits a horrible crime. The man of commanding genius becomes hopelessly insane. Yet these terribly real "unknown quantities" do not render either vital, or mental, or moral science impossible, still less do they place these studies altogether outside of science and in antagonism to it.

Again, as regards the impossibility of any science, as the critic alleges, where *will* intervenes, we have the human will as a constant factor in sociology, in anthropology, in ethical science, in history, in psychology, yet no one maintains that all these studies are opposed to science even if they have, as yet, no claim to rank among established or exact sciences.

Now, so far as we know, the *will* of spirits is no more erratic in its manifestations than the will of living men. It appears to be equally subject to general laws and influences, and, on the average, no more affects the orderly sequence of Spiritualistic phenomena than do the individual wills of human beings affect the orderly sequence of mental, social, or moral phenomena. It is a great mistake to impute all the uncertainty of phenomena with mediums to the erratic *will* of the spirits concerned. Very little is probably due to this cause, while the greater part is certainly owing to what may strictly be termed terrestrial conditions. We know something of these conditions already, and when we know more we have every reason to believe that much of the uncertainty will cease. Not less unsatisfactory is the remark with which our critic concludes this part of the subject:—

> To harmonise science with Spiritualism it will then be, in the first place, necessary to discover the limits of power of spirits, under what conditions it is exerted, and how it may be combated when and where it is desirable.

But in all these respects Spiritualism is fully as advanced as is science itself. We know, practically, the limits of the power of spirits on this earth at the present day, and under ordinary conditions, quite as well as we know the limits of the power of earthquakes and volcanoes, of disease, of insanity, and of the human intellect, and we know how to combat their evil effects quite as well in our domain of observation as do men of science in theirs.

Then we have the bugbear of the "creation or destruction of energy" in Spiritualistic phenomena brought forward, and we are told that scientific men will seek for "precise answers" to the question where the power comes from "before they can accept the Spiritualist theories." But nobody asks them to accept Spiritualist theories before they have investigated the Spiritualist facts.

It has usually been the boast of science that it accepts, and co-ordinates, and studies *all* the facts of nature in order to explain them; but with respect to our facts it applies a different rule and asks for a complete theory—a "precise explanation," before it will even begin to study them. We are informed that, in order—"To establish a harmony between Spiritualism and science it will be necessary, I submit, to show the origin of the energy which is at the disposal of spirits." But science itself does not yet know the "origin of the energy" of gravitation, yet the theory of gravitation is its proudest boast. Science only guesses at the "origin of the energy" of the magnet; and in tracing all terrestrial energy to the sun it only removes the difficulty one step, and cannot do more than make more or less probable guesses as to where the energy of the sun comes from. It is surely not scientific to demand of a new and very difficult science the complete solution of its most fundamental problems as a preliminary to recognising its existence, yet this is how the writer in the *Journal of Science* proposes to treat the students of Spiritualism.

The last passage I shall refer to is that in which the critic considers that Swedenborg was the victim "of delusion or imposture," because, while describing Jupiter and Saturn he said nothing about Uranus or Neptune. The assumption underlying this argument is, that if spirits exist and communicate with men they *must necessarily* know more of the material universe than men do, and *must* communicate their superior knowledge to us. This extraordinary misconception well illustrates the tone of mind of the writer, who has evidently given very little attention to the theories and conclusions of the more advanced of modern Spiritualists. He has yet to learn that the facts of Spiritualism are one thing, the value of the information obtained from Spiritualistic sources quite another thing. It is marvellous that so many people who deny that we have any evidence whatever of the existence of spirits, yet claim to know *à priori* exactly what spirits ought to know and ought to tell us, if they do exist!

* * * * *

Light *itself carried a criticism of the original essay the next week. Wallace's answer to this second analysis appeared in the 25 July 1885 number of that journal:*

Harmony of spiritualism and science (ref 383)

I find some difficulty in comprehending the exact position of Mr. Frederick F. Cook in his elaborate 'Rejoinder' to my article, but with your permission I will briefly notice his direct criticisms of my views, because they have a certain amount of plausibility owing to the extremely condensed form in which I was compelled to express myself in the space that was allowed me.

Mr. Cook first objects to my proposition that—"man consists essentially of a spiritual nature or mind intimately associated with a spiritual body or soul, both of which are developed in, and by means of, a material organism." This, he says, is a case of spiritual suicide, and is directly opposed to my previous statement that—"mind is the cause of organism and perhaps even of matter itself." But surely, it is clear that in the last quoted passage I am speaking of mind in the abstract or as a fundamental principle, while in the former I am dealing with mind as individualised in the human form. There is, I conceive, no contradiction in believing that mind is at once the cause of matter and of the development of individualised human minds through the agency of matter. And when, further on, he asks, "Does mortality give consciousness to spirit, or does spirit give consciousness for a limited period to mortality?" I would reply, "Neither the one nor the other; but, mortality is the means by which a permanent individuality is given to spirit."

His next serious objection is to my supposition that, "it may well be that evolution is a fundamental law of the universe of mind as well as that of matter." This, he says, is a purely materialistic thought. But here again it is clear by the context that I am referring solely to the development of individualised human minds, of which alone we know, or can know, anything, not to mind in the abstract, of which we know absolutely nothing; and I see no materialism in the supposition that such finite individualised minds can only be produced under some law of evolution.

The last special criticism refers to my belief that "progress towards a nobler and happier existence in the spiritual world" is dependent on the cultivation of our higher moral feelings here. My critic says that this is an utter denial of justice or equality, because our moral nature, as well as our environment, is imposed upon us; but he does not say whether he accepts the alternative position, that all are to be at once good and happy in the future state, and that the most selfish, vicious, and sensual are to make equal progress with the benevolent, self-sacrificing, and virtuous. It seems to me that this latter condition of things would be the most opposed to justice, and even to possibility, and would render the present world, with all its trials, a hopeless and insoluble mystery, while it is certainly opposed to the whole body of information and teaching which we receive from spiritual sources.

It seems to me that my critic, throughout, confuses together the general with the special, the universal with the individual, in discussing the relations of spirit and matter, while he equally confounds proximate with ultimate results in his remarks on the spiritual world. My observations and reasonings have been confined throughout to the nature and relations of individualised human minds and their proximate condition in the spirit world. Speculations on the nature or origin of mind in general as well as those on the ultimate states to which human minds may attain in the infinite future, I look upon as altogether beyond the range of our faculties, and to be, therefore, utterly untrustworthy and profitless.

* * * * *

As we shall see in Section Three, Wallace was a vocal advocate of progressive forms of social change. At the base of this interest was his insistence that society's primary function should be to sponsor the individual's intellectual and moral growth, a position fully in harmony with (but by no means inseparable from) his belief in spiritualism. His ideas on the responsibilities of spiritualism vis-à-vis *social change are well summed up in the following address, presented at the International Congress of Spiritualists in 1898:*

Spiritualism and social duty (ref 545)

For the last ten years my attention has been given to other subjects than Spiritualism; and less than three years ago, in a new edition of my writings on the subject [i.e., *Miracles and modern spiritualism*, 3rd edn.], I have restated my firm conviction as to the reality and importance of our inquiry and the worthlessness of all the arguments of our opponents. I have, therefore, nothing whatever to say to you as to Spiritualism itself. But I will take the present opportunity of laying before you a few observations on what appears to me to be the relation of the beliefs we hold as Spiritualists to that subject which now mainly occupies my thoughts—how to raise the bulk of our people out of that terrible slough of destitution, grinding life-long labour for bare subsistence, and shortened lives, uncheered by any of those refinements of art or enjoyment of Nature which are essential to the development of what is best in humanity. In a work published a few weeks since [i.e., *The wonderful century*], I have given ample proof that such is, to-day, the condition of a large proportion of our people, notwithstanding an increase of wealth and of wealth-creating power unequalled in the history of the world, and adequate, if properly utilised, to give not only abundance of all necessaries, but comforts, luxuries, and ample leisure to all. On these matters, however, I will now say no more, but will ask your attention to a few remarks on what I consider to be the relation of Spiritualism to Social Duty.

The old doctrine as to the nature of the future life was based upon the idea of rewards and punishments, which were supposed to be dependent upon *dogmatic beliefs* and ceremonial *observances*. The atheist, the agnostic, even the Unitarian, were for centuries held to be of the first importance; disposition, conduct, health, and happiness were of no account.

The new doctrines—founded almost wholly on the teachings of Modern Spiritualism, though now widely accepted, even among non-Spiritualists—are the very reverse of all this. They are based upon the conception of mental and moral continuity; that there are no imposed punishments; that dogmatic beliefs are absolutely unimportant, except so far as they affect our relations with our fellows; and that forms and ceremonies, and the complex observances of most religions, are equally unimportant. On the other hand, what are of the most vital importance are motives, with the actions that result from them, and everything that develops and exercises the whole mental, moral, and physical nature, resulting in happy and healthy lives for every human being. The future life will be simply a continuation of the present, under new conditions; and its happiness or misery will be dependent upon how we have developed all that is best in our nature here.

Under the old theory the soul could be saved by a mere change of beliefs and the performance of certain ceremonial observances. The body was nothing; happiness was nothing; pleasure was often held to be a sin; hence any amount of punishment, torture, and even death were considered justifiable in order to produce this change and save the soul.

On the new theory it is the body that develops, and to some extent saves, the soul. Disease, pain, and all that shortens and impoverishes life, are injurious to the soul as well as to the body. Not only is a healthy body necessary for a sound mind, but equally so for a fully-developed soul—a soul that is best fitted to commence its new era of development in the spirit world. Inasmuch as we have fully utilised and developed all our faculties—bodily, mental, and spiritual—and have done all in our power to aid others in a similar development, so have we prepared future well-being for ourselves and for them . . .

It seems to me that, holding these beliefs as to the future life and what is the proper and only preparation for it, we Spiritualists must feel ourselves bound to work strenuously for such improved social conditions as may render it possible for *all* to live a full and happy life, for all to develop and utilise the various faculties they possess, and thus be prepared to enter at once on the progressive higher life of the spirit-world. We *know* that a life of continuous and grinding bodily labour, in order to obtain a bare existence; a life almost necessarily devoid of beauty, of refinement, of communion with Nature; a life without adequate relaxation, and with no opportunity for the higher culture; a life full of temptation and with no cheering hope of a happy and peaceful old age, is as bad for the welfare of the soul as it is for that of the body.

If the accounts we get of the spirit world have any truth in them, the reclamation and education of the millions of undeveloped or degraded spirits which annually quit this earth, is a sore burden, a source of trouble and sorrow to those more advanced spirits who have charge of them. This burden must, for a long time to come at all events, *necessarily* be great, on account of the numbers of the less advanced races and peoples still upon the earth; but that *we*, who call ourselves *civilised*, who have learnt so much of the secret powers and mysteries of the universe, who by means of those powers could easily provide a decent and rational and happy life for our whole population—that *we* should send to the spirit world, day by day and year by year, millions of men and women, of children, and of infants, all sent there before their time through want of the necessary means of a healthy life, or by the various diseases and accidents forced upon them by the vile conditions under which alone we give them the opportunity of living at all—this is a disgrace and a crime!

I firmly believe—and the fact is supported by abundant evidence—that the very poorest class of our great cities, those that live constantly below the margin of poverty, who are without the comforts, the necessaries, and even the decencies of life, are, nevertheless, as a class, quite as good morally, and often as high intellectually, as the middle and upper classes who look down upon them as in every way their inferiors. Their condition, socially and morally, is the work of society; and in so far as they appear worse than others they are made so by society. What should we ourselves have been if we had had no education, no repose, no refined or decent homes, no means of cleanliness, which is not only next to, but is a source of, godliness; surrounded by every kind of temptation, and not unfrequently forced into crime? And a direct consequence of the millions who are compelled to lead such lives are the millions of infants who die prematurely—a slaughter a thousand times worse than that of Herod, going on year by year in our midst; surely their innocent blood cries out against our rulers, against all of *us*, who choose such rulers; and more especially against us Spiritualists, who know the higher law, if we do not work with all our strength for a radical reform.

As many of my friends here know, I myself, against all my early prepossessions, have come to believe that some form of Socialism is the only complete remedy for this state of things; and I define Socialism as simply the organisation of labour for the highest common good. Just as the Post Office is organised labour in one department for the benefit of all alike; just as the railways might be organised as a whole for the benefit of the community; just as numbers of vast industries are organised, more especially in America, for the exclusive benefit of rings of capitalists—so all necessary and useful labour might be organised for the benefit of all.

I ask you to think over this question; and above all things, I ask you to consider the necessity for real and fundamental remedies, not mere palliatives, which have been tried with ever-increasing energy and good-will throughout the century, and have absolutely failed. The evil has grown, just as if no such remedies had been applied at all. Charity has increased enormously, and has failed. Now it is time for us to try Justice.

A few years since a talented writer used, and at once popularised, a new term—"equality of opportunity." It expresses briefly and forcibly what may be termed the minimum of social justice. The same idea had been urged by other writers, especially by Herbert Spencer in his volume on *Justice*, when he declared that justice requires every man to receive "the results of his own nature and consequent actions"—this and this only. Fundamentally, the two ideas are the same, but "equality of opportunity" is the more simple and intelligible expression of it.

To Spiritualists, who realise that every child born into this world is a living soul, come here to prepare itself for the higher life of the spirit world, it must appear a crime against that world and against humanity not to see that every such child has the best possible nurture and training, at the very least till it arrives at the adult age and becomes an independent unit of the social organism. And if to each is due the best, then none can have more than the best, and we come thus again to EQUALITY OF OPPORTUNITY.

Of course, many of you will say, "This is impossible. How can we possibly give this equality of nurture and education to every child?" I admit that it is difficult—by no means impossible. It must, of course, be brought about gradually; and where there is a will there is a way. As Herbert Spencer said of another matter—the nationalisation of the land—"justice sternly demands that it be done;" and if we, boasting of our civilisation, declare that it cannot be done, then so much the worse for us and for our false civilisation. But it wants only the will. And it is our duty, as Spiritualists, to help to create that will.

But again, you will say, "Where are the means of doing this? We are already taxed as much as we can bear." True, we are shamefully over-taxed, but, instead of increasing the taxes, there is a necessary corollary of "equality of opportunity" which will not only give us ample funds to bring it about, but will at the same time greatly reduce taxation and ultimately abolish it altogether. For, if every child is given equality of opportunity, and every man and woman receives only "the results of their own nature and consequent actions," then it is evident that there must be no *inequality* of inheritance; and to give equality of inheritance, the State, that is, the community, must be the universal inheritor of all wealth. At first, of course, it would only be needful to take surplus wealth above a fixed maximum; and, so far from this being an injury to the heirs of a millionaire, it would be a great benefit; for it is admitted that nothing has so demoralising an effect on the young as the certainty of inheriting great wealth; and examples of this come before us every year and almost every month. This is the real teaching of the parable of Dives and Lazarus; this gives us the true meaning of Christ's saying that a rich man shall hardly enter into the kingdom of heaven.

Now, many who dislike the idea of Socialism—chiefly, I think, through not understanding what it really implies—will perhaps look more favourably on this great principle of "equality of opportunity," since it would leave individualism untouched, would in fact render it far more complete and effective than it is now. For our present state of society is *not* true individualism, because the inequalities of opportunity in early life are so great that often the worst are forced to the top, while many of the best struggle throughout life without a chance of using their highest faculties, or developing the best part of their nature. Even Tennyson, whose mind was of an aristocratic bent, could say:

> Plowmen, shepherds, have I found, and more than once, and still could find,
> Sons of God and kings of men in utter nobleness of mind;
> Truthful, trustful, looking upward to the practised hustings-liar;
> So the Higher wields the Lower, while the Lower *is* the Higher.
> Here and there a cotter's babe is royal-born by right divine;
> Here and there my lord is lower than his oxen or his swine.

Equality of opportunity would put all this right; everyone would be able to show what power for good he possessed, and society would be enormously benefited in consequence. At the same time, there would be all the stimulus to be derived from individual effort. The man who could surpass his fellows under such equal and fair conditions would be truly great. Some would achieve honour, some would acquire wealth; but it would be all due to their own "nature and consequent actions," and neither the honour nor the wealth would be handed on to individuals who might not be worthy of the one or be able to acquire the other.

I believe myself that such a perfectly fair competition, in which all started on equal terms socially, would be an admirable training, and would be sure to lead, ultimately, to a voluntary co-operation and organisation of labour which would produce most of the best results of Socialism itself. But whether it would or not, I claim that it embodies a great and true principle—Social Justice; and that it affords the only non-socialistic escape from the horrible social quagmire in which we find ourselves. As Spiritualists we must uphold justice; and equality of opportunity for all is but bare justice. Knowing that the life here is the school for the development of the spirit, we must feel it our duty to see that the nascent spirit in each infant has the fullest and freest opportunity of developing all its faculties and powers under the best conditions we can provide for it. And I have ventured to bring this subject before you because it is the one hope nearest to my heart; and I am sure that if the great and rapidly-increasing body of Spiritualists can be brought to consider it, and to feel that the misery and degradation around them *must* be and *can* be got rid of, and that it is especially *their* business and *their* duty to help to get rid of it, the great work will soon be taken in hand.

What we want, above all things, is to educate the people and create a public opinion adequate for the work. In this movement for justice and right, Spiritualists should take the lead, because they, more than any other body, know its vital importance both for this world and the next. The various religious sects are all working, according to their lights, in the social field; but their forces are almost exclusively directed to the alleviation of individual cases of want and misery by means of charity in various forms. But this method has utterly failed even to diminish the mass of human misery everywhere around us, because it deals with symptoms only and leaves the causes untouched. I would not say a word against even this form of charity, for those who see no higher law; but we want more of the true charity of St. Paul—the charity that thinketh no evil, that suffereth long and is kind, that rejoiceth in the truth—not only the lesser and easier charity which feeds the poor out of its superfluity, an action which St. Paul does not allow to be charity at all.

But let us Spiritualists take higher ground. Let us demand Social Justice. This will be a work worthy of our cause, to which it will give dignity and importance. It will show our fellow-countrymen that we are not mere seekers after signs and wonders, mere interviewers of the lower denizens of the spirit-world; but that our faith, founded on knowledge, has a direct influence on our lives; that it teaches us to work strenuously for the elevation and permanent well-being of all our fellow men. In order to do this our watchword must be—NOT CHARITY ONLY BUT JUSTICE.

* * * * *

Wallace's promotion of spiritualism extended to examinations of specific aspects of spiritualistic research as well. In one such work, printed in The Arena *of February 1891, Wallace criticized the theory that telepathy explained the phenomenon of 'phantasms' (i.e., apparitions):*

What are phantasms, and why do they appear? (ref 434)

The theories which have been suggested by the more prominent members of the Society for Psychical Research in order to explain the phenomena of phantasms or apparitions of various kinds, are all founded on telepathy, or thought-transference, the facts of which have been demonstrated by a long series of experiments. It is found that many persons are more or less sensitive to the thoughts or will-powers of others, and are able to reproduce, more or less closely, any definite mental images sought to be conveyed to them. It is urged that those who experience phantasmal sights or sounds are a kind of thought-readers, and are so powerfully affected by the thoughts of friends who are in certain excitable mental states or physical crises, especially at periods of imminent danger or when at the point of death, as to externalize those thoughts in visual or auditory hallucinations either in the waking state or as unusually vivid dreams.

This telepathic theory is held to receive strong support, and in fact to be almost proved, by the curious phenomena of the doubles, or phantasms, of living persons being seen by certain sensitive friends, when those persons strongly *will* that they shall be so seen. Such are the cases of a friend appearing to Mr. Stainton Moses at a time when this friend had fixed his thoughts upon him before going to bed; and those of Mr. B____ who several times appeared in the night to two ladies, on occasions when he went to sleep with the express wish and intention of appearing to them (*Phantasms of the Living*, Vol. 1, pp. 103–108). There are, however, difficulties in these cases. The supposed agent does not usually decide exactly how he will appear or what he will do. In one case Mr. B____ appeared, not to the ladies he was thinking of, but a married sister, hardly known to him, who happened to be occupying their room . . . Now it is an assumption hardly warranted by the facts, that the mere wish or determination to be in a certain part of a house at a certain time could cause a phantasm to appear to a person who happened unexpectedly to be there, and cause that phantasm to perform, or appear to perform, certain acts which do not appear to have been *willed* by the supposed agent. This is certainly not telepathy in the usually accepted sense; it is not the transference of a thought to an individual, but the production of what seems to be an objective phantasm in a definite locality. It is altogether inconceivable, that a mere wish could produce such a phantasm, unless, indeed, we suppose the spirit of the sleeper to leave the body in order to go to the desired place, and that it possesses the power to render itself visible to anyone who happens to be there. Let us then see whether there are any other facts concerning doubles which may throw some light on this question.

Mr. Fryer, of Bath, England, heard his name distinctly called in the voice of a brother who had been some days absent from home. At the same moment, as near as could be ascertained, the brother missed his footing and fell on a railway platform, calling out his brother's name as he fell (*Proc. Soc. Psych. Res.*, Vol. 1, p. 134). Similar in character is the case of Mrs. Severn, who, while in bed one morning felt a violent blow on her lip so real that she put her handkerchief to it, expecting to find it bleeding. At the same time Mr. Severn, caught by a squall in a boat, received a violent blow on the same part of his mouth from the tiller. In the first case, Mr. Fryer's brother had no conscious wish to be heard by him; and in the other case, Mr. Severn certainly did not wish his wife to feel the blow . . . In both these cases, if the supposed agents had anything to do with the actual production of the phantasmal voice and sensation, it was by some unconscious or automatic process. But the experimental evidence for telepathy shows it to be produced by the conscious and active will-power of the agent or agents, and would therefore prove, if anything, that in both these cases there was some third party who was really the agent in willing and producing the telepathic effect. This is rendered still more probable by other cases of "doubles" and of warnings, of which the following is one of the most remarkable.

Mr. Algernon Joy, an engineer employed on the Penarth Docks, at Cardiff, South Wales, was walking in a country lane near the town, absorbed in a calculation connected with the Docks, when he was attacked and knocked down by two young colliers. His thoughts were then immediately directed to the possible cause of the attack, to the possibility of identifying the men, and to informing the police. He is positive that for about half an hour previous to the attack and for an hour or two after it, there was no connection whatever, direct or indirect, between his thoughts and a friend in London. Yet at almost the precise moment of the assault, this friend recognized Mr. Joy's footstep in the street, behind him, then turned and saw Mr. Joy "as distinctly as ever he saw him in his life," saw he looked distressed, asked what was the matter, and received the answer, "Go home, old fellow, I've been hurt." All this was communicated in a letter from the friend which crossed one from Mr. Joy, giving an account of the accident (*Phantasms of the Living*, Vol. 2, p. 524). In this case, whether the "double" was an audible and visual veridical hallucination, or an objective phantasm, it could not have been produced without some adequate cause. To assert that Mr. Joy was himself the unconscious cause cannot be looked upon as an explanation, or as in any way helping us to a comprehension of how such things can happen. We imperatively need a producing agent, some intellectual being having both the will and the power to produce such a veridical phantasm.

The next case still more clearly demands an agent other than that of any of the parties immediately concerned. Mr. F. Morgan, of Bristol, a young man who lived with his mother, was attending a lecture in which he was much interested. On entering the lecture room he saw a friend, with whom he determined to walk home after the lecture. About the middle of the lecture he noticed a door at the side of the platform farthest from the entrance to the hall, and he suddenly, without knowing why, got up and walked half the length of the hall to see if the door would open. He turned the handle, entered, and closed the door behind him, finding himself in the dark under the platform. Noticing a glimmer of light he went towards it, got into a passage which led again into the hall, the end of which he crossed to the entrance door, without any thought of the lecture which was still going on, or of the friend with whom he had meant to return, and then walked home quietly, without any excitement or impression of any kind, and quite unconscious till long after that he had done anything unusual. When he got home, however, he found that the house next to his was on fire and his mother in great alarm. He instantly removed his mother to a place of safety, and then had two or three hours' struggle with the flames . . .

Mr. Morgan states that his character is such that had he felt any impression that there was a fire, or that his mother was in danger, he should probably have shaken it off as mere fancy and refused to obey it. His mother simply wished for his presence, but exerted no will-power towards him. What agency, then, was it that acted upon his mental organization, at first apparently through simple curiosity, in such a strange yet effectual way, bringing him home so promptly, and yet without his feelng that he was in any way being influenced or guided in his actions, which seemed to himself to be perfectly voluntary and normal? We cannot avoid seeing in this case the continuous exercise of some mental influence, guided by accurate knowledge of the character of the individual and of his special surroundings at the moment, and directed with such care and judgment as to avoid exciting in him that antagonism which would have been fatal to the object aimed at. We see then that, even confining ourselves to undoubted phantasms of the living, or to impressions not connected with death, the facts are totally inexplicable on any theory of telepathy between living persons, but clearly point to the agency of preter-human intelligences—in other words, of spirits. The prejudice against such a conception is enormous, but the work of the Psychical Research Society has, it is to be hoped, somewhat undermined it. They have established, beyond further dispute for all who study the evidence, that veridical phantasms of the dead do exist; and the evidence itself—not ignorant or even scientific prejudice—must decide whether these phantasms which, as we have seen in my last article [ref 430], are often objective, are the work of men or of spirits.

Before adducing further evidence on this point, it will be well to consider briefly, the extraordinary theory of the "second self" or "unconscious *ego*," which is appealed to by many modern writers as a substitute for spirit agency when that of the normal human being is plainly inadequate. This theory is founded on the phenomena of dreams, of clairvoyance, and of duplex personality, and has been elaborately expounded by Du Prel in two volumes 8vo, translated by Mr. C. C. Massey. As an example of the kind of facts this theory is held to explain, we may refer to the experiments of the Rev. P. H. Newnham and Mrs. Newnham with planchette. The experiments were conducted by Mrs. N____ sitting at a low table with her hand on the planchette, while Mr. N____ sat with his back towards her at another table eight feet distant. Mr. N____ wrote questions on paper, and instantly, sometimes simultaneously, the planchette under Mrs. N____'s hand wrote the answers. Experiments were carried on for eight months, during which time three hundred and nine questions and answers were recorded. All kinds of questions were asked, and the answers were always pertinent to the questions though often evasions rather than direct answers. Great numbers of the answers did not correspond with the opinions or expectations of either Mr. or Mrs. N____, and were sometimes beyond their knowledge . . .

Now all this, and a great deal more equally remarkable, is imputed to the agency of Mrs. Newnham's "unconscious self," a second independent, intelligent personality of which Mrs. Newnham herself knows nothing except when it "emerges" under special conditions, such as those here described. In the same way Du Prel explains all the phenomena of clairvoyance, of premonitions, of apparent possession, and of the innumerable cases in which sensitives exhibit knowledge of facts which in their normal state they do not possess, and have had no possible means of acquiring.

But is this so-called explanation any real explanation, or anything more than a juggle of words which creates more difficulties that it solves? The conception of such a double personality in each of us, a second self which in most cases remains unknown to us all our lives, which is said to live an independent mental life, to have means of acquiring knowledge our normal self does not possess, to exhibit all the characteristics of a distinct individuality with a different character from own is surely a conception more ponderously difficult, more truly supernatural than that of a spirit-world, composed of beings who have lived, and learned, and suffered on earth, and whose mental nature still subsists after its separation from the earthly body. We shall find, too, that this latter theory explains *all* the facts simply and directly, that it is in accordance with *all* the evidence, and that in an overwhelming majority of cases, it is the explanation given by the communicating intelligences themselves. On the "second self" theory, we have to suppose that this recondite but worser half of ourselves, while possessing some knowledge we have not, does not know that it is part of us, or if it knows, is a persistent liar, for in most cases it adopts a distinct name, and persists in speaking of us, its better half, in the third person.

But there is yet another, and I think a more fundamental objection to this view, in the impossibility of conceiving how, or why, this second-self was developed in us under the law of the survival of the fittest. The theory is upheld to avoid recourse to any "spiritual" explanation of phenomena, "spirit" being the last thing our modern men of science "will give in to." (This was Sir David Brewster's expression after witnessing Home's phenomena. See Home's *Incidents of my Life*, Appendix, p. 245.) But if so—if there is no spiritual nature in man that survives the earthly body, if man is but a highly intellectual animal developed from a lower animal form under the law of the survival of the fittest, how did this "second-self," this "unconscious *ego*," come into existence? Have the mollusk and the reptile, the dog and the ape, "unconscious *egos*"? And if so, why? And what use are they to these creatures, so that they might have been developed by means of the struggle for existence? Darwin detected no sign of such "second-selves" either in animals or men; and if they do not pertain to animals but do pertain to men, then we are involved in the same difficulty that is so often urged against spiritualists, that we require some break in the law of continuous development, and some exertion of a higher power to create and bring into the human organism this strange and useless "unconscious *ego*"—useless except to puzzle us with insoluble problems, and make our whole nature and existence seem more mysterious than ever. Of course this unconscious *ego* is supposed to die with the conscious man, for if not, we are introduced to a new and gratuitous difficulty, of the relation of these two intelligences and characters, distinct yet bound indissolubly together, in the after life.

Finding, therefore, that the theory of duplex personality creates more difficulties than it solves, while the facts it proposes to explain can be dealt with far more thoroughly on the spiritual hypothesis, let us pass on to consider the further evidence we possess for the agency of the spirits of the dead, or of some other preter-human intelligences.

. . . Mrs. Storie, of Edinburgh, living at the time in Hobart Town, Tasmania, one night dreamed a strange, confused dream, like a series of dissolving views. She saw her twin brother sitting in the open air, in the moonlight, sideways, on a raised place. Then he lifted his arm saying, "*The train, the train!*" Something struck him, he fell down fainting, a large dark object came by with a *swish*. Then she saw a railway compartment, in which sat a gentleman she knew, Rev. Mr. Johnstone. Then she saw her brother again. He put his right hand over his face as if in grief. Then a voice, not his voice, telling her he was going away. The same night her brother was killed by a train, having sat down to rest on the side of the track and fallen asleep. The details in the dream, of which the above is a bare abstract, were almost exactly as in the event, and the Mr. Johnstone of the dream was in the train that killed her brother. Now this last mentioned fact could not have been known to the dead man during life, and the dream-picture of the event must, therefore, have been due to the telepathic power of the dead man, or of some spirit-friend acquainted with the facts, and wishing to give a proof of spirit-life.

Dreams giving the details of funerals at a distance are not uncommon. As an example we have one in which Mr. Stainton Moses was invited to the funeral of a friend in Lincolnshire, but could not go. About the time of the funeral, however, he fell into a trance, and appeared to be at the ceremony, and on again becoming conscious, wrote down all the details, describing the clergyman, who was not the one who had been expected to officiate, the churchyard, which was at a distance in Northamptonshire, with a particular tree near the grave. He then sent this description to a friend who had been present, and who wrote back in astonishment as to how he could have obtained the details (Harrison's *Spirits before our Eyes*, p. 148). This may be said to be mere clairvoyance; but clairvoyance is a term that explains nothing, and is quite as mysterious and unintelligible if supposed to occur without the intervention of disembodied intelligences as if with their help. These cases also merge into others which are of a symbolical nature, and which clairvoyance of actual scenes at a distance cannot explain. A well-attested case of this kind is the following:

Philip Weld, a student at a Catholic College, was drowned in the river at Ware, Hertfordshire, in the year 1846. About the same hour as the accident, the young man's father and sister, while walking on the turnpike road near Southampton, saw him standing on the causeway with another young man in a black robe. The sister said, "Look, papa, there is Philip." Mr. Weld replied, "It is Philip indeed, but he has the look of an angel." They went on to embrace him, but before reaching him a laboring man seemed to walk through the figures, and then with a smile both figures vanished. The President of the College, Dr. Cox, went immediately to Southampton, to break the sad news to the father, but before he could speak, Mr. Weld told him what he had seen, and said he knew his son was dead. A few weeks afterwards, Mr. Weld visited the Jesuit College of Stonyhurst in Lancashire, and in the guest-room saw a picture of the very same young man he had seen with his son, similarly dressed, and in the same attitude, and beneath the picture was inscribed "St. Stanislaus Kotska," a saint of the Jesuit order who had been chosen by Philip for his patron saint at his confirmation (Harrison's *Spirits before our Eyes*, p. 116).

Now, here is a case in which phantasms of the son and of another person appear to two relatives, and the presence of an unknown person was eminently calculated, when his identity was discovered, to relieve the father's mind of all fear for his son's future happiness. It is hardly possible to have a clearer case of a true phantasm of the dead, not necessarily produced either *by* the dead son or the Jesuit saint, but most probably by them, or by some other spirit friend who had the power to produce such phantasms, and so relieve the anxiety of both father and sister. It is not conceivable that any living person's telepathic action could have produced such phantasms in two percipients, the only possible agent being the President of the College, who did not recognize by

Mr. Weld's description, the dark-robed young man who appeared with his son.

This introduces a feature rather common in phantasms of the dead some indication of happiness, something to take away any feeling of gloom or sorrow. Thus, a young man is drowned by the foundering of the La Plata telegraphic ship in December, 1874; and, just before the news arrived, his brother in London dreamed that he was at a magnificent fête, in a spacious garden with illuminated fountains and groups of gentlemen and ladies, when he met his brother in evening dress, and "the very image of buoyant health." He was surprised, and said: "Hallo! D____, how are you here?" His brother shook hands with him and said: "Did you not know I have been wrecked again?" The next morning the news of the loss of the ship was in the papers (*Proc. Soc. Psych. Res.*, Part XIV, p. 456). Here, whether the phantasm was caused by the dead man himself, or by some other being, it was apparently intended to show that the deceased was as cheerful and well off after death as during life.

Another characteristic feature of many of these dreams or waking phantasms is that they often occur, not at the moment of death but just before the news of the death reaches the percipient, or there is some other characteristic feature that seems especially calculated to cause a deep impression, and give a lasting conviction of spiritual existence. Several cases of this kind are given or referred to in the *Proceedings of the Society for Psychical Research* (Pt. XV, pp. 30, 31). A most extraordinary example is that of Mr. F. G., of Boston, then of St. Louis, Mo., who, when in St. Joseph, Mo., fully occupied with business, saw a phantasm of his only sister, who had been dead nine years. It was at noonday while he was writing, and she appeared close to him and perfectly life-like, so that for a moment he thought it was really herself, and called her by her name. He saw every detail of her dress and appearance, and particularly noticed a bright red line or scratch on the right hand side of her face. The vision so impressed him that he took the next train home, and told what he had seen to his father and mother. His father was inclined to ridicule him for his belief in its being anything supernatural, but when he mentioned the scratch on the face his mother nearly fainted, and told them with tears in her eyes, that she had herself made that scratch accidentally, after her daughter's death, but had carefully hidden it with powder, and that no living person but herself knew of it. A few weeks after, the mother died happy in her belief that she would rejoin her daughter in a better world (*Proc. Soc. Psych. Res.*, Part XV, pp. 17, 18). Here we clearly see an important purpose in the appearance of the phantasm, to give comfort to a mother about to die, in the assurance that her beloved daughter, though mourned as dead, was still alive.

A case which illustrates both of the characteristics just alluded to, is that of the Rev. C. C. Wambey of Salisbury, England, who, one Sunday evening, was walking on the downs, engaged in composing a congratulatory letter to a very dear friend so that he might have it on his birthday, when he heard a voice saying, "What, write to a dead man; write to a dead man!" No one was near him, and he tried to think it was an illusion, and went on with his composition, when again he heard the voice saying more loudly than before, "What, write to a dead man; write to a *dead* man?" He now understood the meaning of the voice, but, nevertheless, sent the letter, and in reply received the expected intelligence that his friend was dead. Surely, in this case no living agent could have produced this auditory phantasm, which was strikingly calculated to impress the recipient with the idea that his friend was, though dead as regards the earthly life, in reality very much alive, while the spice of banter in the words would tend to show that death was by no means a melancholy event to the subject of it.

In view of the examples now given of phantasms appearing for a very definite purpose, and being in most cases perfectly adapted to produce the desired effect—examples which could be very largely increased from the rich storehouse of the publications of the Society for Psychical Research—the theory put forth by Mr. [Frederick] Myers, that phantasms of the dead are so vague and purposeless as to suggest mere "dead men's dreams" telepathically communicated to the living, seems to me a most extraordinary one. No doubt the range of these phenomena is very great, and in some cases there may be no purpose in the appearance so far as the percipient is concerned. But these are certainly not typical or by any means the best attested or the most numerous; and it seems to me to be a proof of the weakness of the telepathic theory that almost all the cases I have adduced, and many more of like import, have been passed by almost or quite unnoticed by those who support that view.

We have one more class of evidence to notice,—that of premonitions. These are of all kinds from those announcing very trivial events, to such as foretell accidents or death. They are not so frequent as other phantasms, but some of them are thoroughly well attested, and it is difficult to avoid the conclusion that they are realities, and that they are due, generally speaking, to the same agencies as objective veridical phantasms. One or two examples may be given.

A striking case is that of Mrs. Morrison, who was living in the Province of Wellesley, Malay Peninsula, in 1878, and one morning, when awake, heard a voice distinctly say, "If there is darkness at the eleventh hour, there will be death." On starting up in bed the same words were slowly and deliberately repeated. A week afterwards her little girl was taken seriously ill, and some days later, after a week of cloudless weather, a storm came on one morning, a few minutes before eleven, and the sky became black with clouds. At one o'clock the same day the child died (*Proc. Soc. Psych. Res.*, Part XIII, p. 305). The unusual character of the warning renders this case a very remarkable one.

Still more extraordinary is the case of the Yorkshire vicar, who, when a young man of nineteen, was at Invercargill, in New Zealand, and there met a man he knew as a sailor on the ship he had come out in, and agreed to go with him and several others on an excursion to the island of Ruapuke, to stay a day or two for fishing and shooting. They were to start at four the next morning, in order to cross the bar with the high tide, and they agreed to call the vicar in time. He went to bed early with the fullest intention to go with them, and with no doubt or hesitation in his mind. The thing was settled. On his way upstairs to bed he seemed to hear a voice saying, "*Don't go with those men.*" There was no one near, but he asked, "Why not?" The voice, which seemed inside of him, said with emphasis, "You are *not* to go;" and on further question these words were repeated. Then he asked, "How can I help it? They will call me up." And, most distinctly and emphatically, the same voice said, "*You must bolt your door.*" When he got to the room, he found there was a strong bolt to the door, which he had not remembered. At first he determined he would go, as he was accustomed to take his own way at all hazards. But he felt staggered, and had a feeling of mysterious peril, and after much hesitation finally bolted the door, and went to sleep. In the morning about three he was called, the door violently shaken and kicked but though awake he did not speak, and finally the men went away cursing and shouting. About nine o'clock he went down to breakfast, and was at once asked if he had heard what had happened, and was then told that the boat with the party for Ruapuke had been upset on the bar, and *every one of them drowned*. Some of the bodies were washed up on the beach that day, and the others a day or two later, and he adds: "If I had been with them, I must have perished beyond a doubt."

Now what are we to say of this determined, warning voice that insisted on being heard and attended to? Who and what was the being that foresaw the catastrophe that was to happen, and saved the one that it could save? Du Prel would say that it was the second self, the unconscious *ego*, that produced this inner voice; but, as we have shown, this purely hypothetical explanation is both unintelligible and inconceivable, and explains nothing, since the suggested cause has not been proved to exist, nor can it be shown how the knowledge exhibited had been acquired. But phantasms of the dead, manifesting themselves in a way to prove their identity, or exhibiting knowledge which neither the percipient nor any conceivable living agent possesses, afford strong proof that the so-called dead still live, and are able in various ways to influence their friends in earth-life. We will, therefore, briefly summarize the evidence now adduced, and see how the spiritualistic theory gives a consistent and intelligible explanation of it.

It is evident that any general theory of phantasms must deal also with the various cases of "doubles," or undoubted phantasms of the living. The few examples of apparent voluntary production of these by a living person have been supposed to prove the actual production by them, or by their unconscious *egos*; but the difficulties in the way of this view have been already pointed out. In many cases there is no exercise of will, sometimes not even a thought directed to the place or person where, or to whom, the phantasm appears; and it is altogether irrational to ascribe the production of so marvellous an effect as, for example, a perfectly life-like phantasm of two persons, a carriage, and a horse, visible to three persons at different points of its progress through space (as described in my first article [i.e., ref 430]), to an agent who is totally unconscious of any agency in the matter. What is termed the agent, that is the person whose "double" is produced, may be a *condition* towards the production of the phantasm without being the *cause*. I write a telegram to a friend a thousand miles away, and that friend receives my message in an hour or two. But the possibility of sending the message does not reside in me, but in a whole series of contributory agencies from the earliest inventors of the telegraph, down to the clerks who transmit and receive the message.

The clue to a true explanation of these very puzzling "doubles," as of all the other varied phenomena of phantasms and hauntings is, I believe, afforded by the following passage by one of the most thoughtful and experienced of modern spiritualists, Dr. Eugene Crowell:

> I have frequently consulted my spirit friends upon this question, and have invariably been told by them that a spirit while in mortal form cannot for an instant leave it; were it to do so, death would at once ensue; and, that the appearance of one's self at another place from that in which the body at the moment is, is simply a personation by another spirit, who thus often accomplishes a purpose desired by his mortal friend, or some other useful purpose is accomplished by the personation. I am informed, and believe, that in cases of trance, where the subjects have supposed that their spirits have left their bodies, and visited the spheres, their minds have been psychologically impressed with views representing spiritual scenes, objects, and sounds, and many times these impressions are so apparently real and truthful that the reality itself barely exceeds these representations of it, but these are all subjective impressions, not actual experiences. (*Primitive Christianity and Modern Spiritualism*, Vol. II, p. 109)

Accepting, then, as proved by the various classes of phantasms and the information conveyed by them, that the spirits of the so-called dead still live, and that some of them can, under special conditions, and in various ways, make their existence known to us, or influence us unconsciously to ourselves, let us see what reasonable explanation we can give of the cause and purpose of these phenomena.

In every case that passes beyond simple transference of a thought from one living person to another, it seems probable that other intelligences co-operate. There is much evidence to show that the continued association of spirits with mortals is in many cases beneficial or pleasurable to the former, and if we remember the number of very commonplace people who are daily and yearly dying around us, we shall have a sufficient explanation of those trivial and commonplace yet veridical dreams and impressions which at first sight seem so unintelligible. The production of these dreams, impressions, and phantasms, may be a pleasurable exercise of the lower spiritual faculties, as agreeable to some spirits as billiards, chemical experiments, or practical jokes are to some mortals.

Many hauntings, on the other hand, seem to show one mode of the inevitable punishment of crime in the spirit world. The criminal is drawn by remorse or by some indefinable attraction, to haunt the place of his crime, and to continually reproduce or act over some incidents connected with it. It is true that the victim appears in haunted houses, as often as the criminal, but it does not at all follow that the victim is always there, unless he or she was a participator in the crime, or continued to indulge feelings of revenge against the actual criminal.

Again, if there be a spiritual world, if those whose existence on earth has come to an end still live, what is more natural than that many spirits should be distressed at the disbelief, or doubt, or misconception, that so widely prevail, with respect to a future life, and should use whatever power they possess to convince us of our error. What more natural than that they should wish, whenever possible, to give some message to their friends, if only to assure them that death is not the end, that they still live, and are not unhappy. Many facts seem to show us that the beautiful idea of guardian spirits is not a mere dream, but a frequent, perhaps universal reality . . . The numerous cases in which messages are given from those recently dead, in order to do some trivial act of justice or of kindness, are surely what we should expect; while the fact that although indications are frequently given of a crime having been committed, it is but rarely that the criminal is denounced, indicates, either that the feeling of revenge does not long persist, or that earthly modes of punishment are not approved of by the denizens of the spirit world.

The powers of communication of spirits with us, and ours of receiving their communications, vary greatly. Some of us can only be influenced by ideas or impressions, which we think are altogether the product of our own minds. Others can be so strongly acted on that they feel an inexplicable emotion, leading to action beneficial to themselves or to others. In some cases, warning or information can be given through dreams, in others by waking vision. Some spirits have the power of producing visual, others audible hallucinations to certain persons. More rarely, and needing more special conditions, they can produce phantasms, which are audible or visible to all who may be present—real entities which give off light or sound waves, and thus act upon our senses like living beings or material objects. Still more rarely these phantasms are tangible as well as visual—real though temporary living forms, capable of acting like human beings, and of exerting considerable force on ordinary matter.

If we look upon these phenomena not as anything supernatural, but as the perfectly natural and orderly exercise of the faculties and powers of spiritual beings for the purpose of communicating with those still in the physical body, we shall find every objection answered, and every difficulty disappear. Nothing is more common than objections to the triviality or the partiality of the communications alleged to be from spirits. But the most trivial message or act, if such that no living person could have given or performed it, may give proof of the existence of other intelligences around us. And the partiality often displayed, one person being warned and saved, while others are left to die, is but an indication of the limited power of spirits to act upon us, combined with the limited receptivity of spirit influence on our part. In conclusion, I submit, that the brief review now given of the various classes of phantasms of the living and of the dead, demonstrates the inadequacy of all the explanations in which telepathy between living persons, or the agency of the unconscious *ego* are exclusively concerned, since these explanations are only capable of dealing with a small proportion of the cases that actually occur. Furthermore, I urge, that nothing less fundamental and far-reaching than the agency of disembodied intelligences acting in co-operation with our own powers of thought-transference and spiritual insight, can afford a rational and intelligible explanation of the whole range of the phenomena.

* * * * *

It is fully impossible to understand the overall program of Wallace's life without coming to grips with the level of reasoning that underlay his involvement with spiritualism. For example, while it is true that as he aged he took an increasingly purposeful view of evolution, he never held that it had deterministic *qualities. There is not a shred of evidence to suggest that he supported the idea that any* specific *adaptation had been pre-ordained; he did (as we shall see in Section Six), however, believe that the 'law' of natural selection gave rise to an increasing diversification of function within the living world, and that such diversification* in general *served the needs of evolving mankind by offering us the greatest possible range of alternatives of action. Regardless, the* immediate *causes of the ongoing process of diversification could be determined without recourse to teleology. Within this view, spiritual evolution merely represented one aspect of the overall process. As such, it was to be considered no more—nor less—a function of structured, probabilistically occurring, predictable, and therefore understandable relationships than were its more easily investigated biological and physical counterparts.*

One should probably conclude, in fact, that Wallace's adoption of spiritualism about 1865 neither demonstrated nor precipitated a departure from his already operating system of investigation of nature. It is clear that by the time he wrote The advantages of varied knowledge *(ref 1)—over twenty years earlier—he was already committed to the idea that a continuing, dispassionate, and many-sided examination of 'facts' gave rise to an intellectual and moral diversification essential to the long-term well-being of the individual. It is thus not surprising that he began to take spiritualism seriously once he had studied its writings closely enough to perceive their core support for the very same programme of behaviour.*

*Nor is there reason to believe that his post-1865 association with spiritualism caused his subsequent scientific output to 'suffer,' as has been implied by George (1964) and Schwartz (1984*a*). Even a casual glance at the Bibliography compiled for this work will quickly dispel this notion. It is true, of course, that 'non-scientific' subjects formed an increasing* proportion *of Wallace's output as the years passed, but three important considerations must be taken into account before arriving at a hasty conclusion: (1) he had no institutional or other affiliations (or private wealth) to sustain a programme of scientific research at any time after his return from the East Indies in 1862; (2) his timely involvement in subjects of general interest increased the public's awareness of his work and name, thus in many derivative ways ensuring a continuing income (the strategy worked: judging from period newspaper accounts, it appears that in his later years Wallace was not only the world's most celebrated student* *of natural science, but one of its best known figures overall); and (3) it is a rather narrow point of view to equate 'non-scientific' with 'less important' to begin with.*

All of which argues that Wallace's connection with spiritualism is neither to be taken lightly, nor as a personal or professional inconsistency. His acceptance of the belief was predicated on cosmological views he had held for quite some time (see Section Six); natural selection and spiritualism fit into these views equally well, and there is little justification for the idea that Wallace's endorsement of the latter 'overrode' his position on the former. Rather, he simply came to an understanding featuring both.

Related writings

first-hand accounts of seances: refs 118, 126, 132, 137, 219, 243, 255, 258, 259, 261, 271, 273, 282, 283, 396, 398, 400, 416, 452, 467, 637, 717, 729, 739

concerning mediums and spiritualistic phenomena: refs 118, 149, 152, 174, 207, 208, 209, 243, 258, 259, 261, 263, 264, 265, 270, 271, 281, 283, 356, 381, 382, 383, 396, 398, 409, 411, 430, 434, 435, 436, 451, 452, 467, 474, 478, 522, 530, 561, 565, 568, 582, 592, 637, 717 ,726, 729, 742, 745

on the moral, philosophical and historical teachings of spiritualism: refs 118, 243, 379, 383, 398, 451, 506, 545, 717, 727

on science and spiritualism: refs 118, 174, 191, 206, 219, 243, 258, 270, 283, 379, 382, 383, 398, 478, 519, 717, 726, 729, 742, 748

on spiritualism and social responsibility: refs 506, 545, 546, 591, 626, 727

on the criticisms of W. B. Carpenter: refs 206, 263, 264, 270, 276, 277, 281, 283, 285, 287

on miscellaneous topics: refs 118, 174, 204, 243, 251, 258, 409, 451, 478, 519, 530, 546, 582, 611, 623, 624, 627, 662, 664, 675, 685, 699, 701, 705, 706, 717, 726, 729, 732, 738, 739, 740, 741, 744, 746, 750, 753

book reviews and notices: refs 207, 209, 270, 490

secondary literature: Kottler (1974), Malinchak (1987), Oppenheim (1985), Schwartz (1984a), R. Smith (1972), Vorzimmer (1970)

§ 3 · LAND NATIONALIZATION AND OTHER SOCIAL ISSUES

> ... The present writer had his attention forcibly drawn to this great question about eighteen years ago, by the perusal of Herbert Spencer's demonstration (in his *Social Statics*) of the immorality and impolicy of private property in land, and since that time he has endeavoured to make himself acquainted with what has been written on the subject, and by means of constant thought and discussion to arrive at the true solution of the problem. This he believes he has at length done. The difficulties that surrounded the subject were many and great. It was necessary, firstly, to find a means of transferring the ownership of the land from individuals to the State without taking anything away from existing owners, or infringing any right, real or sentimental, which they actually possess; secondly, to devise a new tenure of land which should combine all the incalculable advantages of safe possession and transmissible ownership, together with the full benefit of every improvement and increase of its value, while guarding against the recurrence of unlimited landed estates, absentee landlords, life-interests, subletting, building leases, restriction on improvements, and all other evils which accompany our present system; thirdly, to avoid the dangers which have been hitherto believed to be inherent in State-Landlordism—jobbery, favouritism, waste, and the creation of a vast addiction to State patronage; and, lastly, to render the land a productive and practically inexhaustible source of national income, and to bring all these changes about in a gradual and almost imperceptible manner, by the action of a few simple principles embodied in law, so that society may have ample time to adapt itself to the new conditions; and, during the process of adaptation, successive generations may grow up to whom they will bear the aspect of being as natural, as orderly, and as beneficial, as private ownership does to most of the present generation . . .
>
> ('How to nationalize the land,' 1880: ref 329, p. 735)

FROM Sections One and Two we have seen why the eventuality of Wallace's becoming more and more interested in the study of social problems was predictable: he viewed social change as being no more nor less a part of evolution than biological change, the only fundamental difference being that human beings, unlike other living things, were capable of recognizing toward what end the overall process was leading, and consciously modifying their actions accordingly. All his excursions into social criticism and planning can be traced to derivative concerns—especially as related to the immediate objective of making 'equality of opportunity' (a term borrowed from the writings of Benjamin Kidd—see ref 496) society's byword.

In truth, the matter of consistency of vision aside, Wallace simply could not tolerate one-sidedness, whether this took the form of misrepresentation in a scientific dialogue or what he felt to be any kind of inequity in existing social policy. He had believed in the inherent perfectibility of man since his early teen years, and came to conclude that individual misdeeds were largely the fault of a social system that was 'rotten at the core' (ref 427, p. 330). None the less, he was not pessimistic, viewing social unrest as a positive sign: it showed people were not complacent about society's problems, and were therefore still striving for improvements.

Wallace envisioned a utopian society whose eventual emergence would be assured as soon as a socially universal 'equality of opportunity' was realized. Accordingly, his attacks on existing social structure focused on those institutions he felt were impeding progress in that direction. Far from being the complaints of a crank, the remedial programs he proposed were as fully derived from his 'theory of evolution' as natural selection was. Many of the changes he argued for have in fact come to pass in one form or another. (This is not to say that he would not think there is yet much left to do!)

Wallace believed that the most enduring social reforms were those brought about non-violently. The most effective instrument of change was, simply, the vote. His confidence in the voter was expressed in an ingeniously argued letter that appeared in Reader *in 1865:*

Public responsibility and the ballot (ref 110)

In the review, headed as above, in your last number, and signed "J. S. M." [i.e., John Stuart Mill], the well-known writer does not appear to me to have been quite successful in answering the arguments of the pamphlet he criticizes. Indeed, on the most important point, his own reasoning seems equally applicable to the opposite side of the question. I beg leave, therefore, to make a few observations on what appears to me, as I doubt not it must to many of your readers, a very inconclusive part of his article.

Mr. Mill truly says, that a voter is rarely influenced by "the fraction of an interest, which he as an individual may have, in what is beneficial to the public," but that his motive, if uninfluenced by direct bribery or threats, is simply "to do right," to vote for the man whose opinions he thinks most true, and whose talents seem to him best adapted to benefit the country. The fair inference from this seems to be, that if you keep away from a man the influences of bribery and intimidation, there is no motive left but to do what he thinks will serve the public interest—in other words, "the desire to do right." Instead of drawing this inference, however, it is concluded that, as the "honest vote" is influenced by "social duty," the motive for voting honestly cannot be so strong "when done in secret, and when the voter can neither be admired for disinterested, nor blamed for selfish conduct." But Mr. Mill has not told us what motive there can possibly be to make the man, voting in secret, vote against his own conviction of what is right. Are the plaudits of a circle of admiring friends necessary to induce a man to vote for the candidate he honestly thinks the best; and is the fear of their blame the only influence that will keep him from "mean and selfish conduct," when no possible motive for such conduct exists, and when we know that, in thousands of cases, such blame does not keep him from what is much worse than "mean and selfish conduct," taking a direct bribe?

Perhaps, however, Mr. Mill means (though he nowhere says so) that "class interest" would be stronger than public interest—that the voter's share of interest in legislation that would benefit his class or profession, would overbalance his share of interest in the welfare of the whole community. But if this be so, we may assert, first, that the social influence of those around him will, in nine cases out of ten, go to increase and strengthen the ascendency of "class interests," and that it is much more likely that a man should be thus induced to vote for class interests as against public interests, than the reverse. In the second place, we maintain that any temporary influence whatever, which would induce a man to vote differently from what he would have done by his own unbiassed judgment, is bad—that a man has a perfect right to uphold the interests of his class, and that it is, on the whole, better for the community that he should do so. For, if the voter is sufficiently instructed, honest, and far-seeing, he will be convinced that nothing that is disadvantageous to the community as a whole can be really and permanently beneficial to his class or party; while, if he is less advanced in social and political knowledge, he will solve the problem the other way, and be fully satisfied that in advancing the interests of his class he is also benefiting the community at large. In neither case, is it at all likely, or indeed desirable, that the temporary and personal influence of others' opinions at the time of an election, should cause him to vote contrary to the convictions he has deliberately arrived at, under the continued action of those same influences, and which convictions are the full expression of his political knowledge and honesty at the time?

It seems to me, therefore, that if you can arrange matters so that every voter may be enabled to give his vote uninfluenced by immediate fear of injury or hope of gain (by intimidation or bribery), the only motives left to influence him are his convictions as to the effects of certain measures, or a certain policy, on himself as an individual, on his class, or on the whole community. The combined effect of these convictions on his mind will inevitably go to form his idea of "what is right" politically, that idea which, we quite agree with Mr. Mill, will in most cases influence his vote, rather than any one of the more or less remote personal interests which have been the foundation of that idea. From this point of view, I should be inclined to maintain that the right of voting is a "personal right" rather than a "public duty," and that a man is in no sense "responsible" for the proper exercise of it to the public, any more than he is responsible for the convictions that lead him to vote as he does. It seems almost absurd to say that each man is responsible to every or to any other man for the free exercise of his infinitesimal share in the government of the country, because, in that case, each man in turn would act upon others exactly as he is acted upon by them, and thus the final result must be the same as if each had voted entirely uninfluenced by others. What, therefore, is the use of such mutual influence and responsibility? You cannot by such means increase the average intelligence or morality of the country; and it must be remembered, that the character and opinions, which really determine each man's vote, have already been modified or even formed by the long-continued action of those very social influences which it is said are essential to the right performance of each separate act of voting. It appears to me that such influences, if they really produce any fresh effect, are a moral intimidation of the worst kind, and are an additional argument in favour of, rather than against, the ballot.

Two other questions remain. Is the ballot necessary to prevent bribery and intimidation? Is it so injurious to independence of character as to overbalance its undoubted utility? I think Mr. Berkeley's letter in the *Times* in reply to Mr. Mill, and the experience of every general election, are sufficient to answer the first question in the affirmative. The answer to the second entirely depends upon the state of civilization and independence to which we have arrived; and it seems to me that in the days of standing armies, of an elaborate Poor Law, of State interference in education, of the overwhelming influence of wealth and the Priesthood, we have *not* arrived at the stage of general advancement and independence of thought and action in which we ought to give up so great and immediate a benefit to thousands as real freedom of voting, for the infinitesimal and advantage to the national character which might be derived from the independent and open voting of the few who would feel it compatible with their duty to their families to struggle against unfair influence and unjust intimidation.

* * * * *

The single non-natural science subject to which Wallace devoted the most attention was land nationalization. Although his writings on related issues were limited before 1880, he produced dozens of works promoting this cause after that date. The philosophy and theory of land occupation was, however, a matter he had been turning over in his mind since his very early years. As a postscript to his Presidential Address to the Land Nationalisation Society at their annual meeting of 1885 (ref 380), he summarized the history of his interest in the following fashion:

. . . In a lecture on the Land Question at Newton Hall in February, 1884, Mr. [Frederic] Harrison is reported to have said that—"he thought that Mr. Wallace knew as little about the English Land Question and the management of English estates as Mr. [Henry] George, and his studies in the Eastern Archipelago had not assisted him."

Now, as my practice—in this as in all other enquiries on which I have been engaged—is never to give my personal opinions as of any value, but always to set forth or refer to the facts and arguments which have led me to form the opinion, the above-quoted belief as to my "knowledge" is of little real importance. The question really is, are the facts we allege true, and are the conclusions we draw from those facts sound? But there are many people who will not, or cannot, follow out a train of reasoning founded on an elaborate statement of facts; they prefer relying on authority, and of course they like to know on what grounds any particular person is to be considered an authority. For that reason and that reason only, I beg leave to give you a short autobiographical sketch of my personal relations with the land and the land question.

I left school at the age of 14, and my real education then commenced. I joined an elder brother who was a *land-surveyor*, with considerable knowledge also of engineering and architecture. With him I spent seven years, in various parts of England and Wales, constantly engaged professionally with *land*. We lived almost the whole of this time in small towns or villages, in farmhouses or labourers cottages. We were engaged in surveying for the Tithe Commutation, for the Enclosure Commissioners, or for private estate-owners. We assisted in the process of valuation for the Tithe Commutation, and for the Enclosures of Commons. Living thus almost constantly on the land and among farmers and country people, I soon took a great interest in agriculture. I studied the works of Sir Humphrey Davy and Baron Liebig, at that time the great authorities on agricultural chemistry. When living in village inns I was often present at the meetings of the Farmers' Clubs, and heard agricultural matters discussed, and I really believe that at that period of my life I could have passed a very fair examination in theoretical and practical agriculture. Soon after I came of age, my brother died, and I succeeded him in a small connection as Surveyor and Architect, which occupied me till I went abroad five years later.

Thus, for the twelve most impressionable and most active years of my life I was living in the closest possible relation to the land and its cultivators, and I obtained a knowledge of peasant life and an interest in agriculture which has been a permanent acquisition to me. After returning home from abroad in 1862 I lived a few years in London, but since then have always resided in the country, and having acquired from Herbert Spencer the great principle that private property in land was absolutely wrong, I ever kept the subject in mind, seeking out a mode by which this wrong might be practically and equitably abolished. About eight or ten years ago I began to see my way, and as soon as I had finished the various scientific works which were the result of my twelve years of tropical exploration, I put my ideas in order and wrote the article 'How to Nationalise the Land' [ref 329], which appeared in the *Contemporary Review* in Nov., 1880, and which I believe led to the formation of this Society.

I trust you will excuse the apparent egotism of this narration, but it may perhaps be useful by showing that my whole life has been closely connected with the land, that I have had great opportunities for acquiring knowledge both of it and of those who cultivate it, and that the views I now hold and advocate are not . . . the result of a hasty and superficial literary study, but are, to some extent, the outcome of the practical experience of a lifetime.

* * * * *

As suggested by the preceding sketch, Wallace's Presidency of the Land Nationalisation Society came about as a function of his 1880 article 'How to nationalise the land'. Again, however, he had been thinking about related subjects earlier, as is evident from the following letter, contributed to The Daily News *in 1873:*

Free-trade principles and the coal question (ref 231)

It has now become an axiom with all liberal thinkers that complete freedom of exchange between nations and countries of the various products each has in superabundance and can best spare, for others which it requires, is for the benefit of both parties; and this principle is thought to be so universally applicable, that even when it produces positive injury to ourselves and is certain to injure our descendants, hardly any public writer who professes liberal views ventures to propose a limitation of it. It seems clear, however, that there are limitations to its wholesome application, and that there are certain commodities which we have no right to exchange away without restriction, for others of more immediate use to the individuals or communities who happen to be in possession of them. These commodities may be briefly defined as those natural products which are practically limited in quantity, and which cannot be reproduced. What is meant may perhaps be best explained by taking what may be considered a very extreme case as an illustration. Let us suppose, for instance, a country in which the springs or wells of water were strictly limited in number, but sufficiently copious to supply all the actual needs of the community who had always had the use of them, on making a nominal payment to the owners of the land on which they were situated. Acting on the principles of unrestricted free trade, and anxious to increase their wealth, one after another of the landowners sold their springs to manufacturers, who used up all the water except that required to supply the wants of their own workpeople, thus rendering the remainder of the country almost uninhabitable. A still more extreme case, but one rather more to the point, would be that of a country possessing a surface soil of very moderate depth, but of extreme fertility, and supporting a dense population on its vegetable products. The landowners might find it very profitable to them to sell this surface soil to the wealthy horticulturists of other countries; and if the principle of free trade is unlimited, they would be justified in doing so, although they would permanently impoverish the land, and render it capable of supporting a less numerous and less healthy population in long future ages.

Most persons will admit that in both these cases the exercise of the unrestricted right of free trade becomes a wrong to mankind, and should on no account be permitted; and it will perhaps be said that such cases could never occur in a civilized community as public opinion would not allow the landowners to act in the manner indicated even were they disposed to do so. I believe, however, it may be shown that, under circumstances far worse than those here supposed, the landowners in the most civilized community on the globe do act in a very analogous manner, and, moreover, are not yet condemned by public opinion for doing so. Let us first, however, deduce from such supposed cases as those above given a general principle determining what articles of merchandise are and what are not the proper subjects of free trade. A little consideration will convince us that most animal or vegetable products or manufactured articles, the reproduction and increase of which are almost unlimited in comparatively short periods, are those whose free exchange is an unmixed benefit to mankind; the reason being that such exchange enriches both parties without impoverishing either, and, by leading to improved modes of cultivation and an increased power of production, adds continually to the sustaining power of the earth, and benefits future generations as much as it does ourselves. On the other hand, all those articles of consumption which are in any way essential to the comfort and well-being of the community, and which are, either absolutely or practically, limited in quantity and incapable of being reproduced in any period of time commensurate with the length of human life, are in a totally different category. They must be considered to be held by us in trust for the community, and for succeeding generations. They should be jealously guarded from all waste or unnecessary expenditure, and it should be considered (as it will certainly come to be regarded) as a positive crime against posterity to expend them lavishly for the sole purpose of increasing our own wealth, luxury, or commercial importance. Under this head we must class all mineral products which are extensively used in domestic economy, the arts or manufactures, and which are in any way essential to the health or well-being of the community, and more especially those which from their bulk, weight, and extensive use could not be imported from distant regions without a very serious addition to their cost, such as is pre-eminently the case with coal and iron.

Now, it will be seen that we have here to deal with a case quite as extreme in reality as those supposititious cases with which we commenced this inquiry. For coal and iron are almost as much necessaries of life to the large population of this country as are abundance of water and a fertile soil; but there is this difference, that the water might be restored to its legitimate use, and the soil might be renewed by a sufficient period of vegetable growth; whereas coal burned, and iron oxydised, are absolutely lost to mankind, and we have no knowledge of any restorative processes except after the lapse of periods so vast that they cannot enter into our calculations. It may be replied, that the quantity existing on the globe is vast enough for the necessities of mankind for any periods we need calculate on; but even if this be so (of which we are by no means certain), it may none the less be shown that numerous and wide-spread evils result from our present mode of recklessly expending the stores in certain countries, while the same products remain totally unused in many of the countries they are exported to. For a number of years we have been increasing our production of coal and iron at an enormous rate, and sending vast quantities of both to all parts of the world, civilized and uncivilized, and have thereby produced, so far as I can see, little but evil in various forms, some of which have hitherto received little attention.

Briefly to state these:—In the first place, we have seriously, and perhaps permanently, increased the cost of one of the chief necessaries of life in so changeable a climate as ours—fuel. This is in itself so great and positive an evil that no considerations of mere convenience to remote nations, such as the construction of railways in New Zealand or in Honduras, ought to be mentioned as an excuse for it. Coal in winter is a question of comfort or misery, even of life or death, to millions of the people whose happiness it is our first duty to secure; and shall we coolly tell them that the Antipodes must have railroads, and that landowners, coalowners, and contractors must make fortunes, although the necessary consequence is the yearly increasing scarcity of one of their first necessaries and greatest comforts?

In the second place, by destroying for ever a considerable and ever-increasing proportion of the mineral wealth of our country, we have rendered it absolutely less habitable and less enjoyable for our descendants, and we have not done this by any fair and justifiable use for our own necessities or enjoyments, but by the abuse of increasing to the utmost of our power the quantity we send out of the country, never mind for what purpose, so that it adds to the wealth of our landowners, capitalists, and manufacturers.

In the third place, we have brought into existence a large population wholly dependent on this excessive production and export of minerals, and therefore not capable of being permanently maintained on our soil. In proportion as other nations make use of their mineral productions, and as our own minerals, from the increasing difficulty of procuring them, become necessarily more costly, so must our excessive exports diminish, and with it must diminish our power of maintaining our present abnormal population. A period of adversity will then probably set in for us, only faintly foreshadowed in intensity and duration by those arising from mere temporary fluctuations in the demand for minerals and their manufactured products.

Fourthly, we not only injure ourselves and our successors by thus striving to get rid of our mineral treasures as fast as possible, but we probably do more harm than good to the nations to whom we export them; for we prevent them from deriving the various social and intellectual benefits which would undoubtedly arise from their being compelled to utilise for their own purposes the mineral products of their own lands. The working of mines and the establishment of manufactures bring into action a variety of the mental faculties, and so well vary and supplement the labours and the profits of agriculture or trade, that a people who wholly neglect these branches of industry can hardly be said to live a complete and healthy national life. By considering our rich stores of coal and iron as held in trust by us for the use of the present and future populations of these islands, we should probably stimulate and advance a healthy civilization in many countries which the most lavish expenditure of our own minerals, aided by our capital and engineering skill, fail to benefit.

Lastly, I would call attention to the way in which the lavish production of minerals disfigures the country, diminishes vegetable and animal life, and destroys the fertility (for perhaps hundreds of generations) of large tracts of valuable land. It would be interesting to have a survey made of the number of acres of land covered by slag-heaps and cinder-tips at our iron and copper works, and by the waste and refuse mounds at our various mines and slate quarries, together with the land destroyed or seriously injured by smoke and deleterious gases in those "black counties" which it pains the lover of nature to travel through. The extent of once fertile land thus rendered more or less permanently barren would, I believe, astonish and affright us. How strikingly contrasted, both in their motive and results, are those noble works of planting or of irrigation which permanently increase both the beauty and productiveness of a country, and carry down their blessings to succeeding generations.

This brief sketch of some of the more salient features of the subject of mineral export will serve to show how many and various are the evil results which flow from allowing these invaluable treasures to be wasted at the dictates of mad speculation and the eager race for wealth. These considerations have a very practical bearing at the present time. The recent enormous rise in the price of coal has brought up the question of the advisability of an export duty upon it. The press, almost without exception, have opposed this as being "contrary to the principles of free trade;" and it has further been argued that such a duty would have little or no effect, because the real cause of the high price of coal is that so much is used in the excessive manufacture of iron. But it is evident, from the considerations here set forth, that the export both of coal and iron requires to be regulated or forbidden, and for the same reasons; and if the "principles of free trade" are opposed to this, so much the worse for those "principles," since they will be opposed not only to the true economy of human progress, but also to the clearest principles of social and national morality. Many persons will now ask whether those can be true principles which lead to the exhaustion of our coal-fields for the purpose of lighting South American cities with gas or building railways in every insolvent South American Republic, while our own hard-working population has to suffer the pangs of cold in winter, in consequence of the high price of coal which such reckless projects tend to cause. And the fact that all parties concerned—landowners, colliery proprietors, speculators, and legislators—are so far from seeing anything wrong in what they are doing that their one aim at the present moment is to secure a larger annual output, and an increased export, will be to many an additional argument for taking the property in land altogether out of private hands. Waiving that question, however, for the present, I maintain that it is a wrong to our own population, and a still greater wrong to the next generation, to permit the unlimited export of those mineral products which are absolute necessaries of life, but which once destroyed we can never reproduce. To do so is to sell and alienate for ever a portion of our land itself, and should no more be permitted to private individuals than the selling of the land surface to a foreign State.

Whether or not the period of the total exhaustion of our coal-fields can be approximately estimated, it is clear that the present vast and increasing rate of consumption must be stopped. The numerous evils of the present system I have briefly indicated—where are the benefits which counter-balance them? And the benefits, if they exist, must be large and clear and positive indeed to justify us in recklessly scattering over the whole world the mineral products of our land. It is to their possession that we attribute much of our wealth and power and national prosperity, yet we are doing our best to deprive future generations of any of the advantages we have derived from them.

It appears, then, to be clearly our duty to check the further exhaustion of our coal supplies by at once putting export duties on coal and iron in every form, very small duties at first, so as not to produce too sudden a check on the employment of labour, but gradually increasing, till, by stimulating an increased production in other countries, they may no longer be required. If other nations should see the wisdom and justice of following our example, each may in future develop and enjoy its own mineral products, may help to supply what is necessary to the welfare of those countries which do not possess these natural gifts, and may still leave an ample supply to their descendants.

* * * * *

Wallace's outstanding original contribution to the land nationalization movement was his novel approach to the geography of rent. His arguments began with the idea that the value of a parcel of land rested on two variables that could be both conceptually and practically distinguished from one another. These variables were: (1) land value per se, *as determined by a given parcel's physical qualities* and location relative to the pertinent functions of established society, *and (2) value* added *to the parcel by improvements made by those in (temporary) possession of it. At that time, land reform was a major issue in Great Britain; a very large proportion of the land was in the hands of a relatively small number of owners, and it was becoming increasingly apparent that major social and economic problems were being created by these circumstances. Wallace felt that his solution, easily relatable to his feelings as to what constituted 'progressive' social evolution, went to the very root of the problem.*

In a position paper published in The Westminster Review *in 1890, Wallace discussed the subject of rent:*

The instability of peasant-proprietorship—the necessity of rent (ref 422)

Politicians of to-day, no longer able to withstand the ever-growing public opinion in favour of the radical reform of our land system, profess themselves willing to favour in every possible way the creation of peasant-proprietors; and even the present Tory Government has introduced a Bill, which, if carried, must logically be extended so as to transfer the fee-simple of the entire agricultural land of Ireland to existing occupiers. And the principle of this measure is accepted by both parties, the only difference of opinion being as to *how*, and *when*, and by *whom* it ought to be carried into effect. But not a single voice has yet been raised, in Parliament or out, to proclaim the utter futility of such a proceeding on account of the absence of the equalising agency of rent, an absence which must certainly lead to the failure of some of these new landowners and the aggrandisement of others, till, in a comparatively short period, we shall again have a body of wealthy landlords and rack-rented tenants all over the country.

In order more clearly to see how this result must be produced, let us suppose we have arrived at the period, about half a century hence, when all the land of Ireland has become the property of the tenants and nobody pays any rent. We shall then have a compact body of peasant proprietors holding small farms of very different values, some holding land worth but five or ten shillings an acre, while that of others is worth three or four pounds. Now, it is quite clear that the man with good land and no rent to pay can afford to sell his produce lower than the man who has poor land equally rent-free, and wherever there is competition between them he will do so. When seasons are bad or prices low, the latter will be ruined by this competition, will have to borrow money on his land from his richer neighbour, and will inevitably, sooner or later, have to sell his land, which will be added to the richer land adjoining and be worked together with it. It is to avoid this inevitable result that, almost everywhere on the Continent, the land has been divided up into small detached plots so that each holding consists of a similar proportion of all the different qualities of land in the parish or commune—heavy or light soil, pasture meadow or coppice—a farm of ten or fifteen acres often consisting of twenty or thirty separate patches, all completely isolated and unfenced, and often scattered over a square mile of ground. This, of course, is a dreadfully inconvenient and wasteful mode of cultivation, but it serves rudely to equalise the different holdings; and it is this equalisation which has caused it to be upheld so tenaciously by the peasant proprietors of many different countries.

To understand how peasant-proprietorship would work with us, we may suppose that one half of the cotton manufacturers of England used the old-fashioned machines of thirty or forty years ago, whilst the other half used the very newest and most improved machinery. Is it not absolutely certain that the former would soon be undersold by the latter and would become bankrupt, unless all were taxed exactly in proportion to the benefit derived by the various qualities of the machinery employed? But the land itself is to be the cultivator what machinery is to the manufacturer, and it is permanently and necessarily as different in value as would be the machinery of various periods during the last hundred years if brought into competition to-day. In order to equalise this difference in land value there are two methods in use—the wasteful and imperfect continental method of each cultivator having small detached plots of the different qualities of land, and the far more economical and complete method of *Rent*, by which the advantages of various soils and situations are equalised, and every occupier is able to compete on fair terms with all other occupiers.

Rent, then is a necessary factor in successful agriculture by small farmers, the only question being as to who shall receive the rent and what shall be the conditions of the occupation. The present method of private landlords and rack-rents we nationalisers hold to be the very worst method possible. That of permanent and secure occupation under the State, with the payment of an economic rent, revisable at long intervals and only on changes of value produced by general causes—that is, by the growth or advancement of the whole community—with perfect freedom of action by the cultivator who will be the owner of all improvements of whatsoever kind, to be the very best.

It may indeed be urged that, if by ownership of the land food can be produced and sold cheaper than by tenancy, it must be better for the whole community who are the consumers of food. But this cheapness would be only temporary, because so soon as the land became again the property of the few, owing to the failure of the owners of the poorer lands, it would be let out in farms as now, rent would be paid to equalise the various values of the land, and we should return again to the existent system of landlord and tenant. Under State ownership, however, the rent paid would ultimately take the place of all other taxes, and thus the whole community would benefit far more than by a temporary cheapness of food accompanied by the ruin of a considerable portion of the poorer cultivators.

Rent, therefore, is essential to the stability of any system of the occupation of land. Rent paid to the community, through State or municipal authorities, is the only system which is beneficial to the whole community.

* * * * *

Further perspective on the theory of land nationalization can be gained by considering two short selections (from refs 438 and 584, respectively) that show how Wallace dealt with the key economic concepts 'capital' and 'wealth':

. . . It will be necessary to make a few preliminary observations on the nature of capital, what it is and what it is not. And first, money is not capital. A man may have a houseful of money, but it is not capital till it is converted into tools or raw materials, and even then it will not be productive capital unless it can command both labor and intelligence to utilise it. We often hear the threat that if workmen strike for higher wages capital will go out of the country. That is an idle threat. True capital—railroads, factories, mills, etc., will not pay to carry away; and as for rich people taking money out of the country, that would hurt no one even now, and if we had possession of the land it would be an absolute benefit to us.

Again, few persons realise how perishable and fugitive is capital. It wants constant looking after and constant renewal. Every few years new and improved machinery replaces the old, and even buildings have to be constantly repaired, and at longer or shorter intervals to be altogether re-built. We see how comparatively unimportant is the value of capital existing at any one time, when we observe the rapid renewal of that which has been destroyed by a great fire, like that of Chicago, or by a foreign invasion like that of France by Germany. In a very few years everything is replaced and everybody seems as well off as before. We also see how rapidly capital is created by the fact that so many wealthy men who now own mills and houses by the score began life with nothing. Capital, like all wealth, is created and grows solely by means of labor and intelligence applied to land or to the products of the land. It is one of the misleading errors of political economists that the three factors of wealth are land, labor and capital. The true factors, as long ago pointed out by our vice-president, Mr. Volckman, are land, labor, and intelligence. These existed before capital was created; and had capital been a *necessary* factor in the production of capital, then it could never have come into existence at all. It follows that, when the workers have free access to the *land*, and as the *labor* is certainly all their own, they only need the *intelligence* to produce in a very short time all the capital that is needed. It would be an insult to the working men of

England to suppose that they have not the necessary intelligence;—it would moreover be contrary to all experience and all history, for whence has usually come the intelligence that has created the wealth of England, if not mainly from the ranks of the workers? Hargreaves and Arkwright, Watt and Stephenson, were not landlords or capitalists, though these last have derived much of their wealth from their inventions.

Under present conditions the value of both capital and land to their respective owners, depends entirely on the amount of labor they can command. Both land and machinery are worthless, if left unused, while the former sometimes, and the latter always rapidly deteriorates in value. The landlord and the capitalist are therefore absolutely dependent upon the laborer, and if the laborer can be put in such a position as to be independent of them, he really becomes their master, instead of being, as now, their slave, and they will have to come to him and beg him to enable them to make something, however little, out of their property . . . [from a Presidential Address given to the Land Nationalisation Society in 1891: ref 438, pp. 19–20]

* * * * *

Real and fictitious wealth

Nothing is more certain than that wealth—real wealth—is continually used up and destroyed, and as continually reproduced by fresh labour. All articles of food, of clothing, of furniture, and even tools, machines, dwellings, books, works of art, and ornaments, are either wholly or partly used up day by day or year by year, and as continually reproduced; and it is those which are most continually consumed and reproduced which are, pre-eminently, beneficial wealth. If, then, a man acquires a large surplus of this real wealth beyond what he can consume himself, it must be either profitably consumed by others or be wasted by natural decay. In either case it soon ceases to exist. But our fiscal and legislative arrangements enable a man to change this *perishable* wealth into securities which bring him a *permanent* income—an income supposed to be perpetual, but at all events lasting long after the wealth of which it is the symbol, and which is supposed to produce it, has totally disappeared. Incomes thus derived constitute a tax or tribute on the community, for which it receives nothing in return, and may truly be termed fictitious wealth.

To show that this is so, and at the same time to exhibit clearly not only the evil but the inherent absurdity of these arrangements, let us consider a few indisputable facts. Every year much surplus wealth is accumulated by individuals and is invested as reproductive capital. This invested capital goes on producing an income which, it is supposed, is and ought to be permanent; and as more and more surplus wealth is continually produced and invested, it is evident that the permanent incomes thus derived will continually increase, and thus in each successive generation a larger and larger number of persons will be enabled to live in idleness on incomes derived from this invested capital. But this is a state of things that evidently carries with it its own destruction, since a time must come when the number of idle persons living on "independent incomes," and the aggregate of those incomes, will be so great, that the working portion of the population will be ground down to penury in the attempt to pay them, and this will more quickly come about from the tendency of a large idle and luxurious class to withdraw labour from beneficial work to the production of useless luxuries. The end will inevitably be the worst kind of revolution, brought about by the determination of the labouring poor no longer to support the burthen of an ever-increasing class of idle rich . . . [from 'Interest-Bearing Funds Injurious and Unjust': ref 584, pp. 256–257]

* * * * *

Another cornerstone of Wallace's arguments vis-a-vis *land reform had been laid in an essay published in* Contemporary Review *in 1873:*

Limitation of state functions in the administration of justice (ref 236)

. . . Our system of law is, in great part, the product of times when the security of property was held to be of more importance than protection to the person. The legislators being almost always the great landowners, a large part of the law was adapted to secure them the power of dealing with the land (the most important of all property) in any imaginable way; and in their bungling attempts to do this, they have produced a system of law of real estate of almost unimaginable intricacy. To interpret and carry out this and other branches of the law of property, occupies a large and influential portion of the legal profession. Lawyers exist upon the complexity of the law. It is not to their interest that we should be able to obtain cheap and speedy justice; nor is it their interest to reduce the number of suitors at the courts. We cannot reasonably expect them to do either of these things, which are yet of vital importance to us who are not lawyers. They may, indeed, so modify, and to some extent simplify, procedure as to take away a portion of the terrors of "going to law" in the estimation of aggrieved parties, and so induce a larger number than before to seek their aid against oppression and wrong;—but they will never make any radical reform, or attempt to do what every intelligent suitor knows might be done. Our interests are directly opposed to theirs, and it is mere madness to expect any thorough simplification of the law from lawyers. Such a reform requires the common sense of minds untrammeled by legal technicalities or legal interests. The people must be shown that such a reform is possible—nay, easy—and they will then demand that this matter shall be taken altogether out of the hands of lawyers. It is in the hope of showing how one great branch of this much-needed reform may be made, that the present writer ventures to attack a problem generally considered far beyond the reach of laymen.

A first step, and a very important one, towards rendering cheap and speedy justice possible for every man is, so to simplify the law of property as to free the courts from a large proportion (perhaps one-half, perhaps much more than one-half) of the cases which now occupy them. This would not only render it far easier to dispose promptly of the much simpler cases—which, however, are those which are often of more real importance to the parties affected—but it would allow of the whole method of procedure being altered to suit those simpler cases which would then form the bulk of the business of the courts. Now, this great diminution of cases can be effected without denying redress for any grievance, or a remedy for any wrong, by simply putting out of court a host of matters which ought never to have been taken cognizance of by the law. Here, as in so many other instances, it will be found that reform must begin by a "limitation of State functions;" and that it is because Governments have undertaken to do much that is unnecessary and even injurious, that they are not able to fulfil one of their first and plainest duties—that of giving free, speedy, and substantial justice to the weakest and most indigent, as well as to the most powerful and most wealthy, of their subjects.

The first, and perhaps the largest, group of cases which ought to be taken out of the cognizance of our courts of law, are those which may be comprised under the general term of "trusts." At present any one may place property in the hands of another, either during his own life or to take effect after his death, for certain specified purposes, and if these purposes are neither illegal nor positively immoral, the law will compel the trustee to carry out these purposes to the very letter. They may be trivial, or absurd, or even injurious, but the man who once gets a trustee to accept a trust (and even this is not necessary when it is created by a will) becomes thereby an absolute potentate, who has at his command the whole power of a great State employed to see that his most minute directions are carried out. The number of cases of this kind is enormous, including all those which involve the interpretation and carrying into effect of the provisions of trust-deeds, settlements, and wills; so that a considerable portion of our machinery for administering justice is devoted to ascertaining and giving effect to the whims of individuals for years, and often for scores of years, after they are dead. Under the same general head may be included the power of determining by deed or will the contingent succession to property, and of creating any number and kinds of disqualifications with regard to it. The supposed necessity for providing for every imaginable exercise of this power, has led to such endless complications in the law relating to the transfer of land in all its forms and modes, that years of study are required to comprehend them. They furnish the materials for perhaps the majority of the cases that come before our civil courts, and give occupation to a very large section of the legal profession.

But in the whole group of cases here referred to there is no question of administering justice. For a Government not to carry out a man's wishes after his death is not a wrong, but quite the reverse, since it may with much reason be maintained that, for any Government to occupy itself with carrying out the whims of every man (whether he be sage or fool) who may wish to make his relations or successors subject to his orders in the application of property no longer his, is a positive wrong to the community, inasmuch as it is incompatible with the performance of duties of a paramount nature. What the law should do, and all that it should do, is, to recognize and enforce gifts or transfers of property of all kinds, to living individuals, absolutely. It should utterly refuse to recognize any desires, whims, or fancies of individuals as to the applications of the property, or any limitation to the future owner's absolute possession of it. It should not even recognize any alternative applications of the property in the case of the death of the legatee before that of the testator, who could in that case have altered his will, and if he has not done so the legacy should pass to the legal representatives of the legatee. Property should always be considered by the law to be in the possession of some person absolutely, who can transfer it to another person absolutely, but cannot enforce any stipulations whatever as to the use of it on the next owner. Life interests in landed and other property, with all their attendant evils, would thus never exist.

The wishes of the donor or testator of property, although not a proper subject for the interference of the law, could be in many cases carried out by means of what may be termed a voluntary and amicable trust. The trustee (who would be really the legatee) would be chosen on account of friendship, integrity, and sympathy with the objects and desires of the testator, and he would give just so much effect to those desires as his reason and his conscience impelled him to give. The law would consider him only as the owner of the property, and would in no way interfere with the manner in which he thought proper to interpret the wishes of his friend. To provide for children and minors, property might be either left absolutely to their nearest relative or friend to stand to them *in loco parentis*, or it might be left to themselves, in which case an officer of the court would be their official trustee, and would prevent any misappropriation of their property by relations or guardians till they came of age. We should in this way greatly simplify wills, and almost abolish will-cases, while the courts would be relieved from that great mass of causes of the most tedious kind, in which trust-deeds, settlements, legal estates, shifting uses, entails, and trustees bear a prominent part.

It has been so long and so universally the practice in civilized countries for the law to recognize and enforce the wishes of individuals as to applications of their property other than the simple transfer of it to individuals, that to many, perhaps to most persons, it will at first seem to be a positive injustice to take away from them the power to do so. Yet the law itself recognizes that the practice is beset with evils, and from a very early period legislative restrictions have been applied to it. Hence the laws of mortmain, and the long series of amendments, relaxations, or restrictions of those laws; as well as the limitation of the power of entailing estates for any longer period than a life in being and twenty-one years afterwards. These restrictions prove that the unlimited power of disposition of property has been held to be a law-given custom, not an inherent right; for if the latter, every restriction of its exercise must be a wrong to the parties restricted, which it has never been held to be. The whole question is, however, so very important, and has so many and such wide applications, that it deserves a somewhat fuller discussion.

The establishment of the Endowed Schools Commission has struck the first real blow at the system of a perpetual and blind submission to the wills of dead men; but the new principle, even in its limited application to endowed schools and charities, often excites much opposition. Many liberal and intelligent men still look upon the "intentions" of those who in past ages endowed churches, schools, hospitals, almshouses, and other institutions, as something sacred, which it is almost impious to ignore, and which it is our plainest duty to carry out with only such slight modifications as the changed conditions of society absolutely necessitate. But it is here contended that this notion is not founded on any true conception, either of what is just or what is politic, but that it is, on the contrary, altogether erroneous in principle and mischievous in practice; whence it follows that the sooner it can be got rid of the better for society.

Let us, then, seriously ask, what sufficient reason can be adduced why the State should interfere to carry into effect the desires, whims, or superstitious fancies of any man, for generations or perhaps for centuries after his death? Why should the more enlightened future be bound by the behests of the less enlightened past? Why should we allow, and even encourage, men to hold and administer property after they are dead? For it really comes to that. A man may, justly and usefully, be allowed full liberty (within the bounds of law and order) to use his property as he pleases *during his life*; but why should we go out of our way and make complex arrangements enabling him to continue to do the same after he is dead? During a man's lifetime he can *give* property to whom he thinks fit, or he can apply it to any purpose that he has at heart, without the State's interference; but he absolutely requires the State's assistance in order that his property may continue to be applied precisely in accordance with his ideas of what is best, after his death. The question is, *why* the State should take any cognizance of the matter? It is here contended that this is one of those things quite beyond the proper functions of a Government, and that it has produced, as such excess of authority always does produce, a vast amount of evil. When a man dies he generally has what may be termed natural heirs, that is, children or relatives dependent upon him for a provision in life. For these he is first morally bound to provide, and any surplus beyond their needs, and beyond what the law may give to the State, he may rightly claim the power of bequeathing to any living individuals; and the State on its part is bound to exercise that minimum of interference necessary to secure the property to the respective persons indicated by him. But on what grounds can the testator claim the interference of the State for the purpose of compelling the recipients of the property to do with it what *he* pleases? claim—that is—that *he* shall still be considered to be the real owner of the property *after he is dead*? The thing is so intrinsically absurd, and perhaps even immoral, that nothing but long and universal custom could blind us to the absurdity of it.

What a man may do, and ought to be enabled to do, either during his life or at his death, is to *give* property, and *recommend* (not command) what use he wishes to be made of it. If his morals and his intellect are both good and his judgment sound, his chosen legatees will, at their discretion, carry out his wishes. But to compel them to do so absolutely is monstrous. It implies that the *right* to property continues after death, and that when a man can no longer use it himself he *ought* to be enabled to restrict the freedom of others in the use of it. It implies also that a man with much property to leave is necessarily wise, so wise as to know what will be best for people years after his death. A living agent can modify or supplement his plans as occasions arise or as circumstances require, and he generally does see reason to modify them after a few years' experience. Even acts of parliament, the concentrated essence of the nation's wisdom and foresight one year, often require alteration in the next. But that every man who chooses to do so should be encouraged to make his little "act" before he dies, minutely directing what shall be done with his property for years after his death, and that this "act" should be held to be a fixed law, against which there can be no appeal, all changes of circumstances notwithstanding, and should be enforced by the whole power and authority of the State, is a circumstance which will one day be looked back upon as an amazing anachronism, since it would seem only fitted to exist in a country where the established religion was the worship of ancestors.

We English are wisely jealous of too much government interference in the details of our social life; yet our rulers are living men, imbued with all the ideas and habits and feelings and passions of the age, and are often men of high intellectual attainments, and far in advance of the average of the community. Such a government interferes, at all events, with full command of the most recent knowledge, and with open eyes; yet we will not submit to such interference. But, strange to say, we do submit, and almost pride ourselves in submitting, to have various important social matters determined for us by self-chosen dead men, who are therefore necessarily *behind the age*, and who were sometimes too ignorant, conceited, or superstitious to be up to the intellectual level even of the age in which they lived. It is by such blind guides that we to this day submit to be, in great part, governed in the all-important matters of religion, education, and the administration of charity; and in submission to the immutable laws of these dead rulers we have allowed vast wealth to be misemployed or wasted in the hands of irresponsible and antiquated corporations, which, well bestowed, might have enlightened our people or beautified our land. Who can doubt that the nation would have greatly benefited had our churches and colleges, our schools and charities, our guilds and companies, been free to develop, from age to age, in accordance with the wants and feelings of the living, untrammelled by any slavish adherence to the expressed or implied wishes of the dead?

From the considerations now adduced, it will be evident that the cessation of State interference in the way here objected to, would produce other beneficial results besides that of facilitating the administration of justice. These may be briefly summarized as follows:—

It would take away one of the existing inducements to a life-long devotion to the pursuit of wealth, for if a man could neither make use of it himself nor enjoy the sense of power felt in directing absolutely how it should be employed by others, he would pause in his career of accumulation, and perhaps endeavour to do something useful with it during his own lifetime rather than run the risk of having it all go entirely beyond his control.

It would have the effect of inducing many who now leave their wealth for charitable and philanthropic purposes at their death, to found such institutions as they wished to have established, during their own lifetime, in order to see the working of them, and so adapt them to the fulfilment of an admitted good end as to ensure that they would be preserved by future generations. This active charity or philanthropy would have a most beneficial effect on character, and would undoubtedly lead to more good results than the mere passive bequeathing of money to be employed in some fixed, but often ill-considered and comparatively inefficient manner.

It would prevent the establishment of institutions not adapted to the requirements of the age, and would thus abolish a great bar to mental and moral progress. For the notion of "sacredness" attached to the wishes and commands of the founders of religious, educational, and charitable institutions has done a vast amount of evil, in confusing our notions of what is right and what is useful, and in keeping up the obsolete ideas and practices of a bygone age, long after they have become out of harmony with a more advanced state of society.

There is no fear, as some may imagine, that under the modification of the law here suggested, such institutions would want stability and would be subject to constant fundamental changes in accordance with the ideas of each successive body of governors, for the conservative tendencies of mankind in general, and especially of all governing bodies, are very strong, and customs or practices, even when pernicious or absurd, seldom get changed till long after their hurtfulness or foolishness are universally acknowledged. In proof of this we may adduce the case of our own representative government, which attaches no idea of sacredness to old laws, and is subject to the powerful influence of public opinion; yet we do not find any dangerous instability in our legislation, but rather a slow, many think far too slow, march onward in a tolerably well-defined course of reform.

The change here advocated would also be beneficial, by helping to rid us of the notion that a man can infallibly prescribe what is good for his successors, or that even if he could, he ought to be allowed so to prescribe; for the next generation will be quite as well able to attend to its own affairs as the last was, and will certainly not be benefited by being debarred from the freest action. Once this notion is abolished, our truest philanthropists would be more willing than heretofore to devote their wealth to public purposes, because they would feel confident of its being permanently useful. They would know that each succeeding generation would watch its application critically, and insist that no obsolete customs or erroneous teachings should be perpetuated by means of it,—that it should never become a drag on the wheels of progress, as has been the case with many such institutions, but rather resemble a powerful engine capable of helping on the necessarily slow march of society towards a higher civilization.

If the main principle here advocated—namely, that it is intrinsically absurd and morally wrong that a dead man's will or intention should have power to determine the mode of application of property no longer his—be a sound one, it will have a most important bearing on a question that is now much discussed, as to how far endowments of the National Church by private individuals may be properly claimed by the State. Even writers of very liberal views see in this a stumbling-block to the complete disendowment of the Church of England, because they cannot get rid of the notion that it is something like a robbery to take property given for one purpose and apply it to any other purpose. It is, therefore, a maxim with them, that when any change in the application of such a fund is demanded by public policy, it should still be kept as near as possible to the intentions of the original donor. It is, however, to be remarked, that when the property in question has already been forcibly applied to other uses than those originally intended, the most scrupulous do not propose that it should be brought back to its ancient use; and this seems to imply a doubt of the soundness of their principle. A large part of the existing endowments of the Church of England, for example, were certainly intended to maintain the teaching and services of the Roman Catholic religion. If the donor's intentions are "sacred," these should be given back to the Roman Catholic Church. If it be said that the intention was to maintain the religion of the country, whatever that might be, then the revenues should be fairly divided among all existing sects for the time being,—but that is "concurrent endowment," and is almost universally repudiated. The only consistent, and it is maintained the only true, view, is, that dead men should have no influence (beyond their personal influence on their friends) other than what is due to the intrinsic value of their opinions; and that property cannot be left in trust to carry out dead men's wishes, on the common-sense ground, that the living know better what is good for themselves than the dead can do, and that the latter have no just or reasonable claim to coerce a society to which they no longer belong . . .

The dead are not truly honoured by sacrificing the interests of the living to their old-world schemes; and if, as we may reasonably suppose, the future state is one of progress, at least as rapid as that which obtains on earth, it may be that they are afflicted with unavailing regrets at our blindness, in insisting on being guided by the feeble and uncertain light which they once had the presumption to imagine would for ever be sufficient to illuminate the world.

* * * * *

After the Land Nationalisation Society was established Wallace's study of the land reform problem intensified, culminating in his most important treatment of the subject, the monograph Land nationalisation *(ref 722), in 1882. In 1883 he followed this up with the essay 'The "why" and the "how" of land nationalisation,' published in* Fortnightly Review*:*

The "why" and the "how" of land nationalisation (ref 365)

. . . In order to place . . . [the question of land nationalisation] . . . fairly before my readers within the limits here assigned to me, it will be necessary to omit the consideration of some of its aspects altogether, and to treat others very briefly. The fundamental question undoubtedly is, the right or the wrong, the justice or the injustice, of private property in land. And then follows the question of results; right and justice lead to good results—to happiness and general well-being; wrong and injustice as surely lead to bad results, and their fruits are moral evil and physical suffering. We have to inquire, then, what are the actual results of modern landlordism? and thus confirm or modify the conclusions we have reached from general principles. Finally, we have to consider how we can best carry into effect right and just principles so as most certainly to reap the reward of moral and physical well-being. This really exhausts the subject. The historical inquiry—how private property in land arose, what changes it has undergone, the results of legislation by landlords and usurpation by kings, the story of royal grants, confiscations, and inclosures, are all exceedingly interesting, and will be found to support and strengthen at every point the argument from principle and from results; but it is not essential to a comprehension of the main question, and we shall therefore omit it here . . .

First, then, we have to inquire whether private property in land is just and right; and here we find ourselves at once in conflict with the great body of Liberals and land-law reformers, who advocate, as their sole panacea, free trade in land. For the foundation of their doctrine is, that land should be treated as merchandise; that it is *right* for individuals to own it absolutely and in any quantity; that it is *good* for great capitalists to add farm to farm, and to build up great estates; that land should be bought and sold as easily as iron or railway shares. We nationalisers, on the other hand, say that all this is fundamentally wrong. We maintain that land should not be treated as merchandise, for the following reasons:—

1. Because it is absolutely essential to the exertion of all industry, while it is the first necessity of human existence; therefore those who own it will possess absolute power over the happiness, the freedom of action, and the very lives of the rest of the community.

2. Because it is limited in quantity, and tends therefore to become the monopoly of the rich—a monopoly which will surely be intensified by free trade, which will render it easier than now to accumulate large estates, and will thus make the landless people still more the virtual slaves of the landlords than they are now.

As Professor [Henry] Fawcett and Mr. Shaw-Lefevre have both shown, the land hunger of the rich is insatiable; and, as is well put by the *Edinburgh Review*—"It stands to reason that, if the sale and purchase of land were perfectly easy and free, those persons would buy most land and give the best price for it who had the most money to buy it with."

To determine whether private property in land is right and just, and compatible with the well-being of the whole community, it will be well to glance briefly at the true foundations of property, and the admitted rights of a free man. Property is, primarily, that which is obtained or produced by the exertion of labour or the exercise of skill. In this a man has a right of property, to use, to give away, or to exchange. This is a universally admitted right which forms part of the very foundations of society, and many eminent writers maintain that it is the only way in which private property can justly arise.

Property is, however, usually admitted in any natural product *found* by an individual and obtained without labour; but this kind of property has never the absolute character of the former kind, since if the thing found is not abundant and is essential to life or well-being, the individual right to its exclusive possession is not admitted. The single good spring of water on an island, a single group of fruit or other useful trees, a single pond or stream containing abundance of fish, are not allowed to be appropriated by the first discoverer to the exclusion of his fellow men.

Property in the results of a man's labour has no such limitations; it is usually hurtful to no one, and with free access to natural agencies and products, and freedom of exchange between man and man, is beneficial to all. There is no other natural and universal source of private property but this—that every man has a right to the produce of his own labour; and hence, as land is not produced by man it cannot equitably become private property.

Let us next look at the question from the point of view of the rights of individuals, as members of a society which upholds freedom as a fundamental principle of its existence. In such a society it will surely be admitted that every man has an equal right to live. Not, be it observed, a right to be kept alive by others; not a right to claim any part of the produce of others' labour, but, simply, freedom to support himself by labour, freedom from all obstructions by his fellow-men of his own freedom to labour. Not to have *this* freedom of action is to be a slave; and to this extent at least it will be admitted that all men are, or should be, equal.

But man cannot live without access to the natural products which are essential to life—to air, to water, to food, to clothing, to fire. If the means of getting these are monopolised by some, then the rest are denied their most elementary right—the right to support themselves by their own labour. But neither pure air, nor water, neither food, clothing nor fire, can be obtained without land. A free use of land is, therefore, the absolute first condition of freedom to live; and it follows, that the monopoly of land by some must be wrong, because it necessarily implies the right of some to prevent others from obtaining the necessaries of life.

Another consideration which shows the private ownership of land to be unjust is the fact (admitted by all economists), that the whole commercial value of land is the creation of society, increasing just as population and civilisation increase. If one man had a grant of an uninhabited island or country, the size of Britain, the value per acre of all the land which he did not use himself would be *nil*. Rather than live alone he would give land to any one who would settle near him. And when others came he would sell them land, as it is sold in all new countries, for a mere trifle, while he could never enforce his rights over those who took possession of remote parts of his territory. But, just as the population increased the land would rise in value; till, when towns and cities had sprung up, and all the arts of civilised life were practised, and communications were established with every part of the world, a single acre might sell for 1,000 pounds or 10,000 pounds. Who created this value? Not the original settler, but society. And this shows the absurdity of comparing, as some do, the occasional increase in the value of other property with that of land. In the case of everything which is the product of human labour, the tendency is for it to become cheaper as population increases and civilisation advances. When the reverse occurs it is usually owing to exceptional conditions, or to the influence of some kind of monopoly. But with land the increase of value is universally coincident with and due to the growth of society, and the only fluctuations in this constant rise are owing either to monopoly and speculation forcing the price at a certain epoch above its natural value, or to restrictions on its free use by the people. Here again we bring out a broad distinction between the products of a man's labour which are and should be private property, and land, the gift of nature to man and the first condition of his existence, which should ever remain the possession of society at large.

One other consideration remains, and perhaps the most important of all as affording a demonstration of the necessarily evil results of unrestricted private property in land. If a portion of the community is allowed to appropriate the whole of the land for its private use and benefit, this appropriation necessarily carries with it the right and the power to appropriate the bulk of the products of the labour of the rest of the community, while it keeps down wages to a minimum rate just sufficient to maintain physical existence. Carlyle recognised this truth when he speaks of the poor widow boiling nettles for her only food, and the perfumed lord in Paris extracting from her every third nettle as rent. The late Professor Cairnes in many of his writings dwelt upon this consequence of landlordism. In one of his essays in the *Fortnightly Review*, he says:—

> The soil is, over the greater portion of the inhabited globe, cultivated by very humble men, with very little disposable wealth, and whose career is practically marked out for them by irresistible circumstances, as tillers of the ground. In a contest between vast bodies of people so circumstanced, and the owners of the soil—between the purchases without reserve, constantly increasing in numbers, of an indispensable commodity—the negotiation could have but one issue, that of transferring to the owners of the soil the whole produce, *minus* what was sufficient to maintain in the lowest state of existence the race of cultivators.

But this result has been most clearly and forcibly demonstrated by Mr. Henry George, who makes it the very key-note of his book [i.e., *Progress and Poverty*], and illustrates it by a wealth of illustration and a force of argument which must be carefully studied to be appreciated. I can here only find space for an abstract of one of his illustrations.

Let us suppose an island, with no external communications, and but moderately peopled, in which the land was equally divided among all the inhabitants; and let us suppose that there was free trade in land as in everything else, just as desired by our most advanced politicians. After fifty or a hundred years let us look again at this island, and we shall certainly find the land most unequally divided; some will be very rich and have large landed estates, many will be very poor and have sold or otherwise parted with all their land. We may suppose there to be no wars, a pure government, few taxes, no state church, no hereditary nobility; yet inequality in ownership of land will have caused pauperism and virtual slavery. For, all must live on the land, and from the products of the land; therefore those who do not own land can only have the use of it or obtain its products on the terms of those who do own it. They are really slaves; for, in order to live, they must accept the landlord's terms and do as he bids them.

If landlords are rather numerous it will not *seem* like slavery, because the *forms* of free contract will be observed. But there can really be no free contract, because the landowners can wait, the landless cannot. They must work on the landowner's terms or starve. And thus, just in proportion as population increases, and the competition for land and its products, especially for bare food, becomes keener, the landowners will obtain a larger and larger share of the products of the soil—in other words, rents will rise, and wages will fall or remain stationary. Now let us introduce a fresh element—labour-saving machinery. This will enable more wealth to be produced with the same labour or with less; but it will not decrease the dependence of labour upon land. All the increased production of wealth will go to the wealthy—the landlords, the landless remaining as poor as before. As the climax of this argument, Mr. George supposes the case of labour-saving machinery to be brought to absolute perfection, so that all wealth will be produced by various forms of automata, without human labour. Then all wealth will belong to the landowners, for even standing-room for houses and machinery cannot be obtained except on their terms, and the landless multitude must necessarily starve in the midst of plenty, or live as servile dependents on the landlord's bounty. Thus, private property in land—even were all other social and political evils removed—necessarily makes the many poor that the few may be rich; for it prevents free access to those natural elements without which man cannot live, and thus directly causes *poverty* and *pauperism*, and the long train of miseries and crimes that spring therefrom.

By this preliminary inquiry we have shown—(1) That private property in land can never justly arise, because land is not a product of human labour. (2) That the monopoly of the land by a class is inconsistent with the fundamental rights of individuals in a professedly free country. (3) That the whole commercial value of land is the creation of society, not of landlords and tenants, and should therefore belong to the community. (4) That private property in land necessarily leads to the poverty and subjection of the many for the benefit of the few.

I therefore claim to have completely answered the fundamental question with which I started, and to have demonstrated that, as a matter of principle, our present land-system is absolutely wrong, cruelly and perniciously unjust . . . We now come to the important practical question, What is the outcome of modern landlordism? We have seen, that, from various distinct points of view, it is wrong in principle . . . [works that can be] . . . referred to on the history of the subject show that it had its origin in force, and has since been largely maintained by confiscation and by unjust legislation; but we are so practical a people, that, if it can be shown that its results are good we should care little about principles or about history, but would be quite content to maintain a system which works tolerably well. And people actually do say that it works well. Press and Parliament are never tired of exclaiming—"See how rich we are! What a trade we do with all the world! Our system, which produces such results, *must* be all right!" But along with our great riches we have a mass of terrible poverty, and it is the opinion of disinterested writers that we are the most pauperised country on the globe . . .

Our public men continually assure us that pauperism is diminishing; or that at the worst it is stationary, while our population is increasing rapidly, and that it is therefore proportionally diminishing; and they base their statements on the official statistics of pauperism. I shall show, however, that these are not trustworthy guides, and that there is good reason to believe that, during the very periods in which our aggregate wealth has increased most rapidly, pauperism has increased also in positive amount, and perhaps even in greater proportion than the increase of population.

If we take the official statistics of pauperism in England and Wales for the last thirty years we find great fluctuations, but nothing like a regular diminution. Between 1849 and 1880 the numbers were lowest in the years 1853 and 1876–78, while they were highest from 1862 to 1873. The only years in which the numbers rose above a million were 1863–64 and 1868–71, and this was the very period when our commerce was increasing so rapidly as to excite the enthusiasm of our legislators, and when our prosperity was supposed to be greatest. The extremely irregular fluctuations of official pauperism render it possible almost always to choose some year, twenty, thirty, or forty years back, when it was higher than now, and thus show an apparent decrease of numbers; but if we take the whole period from 1849 to 1880 as one during which our commerce and wealth increased enormously, and all the industrial arts and means of communication made the most rapid strides, and take the average pauperism of the first twelve and last twelve of these years we find them almost exactly the same . . . Between the middle points of these two periods (1854 to 1874) the population increased about 23 per cent., and thus the proportional pauperism *appears* to have decreased considerably, though not at all in proportion to the increase of our aggregate wealth, which was at least doubled during the same period.

Several causes have, however, been in operation during this period which have led to the numbers of officially recorded paupers forming a less and less adequate indication of the total mass of pauperism in the country, so that even the small comfort derived from its supposed decrease in proportion to population may be denied us. In the first place there can be no doubt that the extent and efficiency of private charity all over the country have been steadily increasing, and that by its generous aid large numbers have been saved from becoming paupers. Not only have old charities been better administered, but many societies have been formed for the systematisation of private charity; while all over the country the clergy, and an ever increasing army of lady visitors, have aided the poor with advice and timely relief. It is impossible to estimate the amount of these various agencies, but it seems not impossible that they may have relieved the ratepayers from an amount equal to that due to increase of population during the same period; and this is the more probable when we consider the enormous increase of the wealthy middle class, and the increasing feeling that the poor have some moral claim upon the rich, leading to more and more liberality in every case of undeserved misfortune.

But yet more powerful agencies have been at work tending to decrease the numbers of official paupers without any corresponding decrease of poverty and pauperism. For many years there has been a growing disposition to diminish out-door relief, and apply more generally the "workhouse test" which is the fundamental principle of our poor law. It is well known that there is such a wide-spread dislike and dread of the workhouse among the more respectable poor that many will rather starve than enter it, and as a matter of fact many *do* starve who might have been well fed within its walls. Now in the *Daily News* of April 18th of this year, there was a remarkable article giving an account of the results of this change of system in some London parishes. It states that ten years ago a severe reduction of out-door relief was commenced in Whitechapel and other parishes, till at the present time there is no out-door relief given in that parish, nor in Stepney and St. George's in the East, while the same process is going on in Marylebone and other parts of London, and to a less extent all over the kingdom. The article in question states the remarkable fact that this great change has produced practically *no increase of indoor paupers*. In Stepney, for example, out-door relief has been reduced by 7,000 in the last ten years, and there is no increase whatever of indoor paupers, the reason being (as expressly stated) that *organised private charity has taken the place of out-door relief*. Now if there has been a reduction of 7,000 official paupers in one London parish without any proof of a corresponding decrease of real want and destitution, how utterly unmeaning and even misleading becomes the quotation of these official statistics as showing any real decrease of our pauperism. If we take as our guide the fact, that, in one of the worst and most poverty-stricken districts of the metropolis, organised private charity has been able to take the place of the relieving officer when a *total* cessation of out-door relief has been effected, we may be sure that it has been found quite equal to the much easier task of relieving those thrown on its hands by a very partial application of the same methods of dealing with the poor in other parts of the country. We may, therefore, fairly assume that the diminution of out-door paupers over the whole country during the last thirty years has been largely due to the stricter application of the workhouse test, and that those thus refused relief by the guardians have been aided and kept alive by more extensive and better organised private charity. If this is the case, the only official test of pauperism as actually increasing or decreasing will be found in the records of indoor relief, and these show numbers steadily increasing at a much greater rate than population! Thus in the thirty years from 1852 to 1882 the number of indoor paupers in England has continuously increased, from 106,413 in the former year to 188,433 in the latter, an increase of 83 per cent., while in the same period population only increased 45 per cent. The plain inference is, that confirmed pauperism—that which includes all the most degraded and the most hopeless of our poor—has been steadily increasing at a greater rate than our population, during a period in which our aggregate wealth has been doubled, and our commerce, of which we are so proud, has increased three-fold!

Before quitting this subject, it is well to point out that the way in which the number of paupers is estimated is most misleading, and gives no adequate idea of the real numbers. The tables show only the numbers relieved on the 1st January in each year, but it is estimated that the actual number of persons receiving relief during the year is nearly two and a half times this number, or about an average of two millions for England and Wales, or two and a half millions for the United Kingdom (for details of this estimate see the present writer's *Land Nationalisation*, p. 3). If we add to this latter number those who receive relief in the casual wards (which are not included in the official tables), and the very large numbers who depend wholly or partially on private charity for support, we shall perhaps bring the figures up to three and a half millions. But beyond this number of actual paupers loom a vast host of the poor who ever live on the verge of pauperism, and from whom the ranks of the actual paupers are constantly recruited, including whole populations, like the cottiers of the west of Ireland, and the crofters of the Highlands and Islands of Scotland, living in such a condition of perennial want that it only requires that most certain of periodical events—a bad season—to produce actual famine. If we add only one million for all these, we bring up the number of actual or potential paupers in this civilised, Christian, and pre-eminently wealthy country to about four millions and a half, or about one in seven of the whole population!

This dreadful failure to distribute with any approach to fairness among our workers the enormous wealth which they alone produce, is rendered more disgraceful when we take account of the vast extension of labour-saving machinery during the epoch we are considering.

It is calculated that we now possess steam-engines of about ten million horse power, equal to a hundred million men always working for us. Reckoning six million families in the United Kingdom, we may say that every family has the equivalent of sixteen hard-working slaves, who are never idle and always do a full day's work. What ought to be the result of all this labour, in addition to the grinding labour of all our working men and women? Should we not expect abundance of food and clothing for all, and ample leisure for the cultivation of the mind, and the enjoyment of the beauties of nature and of art? Instead of this, we have wide-spread, ever-present pauperism; crowded cities reeking with squalor, filth, drunkenness, and vice; a depopulated country; and, as a direct consequence of these two factors—streams polluted with wasted fertilising matter, destroying at once natural, valuable fish-food, beauty, and human life. Everywhere we find wealthy people enjoying all the luxuries and refinements of a high civilisation; but amidst them we also find masses of human beings living more degraded lives than most savages, and working harder and more continuously than most slaves.

In our preliminary inquiry we have shown that some such result as this *ought* to arise from absolute private property in land. It is surely a remarkable coincidence (if it be only a coincidence) that these results should occur in such extreme and painful development in the country where land is concentrated in the fewest hands, where the legal rights of landlords are the most absolute, and where, owing to the enormous aggregation of wealth, the divorce between those who own and those who cultivate the soil is the most complete. Let us endeavour to throw further light upon this question by an examination of the effects of our system in special cases.

In Ireland we have the spectacle of landlords doing what they like with their own for three centuries, backed up by a landlord parliament which made any laws they thought necessary; and the result has been a country in continual rebellion and a people ever on the verge of starvation.

This chronic starvation has been imputed to any and every cause but the real one—to over-population, to idleness, to potatoes; the real and all-sufficient cause being that the mass of the population are crowded on small and utterly insufficient holdings of the worst lands at extravagantly high rents, which means, that everything they raise besides enough potatoes to support life goes as tribute to the landlords.

Under such conditions, no population however limited, no industry however great, no agriculture however perfect, no soil however fertile, could save a people from poverty and recurring famines.

As Mr. DeCourcy Atkins well puts it:—

> Less than 2,000 persons own two-thirds of the land in Ireland, and out of its five or six million inhabitants there is no man of those who have tilled it and given it all its present value who owns one sod of its soil. For the land owned by these two thousand persons, many of whom are absentees, five hundred thousand families are competing, as the sole stay between them and starvation.

The Devon Commission, thirty-five years ago, declared authoritatively that in Ireland everything on the land which gives it value—houses, buildings, fences, gates, drains, etc. have been made by the tenants, and are undoubtedly their own property; yet from that day till two years ago our Parliament has allowed and even encouraged the Irish landlords to rob the tenants of this property by forms of law, and thousands and tens of thousands of Irish tenants have been robbed accordingly! Yet more. In the four years succeeding the great famine, there were over two hundred thousand evictions; whole town-lands were depopulated, and their human inhabitants driven off to make room for cattle and sheep—houses, schools, churches, everything being destroyed. The results of this are still to be seen over a large part of Ireland, where the traveller seems to be passing through a land bereft of human inhabitants, but marked by abundant ruins. The *Daily News* special commissioner, writing from Mayo, in October 1880, says:—"Tradesmen, farmers, and all the less wealthy part of the community still speak sorely of the evictions of thirty or forty years ago, and point out the graveyards which alone mark the sites of thickly populated hamlets abolished by the crowbar." The lands thus cleared were let in blocks of several square miles each to English or Scotch farmers for grazing farms, in order, as he tells us, that landlords "might get their rents more easily and more securely," even though they were sometimes less than those paid by the former inhabitants. And what became of these inhabitants? Let the Devon Commission, appointed by Parliament, and consisting mostly of landlords, answer the question:—

> It would be impossible for language to convey an idea of the state of distress to which the ejected tenantry have been reduced, and of the disease, misery, and even vice, which they have propagated in the towns wherein they have settled; so that not only they who have been ejected have been rendered miserable, but they have carried with them and propagated that misery. They have increased the stock of labour, they have rendered the habitations of those who received them more crowded, they have given occasion to the dissemination of disease, they have been obliged to resort to theft and all manner of vice and iniquity to procure subsistence; but, what is perhaps the most painful of all, a vast number of them have perished of want. (*Parl. Rept.* 1845, vol. xix, p. 19)

Now, consider these horrible results produced in four years to a million of people; consider further, that the same kind of eviction, with its consequent misery and vice, has been going on in Ireland in varying degrees down to this very year, and that all this untold wretchedness, this cruel, heart-rending wrong, this vice, and crime, and pauperism, this disease and death, have been caused—not in a great war between nations struggling for supremacy, not to maintain any great principle of religious or civil liberty, but, "in order that landlords may get their rents more securely and more easily!" And now, in this year, 1883, when the people of Ireland, crowded into towns and on the poorest lands of the west coast, are again starving, the only remedy our landlord legislators can propose is to ship them off by thousands to other countries, and thus increase and intensify that widespread hatred of English rule which is the natural and just punishment we are receiving for persistent injustice. These deserted villages are not to be again repeopled; the cattle and sheep must be still allowed to displace Irishmen; the "easy collection" of the landlord's rents must on no account be endangered; let everything go on as before, and when our consciences or our fears are aroused by the cry of too many starving Irishmen, let subscriptions be got up, and let the English people be taxed to ship off a few thousands of surplus paupers to New York, and all will be well!

Here we see pure landlordism having its own way, and working out its natural and inevitable results, in the extreme case of ownership of the soil of a country for the most part by absentees and by aliens in race and religion. About this there can be no dispute. And if the absurd and totally unfounded cry of over-population is always to be followed by more emigration, there can be no end to the process. For even were the population reduced to one million of Irish peasant cultivators, that million would continue in exactly the same condition of misery and destitution as the present population, if they were confined to limited areas and were subjected to the extortions of the agents of absentee or alien landlords. Even before the famine the exorbitant rents and high taxes were paid chiefly by means of exported *food*, showing that the land of Ireland was able to support many more than the eight millions which then inhabited it. Yet now, when the population has been reduced to five or six millions, the same cry of over-population is raised—as it was a century ago when there were only two millions; and whether there be two or five or eight millions in the country, there will certainly be starvation and local over-population if the people are forbidden the free use of their native land, are confined to the least productive districts and to insufficient holdings, and all the surplus produce above a bare supply of potatoes in average years is exported to pay rent! That more starving Irishmen should be expatriated while millions of acres which they once tilled are given up to cattle and sheep, is the condemnation of landlord government. That the chronic famine which has prevailed in Ireland for a century should still devastate it, is the condemnation of landlordism itself.

Let us now turn to another country, where the landlord power has had complete sway for a century, unfettered by any of the difficulties which are often alleged as the reason for its terrible failure in Ireland. In the Highlands of Scotland there has been no religious difficulty, and there has been no antipathy of race; the people have not depended wholly on potatoes, and the country has certainly never been over-populated. Neither has there been any rebellion against authority; but the universal testimony of all who know them best is, that in the whole British dominions there exists no more intelligent, religious, peaceable and industrious people than the Highland peasantry. Yet here too, under the most favourable conditions, we find perennial destitution and famine, and this very year sees another of the long series of Royal Commissions seeking out that which is plain as the sun at noonday, the causes of want and misery among the tenantry of an enormously wealthy, and in their own territories almost omnipotent, body of landlords. The causes are, simply, that the native inhabitants have been driven from the inland valleys to the sea-coast to make room for sheep and deer. The terrible history of the Highland clearances is too long to go into in this place. Suffice it to state, that two millions of acres once inhabited by human beings are now devoted to deer only, from which the noble Highland chieftains or their successors get much sport or large rentals. Seventy men at this day own half Scotland, and if they choose to complete what they have begun, and turn more fields and meadows into hunting-grounds, our existing law permits them to do so; while no amendment of the law as yet proposed by Liberal politicians would place the slightest check upon this iniquitous power . . .

Let us now see what are the conditions under which the Highlanders live at the present day, and consider whether under such conditions anything but poverty, discontent, and famine is possible. We learn from the various reports that have appeared in the daily papers, confirming the testimony of all previous impartial writers, that the Highland crofters are confined to miserably small holdings—the largest croft in Skye, for example, being seven acres; that the land is poor and the rent very high; that the landlords have continually encroached on the commons and mountains, the use of which for grazing is essential to the crofter's existence. These have been usually taken from them, without compensation, to make either large sheep-farms or deer-forests; and in many cases they suffer without redress from the incursions of the deer which eat their crops, while they are not allowed to keep dogs to mind their own sheep (when they have any) for fear of disturbing these sacred deer, whose well-being and due increase are carefully attended to even though it entails starvation on men and women.

Then, again, these great estates, often as large as continental kingdoms or dukedoms, are managed by agents and factors who represent an unknown and unseen landlord, and who are really despotic rulers, carrying out their own decrees under penalty of eviction—a penalty as severe as that imposed by the law of England on hardened criminals. It was stated in the *Daily News*, a paper which is celebrated for the careful accuracy of its information, that on one estate (a generation back) a whole body of crofters were removed because they had good land which the factor wanted; and this is the more credible because many other cases are recorded in which the factors take farms from which the former holders have been evicted. Under the rule of the factors the people may be oppressed and pauperised, even with the most benevolent of landlords. Take the case of the late Sir James Matheson, who bought the extensive island of Lewis (as large as an average English county) about forty years ago, and who is universally admitted to have been personally most benevolent and liberal. Yet under his paternal government tenants were ejected at the will of the factor, and extensive tracts turned into sheep farms and deer forests, and such cruel injustice was perpetrated for years that the people at length rebelled, and then only did their landlord know they had anything to complain of. The *Quarterly Review* declares that Sir James Matheson made wonderful improvements in Lewis, "pouring out money like water," and spending over 100,000 pounds there, besides giving largely in charity; and the result is that in this very year there is famine in Lewis, and the representatives of this wealthy and benevolent landlord are obliged to beg for subscriptions in the city of London to save the people from starvation! Nothing is said about the sheep and deer of the island; no doubt they are fat and flourishing and give handsome returns, whereas men and women are encumbrances and have to be kept alive by charity! What a cruel satire is this. An enormously rich country, which taxes its people heavily under the pretence that it dispenses justice and gives protection to all; which is highly civilised and highly religious; and which yet upholds a system under which large masses of its subjects have no right to live but by the permission of landlords and their irresponsible agents!

I cannot here go further into this distressing subject, but must again refer my readers to . . . easily accessible sources of information . . . .

In the first part of this article I have given a brief outline of the effects of landlord rule in England, Ireland, and Scotland, and I cannot but express my amazement that a writer like Professor Fawcett [in the July 1883 issue of *Macmillan's Magazine*], who must be fully acquainted with the whole of the terrible facts I have here only been able to hint at—who nearly twenty years ago wrote so strongly on the pitiable condition of the British labourer—a condition since changed if anything for the worse, since he has become more than ever divorced from the soil and is tending to a nomadic life—and who has depicted so forcibly the land-hunger of the rich, should yet, in his latest teaching, have no other remedy to propose for the evils due to unrestricted private property in the first essential of human existence, the soil, than to make that property still more easily acquired by the wealthy and still more absolute, by means of complete free trade in land!

It has repeatedly been said, and will no doubt be said again, that many of the admitted evils here pointed out are not due to our system of landlordism, but to various other causes, such as improvidence, over-population, and idleness. To this I reply, that even where these causes do exist they are but secondary causes, and are themselves due to the landlord system. Improvidence is the inevitable result of insecurity for the produce of a man's labour; over-population is always local, and is produced by men being forcibly crowded together and not allowed freely to occupy and cultivate the soil; while idleness is the last accusation that should be made against any of the inhabitants of these islands, whose fault rather is a too great eagerness for work whenever they can work for fair wages or with full security that they will reap the produce of their labour. I reply, further, that the close correspondence between the theoretical and the actual results, between deduction and induction, must not be ignored. An irresistible logic assures us that the possession of the soil by some must make those who have to live by cultivating the soil virtual slaves; that it must keep down wages and enable the landlords to absorb all the surplus wealth produced by the labourer beyond what is necessary for a bare subsistence; while it indisputably prevents all who are not landowners from any use or enjoyment of their native soil except at the landowner's pleasure—a fact everywhere visible in the unnatural mode of growth of our towns and villages, where, with ample land suitable for pleasant and healthy homes all around them, people are forced to live in close-packed houses with the view of nature's beauties shut out and much of the discomfort and insalubrity of large towns carried into the country. The best way, however, to prove that these and many other evils are directly due to landlordism is to show by actual examples how constantly they disappear whenever the land belongs to those who cultivate it. This is the case in many parts of Europe; and although climate, race, laws, and the character and habits of the people may widely differ, we always find an amount of contentment and well-being strikingly contrasted with what prevails under the system of landlordism.

As regards Switzerland, Sismondi, in his *Studies of Political Economy*, gives full information. He declares that here we see the beneficial results of agriculture practised by the very people who enjoy its fruits, in "the great comfort of a numerous population, a great independence of character arising from independence of position, and a great consumption of goods the result of the easy circumstances of all the inhabitants." Many other writers confirm the accuracy of these statements . . .

The late William Howitt, writing on the rural and domestic life of Germany, says:—

> The peasants are not, as with us, for the most part totally cut off from the soil they cultivate—they are themselves the proprietors. It is, perhaps, from this cause that they are probably the most industrious peasantry in the world. They labour early and late, because they feel that they are labouring for themselves.

> The German peasants work hard, but they have no actual want . . . The English peasant is so cut off from the idea of property that he comes habitually to look upon it as a thing from which he is warned by the laws of the large proprietors, and becomes, in consequence, spiritless and purposeless . . . The German *Bauer*, on the contrary, looks on the country as made for him and his fellow-men. No man can threaten him with ejection or the workhouse so long as he is active and economical . . .

And Mr. Baring Gould, although showing how poor the peasant of North Germany often is, owing to the miserable system of each farm being cut up into scores or hundreds of disconnected plots, and his cruel subjection to Jew money-lenders, nevertheless thus compares him with the journeyman mechanic:—

> The artisan is restless and dissatisfied. He is mechanised. He finds no interest in his work, and his soul frets at the routine. He is miserable and he knows not why. But the man who toils on his own plot of ground is morally and physically healthy. He is a freeman; the sense he has of independence gives him his upright carriage, his fearless brow, and his joyous laugh.

In Belgium the most highly cultivated part of the country is that which consists of peasant properties; and even M'Culloch, the advocate of large farms, admits that,—

> in the minute attention to the qualities of the soil, in the management and application of manures of different kinds, in the judicious succession of crops, and especially in the economy of land, so that every part of it shall be in a constant state of production, we have still something to learn from the Flemings.

In France, though the farming may not be what we call good, the industry and economy of the peasant-proprietors is remarkable, while their well-being is sufficiently indicated by the wonderful amount of hoarded wealth always forthcoming when the Government requires a loan. The connection of peasant-cultivation with well-being is apparent throughout France. Sir Henry Bulwer remarks that by far the greatest number of the indigent is to be found in the northern departments, where land is less divided than elsewhere, and cultivated with larger capitals. Mr. Birkbeck, noticing that in one district the poor appeared less comfortable, found, on inquiry, that few of the peasants thereabouts were proprietors, while, in Anjou and Touraine, Mr. Le Quesne noticed that the houses of the country people were remarkable for their neatness, indicative of the ease and comfort of their possessors, and on inquiry as to the cause was told that the land was there divided into small properties. So when the celebrated agriculturist and traveller Arthur Young noticed exceptional improvement in irrigation and cultivation, he is so sure of the explanation of the fact that he remarks:—"It would be a disgrace to common sense to ask the cause; the enjoyment of property *must* have done it. Give a man secure possession of a bleak rock and he will turn it into a garden; give him a nine years' lease of a garden and he will convert it into a desert." (For these and many other examples see Thornton's *Plea for Peasant Proprietors*.)

It hardly needs to adduce more evidence to prove the intimate connection of the sense of secure possession with industry, well-being, and content. But we must briefly notice one more example at our very doors, and under our rule—the case of the Channel Islands. The testimony of all observers is unanimous as to the happy condition of these islands, and to its cause in the almost total absence of landlordism as it exists with us. The Hon. G. C. Brodrick, in his *English Land and English Landlords*, says:—

> If we judge of success in cultivation by the produce, we find that a much larger quantity of human food is raised in Jersey than is raised on an equal area, by the same number of cultivators, in any part of the United Kingdom. Not only does it support its own crowded population in much greater comfort than that enjoyed by the mass of Englishmen, but it supplies the London market out of its surplus production with shiploads of vegetables, fruit, butter, and cattle for breeding. Even wheat, for the growth of which the climate is not very suitable, is so cultivated that it yields much heavier crops per acre than in England; and the number of live stock kept on a given area astonishes travellers accustomed only to English farming. Nor are these only the results of spade-husbandry, for machinery is largely employed by the yeomen and peasant proprietors of the Channel Islands, who have no difficulty in arranging among themselves to hire it by turns.

Mr. Brodrick, like every one else, attributes this wonderful success to the land system of the country.

Lest it be now said that there is something in the climate or soil of these various localities, or in the character or habits of the people to which these favourable results are to be imputed, I must refer to a few crucial examples in which every other cause but difference of land tenure is eliminated, and which therefore complete the demonstration to which this whole argument tends. The first of these is afforded by Italy, over large portions of which there are still, as in the time of the Romans, *latifundia* or large estates farmed by middlemen, and cultivated by labourers or tenants at will. In a recent work the great economist, M. de Laveleye, speaks of the

> naked and desolate fields, where the cultivator dies of famine in the fairest climate and most fertile soil; such is the result of the *latifundia*. Economists, who defend the system of huge properties, visit the interior of the Basilicata and Sicily if you want to see the degree of misery to which your huge properties reduce the earth and its inhabitants.

Yet in the same country and under the same laws, wherever fixity of tenure or peasant properties exist, the utmost prosperity prevails. Again, let us hear M. de Laveleye:—

> I know of no more striking lesson in political economy than is taught at Capri. Whence come the perfection of cultivation and the comfort of the population? Certainly not from the fertility of the soil, which is an arid rock . . . Before obtaining the crops, it was necessary, so to speak, to create the soil. It is the magic of ownership which has produced this prodigy.

Now let us come back to our own country, where we shall find that exactly similar results are produced by similar causes. On the Annandale Estate in Dumfriesshire, plots of from two to six acres were granted to labourers on a lease of twenty-one years. They built their own cottages, having timber and stone supplied by the landlord, and these little farms were all cultivated by the labourer's family, and in his own spare hours. Now note the result. Among these peasant-farmers pauperism soon ceased to exist, and it was especially noticed that habits of marketing, and the constant demands on thrift and forethought brought out new powers and virtues in the wives. In fact, the moral effects of the system in fostering industry, sobriety, and contentment, were described as no less satisfactory than its economical success.

Again, on Lord Tollemache's estate in Cheshire, plots of land from two and a half to three and a half acres are let with each cottage at an ordinary farm rent, and the results have been eminently beneficial. It is remarked here too that the habits of thrift and forethought encouraged by cowkeeping and dairying, on however small a scale, constitute a moral advantage of great importance . . .

A volume might be filled with similar cases, but more are unnecessary here, for the evidence already adduced or referred to is absolutely conclusive. Wherever there are great estates let on an insecure tenure, we find in varying degrees the evils here pointed out. On the other hand, wherever we find men cultivating their own lands, or lands held on a permanent tenure at fixed rents, we find comparative comfort, no pauperism, and little crime. And as this is exactly what a consideration of the immutable laws of human nature and economic science has demonstrated must be the inevitable result, we have fact and reasoning, induction and deduction supporting each other.

Having now cleared the ground by an inquiry into principles and a survey of the facts, we come to the practical question—Can any adequate remedy be found for these widespread and gigantic evils?

The common panacea of the Liberal party, "Free-trade in land," must surely now appear to my readers to be ridiculously inadequate. It was tried in Ireland by the Encumbered Estates Act, and it so aggravated the disease that a Liberal Government has now been forced to stop free-trade in land altogether in Ireland, and fix rents by act of parliament! It has always prevailed in America, yet many of the evils of land monopoly are beginning to appear even there. It exists in Italy, yet great estates prevail, and the tiller of the soil starves in the midst of abundance as with us. It will do nothing for the poor evicted crofters or the famine-stricken cottiers of Ireland. It will not cause the tracts now occupied by sheep and deer to be given up again for the use of men and women; but it will allow rich men, more easily than now, to make more deer-forests and sheep-farms if they choose. It will not help the labourer, the mechanic, or the shopkeeper, to a plot of land where he requires it. It will not give back the land to the use of the people who want it most, and who, as the universal experience of Europe shows, are always benefited by it both physically and morally. Let us then appeal to first principles and simply follow their teaching.

We have seen (1) that private property in *land* cannot justly arise at all; and (2) that its results, except where small portions are personally occupied and tilled, are always evil. Hence we conclude that the land of a country should be the property of the state, and be free for the use and enjoyment of the inhabitants on equal terms. In order that every one may feel that sense of property in the land he cultivates which is the best incentive to industry, absolute security of tenure is necessary. This given, a man becomes virtually the owner of the land he holds from the state, and can deal with it like a freehold, only that it remains subject to such rents and general conditions as may from time to time be held to be for the good of the community. Another important principle is, that sub-letting must be absolutely prohibited; for if this were allowed the evils of landlordism would again arise, as middlemen would monopolise large quantities of land which they would let out at advanced rents and under onerous conditions, so that the actual cultivators would be no better off than under the present landlord system.

Recurring again to first principles we find, that although *land* itself cannot justly become private property, yet everything added to the land by human labour is truly and properly so; and this leads to the important subdivision of landed property into two parts (as is so common in Ireland), the one represented by the landlord's rent for the use of the bare land, the other the *tenant-right*, consisting of the houses, fences, gates, plantations, drains, and other permanent and tangible improvements there always made by the tenants. Now these improvements, which for purposes of sale or transfer may here also be conveniently termed *tenant-right*, should always be the absolute property of the occupier of any plot of land, the state being the owner of the bare land only; and by this simple and logical division it will be seen that all necessity for state *management*, with the long train of evils which Professor Fawcett so properly emphasises, is absolutely done away with, and the cultivator may be left perfectly free to treat his estate as he pleases. For everything on the land which can be deteriorated by bad farming or wilful neglect is his private property, and its preservation may safely be left to the influence of self-interest; while the *land*, which is the property of the state, is practically incapable of deterioration; for its value depends on such natural causes as geological formation, arterial drainage, aspect, rainfall, and latitude, and on such social conditions as density of population, nearness to towns, to seaports, to railroads, or canals, the vicinity of manufactures, etc., none of which can be changed by the action of the tenant. The state, therefore, will have nothing whatever to *manage*, but need only collect its quit-rent as it now collects the land-tax or house-tax, leaving every land-holder perfectly free to do as he pleases, and only interfering with him by means of general enactments applicable to all holders alike. All arrangements that may be necessary for facilitating the acquisition of land by those who need it, should be in the hands of Local Land Courts, acting on principles determined by general enactment.

Having thus given the main outlines of a just and beneficial system of land tenure, let us consider briefly how to bring it into practical operation. And, first, we will explain how existing landlords are to be dealt with, as this is considered by many to be the real difficulty in the way of land nationalisation. We heartily agree with Professor Fawcett that the idea of confiscating all the landed property of the kingdom, and reducing all who derive their income wholly or mainly from land to a state of destitution, is not only grossly unjust, but would be utterly subversive of the very end for which nationalisation is proposed—the well-being of the whole community. We therefore put this suggestion altogether aside, as one never likely to gain ground in England, unless the blind opposition of the landlords to all just reform brings on revolution and anarchy.

We next come to the supposed alternative, to buy out the landlords with hard cash, or state securities, at the full market price of their lands; and here, too, I have always accepted Professor Fawcett's proof that the thing is financially impossible. But I go further, and say, that even if possible, it would be inadmissible; for, so to pay for it, would be to admit that private property in land is *right* even if it be inexpedient, and would provide for the perpetuation of a class of wealthy idlers, supported out of the produce of land which we have declared rightly to belong to the community. I cannot imagine that any sound economic thinker will declare, that it is either right or expedient to provide by special enactment that the unborn generations of our country shall be permanently divided into two classes as regards the national soil—one class tilling it and paying rent to the state for its use, the other class receiving from the state a large portion of these rents to enable them to live in idleness.

There remains, however, a third course, totally distinct from either of these, which, while doing no material injury to any landlord, or to the expectant heirs of any landlord, yet claims and at once secures the use of the soil for the nation. This method is the simple one of allowing all revenues derived from the land by individuals to die out with their lives, while the land itself is resumed by the state for the public benefit. The advantages of this plan are, that it requires no money to be raised, and therefore creates no financial difficulty; that it is just to the landlords, because it would not diminish their net income, or that of their direct living heirs; that it is just to the people because it obtains for them immediate possession and use of the land, and the reversion to all the rents as existing landlords and their heirs die off.

On the part of the landlords it will of course be said that the selling value of their estates will be diminished. Perhaps it will, though not to a great extent; but that is no real hardship if their incomes remain untouched, and as regards their heirs it will be a real and important advantage that their prospective incomes cannot be squandered away. Besides, this objection can hardly be made by any of those Liberals who supported the last Irish Land Act, since in that case not only was the selling value of most estates largely reduced, but the income from them was so much diminished that many landlords and their dependents have been absolutely ruined. If that extreme measure of confiscation was justified by the public benefits it was calculated to produce, then the proposal here made is more than justified, since the injury is hardly perceptible while the benefits are enormously greater. For in the case of Ireland one class has been benefited at the cost of another, while nothing has been done for the labourer or for the nation at large. But by the plan here proposed all will benefit, and there will accrue to the state a steadily-increasing income from quit-rents which will enable the more injurious taxes to be first remitted and ultimately all taxation whatever to be abolished. An especial advantage that may be claimed for this plan is, that the income derived from the annuities to landlords falling in, would begin very gradually, and go on steadily increasing for a period of sixty or eighty years. This would entirely obviate all the financial and commercial difficulty which would certainly be caused, either by the too sudden removal of many kinds of taxation, or by the too great and sudden accumulation of funds in the hands of Government, if the land revenues of the whole country came into its possession at any one fixed period.

Having thus shown how the land may be acquired by the state for the use of the community without cost, and yet without material injury to any living person, we may proceed to consider how it may best be used for the benefit of all; and here we shall be able to answer Professor Fawcett's questions (which he seems to think unanswerable)—"What principles are to regulate the rents to be charged? Who is to decide the particular plots of land that should be allotted to those who apply for them?" The answer to both is easy. Rents will be fixed in the first place by official valuation, following the precedent of the Irish Land Act; afterwards, probably, by free competition. As to *who* is to have land, it is essential that there should be the greatest freedom of choice compatible with the just rights of existing occupiers. How these two important matters may be settled I will now briefly indicate.

It will first be necessary to determine the value of the improvements on the land as distinguished from that of the land itself, and to facilitate subdivision or rearrangement of farms or holdings. This should be done for each separate inclosure shown on the large-scale ordnance maps. Some general principles being laid down for the guidance of the valuers there will be no real difficulty in making the separation. An old pasture field in which the hedges and gates have been constantly repaired by successive tenants may be considered to be, so far as the landlord is concerned, in an unimproved state. Here the whole value will be *land value*. From this as a datum the separation of *improvements* will take place where the landlord has recently put new gates, or has drained, or has built sheds, bridges, or farm buildings. The valuation, when complete, will show the annual value of the *land* for the state *quit-rent*, the present *annual* value of the improvements, and the present *purchase* value of the improvements to a tenant calculated on a scale determined by their quality and probable duration.

This official valuation being made, it would be only fair that the existing occupier of any farm or other land should have the first offer of it under the new conditions; these being that he should become the owner of the improvements and agree to pay the state quit-rent. If he has capital he can purchase the improvements by a cash payment, but if not he should be entitled to purchase them by means of a terminable rental, as in the case of purchases under the Irish Church Act. This would be done through the Land Courts, which would decide on the annual payments to be made and the period for which they are to run, so as to meet the views of both parties. It is the opinion of good authorities that most farmers now hold too much land in proportion to their capital, and that with perfect security of tenure and absolute freedom of action they would reduce their holdings in order to farm more highly and to be able to effect permanent improvements. Some farms would therefore be divided, and remote fields be detached from others, and these would afford land for small holdings or for gardens and fields for labourers.

But much more than this is needed. The crofters and cottiers who have been ejected from their homes, the labourers who have been driven into towns, and all who have been robbed of their ancient rights by the inclosure of commons, require immediate redress. We have seen what beneficial results invariably follow the grant of small plots of land at fair rents and on a secure tenure, and Nationalisation would not deserve the name did it not place this boon within reach of all who desire it. There is no privilege so beneficial to all the members of a community as to have ample space of land on which to live. Surround the poorest cottage with a spacious vegetable garden, with fruit and shade trees, with room for keeping pigs and poultry, and the result invariably is untiring industry and thrift, which soon raise the occupiers above poverty, and diminish, if they do not abolish, drunkenness and crime. Every mechanic and tradesman should also be able to obtain this great benefit whenever he desires it; and this is far too important a matter for the whole community to be left to the chance of land being offered for sale when and where wanted. It is not sufficiently recognised that the use of land for the creation of healthy and happy homes is far higher than its use as a mere wealth-producing agent, in which latter aspect alone it has hitherto been chiefly viewed. To get the greatest benefit from the land of a country it is essential that every inhabitant should be, as far as possible, free to live where he pleases; and to attain this end the right to hold land for cultivation should always be subordinate to the right to occupy it as a home.

To carry these principles into effect, and to allow population to spread freely over the whole country, it is essential that all who desire a permanent home should have some right of free selection (once in their lives) of a plot of land for this purpose. A limit might be placed to the quantity so taken in proportion to the density of the population—near towns perhaps half an acre, in the country an acre or more. Such choice should, of course, be limited to agricultural or waste land, and at first to such land as borders public roads. Other limitations might be, that not more than a fixed proportion of any one farm should be so taken, and that a plot should never be chosen so near the farmer's house as to be an annoyance to him—questions to be decided by the Local Land Courts. Of course this land would be subject to the usual payment of quit-rent to the state according to the official valuation, while the improvements would have to be purchased from the farmer with some small addition as compensation for disturbance.

The effects of such freedom of choice in fixing upon a permanent residence would be gradually to check the increase of towns and to re-populate the country districts. Rural villages would begin a natural course of healthy growth, and if the minimum of land to be taken for one house were fixed at an acre (the maximum being four or five acres) these could never grow into crowded towns, but would always retain their rural character, picturesque surroundings, and sanitary advantages. The labourer would choose his acre of land near the farmer who gave him the most constant employment and treated him with most consideration; and besides those who would continue to work regularly at agricultural labour, there would be many with larger holdings or with other means of living, who would be ready to earn good wages during hay or harvest time. With a million of agricultural labourers, each holding an acre or more of land, and at least another million of mechanics doing the same thing, and all permanently attached to the soil by its secure possession, that scandal to our country, the scarcity of milk and the importation of poultry, eggs, and butter from the Continent would come to an end, while the vast sums we now pay for this produce would go to increase the well-being, not only of the labourers themselves, but of all the retail and wholesale dealers who supply their wants. Our most important customers are those at home, and there is no more certain cure for the now chronic depression of trade than a system which would at once largely increase the purchasing power of the bulk of the community.

The question of house property in towns cannot now be discussed, but it may be treated on the same general principles; and I must refer my readers to a new edition of my book on Land Nationalisation, now in the press, where they will find this part of the subject treated in an appendix.

Before concluding this paper, I would wish to point out how easily the principles of land tenure here advocated may be tried on a small scale without interfering with any private rights or interests; and so convinced am I of the soundness of these principles, that I would venture to stake the whole question of the practicability of land nationalisation on the result of such a trial. I would suggest, then, that all crown lands in any degree suitable for cultivation should be thrown freely open to applicants in small holdings for personal occupation, on the tenure which I have just explained; and I would earnestly press some liberal member of Parliament to urge this trial on the Government by means of an annual motion. The result would certainly be a large increase of revenue from these lands, since all expenses of management would be saved; while it is equally certain that the localities would be benefited by the increased well-being of the inhabitants.

In view of such a trial being made, and its further extension being desirable, a resolution should be passed declaring it inexpedient to sell any crown lands or rights over commons; and the next step should be to stop entirely the further inclosure of common lands for the benefit of landlords, a proceeding which the liberal portion of the community has long condemned as legalised robbery of the people. Many of the more extensive commons and heaths far removed from dense centres of population offer the means for a further trial of this system of land-tenure, and creating a considerable body of virtual peasant-proprietors of the best type. For this purpose all manorial rights of individuals should be declared to be (as they certainly are) injurious to the public, and should be at once acquired by the State. Their present owners might either be repaid the purchase money if they had themselves bought them, or be compensated by means of terminable annuities of amounts equal to the actual average net incomes derived from the several manors. Thus would be afforded ample means for a great social experiment, the result of which, if fairly tried, cannot be doubtful.

* * * * *

Although the preceding selection was undoubtedly Wallace's most significant essay on the subject of land nationalization, his thoughts on the problem continued to evolve into the 1890s. He expressed many of his ideas in his yearly Presidential Addresses to the Land Nationalisation Society and 'Tracts' issued by the same organization. Most of these essays represent developments and/or combinations of themes already present in 'The "why" and the "how" of land nationalisation.' One such work was presented as the Presidential Address to the Land Nationalisation Society's annual meeting of 1893:

The conditions essential to the success of small holdings (ref 471)

As there seems to be considerable difference of opinion on this question, and as success or failure in the first steps towards obtaining free access to land for all who desire it is a matter of vital importance for the progress of our cause, I think it will be useful to set forth for the consideration of Land-nationalisers what seem to me to be the principles which should guide us in this matter.

I begin with the proposition, which will probably be generally accepted, that the primary object of giving the people free access to land is, to enable as many as possible to obtain either the whole or a portion of their living by its cultivation, and, by thus withdrawing some of them altogether from competition with wage-earners, and by giving others a temporary alternative to wage labor, to raise the rate of wages of the less remunerative kinds of work all over the country. And as secondary benefits resulting from this, we shall have, in the first place, an enormous increase in the production of food, displacing much that is now imported from abroad; secondly, an increased consumption of manufactured goods by these self-supporting workers; thirdly, a diminution in the poors' rate from the disappearance of paupers and out-of-works; and, lastly, a public revenue arising from the continuous increase in the value of land owing to the growth and increased well-being of the population. These various benefits I conceive to be important, somewhat in the order in which I have given them; at all events the first is undoubtedly the most important, while the last—the money profit-is the least important. It is a valuable incidental result, but is not to be sought after as one of the chief ends in itself to the diminution of the other kinds of benefit.

Now in order that the various good results above enumerated shall be, in their due order, most certainly obtained, it is of the highest importance that the workers who gain access to the soil shall be able, not only to live upon it, but to live well and thrive upon it. We do not want them merely to earn a living as long as they can work, and go to the poor-house in their old age or during sickness; neither do we want them to be ruined by the first bad season, or by any of the chances and misfortunes to which Agriculture is especially liable. On the contrary, it is of the highest importance to the whole community that the land-cultivators should be in such a position that even in bad years they should not lose, while in average years they should obtain such profits, that, either by increase of the size of their holding or the amount of their stock, their condition should steadily improve, and thus insure them against poverty in old age. If land-nationalisation is to be the success we hope it will be, some such result as this must be aimed at, and we must be sure that we do nothing to prevent its attainment.

In order to bring about this result two things are especially required,—first, a certain amount of knowledge, of experience, of prudence, and of industry in those who first obtain small holdings; in the second place, such very moderate rents and such favorable conditions of tenure as to give them not only a chance but almost a certainty of success. Nothing, in my opinion, is more likely to bring about the failure of the first, and therefore the most important experiments, than the methods usually suggested by politicians, and more or less implied in the various Acts of Parliament, whether already passed or proposed, dealing with this matter. It is, for example, almost always proposed or taken for granted, that before the men can have the land a large cost must be incurred in preparing it for them. We hear of road-making, fencing, draining, and house-building, as if these were absolutely necessities; and in addition to all this, it is generally thought that it will be necessary to advance capital to the proposed tenants to enable them to stock and crop their holdings, and support themselves till they get a return from the land. Now I can hardly imagine a more certain way of bringing discredit on the whole system of small holdings—and with it of land-nationalisation—than such a method. In the first place, this work of "laying out" and "improvement," when it is not done by the occupiers themselves, but for them, by persons who have no interest in doing the work economically but often the reverse, and who have besides no personal knowledge of what these small cultivators really require, will often be unnecessary work and will always be done in an unnecessarily costly way, and will thus add to the rent of the land, without proportionately increasing its value. Then again, the advance of money on loan to men who have never, perhaps, had ten pounds to spend at once in their lives, will in many cases lead to its injudicious expenditure, and be antagonistic to that prudence, thrift, and industry, which are vital to success.

There is yet another, and a very important objection to such methods as these. All this expenditure of public money by other people than those who are to directly benefit by it, will certainly lead to wastefulness and jobbery, since they constitute that very "management of the land by public officials," the evils of which form one of the chief objections to land-nationalisation, and which all our proposals and methods have been calculated to avoid. We must therefore never cease to urge that such management is not only unnecessary, but is calculated to defeat the very purpose for which free access to land is required.

If we consult the reports of the various Royal Commissions on Agriculture we shall find numerous cases of laborers, miners, mechanics, and others, who have become successful cultivators of small holdings and sometimes of considerable farms, often having begun with a lease of waste land which they enclosed, improved, and built houses on, entirely on their own resources and through their own industry; and whenever we find a successful small farmer he has usually worked his way up in some such way. It is an extraordinary thing that whenever it is proposed to allow men to obtain small holdings in England, there is always this talk of "improvements" and "house-building," in addition to giving the land at a fair rent and on a secure tenure; while over a large part of Scotland and Ireland all improvements and all buildings have been done by the tenants themselves, with no security of tenure whatever, so that whenever a misfortune prevents payment of rent—usually rent on the tenants' own improvements—the landlord ejects the poor tenant and confiscates his improvements. The Irish cottar and the Highland crofter ask nothing better than a sufficiency of land at a moderate rent and on a secure tenure. All the perennial misery and often-recurring famine of these two countries have arisen from a denial of this very moderate instalment of bare justice; and if we give this easy access to the land to our English workers, they too will ask for nothing else, and will be far more likely to succeed without any attempt to do for them what they will do much better and more economically themselves. But we do undoubtedly require some process of selection of the best men for this great experiment in the regeneration of our country; and the natural, self-acting, and therefore best mode of selection, will arise from the fact that no man can take a holding unless he has saved money to stock and crop it, or has such a character for industry, sobriety, and capacity as to induce some friend to advance him the money; while the certainty that he is risking the loss of his own savings if he fail, will be the best guarantee that he will have some amount of intelligence and some agricultural experience. No artificial mode of selection will compare with this. A man may get testimonials to character, but no testimonials can show that he will spend borrowed money prudently, or be able to make a profit on land burdened with unnecessary and costly improvements, which, for his purpose, will often be no improvements at all.

We have now to consider the second great essential of success, which is, the rental to be paid for the land and the conditions of tenure. Many of our fellow-workers maintain that the competition-rent offered for land is the best, and in fact the only certain way, of determining its value, and therefore what should be paid for it. From a landlord's or speculator's point of view—considering the money income to be got from the land to be everything, the well-being of the tenants nothing—this is undoubtedly the case; but, from our point of view—looking at the cultivation of the land as leading primarily to the well-being and progressive advancement of the cultivator, and through him the similar advancement of all other manual laborers—it seems to me to be the very worst mode possible. Let us therefore consider it a little in detail.

At the present time, wherever there are allotments to let, there we find it to be the rule that agricultural laborers do most willingly hire them at a rental per acre, sometimes double, sometimes four or five times as much as is paid by Farmers for the same quality of land, and there can be no doubt that if arable land were now offered for allotments and small holdings almost anywhere in England, and the quantity thus offered was not in excess of the demand at the time, it would, if let by auction, realise somewhat similar rentals. Many agricultural laborers, as well as village tradesmen and mechanics, find it advantageous to them to have land even at these high rents, and there is sufficient land thus held all over the country to afford a guide to the prices at which such land would let by auction, even in quantities of from one to five acres. Now the reason such high rents have been, and are paid, is, simply, that the laborers' wages have been always so low and his condition so miserable, that anything by which he could add two or three shillings a week to his earnings by means of overtime work and the assistance of his wife and family, was eagerly accepted. It was his low standard of living that rendered him willing to pay a rent which left him some little profit on his labor; and if he were now offered a small holding on which to live he would be willing (if he could not get it cheaper), to pay a rent which would enable him to live in about the same way as he had hitherto done, but with the chance of occasional better luck and with the satisfaction of being his own master. We see this result very prevalent on the continent, especially in Belgium and in parts of France, where the price of land, and consequently its rental, is very much higher than with us, and as a consequence the small holders often work harder and live as near the starvation line as our poorly paid agricultural laborers or our rack rented Irish cottars. It will be said, no doubt, that this arises from the demand for land being greater than the supply, and that if the land were offered in larger quantities, competition would be less keen and prices lower. But is it at all likely that for a long time the supply of land will be greater than the demand, except quite locally, and temporarily? Is it not, on the contrary, almost certain that the demand will, at first and for a long time to come, perhaps always, be greater than the supply? Is it not our contention that the depression of agriculture and the deplorable condition of so many of our workers is due to the denial of access to land, and that when that access is freely given it will bring about the well-being, first of those who cultivate it, and afterwards of all other wage-earners? There will therefore be a constant and ever increasing demand for land; but unless we take care that those who apply for it have it on such terms that they can not only make a good living from it, but also provide for a comfortable old age, the benefits we anticipate will not arise. It is to avoid any such failure, to prevent the recurrence of the miserable spectacle of men being ejected from their holdings for non-payment of rent, to secure for them something better than a struggle ending in dependence on charity in old age, that I urge the fixing of rents by valuation, taking always the amount paid by prosperous farmers in the same district, rather than that of allotment-holders, as the standard of value. The land when purchased by the local authorities, will be purchased at the farm value, and it can be let at that value at first without loss to the community.

The above sketch of the reasons why I object to the system of competition-rents sufficiently exhibits the principle on which, in my opinion, our dealings with the land should be founded. But there is also a practical objection to that system—that it would be very unequal in its results, and also that it can hardly be carried out unless based on a preliminary valuation. I presume the advocates of competition rents to not propose that land should always be let to the highest bidder, without any reserve whatever. For, if so, whenever the intending tenants were less numerous than the lots to be offered, these lots might be let at much less than agricultural rents. No doubt it will be said there must be a reserved price for each lot, or group of lots of the same kind of land; but such a reserve cannot be fixed without a careful valuation by an expert; so that we should require two processes, both involving some expense, first the valuation, then the auction. The result would be, that in some cases, where there happened to be little competition, the land would be let at the reserved rent, while in other cases—perhaps a few months later, and in the same place—similar land would be run up by competition in much higher rents; and this would inevitably lead to dissatisfaction, and inequality in the prosperity of the tenants,—a dissatisfaction which would compel the authorities to adopt the plan of letting all land at the reserved rent, that is at that fair but low rent which should have been adopted at first.

I maintain, therefore, that it is essential to adopt from the first the only just and equable method, which is, the valuation of the lots by an expert, founded on what would be fair rents of similar land to a large tenant farmer. These lots, with the rents thus determined, would then be open to selection, either on the system of "first come first served," or if thought fairer, of a ballot for the order of choice on certain fixed days. By either of these two methods, supposing the valuation to be fairly made, there would be no inequality of opportunities, no feeling that either by chance or through any other cause, some of the tenants were paying higher rents than others.

Another point of some importance is, that men should be allowed to have as much land as they wished, up to a certain limit—say five or ten acres according to circumstances; and also that the land first let should always be that abutting upon roads or lanes, the inner portion of the farms thus let being reserved for some years, so that any man wishing to add to his holding could have it extended by taking a plot or field behind it, thus avoiding the great inconvenience and loss arising from the separation of plots under one holding. In the meantime this central portion of the farm could be let by the year to any adjacent farmer.

One other point arises in connection with this question, the vital importance of security to the occupier and cultivator of land. We want men to be able to form and keep a home; to be practically as secure in that home, so long as they pay the moderate ground-rent for the land, as if they were the actual owners of the freehold, subject only to the payment of taxes. We want the new tenants under land nationalisation to be really free holders in the old sense—free men holding land from the community never to be interfered with so long as they continued to pay the moderate dues and to be law-abiding citizens. To give this full security all the rights of bequest or sale now appertaining to freehold land should appertain to these state tenancies.

It is, I believe, only by some such process as that which I have here indicated that we can possibly obtain the full benefit of land nationalisation or of the first steps which we may be able to make towards it. We must always remember that the community will be benefited just in proportion to the well-being of the cultivators and of those who obtain access to land. If we rack-rent them so that they just make a living out of the land, they will have little influence in raising the wages of other workers or in enabling them to make a successful bargain with capitalist employers. But if, on the other hand, we allow all occupiers of land to have it on such terms and conditions that they are able to make a good living, provide well for their families, and enjoy an old age of secure repose in their own homesteads, this state of well-being will serve to establish a standard of living which will react on the whole working population, and lead to a corresponding rise in the rate of wages throughout the whole industrial world. There appears to me to be no proposition in the domain of political and social science more certain than this. On the other hand, no mistake can be more fatal than to think that the community would be benefited by screwing from 20 to 50 per cent. more rental from the occupiers of land, thus reducing their profits, rendering their position less secure, and lowering their standard of living, and with it that of the whole working population.

In conclusion, therefore, I would urge most strongly, that in all our arrangements or proposals with regard to the land, we should throw aside altogether the idea of getting the highest possible rents, but should always aim at the maximum of well-being for the cultivators. By thus acting we shall best secure the equal well-being of the whole of the industrial community, and shall initiate that progressive improvement, with the diminution and ultimate abolition both of enforced idleness and of undeserved poverty, which is the whole aim and object of land nationalisation.

* * * * *

Perhaps the single most intriguing aspect of Wallace's social theorizing was the degree to which he managed to combine directly socialist and libertarian (usually referred to by Wallace as 'individualist') ideals. Socialism and libertarianism outwardly seem to have little in common; Wallace, however, did not see things that way. His approach to the matter, not surprisingly, stems from his two-level treatment of the evolution of man. Wallace envisioned a process of evolution culminating in beings characterized by sympathetic activities; that is, among whom chains of mutual support and understanding would be established. The material constraints of existence were to be minimized by an efficient society-level plan of resource-sharing and an increasing appreciation that individual goals of material gain were ultimately counterproductive (that is, counter-evolutionary). Consistent with the 'social consciousness' portion of this vision, Wallace supported the generally socialist positions that workers should unite against over-exploitative capitalist employers, that fundamental physical resources such as land, historically significant monuments, and transportation facilities be nationalized, and that basic health, insurance, and education services be made available to all. His sympathy for the plight of the average labourer (though not, at that point, for socialism in general) is evident in an essay published in The Christian Socialist *in 1884:*

The morality of interest—the tyranny of capital (ref 370)

Having read Professor [Francis] Newman's defence of interest and your remarks thereon, I wish to make a few observations on the general question.

Your position, and also that of Mr. [John] Ruskin, appears to be that money should be lent only as an act of benevolence or charity, and that lending it any other way is not only, in most cases economically and socially, injurious, but is also morally wrong. With the first part of this proposition I am very much inclined to agree, but not with the second. Looked at broadly, I believe that the *power* of obtaining interest on capital, however great, with the corresponding *desire* of the owner of capital to obtain interest on it, is, next to the private monopoly of land, the great cause of the poverty and famine that prevail in all the most advanced and most wealthy communities. To prove this would occupy too much space; but I may just notice that bankruptcies, with the widespread misery they inflict—the speculations of promoters and financiers often bringing ruin on hundreds or thousands of deluded investors—and the vast loans to foreign despots, which can only be paid by the sweat and blood of their unfortunate subjects, are the direct, and in the present state of society, the necessary results of the interest-system. Professor Newman says that if it were to cease, business would be lessened by one-third. But only rotten and speculative business would be stopped; commercial men would then *be* what they now only appear to be, and no really necessary business would cease to be carried on. The late William Chambers has stated (in his *Life of Robert Chambers*) that their vast book-selling, printing, and publishing business was established and carried on from first to last without one penny of borrowed capital; and that, as a result, panics and financial crises which brought ruin to some of their competitors, only caused them a little temporary inconvenience. I believe, therefore, that it would be for the benefit of the community if loans of money, or advances of goods on credit, were not recognised by the law, but were made wholly at the risk of the lender; but I do not see that it follows that he who lends, even under these circumstances, and takes interest for his loan, is doing what is wrong. For I cannot perceive any essential difference in principle between lending on interest, and selling at a profit. If I buy a shipload of drugs or any other goods at wholesale price, warehouse them, and sell them in the course of a year at the current market rate, making a profit of, say, 15 per cent. on my money, am I doing that which is morally wrong? Of the amount gained by me, we may put perhaps 1 or 2 per cent. for my personal trouble in the matter, 2 or 3 per cent. for risk of loss, 5 per cent. for interest on capital, and the other 5 per cent. for surplus profit. Is this 10 per cent. illegitimate gain? and am I morally bound to sell my goods at so much below the market rate as to leave me only fair payment for my time and risk? If it is wrong to take interest for the money which, when lent to another man enables him to do this, surely it is wrong to take a larger share in the shape of profit; and this really means that all trade is immoral which returns more than payment for personal labour, and insurance of the capital employed. But if so, it should be so stated, and the question should not be confined to interest on money loans only, and, in fact, Mr. Ruskin does not so confine it. The quotation you make from Mr.

Ruskin does not, however, seem to me at all to the point. You freely lend your friend an umbrella in his need, and you would even do the same to the merest acquaintance or neighbour, but if your neighbour called every day for your umbrella on his way to the City, and other neighbours followed his example, so that you ceased to have the use of your own umbrellas, you would soon have either to refuse to lend, or to charge a rent for the use of them, and if this were convenient to your neighbours, and they were willing to pay you sufficient to cover the wear and tear of umbrellas, your time and trouble in looking after them, and interest on your capital invested in them, it will require arguments very different from any yet advanced to satisfy me that you would be morally wrong in doing so. In like manner, though you lend your friend or neighbour cab-money, or give him a bed for a night on rare occasions when he urgently requires such aid, you would give none of these things repeatedly to a mere acquaintance. Yet, if circumstances rendered such accommodation very useful to a considerable number of persons, and you or someone else found it profitable to supply such accommodation, you would charge rent for your beds, and interest for your loans, and the transaction would differ nothing in principle from that of every tradesman who sells goods at a profit, of the innkeeper who charges beds in his bill, or of the jobmaster who charges for the use of his horses or his carriages. Nothing deserving the name of proof has yet been given that either of these things are immoral. Whether it is a good and healthy state of society, in which large numbers of persons get their living by such means, is another matter altogether.

The difference of opinion on this question of usury arises mainly from the different standpoints of the disputants. Seeing that it is bound up with many of the evils of modern society, and believing that it should have no place in a system of true Socialism, you and Mr. Ruskin denounce it as immoral. Professor Newman, on the other hand, looks at it as a question of modern society, and finds nothing *in its essential nature* contrary to justice, and here he seems to me to have the best of the argument. No doubt, in a more perfect state of society, in which private accumulations of capital were comparatively small, and the land and its products were freely open to the use of all, usury would have little place, because loans of money would rarely be needed; but when they were needed, I cannot see any grounds for maintaining that it would be morally wrong to lend money on interest. On the contrary, such loans would then retain their use without the evils their wide extension now brings. There would be no great capitalists, and if one man lent to another it would be a convenience to the borrower, and certainly some loss to the lender, because, as Professor Newman well puts it, 100 pounds paid a-year in ten years hence is *not* as valuable as 100 pounds paid to-day. To say that it is so is really to say that it has *no value* to-day, for if its payment can be delayed one year without loss it can two, or three, or ten, or a hundred, or a thousand! Where are we to stop? If we suppose a perfect social state, we suppose all men to be producers, and as capital is an aid to production, no man can give up the use of his capital to another without loss. The true solution to the problem is, I believe, to be found in the proposition that all loans should be *personal*, and, therefore, *temporary*; and that, as a corollary, the repayment of the capital should be provided for in the annual payments agreed to be made by the borrower, either for a fixed period (if he live so long), or for the term of his life. This would abolish the idea of perpetual interest, which is as impossible in fact as it is wrong in principle, while it would avoid the injustice of compelling one man, or set of men, to pay the debts of a preceding generation from which they may have received no real benefit.

This question of interest thus becomes involved in the wider question of the tyranny of capital over labour and its remedy. At present civilised Governments act on the presumption that great accumulations of capital are beneficial, and even necessary, to the well-being of the community, and all legislation favours such accumulations. When the people are once convinced that the reverse is the case, and legislation is directed to favour small holders of capital, and to check its inordinate accumulation, most of the evils complained of will cease. To this end the first step would be to get rid of all Government funds, guaranteed loans, railway stocks, etc., which are the main agents and tools by which capital is accumulated and money is made to breed money. This could be done in every case by making such stocks non-transferable after a certain date, and then declaring the payments to be terminable at the death of the holders and their living heirs, just as I propose to do in the case of landlords. The railways should be taken by the State, existing shareholders receiving annuities of the amount of their average dividends, payable in like manner to themselves and their living heirs. The Limited Liability Act should be repealed, because it has served only to foster the worst and most iniquitous speculations, and has deluded the public into the idea that they could safely share in the profits of commercial enterprises of the nature and management of which they are profoundly ignorant. There would remain no safe investments for money, except in some branch of agriculture, manufactures, or commerce in which either the investor or some relation or friend was personally interested, and thus would be brought about the diminutions and practical abolition of usury as a system, and of whole classes living idle lives on the interest of money derived from the accumulations of previous generations. Of course, it will be said that the plan here proposed is wholesale confiscation and repudiation; but a little consideration will show that it is nothing of the kind, and that it is really the best thing that can happen even to the individual holders of the stocks dealt with. In the case of the National Debt, for example, fundholders are now threatened with a reduction of interest of a quarter per cent., and later on of a half per cent.; and *they will be forced to accept it*, because the interest on the public debt regulates that of all other good investments, which will inevitably rise in price enormously if any considerable portion of the amount now invested in the funds seeks other investments. The offer to pay off fundholders at par will, therefore, be illusory, and the vast class who *live* upon their dividends will inevitably have their incomes reduced one-twelfth or one-sixth, while the cost of living goes on continually increasing. Would they not be far better off to have their present incomes secured to themselves and their living heirs? And when they fully realise their position, will they not choose the latter alternative if offered them? If the series of changes here sketched out were effected, the reign of capital as the tyrant and enemy of labour would be at an end. When the tools with which the financier and the speculator work no longer exist, the piling up of great fortunes will be impossible, and much personal care and attention will be required in order to make capital produce a steady return. Industry and commerce will be the sole means of acquiring wealth, and by these means alone—*under the new conditions of society*—very great wealth can never be accumulated by one man. For the land being nationalised, and the use of some portion of it obtainable by all, the minimum of wages will rise far above the starvation point which now prevails, and every village or other community, however small, will consist of small capitalists, who will be ever ready to unite for the safe employment of their capital. Then will arise a variety of industries on a scale adapted to the size and wealth of the district, and calculated to utilise the surplus labour and spare time of the surrounding population; and these small industries will compete successfully with the establishments of individual capitalists, because they will have an ample and a cheap supply of labour, and because most of their labourers, or their relations, will be share-holders, and will thus be working for themselves. The individual capitalist will then find himself paralysed for want of labour, unless he offers great temptations in the form of high wages and participation in the profits. For when a large proportion of the population are settled upon the land, and are able to devote their savings and their spare time to local industries, they will not, as now, be forced to become parts of a huge manufacturing machine in the success of which they have little personal interest.

By the methods here sketched out the labourer will receive, as Karl Marx and other social reformers maintain that he should do, the whole produce of his labour, and he will obtain this general result without any aid from Government, except what consists in remedying injustice, and removing the restrictions on freedom which now hamper him. Without any laws against usury, usury will practically cease to exist. Without any direct restrictions on wealth, those vast and injurious accumulations of wealth which now prevail will be impossible. The "stealers" and the "beggars" who now, as Mr. Girdlestone has shown, are so numerous among us, will steadily give place to "workers," and just in proportion as that happens, poverty will diminish, and will ultimately disappear. Now, a large portion of the working population are employed in the production of useless and often tasteless luxuries and trifles, the direct consequence of the large number of persons who have surplus money to spend after all their reasonable wants and comforts are fully satisfied. It is this, much more than the mere number of idle people, that is the dead weight which keeps thousands starving in the midst of so much wealth. When mere extravagant luxuries are less in demand great masses of labourers will be set free to produce the necessaries and comforts of life; these will be more abundant and cheaper (whatever their money price may be), and if all those who are now idle aid in the production of these necessaries and comforts, it is evident that, with free exchange, none can want.

I would particularly call attention to the fact that the results here indicated would all be brought about by carrying out the true system of *laissez-faire* now so much abused as if it had failed, when really it has never been tried. Labour, the sole source of all wealth and well-being, has been fettered in all her limbs, and harassed in all her actions, and then because she often stumbles or faints by the way, they cry, "See, she cannot do without help!" But first unloose your bonds, and cease to hamper her with your legal meshes, and then see if she will not achieve a glorious success. Let Government do its duty, and no more. Let it secure peace from external foes, and safety from internal violence; let it give free and speedy justice between man and man; let it secure to all alike free access to the land and all natural powers; let it abolish every monopoly of individuals and classes—either the local or central authority having the management of all institutions or industries which are essential to the public welfare, but which in private hands tend to become monopolies; and let it enact that all debts contracted by individuals shall be payable by those individuals only, and those contracted by the municipality or the State be payable by the generation which contracts them, so that they may never remain a burden on the succeeding generation. When it has done all this, then alone will labour be really free, and, being free, it will work out the well-being of the whole community without any Government interference whatever. This is the true *laissez-faire*; and this, I believe, will enable us to realise the best social state which, *in its present phase of development*, humanity is capable of. The distant future will take care of itself; let us try to improve the future that is immediately before us. I have here very briefly and imperfectly sketched out a series of measures which I believe are best calculated to promote this object, and they have the great and inestimable advantage that they all tend to the diminution of governmental interference with labour and industry, instead of that indefinite increase of it which the German Socialists advocate, and which, as the greatest political thinkers maintain, and as all experience shows, must inevitably fail, while in the present condition of civilisation it will probably lead to evils not less grave than those it attempts to cure.

* * * * *

His pro-labour sympathies notwithstanding, Wallace was critical of acts of organized disobedience. These were, he felt, too short-sighted in their objectives. In 1899, he castigated workers for supporting strikes that cut off production, suggesting that another strategy might realize more satisfactory long-term results.

The inefficiency of strikes—is there not a better way? (ref 560)

Has not the time come when the workers should cease to employ so rude, inefficient, and wasteful a method of improving their condition as by means of STRIKES? In most cases a strike effects little or nothing of a permanent nature, nothing but what may be lost within a year or two, nothing that tends to raise the whole body of the workers in any country. The strike may have been an essential weapon in the past—perhaps the only weapon the worker possessed.

Now, however, all the higher grades of workers are better educated, better organised, and have higher ideals. They have learnt the benefits of co-operation and of union; they have accumulated funds which may be reckoned by millions; and to waste those funds in keeping thousands and tens of thousands of men idle during a strike is one of those economic and social blunders which, in their effects, are often worse than crimes. Instead of keeping men idle for months, in order to obtain a small and perhaps temporary advance in wages or reduction of working hours, would it not be wiser to adopt a totally different method, one which would be much more dreaded by the employers, because it would tend to produce a permanent, instead of a temporary, rise of wages. That method is, *competition with the employers instead of strikes against them*; and it is to be effected by saving and accumulating all the money now spent in keeping men *idle*, and, as occasion arises, using it for the purpose of acquiring shops and tools by which the unemployed in each trade may be gradually absorbed and kept at *work*. Then, step by step, wage-earners would be withdrawn from employers' shops or factories to work in those of their union. Even if, at first, some of these shops were not able to pay full rates of wages, still the men would earn *something* instead of *nothing*, and they could hardly earn less than the usual pay during a strike.

Of course this could not be done all at once, but only step by step; each step, however, rendering the next step easier. The great thing is, to adopt the principle of never spending money in keeping men idle when it is by any means possible to keep them at work. The larger trade unions could probably carry out this method themselves after a few years' preparation, but to work it effectively a federation of unions would be needed; and when the need was clearly seen, this would soon be effected.

I venture to submit for the consideration of the workers that the principle of action here advocated is the only sound one. It alone tends in the direction of enabling them to become their own employers; while every step they take on this road, by withdrawing labour of all kinds from the cruel competition which compels men, and women, and children to work for the barest living rather than actually to starve, would inevitably raise the minimum as well as the maximum wage, and thus permanently benefit the whole body of their fellow-workers.

* * * * *

Wallace became a full-fledged socialist about 1889 after reading American writer Edward Bellamy's best-selling novel Looking backward. *In 1905 he offered his opinion as to how a change-over to a socialistic state might be effected in an essay contributed to the socialist newspaper* The Clarion:

If there were a socialist government—how should it begin? (ref 622)

A few weeks back a writer in the *Clarion* asked this question, and declared that he had made the same inquiry of many Socialist friends, and could get no satisfactory answer. None of them, he said, had the least notion of what could be done *first*, and *next*, and *next*—how, in fact, a capitalistic could be changed into a Socialist *régime*, quietly and systematically, without any compulsion, and yet in such a way as to secure before long the assent and co-operation even of the non-Socialist workers and capitalists.

Now, I quite agree with the writer that these questions should be answered, and that, if well done, it would greatly advance the cause of Socialism. He will, therefore, I am sure, be much surprised to hear that it has been *thoroughly well done*, in the clearest and most direct manner, in a book published eight years ago, and which ought to form part of every Socialist's library—Edward Bellamy's *Equality* . . . I propose to give a sketch here of the method indicated, which is, in my opinion, by far the best and simplest method yet suggested, and only needs to be thoroughly understood by all Socialists who really study their subject, in order that it may be adopted when the good time comes, and we have a Socialist majority in the House of Commons and a Socialist Government in power.

A few preliminaries must first be entered into. It is obvious, for instance, that such a Socialist Government will not follow immediately after such a Government as we have now, or even the most advanced Liberal Government we have ever had. There will inevitably be a series of Parliaments in which Radical and Labour members will be present in ever-increasing numbers, with ever-increasing power over legislation, till at length they will be in a majority, and will form a Radical Government. In such a Parliament Socialists will have great power and influence, till *they*, in their turn, obtain the actual majority, and the time will have come when they will have the power to carry their principles into action.

Now, during this period of continuous progress many things will be done, many principles established, which will greatly facilitate the work of the future Socialist Government. First, all the local public services—water and gas supply, all forms of electrical supply, the liquor traffic, milk supply, etc.—will be municipalised, and such general services as railroads and canals, and the working of coal and other minerals, will have been nationalized. And if the land itself has not been actually taken over by the State to be administered by the municipalities, we shall certainly have adopted the principle of one valuation as a basis for taxation or purchase whenever the State or any local authority requires land for any purposes of public utility.

All these important measures will be greatly facilitated by the preliminary adoption of a great ethical principle—a principle which is, in my opinion, the most indisputable, the most far-reaching, and the most beneficial that has ever been proposed as the basis of economic legislation and social reform. This principle is founded on the axiom that the unborn have no exclusive rights to property, and its full development in the proposition that all inequality of inheritance is unjust. Observation and reason alike prove that all inheritance which enables a man to live idly without giving any adequate service in return for his wealth is an injury to him who receives it, that it renders him the centre of a vast circle of evil influence through his numerous parasites and dependents, and that to permit it is one of the greatest of crimes against society. So soon as this great principle is accepted as a rule of conduct—and I have as yet met with no independent thinker who rejects it—all inheritance will be strictly limited to the direct living heirs, at the period of legislation, of all who possess property. Hence, as numbers die every year without such direct heirs, as years pass away more and more such persons will die, and there will pour into the public treasury an ever-increasing stream of wealth, including land, houses, manufactories, etc., so as to gradually replace all forms of taxation, and also provide ample funds for carrying on the various public services above alluded to, and especially such as are required to absorb the whole of the unemployed in productive labour.

As the application of this great principle will not take one penny from any living individual, while by automatically replacing all other forms of taxation it will benefit all alike, not only by adding to their incomes, but by cheapening all articles which now pay duties, and by removing the periodical annoyance of the demands of the tax and rate collector, it must surely enlist the support of the whole body of those who in any way earn what they live upon; while even the various classes of property-owners will hardly oppose it with the same energy that they certainly would oppose any demand upon their own pockets. It is, therefore, pre-eminently a case of working on the line of least resistance. If it is asked, What will become of the children of property-owners born after the passing of such an Act? the reply is that they must be educated by their parents so as to be able to earn their own living, instead of becoming parasites upon society; while, in the case of those children whose parents die during their childhood, they will become wards of the State, will receive the very best training and nurture possible, with adequate provision to enable them to become useful members of the community—will, in fact, be far better off than the younger sons and daughters of many considerable land owners are now.

What the application of this great principle would really do would be to establish absolute "equality of opportunity," the inherent justice of which has been admitted by such opponents of Socialism as Herbert Spencer and Mr. Benjamin Kidd. It would not necessarily lead to Socialism, but rather to a perfect *individualism* under equal and fair conditions, and this would almost certainly, as I have urged elsewhere, bring about universal *co-operation*, which might or might not lead on to Socialism.

What I particularly wish to point out here is that the principle of non-inheritance would so quickly bring in an enormous revenue as to render it possible for the State to take over so vast a business as the railroads, and to reorganise them in the public interest; and also, whenever it was advisable, do the same with all the educational institutions of the country. In all such cases as the railroads, the minerals, etc., there would be no question of purchase—no need to raise a single pound of capital. The State would simply take over their management in the interest of the public, paying to the owners or the shareholders the same dividends or profits as they had received on an average of the last ten or fifteen years, to be continued to their living heirs. Just in the same way the vast National Debt would be automatically extinguished by the dying out of existing fundholders and their direct heirs, thus again furnishing ample funds for carrying on any undertakings which the general interest rendered necessary.

Having thus endeavoured to forecast what should be, and might probably be, the position of a Socialist Government when for the first time it came into power, I will proceed to point out how such a Government might begin its work so as to fulfil the conditions stated in the first paragraph of this article, and in doing so I shall follow the general lines of Mr. Bellamy's forecast.

As a result of the various undertakings already under Government control, there would now be a considerable army of civil servants, which with their families would probably amount to four or five million individuals, and the very first step of a Socialist Government would be to give to these the benefits of organised productive labour. For this purpose public service stores would be established for the supply of all the necessaries and comforts of life, of the best quality, and at cost price. At first these goods would be purchased wholesale; but this would be quickly followed by the establishment of Government or municipal farms and factories, in which all the more important necessaries would be produced. As the quantities annually required would soon be ascertained, there would be no loss from surplus stocks; the whole enormous expense of advertising would be saved; the cost of distribution would be greatly reduced; while the large profits now absorbed by the manufacturer, the middleman, and the retailer would all go to reduce prices to such an extent that the bulk of goods consumed would cost from 20 to 50 per cent. less than under the competitive and largely monopolistic system of production.

When these centres of production and distribution were in full working order, another useful class of services would be established in the form of cooking, laundry, and housework agencies, by means of which all the heavier and more troublesome kinds of household labour would be saved, and food of all kinds prepared in the best and most wholesome manner at far less cost than the majority of individuals had before paid for greatly inferior food and service. The meals would be supplied either at numerous central restaurants or at the home, as desired, in the latter case at fractionally higher rates.

It must be kept in mind that the various preliminary reforms I have alluded to as certain to precede the establishment of a Socialist Government would already have greatly raised the rate of wages of all kinds of labour through the acquisition of land by local authorities and the absorption of all the unemployed in self-supporting colonies. The result would be that when so much additional labour was absorbed by these various public undertakings, the supply of labour for private capitalists would be so much reduced that wages would rise still further, a rise which would, of course, regulate that of all who were employed in the public service, while the latter would enjoy the advantage of greatly reduced prices, which would gradually be extended to almost every article of general consumption.

Throughout the whole of this process there would have been a continuous approach to an equalisation of earnings by all Government and municipal employees, an equalisation which would first have arisen through the principle of "equality of opportunity" being applied to education, when it would be seen that—*all* labour being equally necessary in its proper place, and everyone being, so far as possible, employed in that kind of work for which he was best fitted—there was no justification whatever for any large difference in the payment of different classes of workers; and this principle would, of course, have been carried out still more completely by a Socialist Government. From this approach to equality in education, in speech and manners, and in general requirements, there would arise such a large demand for the better qualities of all goods as to reduce the useless and wasteful variety now made, thus tending still further to reduced cost.

On the other hand, the abolition of inheritance of wealth would gradually lead to the entire disappearance of millionaires and of the greater part of individual wealth, and with this disappearance the demand would cease for that enormous mass of toys, and jewels, and tasteless frippery now made chiefly to tempt the idle rich to expend their unearned money; hence a great mass of hitherto wasted labour would be available for the production of useful and enjoyable products for all.

Another important result of this non-competitive supply of goods to all public employees would be the gradual displacement of metallic currency, and the proof it would afford that any such currency was wholly unnecessary. As the Government and municipal stores would be strictly limited to public servants and to the amount of their wages and salaries, the ordinary currency would not be received in payment, although wages would for some time continue to be paid in money; but every public servant could obtain, in exchange for the whole or any part of his wages, Government coupons to the same nominal amount, which would be received in payment for all their purchases at the stores. By this simple arrangement it would be impossible for the outside public to obtain the advantages of the stores without entering the Government service. The result would be that gold would depreciate in value as compared with these coupons, and for the first time in the history of civilisation it would be seen that paper money *representing productive work* is the only true standard of value and the best instrument for exchange.

When non-competitive supply of goods to State and municipal employees had reached this stage its advantages would be so obvious that great numbers of those still working for private capitalists would also demand State employment, and this would lead to the continuous extension of public factories and workshops. As this went on the capitalist manufacturers would find it impossible to obtain workmen, and would be obliged to close their factories or hand them over to the Government, receiving in exchange an annuity in coupons based upon the actual selling value of their property; while they themselves could, if they wished it, enter the public service as managers.

It is needless here to enter into further details. It is quite evident that the process here described would go on continuously till, practically, the whole population would have entered the ranks of the Co-operative Commonwealth; and, coincidentally with this inflow, gold and silver currency would steadily diminish in value, till at last it would be absolutely worthless as money, retaining only its value as metal useful in the arts or for ornament—a value which would certainly dwindle to a mere fraction of what it is now. And during the whole of this gradual development of the new social organisation not a particle of compulsion would be needed. There would be a free and open trial of the comparative merits of the two systems side by side. No single step in the measures of the Government could be objected to by the anti-Socialist capitalists and workers, since it would be exactly the same kind of competition which they had always carried on with each other, by which the big capitalist had driven out the small one, and the trust or syndicate had swallowed up the individual manufacturer. The great difference would be that while the result of the competitive system had been to create a regiment of millionaires at the top and an army of unemployed at the bottom—mansions and palaces on the one hand, and the slums and cellars of a thousand cities on the other—the adoption of the Socialistic method of co-operation had abolished all these iniquities and horrors, had rendered it impossible for any human being to suffer want in the midst of wealth, or to be driven to suicide after a hopeless struggle for work.

One more word before I close. Some of my readers may make the objection that as soon as the Government began this system of wholesale production large numbers of capitalists and manufacturers would go abroad, and establish factories where they could obtain cheap labour, thus "impoverishing the country." But if they did so it would merely facilitate the measures of the Socialist Government. They could not carry away their factories, and their money as we have seen would not be wanted.

Numerous other details are discussed by Mr. Bellamy, and I strongly recommend such of my readers as are interested in the subject to study his book. I venture to think that I have, as I proposed to do, pointed out at least one very simple and thoroughly effectual method by which a Socialist Government, whenever it comes into power, could proceed at once to give effect to the fundamental principles of Socialism.

* * * * *

Wallace wrote on railway nationalization on several occasions, one being the following letter to the Editor of The New Age *in 1908:*

Nationalisation, *not* purchase, of railways (ref 661)

In your "Notes of the Week" (September 5th) you remark that railway companies are combining "for the sole purpose of selling their lives dearly" in anticipation of nationalisation. As it seems to be a very general idea, even among Socialists, that the two operations—nationalisation and purchase—must go together, will you allow me space to point out that, while nationalisation is in the highest degree advisable, and may be effected at once by a very simple enactment, purchase is equally unadvisable and unnecessary, and had far better be left till a much later period, when in all probability some general method of dealing with similar claims to other forms of nationally produced wealth may be found practicable.

I may take it for granted that every reader of *The New Age* recognises the advantages to the public, as individuals, in the whole of the railways being worked with the sole view of the maximum of use and enjoyment of the people, so far as is consistent with the safety and well-being of the great army of employees, which will itself tend to secure the safety of the public; while to the nation, this complete unity of organisation and management will be of incalculable advantage as a safeguard against foreign invasion.

But these, and many other collateral advantages will accrue, just as certainly, and even more rapidly, by the State taking over the fixed and rolling stock of the whole of the railways, to be managed and worked in the public interest, while continuing to pay to the present owners of the railways—the shareholders and possessors of every kind of railway stock—that proportion of the net profits to which they are now equitably entitled.

It is, I believe, generally estimated that the economies which would be effected by the co-ordination of the whole system would amount to many millions annually, and this great saving would all be expended in reduced fares, better services, higher wages, and shorter hours of work, by which all shareholders and employees, as well as the whole of the public, would greatly benefit.

But the increased facilities to all who use the railways, and the abolition of the needless and often irritating restrictions of most of the existing managements, would certainly lead to a large increase of traffic, and thus render any considerable discharge of existing railway employees unnecessary, while the position of all would be much improved.

It is needless here to go into the question of the exact future status of the shareholders. As one mode of dealing with them, I would suggest that the relative market-value of each kind of railway security having been ascertained, with due regard to the condition of the line and rolling-stock, the holders of these securities should be offered in exchange government annuities for their own lives, and that of the legal heirs, in the direct line, living at the time of the owner's death, these annuities to be for amounts approximately equal to the dividends or interest they had received on an average of the three preceding years. This fixed and certain annuity would be fully equal in value to the less secure and fluctuating railway stocks. Those who declined to accept this mode of payment would receive whatever dividends the Government should declare to have accrued, after full provision for the upkeep of the line, efficiency of the service, and reduction of debt.

I believe myself that a majority of railway shareholders would accept the annuities, and this would lead to the possibility of the railways becoming unencumbered national property in two or three generations. Debenture-holders would, of course, be gradually paid off at par out of profits.

The special advantage of such a mode of nationalising our railways is, that it involves no vast financial operation of valuation and purchase, certain to be disadvantageous to the public—an operation which so many people think an insuperable objection to nationalisation. It secures all the advantages of public management, and at the same time safeguards the equitable interests of the shareholders.

The method is, of course, applicable to the acquisition of every kind of property in the hands of corporations, which should belong to the community.

* * * * *

It must be understood that underlying the rhetoric of Wallace's socialist message is a plea for the liberation of the individual. Believing rampant capitalism promoted class distinctions and emotional subservience to the idea that amassing material wealth was life's primary object, he regarded socialism as a vehicle through which 'equality of opportunity' could be realized, and the individual's dreams of material accumulation eventually refocused in more productive directions. It is significant that despite his very early introduction to socialist concepts, he remained unconvinced as to their practicality for over fifty years. The programme described in Bellamy's book Looking backward *changed his mind. But he would have nothing of social forces that pressed for arbitrary 'improvement' of the human race through directed wholesale elimination of 'inferior' races or individuals—this confused mere physical change with moral (spiritual) evolution. His opinions on the matter were related straightforwardly during an interview in 1912 (ref 750), the source of the following passage:*

. . . But you must not dream that I approve of any of the modern eugenic heresies that are now being advocated. I feel a little sore on this point because in a popular scientific publication that has just been sent to me, I am referred to as spending the evening of my days in furthering the teaching of eugenics. Wherever did I advocate any such preposterous theories? Not a reference to any of my writings; not a word is quoted in justification of this scientific libel. Where can they put their finger on any statement of mine that as much as lends colour to such an assertion? Why, never by word or deed have I given the slightest countenance to eugenics. Segregation of the unfit, indeed! It is a mere excuse for establishing a medical tyranny. And we have enough of this kind of tyranny already. Even now, the lunacy laws give dangerous powers to the medical fraternity. At the present moment, there are some perfectly sane people incarcerated in lunatic asylums simply for believing in spiritualism. The world does not want the eugenist to set it straight. Give the people good conditions, improve their environment, and all will tend towards the highest type. Eugenics is simply the meddlesome interference of an arrogant scientific priestcraft. There are no really bad people; no one absolutely beyond reclaim. That is where our prison system is all wrong. We treat our prisoners as though they were utterly bad. There are none utterly bad, but only different degrees of goodness. When we understand that, we shall give up our absurd ideas of punishing crime, and shall, instead, try to reform the criminal . . .

* * * * *

'Might is right' arguments also made Wallace shudder. The measure of success at the human level of evolution should be, he felt, the degree to which social strategies invited sharing rather than domination (a theme later borrowed, and further developed, by P. A. Kropotkin in his Mutual aid*). Consider the following selections, drawn from letters printed over the period 1898–1900:*

Imperial might and human right (ref 579)

It is to me very distressing to see Mr. Bernard Shaw exercise his great talents and his caustic wit in paradoxes and verbal quibbles calculated to make the enemy rejoice at the dissensions among Socialists; and at the apparent absence of agreement among them even on fundamental principles of politics and of ethics. Passing by Mr. Shaw's doctrine that we Socialists should uphold "robbery under arms" of any land or property which we think we could make a better use of than the present possessors—the doctrine that might is right for Socialists as for burglars—I will now only say a word or two on his statement that "independence and liberty produce not freedom, but slavery." Of course, Mr. Shaw can prove that he is right by taking only one meaning of the words, and that not the generally accepted meaning. Independence in the individual means that he is not the slave or the servant of another man, either directly or indirectly, as a serf, tenant, or wage-thrall. It does *not* mean that he must be absolutely self-suffering, without any help from his fellows either through friendship or co-operation, or social organisation. Yet it is only by adopting this last sense—a sense only justifiable etymologically—that Mr. Shaw's statement has a shadow of truth.

And as regards communities or nations, independence has but one meaning—self-government as opposed to government by an outside power which has annexed, purchased, or conquered some smaller and weaker people. Whether such an enforced government is relatively good or bad, it is still slavery for the weaker people, and, like all slavery, is demoralising to both parties. With all my heart and soul I protest against and condemn the doctrine that we have any right to force our rule upon people who do not want it, under the pretence of better government. I maintain that force is *never* the better way, and that every people should be left to develop their own government, aided by advice and example, but never by compulsion. No truth is, I believe, more certain than that stated (I think) by Mill, that the worst government of a people by themselves is better than the best government by foreign conquerors. To my mind, Socialism can only come about voluntarily. Compulsion, whether of individuals by the majority or of weak nations by stronger ones, is not only ethically wrong, but is antagonistic to all real progress towards the hoped-for Co-operative Commonwealth.

* * * * *

Letter to *The Eagle and the Serpent* (ref 580)

I can understand Dr. [Ragnar] Redbeard's position, though I cannot accept it. If men were only "herds of animals" his view might be the true one. But the mere fact that men, everywhere, and throughout history, have had words and ideas corresponding to *truth*, *justice*, *virtue*, *right*, and that there have always been men who would sacrifice even their lives for these ideas, proves that mankind is *more* than an aggregation of "herds of animals." The mere physical struggle—the rule of the biggest and strongest brutes among men—is not therefore conducive to man's advance, prosperity, happiness.

Again, something *may* be said in favour of the struggle of individual with individual man, as leading to the survival of the best physical types. It may be said that such a struggle is good for humanity. But no such advantage can be predicated of the struggle between communities very unequal in numbers. Forty millions, even though mostly fools and scoundrels, may be able to destroy half a million of far higher average type mentally, morally, and physically. The massacre of St. Bartholomew was such an exercise of *might*; but it was not a benefit to the race, it was opposed to our ideas of justice and morality both now and then, and was therefore not *right* . . .

* * * * *

Letter to *The Eagle and the Serpent* (ref 540)

If Mr. [Thomas] Common's statement of Nietzsche's teaching and the social reforms at which he aims, are accurate, then, even though some of his *methods* of obtaining social reforms may be good, the reforms themselves seem to me to be both impracticable and worthless, if they are not even retrogressions. Mr. Common tells us that Nietzsche is the apostle of "a true aristocracy," and of apportioning "advantages and disadvantages respectively to merits and demerits." If by "advantages" he means material superiority or greater wealth, and that the aristocracy of merit claim this superiority as their right, that alone would, in my opinion, show that they were *not* a true "aristocracy" and that they did *not* really "merit" what they claimed. Again, what is merit, and who is to decide on the merits and demerits of individuals? If it means intellectual, moral, or physical, superiority, or any combination of them, and if these qualities are fully exerted for the benefit of society at large, those who possess and so use their superiority will, under any rational condition of society, receive the greatest reward men can receive—the respect, honour, and affection of their fellows. But such men can only *prove* that they possess such superior qualities and that they are worthy of the honour they will receive, by working and living under *equal* conditions and *equal* advantages with their fellows. Without this absolute "equality of opportunity," there can be no possibility of accurately determining "merit and demerit" as regards society; hence, I maintain that the only object worth working for, as the first and essential stage towards utilising all the best powers and faculties of a nation for the common good, is, to bring about this "equality of opportunity." This, however, is simple justice, as between man and man. It is a fundamental axiom of ethics. It is not an "esoteric" doctrine, and it does not need to be upheld by "falsehood," as apparently does Nietzsche's system of aristocracy—and from falsehood, esoteric teaching, and a ruling aristocracy, nothing that is of permanent good ever has arisen or can arise . . .

* * * * *

Darwinism in sociology (ref 549)

... There is little to object to in the first four principles ... [Mr. Common] ... lays down, except that they imply ... an exclusive application to "physically and mentally superior individuals" instead of to all. This would lead, logically and practically, to a restricted and perhaps hereditary oligarchy of the stronger and less moral individuals, with the slavery or destruction of all who opposed them. This is more plainly indicated in ... [the fifth principle] ... where "greater advantages" are claimed for the "better class of individuals," which really means the stronger, more cunning and more self-seeking, who, believing themselves to be the best, will enforce that belief, as to a considerable extent they always have done and do now. Mr. Common nowhere suggests that these "superior individuals" are likely to be restrained by any moral considerations in their efforts to obtain those "greater advantages" which he declares are their due.

Under such an oligarchy, once established, what chance would there be for the inventors, the thinkers, the artists, the moralists—the real aristocracy of the human race—except so far as they helped to increase the wealth and luxury and power of the "superior individuals." The slavery and tyranny, the assassinations and poisonings, of Greece and Venice, would inevitably be reproduced in any society whose fundamental principle was, that social and economic inequality was right, and that the stronger—physically or mentally—were justified in using their alleged inferiors for their own ends. Such a system would bring back the old bad days of—"Truth for ever on the scaffold, Wrong for ever on the throne"—and I challenge Mr. Common to prove that it could lead to anything else.

One would think that the evils of class-privilege and every other form of aristocracy had been sufficiently demonstrated in the past, to render it certain that this is *not* the road by which the future welfare of humanity is to be attained. True social excellence, to use Mr. Common's own term, can only be reached by utilising all the powers and capacities of the whole community, under the conditions best adapted to develop them; and no other means of doing this has been suggested but that "equality of opportunity" and of economic status which Mr. Common declares to be absurd. But his statement on this point, in his principle 5, shows that he entirely misunderstands the grounds upon which it is advocated. His argument as to agriculturists not giving the same chance to inferior plants and to weeds as to the crops they are raising, is a false analogy, and is entirely beside the question. The true analogy of men wishing to improve men, or rather of communities seeking their own highest well-being, is not that of the farmer and his crops, but of the plant-world seeking to improve itself, and having some power, though limited, of doing so. In that case it is quite certain that many of the best, in our opinion, that is the most beautiful and highly developed in flower and foliage would be exterminated, while the quickest growing trees and the coarsest and most vigorous weeds and climbers would alone survive. They would claim to possess the highest plant-excellence; they would therefore claim greater advantages of situation, soil, etc., and would only permit the existence of what they would term inferior types so far as not to interfere with their own well-being.

Again, when he says that semi-idiots should not have a university education, he equally misunderstands the principle of equality of opportunity. This does not mean the *same* education for all—that is the present absurd plan—but equal facilities for developing all that is best in their natures, so as to enable them to spend happy and useful lives; for there is probably no human being that could not attain to this condition by proper nurture and training . . .

But perhaps the most erroneous and most vicious of Nietzsche's principles, according to Mr. Common, is that enunciated in the last sentence of [his principle] 5—"And it is still more absurd to advocate, . . . that the inferior class should be allowed to breed like vermin, and that their spawn should be supported at the cost of the better classes." In order to understand what this really means we want a tolerably accurate definition of the terms "inferior" and "better." We may take either of two extreme definitions—that the inferior are the criminal classes only, or that they are the working as opposed to the professional and propertied classes. If we take the latter and more popular definition we have to remark first that the facts as to who supports who, are the exact opposite of Mr. Common's statement, since the workers undoubtedly support the non-workers, and the children of the non-workers as well as their own. Secondly, it is generally admitted that the "better classes," as above defined, would either remain stationary in numbers or would even decrease if they were not recruited from the "inferior" class. The more rapid increase of these latter is therefore a benefit to the nation.

Now considering the alternative view, that the criminals are to be considered as the inferior class (compared by Mr. Common with vermin), who are not to be allowed to breed, we shall at once see that this is not only impracticable, but would be probably hurtful to society. It is a mistake to assume that those whom we now punish as criminals are essentially bad in nature, or are even, on the average, at all inferior to the majority of those who lead ordinarily respectable lives. Many highwaymen of the last century and some burglars of to-day, are men of exceptional ability who, under favourable conditions of education and nurture would become esteemed and useful members of society. A considerable proportion of the population of New South Wales, before the gold discoveries, consisted of released convicts and their descendants, and these were in no way inferior to the average of honest emigrants. So, the waifs and strays from the gutters of the East End of London, when rescued and trained by Dr. Barnardo, are found to be quite up to the average of respectable citizens. These examples support the view so strongly urged by Herbert Spencer, that what are termed the lower, the middle, and the upper classes are fundamentally alike, morally and intellectually. Their differences depend wholly on their environment and early education, and there is no reason to believe that the average of inherent vice or passion is greater in one class than in the other. But owing to the great differences in the environment and economic condition of the three classes their vices are exhibited in widely different ways, and those of the upper and middle classes, though equally great, do not so often bring them within the meshes of the law. The aristocratic bankrupts or gamblers, and the middle-class promoters or directors of bubble companies, are often worse morally than the bulk of our criminals, but they are not so often found out . . .

The idea that society can be improved by forcible interference with the increase of the various classes is one of the most wildly impracticable as well as one of the most dangerous and unnecessary of all the panaceas for remedying some of the evils of our social system while leaving the real causes of those evils absolutely untouched. The dread of too rapid increase of the population, always brought forward by the opponents of real social reform, is a mere figment of the imagination, which will altogether vanish under a rational organisation of society. Those who marry earliest and breed fastest, are those who are economically worst off—the casual and unskilled labourers; the higher class, of skilled labourers, marry later; the mercantile and professional middle classes later still; while the landed and capitalist aristocracy marry latest, and increase slowest or not at all. Yet, in face of these palpable facts, the objection is continually made, that if *all* are well-off and free from anxiety as to a comfortable living, population will increase so rapidly that *all* will soon sink to starvation-point. No reason whatever, founded on actual facts, is given for this statement, and some of the most important of the factors that would come into play and which would lead to an opposite result, are left wholly unnoticed. These factors I have set forth in an article on 'Human Selection' [ref 427] in the *Fortnightly Review* . . .

Many other points call for notice, but I need here only mention that Mr. Common is in error when he refers to myself as having shown that "falsehood, imposture, and hypocrisy" play a large part in nature. He quite misunderstands the facts of mimicry and protective resemblances, which are wholly involuntary, and cannot therefore imply falsehood, etc. But this, and all other arguments drawn from the animal or vegetable kingdoms are valueless, because in social and civilised man the mental and moral nature rules over the physical; and . . . a new and high kind of selection will come into action as soon as he learns how to subordinate the latter to the former, and how to so organise his social state as to satisfy the economic requirements of all. His power over the forces and products of nature is now so great, that, not only necessaries and comforts, but all beneficial luxuries could be enjoyed by every human being, with an amount of labour less than half of that now expended in giving hurtful luxury and idleness to some, while millions remain subject to a degree of want and misery greater probably than has ever been seen at any previous era in the world's history. The method of bringing about social excellence here suggested, works with nature, not against her. It depends upon the natural play of the higher qualities of human nature, and will therefore be both self-acting and efficient; while any forcible intervention of authority will be as certain to produce evil as it must be powerless for good.

* * * * *

Having travelled extensively and dealt intimately with many different native peoples in his earlier years—and having thereby gained considerable respect for them—Wallace strongly objected to the way the uncivilized countries of the world were being subjugated by the Western powers. In the following essay, published in 1906, he argued for a more enlightened policy toward colonial dependants:

The native problem in South Africa and elsewhere (ref 630)

It has been well said that no man is good enough to have absolute power over another, and the same may be said with still greater certainty of nations. It almost invariably happens that when any such despotism is attempted both the rulers and the men or nations who are governed by them suffer in character even more than in material results. It is now universally admitted that actual slavery produces these evil consequences to the individuals concerned, while few will deny that they are almost equally clear in the case of nations, often leading to the downfall or deterioration of the ruling state. Assyria and Egypt, Greece and Rome, are examples from the ancient world, while in more recent times we have the disastrous results of her vast American conquests to Spain, of the Napoleonic empire to France, of Ireland, India, North America, and South Africa to England.

In all these cases, just in proportion as despotic rule has been enforced, discontent, rebellion, and devastating wars have been the result; while on the other hand, whenever subject peoples, whether civilised, barbarian, or savage, have been permitted either wholly or partially to govern themselves, peace and contentment have followed to the equal benefit of both parties. As illustrative of such happy results we need only mention Penn and Lord Elgin in North America, Lord Lawrence in the Punjaub, and Sir James Brooke in Borneo. Equally suggestive is the fact, that some of the best governed and most contented parts of India to-day, are to be found among the Native Provinces, ruled by their hereditary or native chiefs with the advice of a British Resident and protection from external enemies by the British Government. So in South Africa, Basutoland and Bechuanaland, governed by native law and custom under their own chiefs, assisted by the sympathetic advice of a British Resident Commissioner, are thoroughly contented; but this happy state of things is only preserved by the strict prohibition of European settlement, and thus of the exploitation and so-called development of the country with its inevitable consequence, the deterioration both of the native and the European.

Surely these examples are of so striking a nature as to deserve our careful consideration, and we may well ask ourselves whether it is not possible to learn from them how to deal with those more complex and more difficult cases where two antagonistic races are more or less intermingled, and where the actual state of things is unsatisfactory to both, ranging from smouldering discontent as in many parts of India, to more open disaffection and resentment sometimes rising to what we term rebellion, as we now see in parts of South Africa.

The fundamental reason of success and failure in these several cases, is, that where we have succeeded we have done so by giving to the inferior race our protection and our advice, through some man of high character in sympathy with those he lives among and who exercises the minimum of compulsion. We have acted as the disinterested friend, not as the mere self-interested conqueror and despotic ruler. However much we may protest to the contrary, the latter is our position in South Africa to-day, and it is also our position over the larger part of India. In both these countries we make laws affecting the coloured races without consulting them in any way, and however much these laws may conflict with their mode of life, their customs, or their prejudices, we enforce them by means of armed police or bodies of troops, and if they resist we dub them rebels and remorselessly destroy them by means of our overwhelming superiority in arms.

In those cases where a large coloured population inhabits the same territory with a much smaller body of whites, we not only enforce against them our complex machinery of law and punishment, but we also often make special laws or regulations against the coloured races, which have the effect of stamping them with the badge of inferiority and serfdom. This treatment is the more galling to them now, because we have for many years striven to convert them to our religion, and have either educated them or assisted them to educate themselves, with the result that considerable numbers of them are now better educated and more truly civilised than a considerable number of the low-class Europeans who despise and ill-treat them.

It is evident that we have here a state of things which cannot go on indefinitely. We have drifted into it without any forecast of the necessary consequences. One set of influences and one trend of opinion, most prevalent in the Colonies, leads to the belief that the natives are the pre-destined servants or serfs of the whites, and sees no reason why they should not for ever be thus treated, kept in their place by force if necessary, and denied all the rights of freemen or of equals. On the other side we have the religious world and the philanthropists at home who have always exerted themselves to convert the natives to Christianity, to educate them, to force upon them our complex civilisation, and to claim for them, though rather half-heartedly, equal political and social rights. These persons almost ignore the antipathies and ingrained prejudices which everywhere manifest themselves where white and dark races are forced to live together, and which even the loudly proclaimed belief in the brotherhood of man, the equality of rights and the teachings of Christianity, are powerless to overcome.

On a calm consideration of the whole problem it must be admitted that the former point of view—that of inherently superior and inferior races—of master and servant, ruler and ruled, is the most consistent with actual facts and perhaps not the less fitted to ensure the well-being, contentment, and ultimate civilisation of the inferior race. It is also by no means incompatible with a just treatment of the native, with sympathetic interest in his welfare, and with the grant of a considerable amount of self-government; and it is for the purpose of suggesting how this latter may be effected that I am venturing to make a slight contribution to this very thorny subject.

The difficulty is of course very much increased in the case of a self-governing colony, composed largely of people who are seeking for wealth through the labour of the natives in agriculture or in various forms of commerce or industry, and whose interests and prejudices largely influence the government. It is therefore unwise, when the proportion of whites to natives is very small, to grant self-government to such a colony unless special reservations are made for the protection of the natives. At the present time there is a strong desire for self-government in Trinidad, and perhaps in some other West Indian islands, but it is asked for by the white traders and planters for the express purpose of obtaining complete power over the coloured labouring classes, free from what they consider the harassing regulations for their protection now in force. But as there is in the island a large population of negroes and mulattoes as well as coolies from India, races very antipathetic to each other, even the greatest friends of self-government consider that in this case it would certainly lead to injurious results.

The same difficulties occur in the case of the new colonies in South Africa and especially in Natal, where besides the very large number of natives there are also Indians in nearly equal numbers with the Europeans. The extreme diversity of ideas, customs, laws, and religion between these two dark races, renders the application of our laws and regulations to both of them, extremely irritating and often unjust; while, when the two coloured races come into conflict, the methods of our ordinary courts of justice are quite unfitted to hold the balance fairly between them. Hence there almost always arises a sense of injustice and oppression, which is increased by the fact that in many cases these people are well educated and have very strong feelings in regard to infringements of their personal freedom or the sanctity of their homes.

If, now, we are to be guided by the varied experience in all parts of the world here briefly referred to, it becomes clear that in handing over the government of Zululand to the Governor of Natal, who is necessarily very largely influenced by local opinion, a great mistake was made, which should be at once retraced and the country administered in the same way as the other protected states under the direct supervision of the home government, and with the same sympathetic consideration for native habits and prejudices as in the other Protectorates. Looking at this matter merely from a Colonial point of view, it would surely be better for Natal to have as a neighbour a contented and prosperous native state, with which a considerable trade would arise, and which would, as the population increased, afford a large amount of native labour when required.

We now approach the main problem, which is, how to provide some amount of self-government for the native population of Natal and of the two new Boer Colonies, so as to satisfy to some extent the just demands of the educated and civilised portion of the natives, and to do this without exciting the strong and very natural opposition to any suggestion of putting the coloured races on a political equality with the whites. What we have to aim at is, in the first place, to diminish the sense of injustice now felt by the educated and christianised natives, at being treated as a subject and degraded race, despotically ruled by aliens who, for the most part, take no account whatever of their feelings and claims as British subjects and fellow Christians. In the second place, we must proceed tentatively so as not to arouse antagonism in the ruling race, our aim being to give the better and higher among the natives an opportunity of freely stating their political and social grievances, so as to influence the legislature towards a more just and sympathetic treatment of them.

The first and most obvious thing to do is to give to the natives in every district of each Colony one or more chiefs or magistrates of their own race, chosen from the native clergy or schoolmasters or any other adequately qualified individuals. These native magistrates should sit with the ordinary magistrates, and in all cases, criminal or civil, where both natives and Europeans were concerned, would act as the official protector or advocate for the native in the interests of justice, and for the purpose of putting the native point of view before the European magistrate or judge, who would alone be responsible for the decision of the court.

In the case of disputes between or crimes by natives, in which no whites were concerned, the native magistrate would hear and decide the matter according to native law and custom, but modified where necessary in accordance with European law. Here too the Colonial magistrate would (at first) preside over the court, giving advice and suggestions to the native magistrate; but except in very difficult or important cases would allow the native magistrate to give the judgment of the court.

In this way it seems to me not only would much hardship and injustice to natives be avoided, but the effect on the native mind generally would be most beneficial, by showing a full recognition of their rights in the very large and complex department of the administration of justice. The association of the two races in this way would also have an important educational influence on both. The native magistrate would learn much from frequent official intercourse with an educated European, while he would gain thereby a dignity and influence among his countrymen which he would value highly, and which would have an important and beneficial influence on his own character and conduct. On the other side the European magistrate would be equally or even more benefited, since by association with educated natives in the hearing of the infinitely varied cases that come before him, he would acquire a knowledge of native character, customs, and feelings which he could obtain in no other way, and which might have an important influence on the welfare of the community, when, in after years, some of these magistrates became members of the legislature.

Another step of very great value and importance would be the introduction into all local authorities, such as education boards, district councils, town councils, etc., of one or more educated natives of each nationality (Kaffir or Indian) chosen to represent their fellow-countrymen, and to express their views and wishes as to any bye-laws or regulations which they found to be oppressive and unjust. Even if these representatives of the coloured races did not have votes, their constant presence at meetings, their right to protest against oppressive or one-sided regulations, often made in ignorance of the effects produced by them, could not fail to be of great value. They would also be able to bring before the authorities all cases of cruel or unfair treatment of natives by the police, a matter of the greatest importance in all countries, and especially in those where there are antagonisms of race.

Here again the educational influence on the ruling class of constant official communication with some of the best among the native races would certainly be very great, and would probably lead in time to a better understanding between them than has been created by the present system of aggressive rule of a superior caste over their serfs.

Perhaps more important still would be the application of the same principle to the Colonial legislature itself in both chambers. The native representatives need be few in number—perhaps three or four in the lower and one or two in the upper house, the object being in no sense to place the coloured race on an equality with the white, but to provide each branch of the legislature with accurate and precise information as to how both existing and proposed laws affect the natives, how and why they feel themselves injured or oppressed by them, and thus enable modifications to be made which, though apparently of trifling importance, may make all the difference between a condition of constant irritation and one of cheerful acquiescence.

As to the mode of election of these native representatives it would be best to leave it largely to themselves to suggest the best method. It would probably be most in accordance with their customs for local chiefs, ministers of religion, teachers, etc., to meet together and nominate perhaps twice the number of members required. From these names the Governor of the Colony, after due inquiry among local magistrates or others well acquainted with the natives, would choose the members, who would thus be not only elected by their countrymen, but nominated by the Governor as the representative of the Crown. As the chief function of these members would be consultative and educational as regards the opinions and feelings of the natives, it would not much matter whether they had votes or not; but I do not think it would be necessary or wise to make any invidious distinction against them, as their very limited numbers would not give them any real power except in cases where public opinion was already nearly equally divided as to the advisability of the legislation in question.

It is, I think, almost impossible to exaggerate the beneficial influence of some such small share in the government of a country as here suggested being awarded to the natives. Not only would it diminish, and in time destroy, the sense of injustice and oppression that now so largely prevails among the native races, but it would greatly extend mutual knowledge and give rise to that mutual respect which generally exists when men of different races and civilisations come to know each other intimately. It is to be hoped that in the constitutions now being elaborated for the two Colonies we have so recently devastated and conquered, some instalment, however small, of such a concession to just principles may be introduced; and, that powers may be reserved by the Crown for a further extension of the principle whenever it seems advisable. If that is done, and the results are found to be beneficial, other self-governing colonies may find it advantageous to adopt the plan.

So long as we possess colonies in which a considerable native population still exists we should, I think, always retain our guardianship of those natives in order to protect them from the oppression and cruelty which always occurs when a young, and mainly wealth-seeking community has absolute power over them. Where these natives are numerous and energetic, and are rapidly acquiring our education, our religion, and the outward form at all events of our civilisation, things cannot remain as they are. What the ultimate condition of such mixed communities may be it is difficult to say, but, whatever the future may have in store for us, it is certain that a method which recognises that the coloured races are men of fundamentally the same nature as ourselves, and which aims at developing the best that is in them, by granting them some at least of the elementary rights of men and citizens, is more likely to bring about a satisfactory solution of this difficult problem, than that system of contemptuous superiority and denial of all political and social claims that has hitherto so largely prevailed.

Having no personal knowledge of the country more particularly referred to in this article, I only put forward my views in a suggestive form. Forty years ago I had the privilege of enjoying the friendship of Sir James Brooke, and, during more than a year's residence in Sarawak, of observing the mode and results of his beneficent and sympathetic rule over antagonistic native races. A little later I spent several months in North Celebes, in Java, and in East Sumatra, where I had full opportunity of noticing the effects of the judicious rule of the Dutch, almost wholly exerted through native chieftains. For nearly twelve years I travelled and lived mostly among uncivilised or completely savage races, and I became convinced that they all possessed good qualities, some of them in a very remarkable degree, and that in all the great characteristics of humanity they are wonderfully like ourselves. Some, indeed, among the brown Polynesians especially, are declared by numerous independent and unprejudiced observers, to be both physically, mentally, and intellectually our equals, if not our superiors; and it has always seemed to me one of the disgraces of our civilisation that these fine people have not in a single case been protected from contamination by the vices and follies of our more degraded classes, and allowed to develope their own social and political organism under the advice of some of our best and wisest men and the protection of our world-wide power. That would have been indeed a worthy trophy of our civilisation. What we have actually done, and left undone, resulting in the degradation and lingering extermination of so fine a people, is one of the most pathetic of its tragedies.

* * * * *

Wallace was, as might be expected, a very vocal critic of militarism. In 1899 he was asked to submit an essay on 'The causes of war, and the remedies' to the French newspaper L'Humanité Nouvelle. *It was shortened for printing in French in that serial, but the full length version appeared in English in* The Clarion *shortly afterward:*

The causes of war, and the remedies (ref 567)

In response to a request from the Editor of *L'Humanité Nouvelle*, I here give my views, briefly, on the questions submitted to me. [These were: (1) Is war among civilised nations still necessary on the grounds of history, right, and progress? (2) What are the effects of militarism—intellectual, moral, physical, economic, and political? (3) What is the best solution of the problems of war and militarism in the interests of the future civilisation of the world? (4) What are the most rapid means of arriving at this solution?]

(1) Under the existing conditions of society in all civilised communities, and as a consequence of the principles and methods of government which prevail in them, war cannot cease to be more or less prevalent among them.

The conditions which almost inevitably lead to war are the existence of specialised ruling and military classes, to whom the possession of power and the excitements and rewards of successful war are the great interests of life. So long as the people permit these distinct and independent classes to exist and—more than this—continue to look up to them as superiors and as necessary for the proper government of the country and for the effective protection of individual and national freedom, so long will these rulers continue to make wars.

All civilised governments, whatever may be their professions, *act* on the principle that extension of territory and the absorption of adjacent or remote lands, so as to increase both the extent of country and the population over which they have sway, is a good in itself, quite irrespective of the consent of the peoples so absorbed and governed, and even when the peoples are alien in race, in language, and religion. Although they may not openly avow their acceptance of this doctrine, yet they invariably *act* upon it, though in some cases they think it necessary to make excuses for their action. They declare that such conquest and absorption is necessary for the national safety, for the increase of trade, and for many other reasons. The majority of the workers, and of educated people who do not belong to the ruling or the military classes, however, do not accept this principle. They more or less decisively hold the opinion that governments can only justly derive their power from the consent of the governed, and that all wars for territory and all conquests of alien peoples are wrong.

The reason of this difference of opinion is very simple. Every addition of territory, every fresh conquest even of barbarous nations or of savages, provides outlets and additional places of power and profit for the ever-increasing numbers of the ruling classes, while it also provides employment and advancement for an increased military class in first subduing and then coercing the subject populations, and in preparing for the inevitable frontier disputes and the resulting further extensions of territory. Wars and conquests and ever-expanding territories are thus found to be essential to their existence and continued power as superior classes. But the people outside these classes derive little, if any, benefit from such extensions, while they invariably suffer from increased taxation, either temporarily or permanently, due to increased armaments, which the protection of the enlarged territory requires. Almost without exception every war of modern times has been a dynastic war—a war conceived and carried out in the interests of the two great governing classes, but having no relation whatever to the well-being of the peoples who have been forced to fight each other. In every case the people suffer by the loss or disablement of sons, husbands, and fathers, by the destruction of crops, houses, and other property, and by increased taxation, due to the increase of armaments that always follows such wars even in the case of the victors. Hence the material and moral interests of the mass of the people of every country are wholly opposed to war, except in the one case of defending their country against invasion and conquest. They are therefore more open to the influence of moral and humane considerations, while they alone feel the numberless evils which war brings upon them. Except in very rare cases, a plebiscite fairly taken would decide against any other than a defensive war.

(2) To discuss the effects of militarism under the various heads suggested in the question would require much space and some special knowledge which I do not possess. That these effects have both good and evil aspects may be admitted. The evil effects have been often set forth and are sufficiently known, both in their vast extent and far-reaching consequences, while the greatest of them—the perpetuation of war and the desire for military glory—has already been alluded to. I will, therefore, confine my remarks to the partial good that undoubtedly exists in this fundamentally evil thing, chiefly for the purpose of showing that whatever good there is in it may be obtained in other ways which are as essentially humane, moral, and beneficial as war is essentially cruel, immoral, and hurtful.

The good that results from militarism arises wholly from the perfection of its organisation, of its training, of the habits of order, cleanliness, and obedience which the soldier soon learns are essentials to efficiency, from the social and brotherly life of the soldier, whether in camp or in the field, from the *esprit de corps* which grows out of its systematic organisation and companionship, leading to generous rivalry and to those deeds of heroism and self-sacrifice which are universally admired. And, further, every soldier learns by experience the marvellous power of organised labour under skilled direction to overcome what to the ordinary man seem insurmountable difficulties. He sees how foaming torrents or broad rivers can be rapidly bridged; how roads can be made over morasses or across mountains; how the most formidable and apparently impregnable defences are attacked and taken; and how a few bold men in a "forlorn hope," by the sacrifice of their lives, often insure the success of the army to which they belong. Many of the finest qualities of our nature are thus called into action by the soldier's training and during his struggle against the enemy; and so greatly has humanity developed among us that it may be fairly argued that these good effects more than balance the evil passions of cruelty, lust, and plunder which even now are to some extent manifested in every great war, though to a far less degree than even 50 years back.

But every one of these good results of militarism could certainly be obtained by any equally extensive and equally skilful organisation for wholly beneficial purposes. If labour, where organised for military ends, is so effective in results and so beneficial as a training, it would be equally effective and equally beneficial when devoted to overcoming the obstacles to man's progress presented by nature to the production of the necessaries of civil life, to sanitary works for the preservation of health, and to everything that facilitates communication and benefits humanity. If the same amount of knowledge, the same amount of energy, and the same lavish expenditure, where absolutely required, were devoted to the training of great industrial armies, to their maintenance in the most perfect health and efficiency, and to their employment in that great war which man is ever waging against Nature, subduing her myriad forces to his service, guarding against those sudden attacks by storm and flood, by violence and earthquake, which he cannot altogether avoid, and in the production of all the essentials of human life and of a true and beneficent civilisation, the good effects on character would surely be much greater than those produced by mere military training, as the objects aimed at and the results achieved would be more beneficial and more calculated to promote the higher interests of man. And if these industrial armies were allowed to reap the full advantages, material as well as moral, which they created, the results would be so striking that almost the entire population, male and female alike, would claim to be so trained and organised for their own physical, moral, and economic benefit. And the enjoyment of life under such a system of voluntary organised labour would be so enhanced that few indeed would wish to escape from it. Labour in companionship for the common good almost ceases to be labour at all . . .

It has been objected to all such proposals for the organisation of industry that it would deteriorate character by destroying individuality; but no such objection is made to the military organisation, while under its best forms the reverse is found to occur. In point of fact, *all* organisation is beneficial to character just in proportion as it rises above slavery. And when it shall have reached the point of being the organisation of social equals, for the equal benefit of all, it will attain to its most beneficial influence. Then, character and merit will alone give authority, and the highest and best will inevitably rise to the highest positions. And, just in proportion as the rank and file became educated, and felt the inspiring influences of comradeship and emulation, they could be left more and more to their own initiative; each one's individuality would have the fullest play, controlled only by the influence and opinion of his immediate fellow-workers, and the whole great organisation would become almost automatic in its harmonious working.

Such is found to be the case in the best military organisations, in which the intelligence and individual action of both commissioned and non-commissioned officers, and even of privates, is cultivated, and becomes of the greatest value, giving to the army in which it most generally exists an undoubted superiority. In any army thus intelligently and sympathetically trained and organised none of the results so dreaded in industrial organisation are found to occur. Men are *not* brought to a dull level of mediocrity; interest in the work they have to do is *not* lost; skulkers, malingerers, and deserters do *not* abound in any appreciable or hurtful proportion; nor is there any indication that men of superior abilities refuse to exercise their talents for the common good because the money rewards of such ability are small as compared with those often obtained in civil life; and lastly, the fact that all are provided with food and clothing, and are thus removed from the influence of economic competition, is not found to have any injurious effect on their effectiveness as workers, fighters, or organisers. And that these effects are not caused by compulsion and the severe penalties of military law is shown by the fact that during the civil war in America, where compulsion and punishment were rarely used, the whole of the opposing armies being practically volunteers, cheerfully submitting to military drill and organisation for the common good, these high qualities were equally manifested.

Yet objections of this class are held to be fatal to any proposal for national industrial organisation for the benefit of all, and the very system of training and co-operation which in the one case is admitted to have beneficial effects on character, and is undoubtedly, even under very unfavourable conditions, attractive in its comradeship and freedom from care, is condemned as being injurious and unworkable when applied industrially. Oh! that some great ruler of men would arise to benefit humanity by organising industrial armies, leading to the elevation and happiness of a whole people, and thus proving that peace may have its victories, far greater and more glorious than those of war!

(3–4) . . . It is, I think, clear that no hope of a complete solution—hardly even of amelioration—is to be expected from the ruling classes, urged on as they are on the one hand by those who are ever seeking for place and power, or for official appointments in newly-acquired territories, and on the other hand by the military class, who ever seek to justify their existence and the enormous burden they are to the nation by obtaining for it extensions of territory or military glory, and with either of these an extension of their own influence. It is, therefore, the *people*, and the people alone, that must be relied upon to banish militarism and war, and for this end every possible effort must be made to educate and enlighten them, not only as to the horrors and iniquity of war, but as to the utter inadequacy and worthlessness of almost all the causes for which wars are waged. They must be shown that all modern wars are dynastic; that they are caused by the ambition, the interests, the jealousies, and the insatiable greed of power of their rulers, or of the great mercantile and financial classes which have power and influence over their rulers; and that the results of war are *never* good for the people, who yet bear all its burthens.

In the course of this education of the people there are certain points that should be specially advocated. For example, nothing is more inconsistent, more foolish, and more wicked than the universal practice of civilised and Christian nations in selling all the most improved weapons and instruments of destruction to semi-civilised, barbarous, or savage rulers, thereby rendering it more difficult—more costly in blood and treasure—to deal with such rulers when their crimes against their own peoples or against humanity become too great to be borne. This practice also renders it ever more and more difficult for advanced nations to disarm, and thus gives to militarism an additional reason for its existence. From every point of view, whether of Christianity, humanity, or human progress, the supply of modern instruments of war to barbarous rulers, for the coercion of their own subjects, and as a standing menace to civilisation, should be absolutely forbidden. For this purpose, and in order that legal enactments to this end may be effective, we must try and create a sentiment of horror against those who continue thus to betray the cause of civilisation, as being not only traitors to their country, but enemies to the human race. In my opinion, men who after due notice, and in spite of its declared illegality, continue to supply these weapons to the possible enemies of their country should be declared outlaws in every Christian or civilised community. Hardly less foolish and wicked is the free trade in these instruments and armaments of war, so that directly one or more of the civilised nations are preparing for war the workshops of all the other civilised nations are at once engaged in supplying every kind of destructive appliance, even though they may in a year or two be used against themselves. The time will surely soon come when this conduct will be looked upon as the very culminating point of combined folly and wickedness that the world has seen. The only rational mode of procedure would be to forbid altogether the private manufacture or sale of war material. War is a national act, and so long as it exists all preparation for it should be kept strictly in the hands of national governments.

This supply of the implements of war is the work of capitalists in their own interests; but even worse if that be possible, is the action of the great civilised governments themselves in allowing their trained officers to engage in the organisation of the armies of semi-barbarous rulers, thus rendering it more difficult to coerce these rulers in the interests of civilisation, and indirectly, yet most certainly, leading to a vast extension of the horrors of war. The entire absence of ethical principle created by militarism is especially shown in the fact that no effective protest has been raised against this most pernicious and suicidal practice. Here, again, the people alone can take effective action, and the people want educating. Common justice, common humanity, even common sense, alike demand that this practice be absolutely forbidden, and that any officer engaging in the organisation of the armies of semi-barbarous or alien rulers should be declared an outlaw by the Government in whose army he was trained, be demanded from the employing Government as a traitor to his country, and the refusal to give him up be followed by an instant declaration of war from *all* the civilised governments.

Yet another point on which the people should be educated is, that they should claim and exercise the right to refuse, as soldiers, to act against their fellow-countrymen or against other countries with whose people they have no quarrel. Accepting the principle that the only just rights of governments rest upon the consent of the governed, what is termed rebellion is not a crime, but is usually the just demand of a community for self-government, a demand which, instead of being repressed by force, should be tested by a plebiscite. And smaller disturbances, termed riots, always arise from some injustice or supposed injustice, and are not proper subjects for massacre by armed soldiers. To use fire arms against a crowd, and kill or maim innocent persons, women and children, as almost always happens, is to authorise murder. Whenever it may be necessary to prevent violence by a mob, and the available force of police is not sufficient, special constables should be enrolled. But a far better plan would be to organise the fire-brigades as coadjutors of the police, since it is certain that no unarmed (or even armed) mob can stand against the jet of a fire-engine or of several fire engines. The mob would instantly disperse, and be rendered ridiculous without endangering life.

Of course, any proposed system of arbitration to settle disputes between nations should be strongly supported; but the existing condition of all the great civilised governments renders it certain that, so long as the ruling and military classes exist, and are allowed to possess the almost absolute powers they now exercise, war, as the ultimate mode of settling national disputes, will not cease.

* * * * *

Wallace's thoughts on the institution of a paper money standard of value, a peripheral but interesting subject he discussed a number of times in various degrees of detail, were set out in a pair of letters to The Clarion *in 1898:*

Is there scarcity or monopoly of money? (ref 553)

. . . Leonard Hall . . . [has said]:

> By money monopoly I, of course, mean monopoly of means of exchange . . . The laws now regulating (restricting) the currency and banking are the last links in the long chain of that private and class monopoly which has limited, taxed, and hocussed the exchange medium from the beginning. Monopoly depends upon maintaining artificial scarcity. The scarcer a thing is made, the easier it is monopolised, cornered, kept out of circulation.

Now, the whole of this I maintain to be erroneous. There is no monopoly of money, and no scarcity; and neither workers nor producers of any kind suffer loss or inconvenience to any appreciable amount from such alleged scarcity. This is a subject which has been so obscured by vague generalisation and a misleading terminology that it is necessary to come down to plain concrete facts in order clearly to see what happens.

If, owing to some continuous increase of business, an increasing amount of money (gold and silver) is required for payments of weekly wages, etc., and the banks have any difficulty in supplying the amounts needed, any of the largest houses who are inconvenienced may purchase gold and send it to the Mint to be coined; but usually the Bank of England does this, because it first feels the scarcity, and the coinage goes on till the supply of coin in circulation is found to be sufficient. So, when silver and copper coin become scarce, and the Bank of England cannot supply the increasing demand of their customers, they apply to the Mint, which then coins more silver and copper till the demand ceases. The supply of metallic money is therefore strictly regulated by the demand; the Mint exists for the purpose of supplying the demand, and no scarcity that can affect producers to any perceptible extent ever occurs.

And on the side of the producers and workers the same result is everywhere seen. Anyone who holds goods or produce of any kind for which there is an efficient demand can always sell it for cash, at retail prices if he takes it to market himself; at wholesale prices if he prefers selling it to a merchant or dealer. No doubt the dealer will sometimes say that he has no money and will offer to buy the goods on credit; but that is not in any way due to there being any real scarcity of money in the country, but to the unsound credit-system on which almost all business is carried on, so that the money the dealer is receiving every day from his customers has to be saved to meet bills periodically falling due. Were there double the amount of money in circulation, the man who carries on his business by credit, and in competition with wealthier and larger dealers, will always be short of money.

People are deceived by the terms "money market," "dear money," "cheap money," etc.; but these terms have no application to the quantity of money in circulation, but solely to the amount of interest charged for loans on personal security and of discounts on bills; and this depends on the general stability of trade, which again is dependent on politics, on the prospects of war, on the amount of speculation, and other similar causes, but has nothing whatever to do either with the quantity of money in circulation, or the amount of wealth and credit in the country. Often, indeed usually, when money is said to be "dear," any amount of money can be had at very low interest on good security, or for sound enterprises.

It must not be supposed that I think our system of money and finance is a good one; quite the reverse. But I maintain that the imperfection of the system does not directly affect workers or producers. The whole amount of gold and silver in our current coin is so much loss to the country, and a sound system of credit-notes might well take its place, but this would make practically no difference to producers. Whether a man receives 40*s*. for his week's labour or a credit-note for the same amount, it is the same to him, if the two sovereigns and the credit-note are alike in purchasing power. If it is said that, by means of co-operative stores and credit-notes, he will be able to purchase *more* for the same nominal amount than if he is paid in money, that has not been proved. With co-operative stores in full working order which would receive his produce at fair retail prices, he would no doubt obtain a considerable advantage; but the advantage would be due to the co-operation, and the small margin of difference between wholesale and retail prices not at all, or if at all, in an infinitesimal degree, to the use of notes instead of money. That the amount saved by not using metallic money would be very small, if perceptible, can be shown in two ways. (1) For a long time, at all events, the stores must have a considerable money capital to buy the various goods not produced by the co-operators. (2) The amount of the money required as permanent cash capital would be very small compared with the whole business done, because every day and hour, on the average, more money would be received for sales than would be paid out for purchases, and thus a very small permanent cash balance would suffice to guard against the purchases in any one day or week exceeding the sales; and this sum, as compared with the total amount of the sales in the year, would be quite insignificant.

It appears, then, that the supply of coined money is always such as to satisfy the demand, acting automatically by the agency of the Bank of England and the Mint. There is, therefore, *never* any scarcity of the circulating medium. For the same reason, monopoly of it is impossible, since the first attempts at a monopoly would lead to increased coinage, and the monopolists would then have to export their hoarded gold or turn it into bullion at a loss.

It is also clear that the actual amount of money in circulation, though absolutely large, is, relatively, exceedingly small, when compared with the amount of work it does. Every sovereign probably buys a hundred pounds' worth of goods in a year, and the very same sovereign may go on buying for fifty, or even a hundred, years, so that although the total amount of coin circulation is enormous, yet it is a very small fraction as compared with the exchanges it facilitates before it is worn out or replaced, and thus the saving effected by the universal use of credit notes might probably not average a shilling a year to each worker. I think I have now shown that there is, as a fact, no monopoly, no artificial scarcity, no restriction, no hocussing of the circulating medium as it affects the workers; while for traders and merchants on a large scale the supply of banknotes, cheques, bills of exchange, etc., etc., is unlimited. In the case of this form of money it is undue inflation, never restriction, that produces evil results.

The use of metallic money as a standard is also disadvantageous on account of fluctuations in the intrinsic value of gold or silver as compared with other commodities; but these fluctuations are certainly not great or rapid in the case of gold, and do not therefore affect the workers, because any changes arising from this cause will affect both wages and prices in the same way. The only real and important evil of our financial system is due, as stated in my former letter [ref 552], to the existence and continual increase of interest-bearing funds, bonds, and shares, which not only encourages that form of gambling termed financial operations, but enables the surplus savings of each year to be permanently invested, and thus increases year by year the number of persons who are able to live in idleness upon the labour of the productive workers, and therefore to their injury and impoverishment. It is for this reason that the continuous increase of our commerce and our wealth is, and must necessarily be, accompanied by a corresponding increase of poverty and starvation. This I have demonstrated by indisputable facts in my recent work *The Wonderful Century* [ref 726], and it is the one thing above all others that should be continually brought before the public, till it at last penetrates the thick armour or optimism with which the middle and upper classes, and especially politicians and the literary, artistic, and scientific cliques, protect themselves against the contemplation of the terrible realities and heartrending miseries which are the necessary results of our barbaric competitive system.

* * * * *

A complete system of paper money (ref 556)

. . . A. P. Hazell asks me to explain how a safe and effective system of credit notes or other form of paper money can be established and worked. I will endeavour to do so as briefly as possible; but to explain the matter fully would require a lengthy article.

A gold currency is supposed to be necessary in order that we may have money which is a measure of value as well as a tool of exchange. It is, however, now admitted that gold is not a permanent and stable measure of value, though I believe it is much more nearly so than is generally supposed. Most of the money specialists believe that for many years past the value of gold has been rising, basing their conclusion on the continual reduction in price of most commodities. But it is evident that the price of goods may be greatly reduced by improved machinery and production on a larger scale, and it seems to me that in the case of most of our manufactured goods this cause alone is sufficient to have reduced prices much more than they have actually been reduced; and in that case gold will have *diminished*, not increased in value, as the enormously increased production during the last half-century would lead us to think it should have done.

The usual objection to paper money is that it will change in value according to the amount issued, as is well seen in all countries where the Governments have tried to raise funds by such over-issues. This is quite true; but it is this very property of paper money that makes it easy to keep its value stationary, and, therefore, renders it, when the issue is properly regulated, a better and more stable measure of value than gold, or than any single commodity whatever. How this stability can be attained, I will now endeavour to explain.

Stability, or equality of purchasing power at different times, can only be known by the same nominal amount of money—say, 100 pounds or 1,000 pounds—being able to purchase the same quantities of all the chief necessaries of life on the average. Luxuries used by the few—ornaments, jewellery, works of art, etc.—may be left out of consideration. As necessaries of life, we may take the four great groups of food, clothing, houses, and fuel; and each of these may be represented by a limited number of the most important items, as bread, meat, potatoes, sugar, tea, and beer, to represent food; timber, iron, bricks, and glass for houses, or a larger number of items if thought advisable by experts. Having fixed upon the list of commodities—perhaps 50, perhaps 100, in all—which are considered to be amply sufficient as the basis of an estimate of the purchasing power of money, the next step will be to estimate the proportionate quantity of each consumed in the whole kingdom, or in some representative part of it, during a year. This is necessary in order to give to each its proper weight in the estimate; for if 100 tons of A and 1,000 tons of B are used per annum, it will lead to very erroneous conclusions if we were to use equal quantities of each in our estimate, and I believe that this very mistake has been made in the estimation leading to the conclusion that gold has for many years been appreciating in value. Having now got our typical list of commodities with the proportionate quantities of each, we next have to get the average price for a series of years—seven, ten, twenty, or whatever number may be fixed upon as the basis on which to calculate the standard purchasing power of our new national currency. All these facts can be got at with sufficient accuracy by means of agricultural and commercial statistics and market prices. When completed, a table will be constructed something in this form:

Proportions of standard products consumed, and their value on the average years—1890–1896.

Bread	10,000 lbs	value 50	pounds
Meat	4,000 lbs	" 200	"
Sugar	1,500 lbs	" 10	"
Tea	500 lbs	" 40	"
Timber	1,000 cub. ft.	" 100	"
Coal	200 tons	" 200	"
etc., etc., etc.,		600	"

These proportions and prices are put down at a mere guess, but when obtained as accurately as possible for the whole of the 50 or more commodities chosen, we shall have, as a result, that these quantities of these commodities have, on the average of the last seven (or 10 or 20) years' cost a certain gross sum. Now, what I maintain is, that paper money (called credit-notes, or anything you like) can be so issued as, for any number of years, to continue to purchase the same quantities of this whole series of commodities for approximately the same nominal amount. Some of the items will, of course, rise in value from one year to another, and others will fall: but the paper currency will always, within very small limits of variation, purchase the same total amounts.

To do this, a Minister, or Commissioner of Currency, with a sufficient staff of clerks, will be appointed, whose duty it will be to have regular returns made of the market prices of the standard commodities week by week, and to have the averages calculated. If during any month or quarter these averages are seen to fall continuously, that is, everything becomes cheaper, he will advise the Treasury to issue more notes which they will bring into circulation (by using them to pay salaries and current expenses) till the fall is checked and the true average reached. When, on the other hand, the standard goods show a rise in price, it indicates that there is a slight surplus of the currency, which is to be checked by cancelling old notes as they come back to the Treasury. This process could be so nicely regulated that, practically, there would be no rise or fall of prices on the average, since either would be remedied before it could possibly be detected by the public.

Here, then, we should have a most useful and portable currency—which could be issued for any amounts in very thin but tough cards about the size of railway tickets, and of different colours for the different denominations—and which would be a stable measure of value as well as a convenient instrument of exchange. And it would have the great advantage of working almost automatically and preserving an unchanged purchasing power by the very act of supplying the demands of the community. And as, with an increasing population, more and more currency would be required, and as many small notes would be lost, burnt, or otherwise destroyed, this currency would be a constant source of revenue to the Government.

During the process of change from metal to paper the gold paid into the Treasury for taxes, duties, stamps, etc., would be accumulated, and form a reserve fund for pressing purchases from other countries in case of war. But the great point is, that by regulating the amount of notes issued in the way above described, this money would become a real *measure of value*, which gold can never be so long as its production is a matter of private speculation, and its cost, and consequent value in exchange, liable to indefinite variation.

* * * * *

Utopian though it may be, there is yet much fertile ground for exploration in Wallace's social theorizing. His 'theory' of social evolution accepts the liberal premise that all individuals are fundamentally amoral, such wrinkles as may invade their character stemming from the imperfections of the social systems into which they are born, and being largely beyond their immediate awarenesses. Wallace's was not, however, a naïvely permissive form of liberalism. Importantly, he had also considered the dynamics of the situation, and concluded that social progress *was at heart the function of a wilful rejection of selfish, materialistic goals: that is, a slow transcendence of the one-cause, one-effect determinism (and thus limitations) of animal existence. Counter-evolutionary social structures could be toppled by public opinion, and as long as these were replaced by ones that* in general*: (1) increased equality of opportunity, and (2) permitted a growth of counter-materialistic intuitions, there was true progress. Land nationalization and socialism were society-level programmes designed to promote equality and reduce individual worry over satisfying fundamental needs; spiritualism, a vehicle for developing non-materialistic libertarian goals.*

Though various flaws can be recognized in the particulars of Wallace's progressive arguments, it is none the less true that his overall model of societal priorities has proved to be an accurate predictor of important trends that have developed since his time. Peace-and-quiet-seeking suburbanites now outnumber urban dwellers by a wide margin, and even fully rural areas are again increasing in population. Capitalists have begun to recognize that an emphasis on the manufacture of cheap goods that disregards the matter of quality can ultimately be counter-productive. Employee buy-outs of troubled companies have become commonplace occurrences. Unenlightened colonial policies have given rise, time and time again, to bloodshed and revolution. Fundamental social and transportation services have been nationalized in most of the leading nations of the world. Trusts, both governmentally and privately organized, continue to save an increasing portion of the land from the materialistic exploitation and degradation that doubtless would have been its fate in the hands of individual owners. The 'average person' has increasingly collectivized to exert pressure on morally questionable social, economic, and military policies. And so on.

Some may feel that Wallace, optimist that he was, was fortunate to die when he did, in late 1913. The events of 1914, which would have confirmed his suspicion that society was 'rotten at the core,' would none the less have given him little satisfaction.

Related writings

on the conservation of natural resources: refs 231, 292, 306, 380, 715, 720, 722, 726, 727

on poverty/the inequitable distribution of wealth: refs 329, 344, 345, 346, 365, 369, 370, 375, 387, 407, 414, 423, 427, 450, 466, 498, 505, 507, 512, 545, 553, 554, 563, 584, 587, 593, 639, 655, 665, 722, 723, 726, 727, 729, 733, 734, 749, 752

on rural depopulation/under-population: refs 341, 344, 345, 346, 365, 375, 380, 387, 407, 423, 688, 722, 723, 727, 729, 734

on inherited wealth and trusts: refs 236, 329, 370, 483, 491, 498, 505, 508, 531, 576, 587, 622, 634, 639, 722, 727, 729, 752

on labour and labour problems: refs 344, 345, 364, 370, 375, 407, 414, 423, 438, 466, 498, 501, 505, 507, 512, 528, 560, 567, 576, 581, 610, 617, 655, 691, 722, 723, 726, 727, 729, 733, 734, 749

on capital/interest/loans/a paper money standard: refs 370, 380, 387, 438, 498, 507, 552, 553, 556, 557, 584, 622, 722, 723, 726, 727, 729, 749

on the regulation and other aspects of trade and commerce: refs 65, 104, 231, 306, 310, 312, 387, 553, 556, 557, 567, 720, 722, 723, 727, 734, 749

concerning land nationalization (landlord–lessee relationships, agricultural economics, the theory of land occupation and use, proposals and experiments and related legislation, etc.): refs 95, 104, 225, 292, 329, 339, 341, 343, 344, 345, 346, 347, 361, 364, 365, 366, 371, 375, 376, 380, 385, 386, 387, 399, 403, 407, 408, 412, 414, 418, 422, 423, 426, 438, 450, 466, 471, 483, 495, 498, 501, 507, 508, 512, 525, 531, 533, 541, 552, 555, 563, 566, 576, 584, 585, 586, 623, 625, 638, 653, 655, 662, 678, 679, 683, 684, 688, 696, 703, 707, 715, 722, 723, 726, 727, 729, 734, 735, 747, 749, 750, 753

on the treatment of commons lands: refs 225, 292, 329, 414, 443, 450, 498, 625, 679, 722, 727 on the nationalization of railways and like services: refs 438, 507, 584, 622, 628, 634, 661, 727, 749

on the social economy/equality of opportunity/social justice concepts: refs 375, 380, 427, 450, 466, 495, 496, 498, 507, 508, 512, 531, 540, 545, 573, 587, 591, 593, 622, 626, 634, 684, 723, 727, 729, 733, 736

concerning socialism: refs 403, 418, 431, 438, 498, 501, 507, 515, 537, 545, 552, 579, 586, 587, 595, 617, 622, 643, 644, 647, 654, 684, 710, 727, 729, 735, 745, 748, 750

on militarism/imperialism/'might vs. right'/warfare: refs 236, 387, 496, 525, 528, 540, 549, 550, 559, 567, 569, 570, 571, 572, 579, 580, 591, 595, 601, 610, 617, 629, 630, 659, 670, 677, 684, 698, 723, 726, 727, 729

on voting, suffrage and institutional politics: refs 110, 157, 158, 225, 236, 438, 454, 491, 498, 572, 576, 630, 632, 635, 647, 671, 715, 720, 722, 727, 729, 735, 747

concerning institutional religion: refs 111, 113, 225, 236, 243, 398, 498, 505, 545, 639, 715, 720, 727, 729

concerning criminality and punishment: refs 549, 576, 626, 684, 715, 726, 729

on eugenics: refs 161, 427, 445, 549, 727, 733, 736, 737, 750

on education (also see Section Eight): refs 157, 225, 292, 427, 445, 549, 586, 630, 684, 715, 720, 726, 727, 729, 733, 734, 736, 737, 747, 750

on miscellaneous topics: refs 104, 431, 515, 543, 552, 554, 562, 563, 564, 566, 581, 590, 623, 639, 641, 690, 691, 693, 710, 715, 720, 722, 723, 726, 727, 729, 733, 734, 747, 748, 752

book reviews and notices: refs 414, 450, 495, 496, 501, 515, 641, 727

secondary literature: Clements (1983), Durant (1979), George (1964), Malinchak (1987), Marchant (1916), Oppenheim (1985), Peirce (1906), R. Smith (1972), Turner (1974)

§4 · THE ANTI-VACCINATION CAMPAIGN

> ... During the course of my examination, I have been asked questions which implied that I had taken up this subject and written upon it without the full and accurate information befitting a man of science. I admit that this is, to some extent, true; but my answer is that I did not take it up as a question of pure science. If it had remained a question of medical science and practice, I should not have troubled myself about it, and certainly not have written on it. But from the moment when, through the great influence of the medical profession, a medical dogma was enforced by penal law, it became a question of personal liberty. When almost every week I read of men fined or imprisoned for refusing to subject their children to a surgical operation which they (and I) believed to be, not only useless, but injurious and dangerous, I felt impelled to aid, if ever so little, in obtaining a repeal of a cruel and tyrannical law . . .
>
> (ref 420, p. 167)

In the context of today's world, the idea that a sizeable portion of the turn-of-the-century population on both sides of the Atlantic opposed smallpox vaccination may seem quite absurd. But such a sentiment did exist, and was based on legitimate concerns.

Jenner's development of vaccination in the early 1800s was a milestone in the evolution of preventive medical treatment. Its significance was quickly recognized, and before long it became mandatory in England and other places to have one's children vaccinated. Over the course of the nineteenth century, incidence of smallpox dramatically declined, and the medical establishment came to believe that the practice was beyond criticism.

This was, however, a period during which the theory of prevention was running rather ahead of implementation practices. Vaccines were neither produced nor administered under the most sanitary of conditions, and it appears that a significant number of individuals died as a direct result of vaccination itself. Just how many did will never be known, as such records as were kept ran afoul of conflict-of-interest biases: obviously, it would not be good for a doctor's reputation were it discovered that his patients were expiring at suspiciously high rates.

There was, further, the fact that social living conditions in general were becoming more sanitary; thus, it was difficult to assess just how much of the reduction in incidence of smallpox was a function of vaccination, and how much was related to environmental causes. By the 1870s, however, there were both enough individual incidents being publicized and enough suspicious statistics emerging to make some—including a number of doctors—question the wisdom of continuing a mandatory vaccination programme. Wallace, who had 'duly' had his own children vaccinated some years earlier, was apparently introduced to the evidence late in that decade, and soon after came to the conclusion that a social injustice was being perpetrated.

Wallace was of course welcomed into the anti-vaccination movement. Although he was not close enough to the legislative superstructure to have any direct influence, both his name and powers of argument were by then well-known. He became one of the campaign's leading voices (his formal connections with it, however, were actually rather limited); in fact, when a Royal Commission was finally appointed to assess its claims, Wallace was offered a seat on the examining board. He declined the invitation, deciding that it would be more productive to serve in the capacity of expert witness during the examination proceedings. (He eventually ended up presenting four full days of testimony, the text and accompanying diagrams of which consumed over fifty pages of the Commission's final report: ref 420.)

Wallace's contribution to the anti-vaccination movement was also methodologically significant, as he became one of the first to organize an entirely statistics-based critique of a public health problem. His three treatments of the problem were published, in pamphlet form, in 1884, 1898, and 1904 (refs 374, 536, and 616, respectively). These went through a total of eight printings. The 1898 work was about a hundred pages long, and was simultaneously incorporated into the social criticism portion of the first edition of The wonderful century *(ref 726). The summary of 1904, however, despite its too-ambitious title, is of a length ideal for present purposes, and is reproduced here in full:*

A summary of proofs that vaccination does not prevent small-pox but really increases it (ref 616)

I. Why doctors are not the best judges of the results of vaccination.

(1) In the first place they are interested parties, both pecuniarily and in a much greater degree on account of professional training and prestige. Only three years after vaccination was first introduced, on the recommendation of the heads of the profession, and their expressed conviction that it would give lifelong protection against a terrible disease, Parliament voted Jenner 10,000 pounds in 1802, and 20,000 pounds more in 1807, besides endowing vaccination with 3,000 pounds a year in 1808. From that time doctors as a body were committed to its support; it has been taught for nearly a century as an almost infallible remedy in all our medical schools; and has been for the most part accepted by the public and the legislature as if it were a well-established scientific principle, instead of being as the historian of epidemic diseases—Dr. Creighton—well terms it, a grotesque superstition.

(2) Whether vaccination produces good or bad results can only be determined by its effects on a large scale. We must see whether, during

epidemics—at different periods or in different places—small-pox mortality is diminished as compared with that from other diseases in proportion to the total amount of vaccination; and this can be done only by the Statistician, using the best materials—in this country those of our Registrar-Generals.

Two of the greatest medical authorities on vaccination, Sir John Simon and Dr. Guy, F.R.S., have declared this to be necessary. The former, in 1857, in a Parliamentary Report on the *History and Practice of Vaccination*, says: "From individual cases the appeal is to *masses of national experience*." Dr. Guy, in a celebrated paper published by the Royal Statistical Society, says: "Is vaccination a preventive of small-pox? To this question there is, there can be, no answer except such as is couched in the language of figures." The language of figures is "Statistics;" hence, statisticians, not doctors, are the only good judges of this question. But the last Royal Commission consisted wholly of doctors, lawyers, politicians and country gentlemen, not one trained statistician! Hence, as I have demonstrated in my *Vaccination a Delusion* [ref 536], they have made the grossest blunders and their Report is absolutely worthless.

II. What is proved by the best statistical evidence available.

(1) The only complete and trustworthy records of mortality and of the causes of death which we possess, are those of the Registrar-Generals for England and Wales, for Scotland and for Ireland; the former from 1838, the two latter from later dates. But for London we have records from a much earlier period—the *Bills of Mortality*, which, though not completely accurate, are yet considered to show the rise and fall of the death-rates from the chief diseases then recognised, with sufficient general accuracy to be very valuable. They are continually appealed to in order to show the enormous improvement in the health of London in the nineteenth as compared with the eighteenth century, and this comparison as regards small-pox is one of the stock arguments of the doctors, and was strongly urged by the Royal Commissioners. It is stated over and over again that, down to the year 1800, small-pox deaths were excessive, but that from the very introduction of vaccination in 1800 they began to decrease, and have been getting less and less ever since. No other disease, it is said, has decreased in the same striking manner.

(2) This being the very foundation of the supposed evidence in favour of vaccination it is necessary to examine it closely, when it will be found to be wholly *worthless*, and to illustrate in a striking manner the complete ignorance of doctors, and also of the Royal Commissioners, of the very elements of statistical enquiry. This requires some little explanation, though it is really a very simple matter.

In order to be able to study the effect of any alleged cause of improved health of the community, we must compare the death-rates before and after the introduction of the supposed cause of improvement (in this case vaccination), and also compare these with the death-rates from other groups of diseases, and from all causes. These facts are given by the Registrar-General in tables showing the number of deaths each year in each million of the population. Now, small-pox, many fevers, cholera, etc. are what are termed epidemic diseases, which attack large populations at irregular intervals with great severity, while at other times they are far less fatal or more local. Hence the yearly death-rates vary enormously. In 1796 more than 4,000 per million died of small-pox in London, while in the next year there were only about 800, and the following year (1798) over 3,000. Again, in 1870 less than 100 per million died of it, while in 1871 there were about 300, and in 1872 about 2,500. Thus the figures go increasing and decreasing so suddenly and so irregularly, that by taking only a few years at one period, and a few at another, you can show an increase or a decrease according to what you wish to prove. Hence it is often ignorantly said that figures can be made to prove anything. But this is quite untrue. They can often be made to *show* anything, which is quite another matter; but if properly exhibited and compared they lead to one conclusion only; they show the *truth*.

(3) There are a few simple rules for getting at the truth in such statistics as we are now discussing. One is that we must take as long *periods of time* as possible; another is that we must use the *largest* populations available. Two other conditions are almost equally important; we must compare, when possible, equal periods before and after vaccination was introduced; and we must also compare the increase or diminution of *small-pox* with those of *other diseases*, in order to discover whether there is anything exceptional in the decrease of small-pox mortality which requires a peculiar cause to explain it.

But the ever-varying figures in long columns are so confusing to most people, that it is impossible to make anything out of them, and to simplify them, averages have to be taken, showing the deaths every five or every ten years, and in other ways, so as to find out what the figures really mean, and even then, by altering the *periods* or beginning *at different years*, a very different result may often be shown.

(4) By far the best way and that usually adopted by statisticians and mathematicians, is to draw out diagrams by which the whole course of the mortality from each disease or group of diseases can be seen and compared at a glance. From the various elaborate tables given in the Reports of the Royal Commission and from the annual reports of the Registrar-General, I constructed twelve diagrams, each showing the comparative rise or fall of small-pox mortality and other diseases in various places and under different conditions; and all these without exception *demonstrate* either that vaccination has no effect whatever, or that it tends to *increase* rather than decrease *small-pox mortality*. These are all given in my little book *Vaccination a Delusion*, which can be had from the National Anti-Vaccination League for 9*d*. a copy, or 10½*d*. by post.

(5) As many people do not understand these diagrams I here give a part of one of them in a simplified form in order to render statistical diagrams intelligible to all, and it will serve to show what is the nature of the evidence against vaccination, and also how I prove that the statements made by the doctors and by the Royal Commissioners are not only misleading but absolutely untrue.

DIAGRAM OF LONDON MORTALITY.

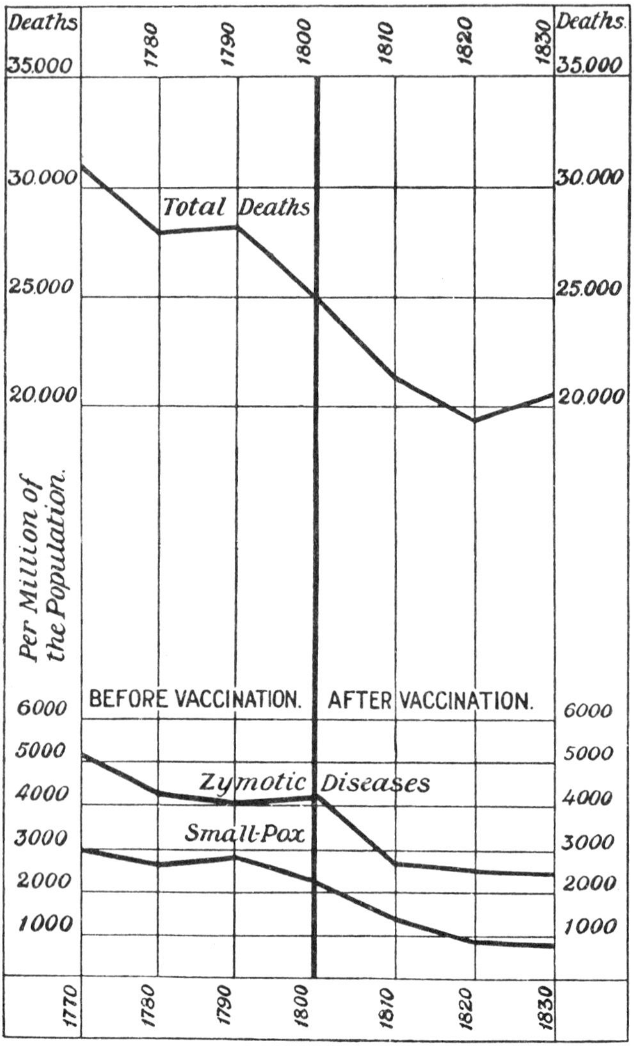

Explanation of the diagram.

(6) The figures on the bottom and top of the diagram show the years, from 1770 to 1830, while those on the right and left show the *proportion of deaths* to population during this period of 60 years; the lower line the *small-pox* deaths; that next above it the deaths from the other *zymotic diseases* (fevers, diphtheria, whooping-cough, etc.); while the top line shows deaths from *all diseases*. These last deaths, being so much more numerous, have had to be drawn out on a smaller scale in order to show them on the same page as the others.

(7) This diagram shows us that small-pox decreased during the ten years before vaccination at very nearly the same rate as it did in the ten years after vaccination. The other zymotic diseases decreased even more than small-pox during the ten years after vaccination. General mortality also decreased after 1800 more rapidly than before 1800. Yet the Royal Commissioners declare that there was *nothing* but vaccination to produce the decrease of small-pox, and that there was no improvement in sanitation in the beginning of the nineteenth century, as compared with the latter part of the eighteenth century, to account for the difference.

(8) Now, in an Appendix to my *Vaccination a Delusion*, I have given an account of a number of improvements affecting health at this very period which are amply sufficient to produce the results shown by the diagram, and I believe it is the most compact and most interesting account of these improvements yet given. The chief of them are (1) That many West-end squares and suburbs were built at this very period, and were inhabited chiefly by city people. (2) That the streets were more systematically cleaned and the roads improved. (3) That the water supply was much improved. (4) That potatoes, tea, and coffee came into more general use; while the better roads allowed more fresh meat, vegetables and milk to be used. (5) Cemeteries were formed outside London and many City graveyards were permanently closed. The result of these five groups of improvements was strikingly shown in the decrease of the death-rate in a number of the most fatal diseases (as recorded in a Table by Dr. Farr, reprinted in the Third Report of the Royal Commission) to fully one-half in 1801–10 as compared with 1771–80; an amount of improvement which has never occurred in any similar period during the whole of the 270 years for which we have official statistics. And yet the Royal Commissioners declare that nothing *but vaccination* can explain the corresponding and very similar decrease in small-pox!

(9) As you will now understand the method of exhibiting statistics by means of diagrams, I will proceed to state the other more important conclusions to be drawn from our national statistics of death-rates. Those who wish to study them more fully must obtain the book itself, and examine the diagrams and the full details there given.

III. London death-rates during registration. 1838–96.

(1) These tables show us that neither the general mortality nor that from zymotic diseases decreased much till about 1868, but from that date there has been a large and continuous decrease. Small-pox had a sudden increase in 1838, in which year the mortality was greater than it had been for the preceding twenty-five years. Then it decreased slowly till 1870, and this decrease is always ascribed by the doctors to vaccination. But in 1871 there was a great epidemic, when the mortality was greater than at any period during the preceding seventy years of constantly increasing vaccination! Since 1871 small-pox has decreased, but only at about the same rate as the other zymotic diseases. The interesting thing to note here is, that the Main Drainage of London was completed in 1865, and about five years later (the time required

for the connection of all the house drainage) the marked diminution in the mortality above-mentioned began to show itself. And if we average the enormous small-pox mortality of 1871 with that of the preceding ten years, we shall find that it will bring the small-pox mortality into almost exact correspondence with that from all other causes, and thus leave nothing to be imputed to vaccination!

(2) In another diagram in my book I show the mortality from the five groups of zymotic diseases taken separately: Fevers, Whooping-cough, Diphtheria and Scarlatina, Measles, and Small-pox, for the same period of Government Registration. All of these diseases show a nearly similar decrease in the latter half of the period, except measles, which shows hardly any diminution; but there is reason to believe that the cause of this is, that, when vaccinated children after a short illness die of small-pox, measles or chicken-pox are often given as the cause of death.

IV. Death-rates in England and Wales during the period of registration.

(1) My third diagram is one of the most instructive and conclusive in my book, because it deals with the whole population of England and Wales and the death-rates from various groups of diseases as in the illustrative diagram. In the first twenty-five years, from 1848 to 1872, there is on the average hardly any decrease either of general mortality, zymotics, or small-pox, since the enormous small-pox mortality of 1871–72 if distributed over the preceding ten years will bring it to correspond closely with the other classes of mortality. But from 1873 to 1895—the last twenty-three years shown—there is a diminution in all three of the diseases to a considerable amount. For the last ten years the diminution in small-pox is the greatest; but this can be proved to be *not* due to vaccination, as I will now explain.

(2) It is only from the year 1872 (after the great epidemic of small-pox) that all vaccinations, private as well as public, have been officially registered, and a table showing their amount has been given in the Final Report of the Royal Commission. From 1872 to 1882 the vaccinations amounted to 95 per cent of the births; practically *all* were vaccinated if we allow for those that died before they could be operated on or very soon afterwards. But from that date the number of vaccinations steadily decreased, till in 1895 they were only 80 per cent of the births, a diminution of 15 per cent in fourteen years. If vaccination were the chief or only preventive of small-pox we ought to have a considerable *increase* of the disease during this period, instead of which it is in this period only that the diminution of small-pox has been *more marked* than that of the other zymotic diseases! Here, then, we have the first distinct proof that it is vaccination which *keeps up the disease*, and that when a larger number of children escape the blood-poisoning lancet small-pox diminishes!

Another and even more conclusive proof is given . . . by Dr. Ruata, M.D. The whole male population of Italy are revaccinated on entering the army. Under the age of 20, men and women are *alike* as regards vaccination; afterwards men have an *enormous advantage*, if vaccination is of any use. Yet, over 20, many *more* men than women die of small-pox, while under 20 the mortality is equal, again demonstrating that vaccination *increases* small-pox mortality!

V. Thirty years of rapidly decreasing vaccination in Leicester, and its teachings.

(1) The great manufacturing town of Leicester, with nearly 200,000 inhabitants, affords the most conclusive proof of the uselessness of vaccination that it is possible to have; and the doctors and government officials carefully avoid dealing with it except to prophecy evils which have never come to pass.

Down to 1872 Leicester was one of the most completely vaccinated towns in the kingdom, the number of vaccinations, owing to alarm after epidemics, several times *exceeding* the number of births. Yet in 1871, at the very height of its good vaccination record, it was attacked by the epidemic with extreme severity, its small-pox deaths during that year being more than 3,500 per million of the population, or about a thousand per million *more* than the mortality in London during the same epidemic. If ever a test experiment existed it is this of Leicester, where an almost completely vaccinated community suffered more than unvaccinated and terribly insanitary London, on the average of the last forty years of the eighteenth century.

But even more conclusive evidence is to come.

(2) That fearful mortality destroyed the faith of Leicester in vaccination. Poor and rich alike, the workers and even the municipal authorities began to refuse vaccination for their children. This refusal continued till, in 1890, instead of 95 per cent the vaccinations reached only 5 per cent of the births! As this ominous decrease of vaccination went on doctors again and again prophesied against it, that once small-pox was introduced it would run through the town like wildfire and decimate the population. Yet it *has* been introduced again and again, but it has never spread; and from that day to this no town in the kingdom of approximately equal population has had such a very low small-pox mortality as this almost completely unvaccinated and—as the doctors say—*unprotected* population! Surely this completes the demonstration that vaccination, instead of preventing, increases the liability to small-pox, and that the only way to abolish the disease is to do as Leicester did, leave off vaccination altogether and devote our energies to sanitation, and the isolation of such rare cases as do occur.

Yet this wonderfully conclusive test experiment was passed over by the Royal Commissioners in 1894, with a few scattered remarks, which are either absolutely untrue or entirely beside the question. (See *Vaccination a Delusion*, p. 277.)

VI. The army and navy: A demonstration of the uselessness of vaccination.

(1) The doctors always claim that, though the effect of vaccination in infancy wears out, yet re-vaccination offers an almost complete protection for the rest

of the person's life. In a circular issued in 1884, and up to the time of the Royal Commission widely distributed *with the approval* of the *Local Government Board*, it is stated that: "Soldiers who have been re-vaccinated can live in cities intensely affected by small-pox without themselves suffering to *any appreciable degree* from the disease." I will now show you that this official statement is *absolutely false.*

(2) All soldiers and sailors are re-vaccinated on entering the service, unless they have recently had small-pox. The reports of the Royal Commission give the small-pox deaths in the Army and Navy from 1860 to 1894. The Registrar-General gives the total mortality from disease in the two services for the same period. I have compared these two mortalities by means of a diagram constructed from the tables, and this is what we find. First, throughout the whole period the total mortality from all diseases in the Army is much higher than in the Navy. Clearly, this is the result of the one class living in barracks, largely in towns and cities, the other in the midst of the pure and bracing sea air.

In the second place, there has been, in both services, throughout the thirty-four years a continuous diminution of mortality, so that it is now only about one-third of what it was thirty-four years ago; and this enormous improvement is stated by the Army and Navy doctors to be due to the much better *sanitation* of ships and barracks, and to the great improvement in the *food* and general *treatment* and *medical* attention in both services.

Thirdly, in both Army and Navy there has been a large decrease in the small-pox mortality throughout the whole period, corresponding closely with that of the general mortality, and certainly due to the same causes—improved sanitation and medical treatment.

Fourthly, in the very same years (1871–2) as the great epidemic in England and on the Continent, there was also a small-pox epidemic both in the Army and the Navy, and taking account of the age of the men and their condition of constant medical supervision, quite as severe as among the general population, who had *not* the alleged *complete protection* of re-vaccination.

Fifthly, this is proved by two comparisons—with Ireland and with Leicester—from tables given in the Reports of the Royal Commission extending from 1864 to 1894. The diagrams formed from these tables show us that Irishmen of about the *same ages* as our soldiers and sailors suffered more during the epidemic of 1872, but for the remainder of the thirty years they had rather less small-pox mortality; while since 1881 they have had *not half the small-pox mortality* of the Army and Navy.

(3) The other comparison is with Leicester, which city, in the period of twenty years (1873–1892), during which they had been growing less and less vaccinated, has had a total of only 16 small-pox deaths per 100,000 of its population, which includes thousands of unvaccinated children and infants; while for the same period the deaths in the Army and Navy amounted to over 70 per 100,000.

And yet we have had the impudently false statement circulated by thousands, under the approval of the Local Government Board, that the re-vaccinated Army and Navy do not, under the worst circumstances, "appreciably suffer!" The Royal Commissioners, on the other hand, shirk the whole matter—make no comparisons with other populations—but state vaguely that "particular classes" who have been "exceptionally" re-vaccinated exhibit "quite exceptional advantages in relation to small-pox,"—a statement which, as regards the only "exceptionally" re-vaccinated large classes of men, is, as their own tables show, the very reverse of the truth, since they suffer much more than the least vaccinated class of about equal population in the whole kingdom.

It is thus absolutely demonstrated that it is the exceptionally *unvaccinated* that possess the exceptional advantages, while the "exceptionally *re-vaccinated*" Army and Navy show quite exceptional *disadvantages*, in a small-pox mortality during the same twenty years, more than *four* times as great as the exceptionally *unvaccinated* town of Leicester!

But the learned men of the Royal Commission never put these two facts side by side, so that the Government and the public might draw their own conclusions from them. So far as their Final Report shows, these gentlemen were ignorant or oblivious of the very existence of these facts, which conclusively prove that Vaccination is not only *worthless* but an *injurious* operation—a Gigantic Medical Imposture!

(4) For the reasons now stated, we call upon voters of all parties to refuse support to every candidate who upholds the legal or other enforcement of vaccination, which, as we have shown, both spreads disease and increases mortality. No government has the right to order healthy infants to be blood-poisoned, under the pretence of protection against a danger that may never arise. The abolition of all laws enforcing or encouraging vaccination is therefore of more immediate and vital importance than any party dogma or any political programme.

VII. How to deal with medical pro-vaccinators.

(1) In my *Vaccination a Delusion* I have given examples of the grossest misstatements of doctors and officials from the time of Jenner down to the present day. They are such as often appear to be incredible, but none of them have ever been disproved. Several have been given here; but there is one more which is so universal that it must be briefly referred to. In all Official Reports of small-pox epidemics the fatality of the unvaccinated is always declared to be enormous as compared with the vaccinated. As an example, Dr. Gayton, in a Table published in the Second Report of the Royal Commission, gives the percentage of deaths to cases as follows: Vaccinated—7.45 per cent; Unvaccinated—43.00 per cent.

But all the medical writers on small-pox during the eighteenth century agree in stating that the average death-rate of small-pox patients was then from fourteen to eighteen per cent. At that time, however, the sanitary state of our towns and hospitals was abominable, while the medical treatment of small-pox was so incredibly bad that it is a wonder any survived. Yet the doctors ask us to believe that *now*, with far healthier conditions and with far better treatment and nursing, more than *twice* as many unvaccinated small-pox patients die as died *then*, when *all* were unvaccinated! The thing is absolutely incredible and absurd; and the belief in it is due solely to the fact that doctors register all deaths from small-pox as "unvaccinated" when they can possibly find any excuse for doing so. One of them has stated that "the mere assertions of patients or their friends that they were vaccinated counts for nothing." The alleged enormous mortality of the unvaccinated is further shown to be erroneous by the fact that the published Reports of three of the largest small-pox hospitals for London from 1876 to 1879 showed that the average small-pox mortality of all patients was about 18 per cent, or a little higher than during the eighteenth century. This may be explained partly by the fact that many of the milder cases do not go to the hospitals, and partly by the weakening of the constitution due to the blood-poisoning operation of vaccination, which, when conditions are alike, renders the vaccinated less able to resist small-pox than the unvaccinated. It has been well asked: "If about 36 per cent of unvaccinated patients die of small-pox while only about 18 per cent died in the eighteenth century who or what kills the other 18 per cent?" It cannot be the general conditions, since the mortality from all diseases has greatly diminished. There remains only the medical treatment. Do doctors accept this?

(2) Now if any one thing brings forward doctor's or official's figures as to the enormous value of vaccination, ask them first the above questions. They will deny the facts. Then, in my book you will find the official authority for these and all the other facts referred to. They will be obliged to say they have never enquired into them, and you may then tell them that they have no right to teach you who *have* enquired into them.

If you have a medical man to deal with, ask him why he does not admit Sir John Simon's statement, that "*the great masses of national experience* can alone prove the value of vaccination." Then show him the diagrams (in my book) which I have here referred to, and ask him to prove that they show "great benefits of vaccination," instead of showing as they do its absolute worthlessness.

(3) As to its terrible dangers, the thousands of lives vaccination has destroyed or ruined as regards health, I have no space to refer to them here, but ample evidence from the Royal Commission Reports is given in my book.

(4) Doctors and Members of Parliament are alike grossly ignorant of the true history of the effects of vaccination. They require to be taught; and nothing is so likely to teach them as to show them the diagrams I have referred to in this short exposition of the subject—those of *London* for thirty years before and after vaccination—of *England and Wales* during the period of official registration—of *Leicester* which has almost abolished small-pox by refusing to be vaccinated for thirty years—and for the *Army and Navy*—which, though thoroughly *re-vaccinated* and therefore (according to the doctors) as well protected as they possibly can be, yet die of small-pox at least as much as badly vaccinated Ireland, and many times more than unvaccinated Leicester.

A doctor who has *not* studied these most vital statistics has no right to an opinion on this subject.

A candidate for Parliament who *will not* give the necessary time and attention to study them, but is yet ready to vote for penal laws against those who know infinitely more of the question than he does, is utterly unworthy to receive a single vote from any self-respecting constituency.

* * * * *

Vaccination a delusion *(ref 536), especially, created a considerable stir, and attracted plenty of criticism. Among the most vocal objectors was one Dr. Francis T. Bond, Secretary of the Jenner Society. One Bond–Wallace exchange appeared in the pages of the London newspaper* The Echo *in mid-1898, the two following letters being Wallace's contribution to the discussion:*

Dr. Bond and Mr. A. R. Wallace. Mr. Wallace replies. (ref 544)

Dr. Bond has now written four long letters without approaching the real point at issue between us, which is, whether the large masses of national statistics exhibited in my diagrams are, as he alleged, "fallacious;" and, if not, whether they prove or disprove the alleged value of vaccination. His first two letters dealt only with the origin and spread of zymotic diseases, as to which I stated what I believe to be the opinion of the best authorities; but Dr. Bond holds different views, and he is entitled to his own opinion; though it is doubtful whether this and other small details were worth discussing at such portentous length.

In his last two letters, however (*Echo*, June 7th and 11th), he comes to a definite though still very small point, disputing the accuracy of a single case I quoted to illustrate the phenomenon of pro-vaccinist "bluff," of which I have given several examples; and he accuses me of having "doctored" a death-rate, of "statistical sleight of hand," and of "suppressing" essential facts. Let us see how the matter really stands.

(1) I quoted from the appendix to Sir John Simon's Parliamentary Paper on the 'History and Practice of Vaccination' (1857), a statement by Dr. T. Graham Balfour as to vaccination and small-pox in the Royal Military Orphan Asylum at Chelsea. Dr. Bond, however, does not consult this original paper, but the reprint of it in the first report, from which reprint Dr. Balfour's statement is omitted, as I stated to the Royal Commission (*Third Report*, Q. 7,211). Hence arises a long series of blunders. Dr. Bond says:—"The quotation of Dr. Balfour by Sir John Simon, on which Mr. Wallace bases his comparison, is to be found on page 72 of the first report of the Royal Commissioners"—and then he quotes what Sir John says about another paper of Dr. Balfour's, not the one in the appendix to his original Parliamentary Paper!

Having thus blundered at the very beginning, he goes on to say that I have "for some unexplained reason inflated 5,744 into 31,706." But Dr. Balfour, in his original article, gives 31,705 as the number of boys during 48 years . . . among whom there were 39 cases and four deaths; and he himself gives the death-rate at 126 per million, a rate which Dr. Bond accuses me of having "doctored." It is true that I added the words "on the average number in the

Asylum," because it is only thus you can compare one death-rate with another. That average number is 660 and a little more than a half over, and whether you calculate the death-rate among 31,705 in 48 years or among 660 each year, the result is the same, 126 per million. Of course, the numbers are too small to have any real statistical value. But they were adduced by Dr. Balfour himself to show the value of vaccination, and he took it for granted that 126 per million was an exceptionally low rate. But I show that among the very imperfectly vaccinated outside population of the same ages the rate was only 94 per million, so that the case, if it proves anything, is dead against vaccination. The number 5,744, given, apparently, in Dr. Balfour's other paper, may probably be the actual number of the boys who passed through the asylum in the 48 years, a number which, for statistical purposes, is unimportant.

(2) Dr. Bond's accusation as to my "suppression" of the alleged fact that the four boys who died, were all unvaccinated because they were "believed already to have suffered from small-pox" is equally unfounded, for that statement is not made in the original article. Dr. Balfour says, most clearly, that every boy was vaccinated unless he had been vaccinated before, or had had small-pox; and he states positively that they were "all protected." It is only Sir John Simon, in his second-hand statement, who hints a doubt by his "believed to have had small-pox;" and the fact of the whole four deaths occurring after previous small-pox is too improbable to be accepted except on the very best first-hand evidence. But, if it were the fact, it is not one to be suppressed by us, but rather by the pro-vaccinists. For they all admit that protection by vaccination can never exceed, and very rarely equal, that afforded by an attack of small-pox: so that if small-pox does not protect against the disease still less does vaccination.

I have now shown that every word and figure of my reference to the Chelsea Asylum case (at pp. 50–51 of my *Vaccination a Delusion*) is strictly accurate; and that my conclusion, which, as Dr. Bond tells his readers, "seems on the face of the evidence to be justifiable," is really so.

To use the very words of my critic against himself, "This is not a bad illustration of Dr. Bond's trustworthiness as a controversialist, as well as of his competency as a statistician." He has misread my very plain reference; he has, therefore, quoted the wrong authority; he has got hold of the wrong figures; he has thus left my facts and my conclusions wholly untouched, and he has occupied a column and a-half of small print in thus convicting himself of incompetence as a critic . . .

* * * * *

Dr. Bond and Mr. A. R. Wallace. Mr. Wallace's final reply. (ref 548)

Dr. Bond's three long letters on the Army and Navy statistics of small-pox really call for no reply from me, since he leaves my main statement, that the best re-vaccination possible has produced absolutely no effect in diminishing their small-pox mortality, entirely unaffected. The crucial fact that, in both

Army and Navy for the last 34 years, the total mortality from all diseases has decreased more rapidly than the small-pox mortality, is an unanswerable proof of this, and neither the Royal Commissioners nor Dr. Bond have attempted to answer it. And the further fact that these two bodies of picked and re-vaccinated men have suffered from small-pox about as much as the whole population of Ireland of the same ages, and more than twice as much as almost unvaccinated Leicester, still further show the absolute inutility of the best re-vaccination the medical staff of the Army and Navy can give. In the face of these two indisputable facts Dr. Bond's various petty criticisms, which never touch the main points, may be neglected.

My general conclusion, founded on the very best and most extensive body of statistics in existence, as given in my essay, remains totally uncontroverted by anything in Dr. Bond's numerous letters. It is, that the long series of records of mortality in London (the largest and most continuous known), the whole modern registration statistics of England, Scotland, and Ireland, the accurate official records of the Army and Navy, and those of the almost unvaccinated town of Leicester, all give the same consistent results . . .

Against this overwhelming consensus of the best evidence available, we have nothing but an array of individual experiences not founded on authoritative evidence, and in any case worth nothing, since an equal array can be and has been adduced on the other side. The broad and consistent teachings of the great masses of national experience we now have at our disposal, combined with that of extensive communities exceptionally treated—as in the case of the Army and Navy and the town of Leicester—cannot be affected by any amount of more limited or personal experience. Evidence must be weighed as well as counted. Yet the bulk of the medical profession and the great majority of our legislators, without paying the slightest attention to these facts, content themselves with again and again declaring their "firm belief in the immense benefits of vaccination." Like the historic Bourbons, they learn nothing and forget nothing, but blindly cling to their early beliefs . . .

* * * * *

At present it is difficult to assess the degree to which Wallace's involvement in the anti-vaccination campaign influenced either its own outcome (compulsory vaccination was abolished in England some years after Wallace's death) or other developments—in either epidemiology or other fields. Although he was no mathematician, Wallace was a skilled interpreter of descriptive statistics. Statistics-based arguments surfaced from time to time in his writings on many subjects, but in the case of the anti-vaccination crusade they provided the essence *of argument. His contribution to the development of the modern 'synthetic' approach to epidemiology (i.e., the understanding that epidemics arise only when several mutually reinforcing conditions are met), however, lies unassessed. In this connection it is suggestive, for example, that the scene of one of the first and most significant of the 'synthetic schools' was the Soviet Union, a country whose pre-Revolution naturalists had demonstrated an obvious empathy for Wallace's views on evolution and 'diffusionist' biogeography, and where to this day epidemiology, medical geography, and biogeography remain closely connected studies.*

Related writings

anti-vaccination statistical arguments: refs 374, 420, 509, 536, 542, 544, 548, 551, 616, 686, 687, 694, 697, 726, 729

concerning vaccination legislation and laws: refs 536, 588, 598, 603, 607, 613, 619, 621, 636, 640, 652, 674, 682, 687, 726, 729

letters of support sent to anti-vaccination conferences: refs 368, 588, 598, 603, 607, 613, 619, 636, 652, 674, 680, 682, 687, 694, 702

anti-vaccination testimony presented: refs 420, 686, 697

on miscellaneous topics: refs 536, 538, 578, 605, 608, 618, 681, 726, 729

secondary literature: Clements (1983), George (1964), Marchant (1916), Peirce (1906)

§ 5 · BIOGEOGRAPHY

. . . Not only does the marvellous structure of each organised being involve the whole past history of the earth, but such apparently unimportant facts as the presence of certain types of plants or animals in one island rather than in another, are now shown to be dependent on the long series of past geological changes—on those marvellous astronomical revolutions which cause a periodic variation of terrestrial climates—on the apparently fortuitous action of streams and currents in the conveyance of germs—and on the endlessly varied actions and reactions of organised beings on each other. And although these various causes are far too complex in their combined action to enable us to follow them out in the case of any one species, yet their broad results are clearly recognisable; and we are thus encouraged to study more completely every anomaly in the distribution of living things, in the firm conviction that by so doing we shall obtain a fuller and clearer insight into the course of nature, and with increased confidence that the 'mighty maze' of Being we see everywhere around us is 'not without a plan.'

(*Island life*, 3rd ed., 1902: ref 721, pp. 544–545)

Before going on to the studies through which Wallace established his scientific reputation—biogeography and evolutionary theory—we need briefly to examine Wallace's general method of doing science. At the outset, it is fair to say that he represents one of history's best examples of the way divergent thinking can be successfully applied to scientific advance. It should be clear from the material presented thus far that it was not Wallace's style to worry about establishment expectations; that he was so often on target is strong testimony to his ability to penetrate to the essential.

He was, of course, basically a theoretician—or, perhaps more accurately, a 'natural philosopher.' Contrary to the image of him that many people have formed on the basis of his Ternate 'fever-inspired discovery' of natural selection, however, he most definitely was not a man given to extemporaneous conjecture. In fact, it appears he almost compulsively committed himself to thorough literature reviews of every subject he took up, ostensibly in an effort to ascertain the range and character of 'facts' pertaining. (Given Wallace's views on 'the advantages of varied knowledge,' this is not surprising.) Thus, much of Wallace's scientific work finds its origins in empiricism: whether based on his own discoveries as a natural history collector or on those of others, whereupon his function became that of interpreter, critic, and synthesizer. Perhaps, had he been afforded the opportunities of wealth or formal training, he might have made a decent experimentalist as well. Many of his works contain suggestions for experiments designed to test particular hypotheses. It seems rather doubtful, however, that he ever would have really excelled in that role, as there is nothing in his writings that conveys any regrets at not having contributed in this way.

In any case, it certainly cannot be doubted that Wallace possessed a considerable talent for isolating significant pieces of factual information, and for relating these to one another. Once the domain of a problem had been established and all pertinent information digested, Wallace would proceed to the preparation of a model accounting for the facts. From the model emerged internal consistencies suggestive of possible causalities; cause and effect were then investigated through whatever concentration of attention his limitations of background permitted. Fortunately for him, he was also an adept analytical thinker, and in consequence was not afraid to extend himself. All things considered, he committed remarkably few significant errors in his many forays, probably because of his attention to fundamentals and general attitude of scientific scepticism.

In many instances the models he worked out dwelled on spatial relationships—again, the reason for considering Wallace as being, fundamentally, a geographer. It is usually little appreciated how strongly natural processes are constrained by the necessity of having to take place in a physically three-dimensional, 'real world' space. The less directly tied a train of thought is to this reality, the generally easier it is to validate in a manner that does not contradict competing arguments (consider, for example, the plethora of psychological theories, or the curious spatial properties of non-Euclidean geometries). One secret of Wallace's success as an innovative scientific thinker was his unusually developed talent for constructing process models explaining spatial patterns and how these change, relative to one another, over time. His conclusions rendered therefrom were of course wrong from time to time, but it seems more accurate in most of these instances to point to incomplete data, rather than poor reasoning *per se*, as being responsible.

Wallace's analytical abilities built him an enviable reputation as critic, reviewer, and popularizer. A number of ideas set out by other workers (notably, the classification of islands, the six-region faunal classification scheme of P. L. Sclater, natural selection, mimicry, the permanence of ocean basins and continents, the causes of the ice age, and the glacial origin of alpine lake basins—to all of which we shall come in due course) were either strongly defended, extended, or demonstrated outright by Wallace, and the way his efforts contributed to their success has in general been but poorly evaluated.

Wallace's skill at spatial analysis is best illustrated by his contributions to biogeography—specifically, by his innovations regarding the historical reconstruction of faunas ('zoological geography') and the spatial aspects of phylogenesis ('geographical zoology'). The wonders of the diversity of life first struck him during the period of his leisure-time collecting activities while working as a surveyor. As a teacher at Leicester, he began to mull over the concept of evolutionary change; a central influence in this regard was his reading of Robert Chambers's *The vestiges of the natural history of Creation* in 1845. At that point, evolution was little more than a hypothesis, and although Chambers helped focus interest on the subject, he was unable to advance a compelling causal framework for the process. Wallace apparently decided early on—probably no later than 1846—that an intensive investigation of the facts of animal and plant distribution might be just what was needed to determine how biological change took place. There were not sufficient distribution data available at that time to permit an attack on the problem based on literature review alone, however, and in any case, Wallace had no means through which to support such study. Investigation through professional collecting seemed a way around both problems, so off he and Bates sailed for South America.

Though many 'facts' were collected during the South American adventure, their interpretation was thwarted by the accident that occurred on Wallace's way back to England. None the less, the expedition was far from being a complete failure. Thanks to the salvage of his palm drawings, Wallace was able to put together the short monograph *Palm trees of the Amazon and their uses* (ref 713) in 1853. He followed this up later the same year with *A narrative of travels on the Amazon and Rio Negro* (ref 714). Both books were reasonably well received, though the information they presented on the distribution and uses of various Amazonian plant and animal forms did more to establish Wallace's reputation as a credible observer than as a thinker.

During the same period Wallace also managed to publish a few shorter works. On pages 109–110 of one of these efforts, 'On the monkeys of the Amazon' (ref 8), a paper presented at a meeting of the Zoological Society of London in late 1852, is recorded his first of many pleas that naturalists ought to give more attention to recording the precise locations at which their specimens were taken:

. . . In the various works on natural history and in our museums, we have generally but the vaguest statements of locality. S. America, Brazil, Guiana, Peru, are among the most common; and if we have "River Amazon" or "Quito" attached to a specimen, we may think ourselves fortunate to get anything so definite: though both are on the boundary of two distinct zoological districts, and we have nothing to tell us whether the one came from the north or south of the Amazon, or the other from the east or the west of the Andes. Owing to this uncertainty of locality, and the additional confusion created by mistaking allied species from distant countries, there is scarcely an animal whose exact geographical limits we can mark out on the map.

On this accurate determination of an animal's range many interesting questions depend. Are very closely allied species ever separated by a wide interval of country? What physical features determine the boundaries of species and of genera? Do the isothermal lines ever accurately bound the range of species, or are they altogether independent of them? What are the circumstances which render certain rivers and certain mountain ranges the limits of numerous species, while others are not? None of these questions can be satisfactorily answered till we have the range of numerous species accurately determined.

During my residence in the Amazon district I took every opportunity of determining the limits of species, and I soon found that the Amazon, the Rio Negro and the Madeira formed the limits beyond which certain species never passed. The native hunters are perfectly acquainted with this fact, and always cross over the river when they want to procure particular animals, which are found even on the river's bank on one side, but never by any chance on the other. On approaching the sources of the rivers they cease to be a boundary, and most of the species are found on both sides of them. Thus several Guiana species come up to the Rio Negro and the Amazon, but do not pass them; Brazilian species on the contrary reach but do not pass the Amazon to the north. Several Ecuador species from the east of the Andes reach down into the tongue of land between the Rio Negro and Upper Amazon, but pass neither of those rivers, and others from Peru are bounded on the north by the Upper Amazon, and on the east by the Madeira. Thus there are four districts, the Guiana, the Ecuador, the Peru and the Brazil districts, whose boundaries on one side are determined by the rivers I have mentioned.

In going up the Rio Negro the difference in the two sides of the river is very remarkable. In the lower part of the river you will find on the north the *Jachus bicolor* and the *Brachyurus couxiu*, and on the south the red-whiskered *Pithecia*. Higher up you will find on the north the *Ateles paniscus*, and on the south the new black *Jacchus* and the *Lagothrix humboldtii*.

Spix, in his work on the monkeys of Brazil, frequently gives, "banks of the river Amazon" as a locality, not being aware apparently that the species found on one side very often do not occur on the other, though the fact is generally known to the natives. In these observations I have only referred to the monkeys, but the same phaenomena occur both with birds and insects, as I have observed in many instances.

* * * * *

In early 1854 Wallace began the collecting expedition that would secure his place in history. By early 1855, he was confident enough of the connections between geography and geology and evolution to prepare his first major analysis of the subject:

On the Law which has regulated the introduction of new species (ref 20)

Every naturalist who has directed his attention to the subject of the geographical distribution of animals and plants, must have been interested in the singular facts which it presents. Many of these facts are quite different from what would have been anticipated, and have hitherto been considered as highly curious, but quite inexplicable. None of the explanations attempted from the time of Linnaeus are now considered at all satisfactory; none of them have given a cause sufficient to account for the facts known at the time, or comprehensive enough to include all the new facts which have since been, and are daily being added. Of late years, however, a great light has been thrown upon the subject by geological investigations, which have shown that the present state of the earth, and the organisms now inhabiting it, are but the last stage of a long and uninterrupted series of changes which it has undergone, and consequently, that to endeavour to explain and account for its present condition without any reference to those changes (as has frequently been done) must lead to very imperfect and erroneous conclusions.

The facts proved by geology are briefly these:—That during an immense, but unknown period, the surface of the earth has undergone successive changes; land has sunk beneath the ocean, while fresh land has risen up from it; mountain chains have been elevated; islands have been formed into continents, and continents submerged till they have become islands; and these changes have taken place, not once merely, but perhaps hundreds, perhaps thousands of times:—That all these operations have been more or less continuous, but unequal in their progress, and during the whole series the organic life of the earth has undergone a corresponding alteration. This alteration also has been gradual, but complete; after a certain interval not a single species existing which had lived at the commencement of the period. This complete renewal of the forms of life also appears to have occurred several times:—That from the last of the Geological epochs to the present or Historical epoch, the change of organic life has been gradual: the first appearance of animals now existing can in many cases be traced, their numbers gradually increasing in the more recent formations, while other species continually die out and disappear, so that the present condition of the organic world is clearly derived by a natural process of gradual extinction and creation of species from that of the latest geological periods. We may therefore safely infer a like gradation and natural sequence from one geological epoch to another.

Now, taking this as a fair statement of the results of geological inquiry, we see that the present geographical distribution of life upon the earth must be the result of all the previous changes, both of the surface of the earth itself and of its inhabitants. Many causes no doubt have operated of which we must ever remain in ignorance, and we may therefore expect to find many details very difficult of explanation, and in attempting to give one, must allow ourselves to call into our service geological changes which it is highly probable may have occurred, though we have no direct evidence of their individual operation.

The great increase of our knowledge within the last twenty years, both of the present and past history of the organic world, has accumulated a body of facts which should afford a sufficient foundation for a comprehensive law embracing and explaining them all, and giving a direction to new researches. It is about ten years since the idea of such a law suggested itself to the writer of this paper, and he has since taken every opportunity of testing it by all the newly ascertained facts with which he has become acquainted, or has been able to observe himself. These have all served to convince him of the correctness of his hypothesis. Fully to enter into such a subject would occupy much space, and it is only in consequence of some views having been lately promulgated, he believes in a wrong direction, that he now ventures to present his ideas to the public, with only such obvious illustrations of the arguments and results as occur to him in a place far removed from all means of reference and exact information.

The following propositions in Organic Geography and Geology give the main facts on which the hypothesis is founded.

Geography

1. Large groups, such as classes and orders, are generally spread over the whole earth, while smaller ones, such as families and genera, are frequently confined to one portion, often to a very limited district.
2. In widely distributed families the genera are often limited in range; in widely distributed genera, well-marked groups of species are peculiar to each geographical district.
3. When a group is confined to one district, and is rich in species, it is almost invariably the case that the most closely allied species are found in the same locality or in closely adjoining localities, and that therefore the natural sequence of the species by affinity is also geographical.
4. In countries of a similar climate, but separated by a wide sea or lofty mountains, the families, genera and species of the one are often represented by closely allied families, genera and species peculiar to the other.

Geology

5. The distribution of the organic world in time is very similar to its present distribution in space.
6. Most of the larger and some small groups extend through several geological periods.
7. In each period, however, there are peculiar groups, found nowhere else, and extending through one or several formations.
8. Species of one genus, or genera of one family occurring in the same geological time are more closely allied than those separated in time.
9. As generally in geography no species or genus occurs in two very distant localities without being also found in intermediate places, so in geology the life of a species or genus has not been interrupted. In other words, no group or species has come into existence twice.
10. The following law may be deduced from these facts:—*Every species has come into existence coincident both in space and time with a pre-existing closely allied species.*

This law agrees with, explains and illustrates all the facts connected with the following branches of the subject:—1st. The system of natural affinities. 2nd. The distribution of animals and plants in space. 3rd. The same in time, including all the phaenomena of representative groups, and those which Professor [Edward] Forbes supposed to manifest polarity. 4th. The phaenomena of rudimentary organs. We will briefly endeavour to show its bearing upon each of these.

If the law above enunciated be true, it follows that the natural series of affinities will also represent the order in which the several species came into existence, each one having had for its immediate antitype a closely allied species existing at the time of its origin. It is evidently possible that two or three distinct species may have had a common antitype, and that each of these may again have become the antitypes from which other closely allied species were created. The effect of this would be, that so long as each species has had but one new species formed on its model, the line of affinities will be simple, and may be represented by placing the several species in direct succession in a straight line. But if two or more species have been independently formed on the plan of a common antitype, then the series of affinities will be compound, and can only be represented by a forked or many-branched line. Now, all attempts at a natural classification and arrangement of organic beings show, that both these plans have obtained in creation. Sometimes the series of affinities can be well represented for a space by a direct progression from species to species or from group to group, but it is generally found impossible so to continue. There constantly occur two or more modifications of an organ or modifications of two distinct organs, leading us on to two distinct series of species, which at length differ so much from each other as to form distinct genera or families. These are the parallel series or representative groups of naturalists, and they often occur in different countries, or are found fossil in different formations. They are said to have an analogy to each other when they are so far removed from their common antitype as to differ in many important points of structure, while they still preserve a family resemblance. We thus see how difficult it is to determine in every case whether a given relation is an analogy or an affinity, for it is evident that as we go back along the parallel or divergent series, towards the common antitype, the analogy which existed between the two groups becomes an affinity. We are also made aware of the difficulty of arriving at a true classification, even in a small and perfect group;—in the actual state of nature it is almost impossible, the species being so numerous and the modifications of form and structure so varied, arising probably from the immense number of species which have served as antitypes for the existing species, and thus produced a complicated branching of the lines of affinity, as intricate as the twigs of a gnarled oak or the vascular system of the human body. Again, if we consider that we have only fragments of this vast system, the stem and main branches being represented by extinct species of which we have no knowledge, while a vast mass of limbs and boughs and minute twigs and scattered leaves is what we have to place in order, and determine the true position each originally occupied with regard to the others, the whole difficulty of the true Natural System of classification becomes apparent to us.

We shall thus find ourselves obliged to reject all those systems of classification which arrange species or groups in circles, as well as those which fix a definite number for the divisions of each group. The latter class have been very generally rejected by naturalists, as contrary to nature, notwithstanding the ability with which they have been advocated; but the circular system of affinities seems to have obtained a deeper hold, many eminent naturalists having to some extent adopted it. We have, however, never been able to find a case in which the circle has been closed by a direct and close affinity. In most cases, a palpable analogy has been substituted, in others the affinity is very obscure or altogether doubtful. The complicated branching of the lines of affinities in extensive groups must also afford great facilities for giving a show of probability to any such purely artificial arrangements. Their death-blow was given by the admirable paper of the lamented Mr. Strickland, published in the *Annals of Natural History*, in which he so clearly showed the true synthetical method of discovering the Natural System.

If we now consider the geographical distribution of animals and plants upon the earth, we shall find all the facts beautifully in accordance with, and readily explained by, the present hypothesis. A country having species, genera, and whole families peculiar to it, will be the necessary result of its having been isolated for a long period, sufficient for many series of species to have been created on the type of pre-existing ones, which, as well as many of the earlier-formed species, have become extinct, and thus made the groups appear isolated. If in any case the antitype had an extensive range, two or more groups of species might have been formed, each varying from it in a different manner, and thus producing several representative or analogous groups. The Sylviadae of Europe and the Sylvicolidae of North America, the Heliconidae of South America and the *Euplaeas* of the East, the group of trogons inhabiting Asia, and that peculiar to South America, are examples that may be accounted for in this manner.

Such phaenomena as are exhibited by the Galapagos Islands, which contain little groups of plants and animals peculiar to themselves, but most nearly allied to those of South America, have not hitherto received any, even a conjectural explanation. The Galapagos are a volcanic group of high antiquity, and have probably never been more closely connected with the continent than they are at present. They must have been first peopled, like other newly-formed islands, by the action of winds and currents, and at a period sufficiently remote to have had the original species die out, and the modified prototypes only remain. In the same way we can account for the separate islands having each their peculiar species, either on the supposition that the same original emigration peopled the whole of the islands with the same species from which differently modified prototypes were created, or that the islands were successively peopled from each other, but that new species have been created in each on the plan of the pre-existing ones. St. Helena is a similar case of a very ancient island having obtained an entirely peculiar, though limited, flora. On the other hand, no example is known of an island which can be proved geologically to be of very recent origin (late in the Tertiary, for instance), and yet possesses generic or family groups, or even many species peculiar to itself.

When a range of mountains has attained a great elevation, and has so remained during a long geological period, the species of the two sides at and near their bases will be often very different, representative species of some genera occurring, and even whole genera being peculiar to one side only, as is remarkably seen in the case of the Andes and Rocky Mountains. A similar phaenomenon occurs when an island has been separated from a continent at a very early period. The shallow sea between the Peninsula of Malacca, Java, Sumatra and Borneo was probably a continent or large island at an early epoch, and may have become submerged as the volcanic ranges of Java and Sumatra were elevated. The organic results we see in the very considerable number of species of animals common to some or all of these countries, while at the same time a number of closely allied representative species exist peculiar to each, showing that a considerable period has elapsed since their separation. The facts of geographical distribution and of geology may thus mutually explain each other in doubtful cases, should the principles here advocated be clearly established.

In all those cases in which an island has been separated from a continent, or raised by volcanic or coralline action from the sea, or in which a mountain-chain has been elevated, in a recent geological epoch, the phaenomena of peculiar groups or even of single representative species will not exist. Our own island is an example of this, its separation from the continent being geologically very recent, and we have consequently scarcely a species which is peculiar to it; while the Alpine range, one of the most recent mountain elevations, separates faunas and floras which scarcely differ more than may be due to climate and latitude alone.

The series of facts alluded to in Proposition 3, of closely allied species in rich groups being found geographically near each other, is most striking and important. Mr. Lovell Reeve has well exemplified it in his able and interesting paper on the Distribution of the *Bulimi*. It is also seen in the Humming-birds and Toucans, little groups of two or three closely allied species being often found in the same or closely adjoining districts, as we have had the good fortune of personally verifying. Fishes give evidence of a similar kind: each great river has its peculiar genera, and in more extensive genera its groups of closely allied species. But it is the same throughout Nature; every class and order of animals will contribute similar facts. Hitherto no attempt has been made to explain these singular phaenomena, or to show how they have arisen. Why are the genera of Palms and of Orchids in almost every case confined to one hemisphere? Why are the closely allied species of brown-backed Trogons all found in the East, and the green-backed in the West? Why are the Macaws and the Cockatoos similarly restricted? Insects furnish a countless number of analogous examples;—the *Goliathi* of Africa, the Ornithopterae of the Indian islands, the Heliconidae of South America, the Danaidae of the East, and in all, the most closely allied species found in geographical proximity. The question forces itself upon every thinking mind,—why are these things so? They could not be as they are, had no law regulated their creation and dispersion. The law here enunciated not merely explains, but necessitates the facts we see to exist, while the vast and long-continued geological changes of the earth readily account for the exceptions and apparent discrepancies that here and there occur. The writer's object in putting forward his views in the present imperfect manner is to submit them to the test of other minds, and to be made aware of all the facts supposed to be inconsistent with them. As his hypothesis is one which claims acceptance solely as explaining and connecting facts which exist in nature, he expects facts alone to be brought to disprove it; not *a-priori* arguments against its probability.

The phaenomena of geological distribution are exactly analogous to those of geography. Closely allied species are found associated in the same beds, and the change from species to species appears to have been as gradual in time as in space. Geology, however, furnishes us with positive proof of the extinction and production of species, though it does not inform us how either has taken place. The extinction of species, however, offers but little difficulty, and the *modus operandi* has been well illustrated by Sir. C. Lyell in his admirable *Principles*. Geological changes, however gradual, must occasionally have modified external conditions to such an extent as to have rendered the existence of certain species impossible. The extinction would in most cases be effected by a gradual dying-out, but in some instances there might have been a sudden destruction of a species of limited range. To discover how the extinct species have from time to time been replaced by new ones down to the very latest geological period, is the most difficult, and at the same time the most interesting problem in the natural history of the earth. The present inquiry, which seeks to eliminate from known facts a law which has determined, to a certain degree, what species could and did appear at a given epoch, may, it is hoped, be considered as one step in the right direction towards a complete solution of it.

Much discussion has of late years taken place on the question, whether the succession of life upon the globe has been from a lower to a higher degree of organization? The admitted facts seem to show that there has been a general, but not a detailed progression. Mollusca and Radiata existed before Vertebrata, and the progression from Fishes to Reptiles and Mammalia, and also from the lower mammals to the higher, is indisputable. On the other hand, it is said that the Mollusca and Radiata of the very earliest periods were more highly organized than the great mass of those now existing, and that the very first fishes that have been discovered are by no means the lowest organized of the class. Now it is believed the present hypothesis will harmonize with all these facts, and in a great measure serve to explain them; for though it may appear to some readers essentially a theory of progression, it is in reality only one of gradual change. It is, however, by no means difficult to show that a real progression in the scale of organization is perfectly consistent with all the appearances, and even with apparent retrogression, should such occur.

Returning to the analogy of a branching tree, as the best mode of representing the natural arrangement of species and their successive creation, let us suppose that at an early geological epoch any group (say a class of the Mollusca) has attained to a great richness of species and a high organization. Now let this great branch of allied species, by geological mutations, be completely or partially destroyed. Subsequently a new branch springs from the same trunk, that is to say, new species are successively created, having for their antitypes the same lower organized species which had served as the antitypes for the former group, but which have survived the modified conditions which destroyed it. This new group being subject to these altered conditions, has modifications of structure and organization given to it, and becomes the representative group of the former one in another geological formation. It may, however, happen, that though later in time, the new series of species may never attain to so high a degree of organization as those preceding it, but in its turn become extinct, and give place to yet another modification from the same root, which may be of higher or lower organization, more or less numerous in species, and more or less varied in form and structure than either of those which preceded it. Again, each of these groups may not have become totally extinct, but may have left a few species, the modified prototypes of which have existed in each succeeding period, a faint memorial of their former grandeur and luxuriance. Thus every case of apparent retrogression may be in reality a progress, though an interrupted one: when some monarch of the forest loses a limb, it may be replaced by a feeble and sickly substitute. The foregoing remarks appear to apply to the case of the Mollusca, which, at a very early period, had reached a high organization and a great development of forms and species in the Testaceous Cephalopoda. In each succeeding age modified species and genera replaced the former ones which had become extinct, and as we approach the present aera but few and small representatives of the group remain, while the Gasteropods and Bivalves have acquired an immense preponderance. In the long series of changes the earth has undergone, the process of peopling it with organic beings has been continually going on, and whenever any of the higher groups have become nearly or quite extinct, the lower forms which have better resisted the modified physical conditions have served as the antitypes on which to found the new races. In this manner alone, it is believed, can the representative groups at successive periods, and the risings and fallings in the scale of organization, be in every case explained.

The hypothesis of polarity, recently put forward by Professor Edward Forbes [note on death of Forbes omitted] to account for the abundance of generic forms at a very early period and at present, while in the intermediate epochs there is a gradual diminution and impoverishment, till the minimum occurred at the confines of the Palaeozoic and Secondary epochs, appears to us quite unnecessary, as the facts may be readily accounted for on the principles already laid down. Between the Palaeozoic and Neozoic periods of Professor Forbes, there is scarcely a species in common, and the greater part of the genera and families also disappear to be replaced by new ones. It is almost universally admitted that such a change in the organic world must have occupied a vast period of time. Of this interval we have no record; probably because the whole area of the early formations now exposed to our researches was elevated at the end of the Palaeozoic period, and remained so through the interval required for the organic changes which resulted in the fauna and flora of the Secondary period. The records of this interval are buried beneath the ocean which covers three-fourths of the globe. Now it appears highly probable that a long period of quiescence or stability in the physical conditions of a district would be most favourable to the existence of organic life in the greatest abundance, both as regards individuals and also as to variety of species and generic groups, just as we now find that the places best adapted to the rapid growth and increase of individuals also contain the greatest profusion of species and the greatest variety of forms,—the tropics in comparison with the temperate and arctic regions. On the other hand, it seems no less probable that a change in the physical conditions of a district, even small in amount if rapid, or even gradual if to a great amount, would be highly unfavourable to the existence of individuals, might cause the extinction of many species, and would probably be equally unfavourable to the creation of new ones. In this too we may find an analogy with the present state of our earth, for it has been shown to be the violent extremes and rapid changes of physical conditions, rather than the actual mean state in the temperate and frigid zones, which renders them less prolific than the tropical regions, as exemplified by the great distance beyond the tropics to which tropical forms penetrate when the climate is equable, and also by the richness in species and forms of tropical mountain regions which principally differ from the temperate zone in the uniformity of their climate. However this may be, it seems a fair assumption that during a period of geological repose the new species which we know to have been created would have appeared, that the creations would then exceed in number the extinctions, and therefore the number of species would increase. In a period of geological activity, on the other hand, it seems probable that the extinctions might exceed the creations, and the number of species consequently diminish. That such effects did take place in connexion with the causes to which we have imputed them, is shown in the case of the Coal formation, the faults and contortions of which show a period of great activity and violent convulsions, and it is in the formation immediately succeeding this that the poverty of forms of life is most apparent. We have then only to suppose a long period of somewhat similar action during the vast unknown interval at the termination of the Palaeozoic period, and then a decreasing violence or rapidity through the Secondary period, to allow for the gradual repopulation of the earth with varied forms, and the whole of the facts are explained. We thus have a clue to the increase of the forms of life during certain periods, and their decrease during others, without recourse to any causes but those we know to have existed, and to effects fairly deducible from them. The precise manner in which the geological changes of the early formations were effected is so extremely obscure, that when we can explain important facts by a retardation at one time and an acceleration at another of a process which we know from its nature and from observation to have been unequal,—a cause so simple may surely be preferred to one so obscure and hypothetical as polarity.

I would also venture to suggest some reasons against the very nature of the theory of Professor Forbes. Our knowledge of the organic world during any geological epoch is necessarily very imperfect. Looking at the vast numbers of species and groups that have been discovered by geologists, this may be doubted; but we should compare their numbers not merely with those that now exist upon the earth, but with a far larger amount (See on this subject a paper by Professor Agassiz in the *Annals* for November 1854.—Editor of the

Annals). We have no reason for believing that the number of species on the earth at any former period was much less than at present; at all events the aquatic portion, with which geologists have most acquaintance, was probably often as great or greater. Now we know that there have been many complete changes of species; new sets of organisms have many times been introduced in place of the old ones which have become extinct, so that the total amount which have existed on the earth from the earliest geological period must have borne about the same proportion to those now living, as the whole human race who have lived and died upon the earth, to the population at the present time. Again, at each epoch, the whole earth was no doubt, as now, more or less the theatre of life, and as the successive generations of each species died, their exuviae and preservable parts would be deposited over every portion of the then existing seas and oceans, which we have reason for supposing to have been more, rather than less, extensive than at present. In order then to understand our possible knowledge of the early world and its inhabitants, we must compare, not the area of the whole field of our geological researches with the earth's surface, but the area of the examined portion of each formation separately with the whole earth. For example, during the Silurian period all the earth was Silurian, and animals were living and dying, and depositing their remains more or less over the whole area of the globe, and they were probably (the species at least) nearly as varied in different latitudes and longitudes as at present. What proportion do the Silurian districts bear to the whole surface of the globe, land and sea (for far more extensive Silurian districts probably exist beneath the ocean than above it), and what portion of the known Silurian districts has been actually examined for fossils? Would the area of rock actually laid open to the eye be the thousandth or the ten-thousandth part of the earth's surface? Ask the same question with regard to the Oolite or the Chalk, or even to particular beds of these when they differ considerably in their fossils, and you may then get some notion of how small a portion of the whole we know.

But yet more important is the probability, nay, almost the certainty, that whole formations containing the records of vast geological periods are entirely buried beneath the ocean, and for ever beyond our reach. Most of the gaps in the geological series may thus be filled up, and vast numbers of unknown and unimaginable animals, which might help to elucidate the affinities of the numerous isolated groups which are a perpetual puzzle to the zoologist, may there be buried, till future revolutions may raise them in their turn above the waters, to afford materials for the study of whatever race of intelligent beings may then have succeeded us. These considerations must lead us to the conclusion, that our knowledge of the whole series of the former inhabitants of the earth is necessarily most imperfect and fragmentary,—as much so as our knowledge of the present organic world would be, were we forced to make our collections and observations only in spots equally limited in area and in number with those actually laid open for the collection of fossils. Now, the hypothesis of Professor Forbes is essentially one that assumes to a great extent the *completeness* of our knowledge of the *whole series* of organic beings which have existed on the earth. This appears to be a fatal objection to it, independently of all other considerations. It may be said that the same objections exist against every theory on such a subject, but this is not necessarily the case. The hypothesis put forward in this paper depends in no degree upon the completeness of our knowledge of the former condition of the organic world, but takes what facts we have as fragments of a vast whole, and deduces from them something of the nature and proportions of that whole which we can never know in detail. It is founded upon isolated groups of facts, recognizes their isolation, and endeavours to deduce from them the nature of the intervening portions.

Another important series of facts, quite in accordance with, and even necessary deductions from, the law now developed, are those of *rudimentary organs*. That these really do exist, and in most cases have no special function in the animal oeconomy, is admitted by the first authorities in comparative anatomy. The minute limbs hidden beneath the skin in many of the snake-like lizards, the anal hooks of the boa constrictor, the complete series of jointed finger-bones in the paddle of the Manatus and whale, are a few of the most familiar instances. In botany a similar class of facts has long been recognized. Abortive stamens, rudimentary floral envelopes and undeveloped carpels, are of the most frequent occurrence. To every thoughtful naturalist the question must arise, What are these for? What have they to do with the great laws of creation? Do they not teach us something of the system of Nature? If each species has been created independently, and without any necessary relations with pre-existing species, what do these rudiments, these apparent imperfections mean? There must be a cause for them; they must be the necessary results of some great natural law. Now, if, as it has been endeavoured to be shown, the great law which has regulated the peopling of the earth with animal and vegetable life is, that every change shall be gradual; that no new creature shall be formed widely differing from anything before existing; that in this, as in everything in Nature, there shall be gradation and harmony,—then these rudimentary organs are necessary, and are an essential part of the system of Nature. Ere the higher Vertebrata were formed, for instance, many steps were required, and many organs had to undergo modifications from the rudimental condition in which only they had as yet existed. We still see remaining an antitypal sketch of a wing adapted for flight in the scaly flapper of the penguin, and limbs first concealed beneath the skin, and then weakly protruding from it, were the necessary gradations before others should be formed fully adapted for locomotion. Many more of these modifications should we behold, and more complete series of them, had we a view of all the forms which have ceased to live. The great gaps that exist between fishes, reptiles, birds and mammals would then, no doubt, be softened down by intermediate groups, and the whole organic world would be seen to be an unbroken and harmonious system.

It has now been shown, though most briefly and imperfectly, how the law that "*Every species has come into existence coincident both in time and space with a pre-existing closely allied species,*" connects together and renders intelligible a vast number of independent and hitherto unexplained facts. The natural system of arrangement of organic beings, their geographical distribution, their geological sequence, the phaenomena of representative and substituted groups in all their modifications, and the most singular peculiarities of anatomical structure, are all explained and illustrated by it, in perfect accordance with the vast mass of facts which the researches of modern naturalists have brought together, and, it is believed, not materially opposed to any of them. It also claims a superiority over previous hypotheses, on the ground that it not merely explains, but necessitates what exists. Granted the law, and many of the most important facts in Nature could not have been otherwise, but are almost as necessary deductions from it, as are the elliptic orbits of the planets from the law of gravitation.

* * * * *

The preceding paper is perhaps best known for its place in the sequence of events leading to the publication of Darwin's On the origin of species *in 1859, but it also represents Wallace's first major treatise on the method of biogeographical analysis, and in many respects the birth of the modern approach to that subject. From that point onward, the historical analysis of biotas became an increasingly synthetic enterprise, relying on diverse sources of information pertaining to phylogeny, geology, past and present geography and climate, and the past and present distribution of life.*

Wallace's first significant application of the methods implied in this paper was recorded in a discussion of the faunal affinities of the Aru Islands, eastern Indonesia, published in The Annals and Magazine of Natural History *(ref 38) in late 1857. From this work, 'On the natural history of the Aru Islands,' I have excerpted the greater portion of its analytical section (pp. 478–483):*

From . . . the preceding observations, it will be seen that many Australian genera and some species occur in Aru; while, considering the very small number of species known from New Guinea, and the necessarily very imperfect exploration of Aru in such a short time, the number of identical species is very remarkable. I believe that nearly one-half of the hitherto-described species of passerine birds from New Guinea will be found in my Aru collections, a proportion which we could only expect if all the species of the latter country inhabit also the former. Such an identity occurs, I believe, in no other countries separated by so wide an interval of sea, for the average distance of the coast of Aru from that of New Guinea is at least 150 miles, and the points of nearest approach upwards of 100. Ceylon is nearer to India; Van Diemen's Land [Tasmania] is not farther from Australia, nor Sardinia from Italy; yet all these countries present differences more or less marked in their faunas; they possess each their peculiar species, and sometimes even peculiar genera. Almost the only islands possessing a rich fauna, but identical with that of the adjacent continent, are Great Britain and Sicily, and that circumstance is held to prove that they have been once a portion of such continents, and geological evidence shows that the separation had taken place at no distant period. We must, therefore, suppose Aru to have once formed a part of New Guinea, in order to account for its peculiar fauna, and this view is supported by the physical geography of the islands; for, while the fathomless Molucca sea extends to within a few miles of them on the west, the whole space eastward to New Guinea, and southward to Australia, is occupied by a bank of soundings at a uniform depth of about 30 or 40 fathoms. But there is another circumstance still more strongly proving this connexion: the great island of Aru, 80 miles in length from north to south, is traversed by three winding channels of such uniform width and depth, though passing through an irregular, undulating, rocky country, that they seem portions of true rivers, though now occupied by salt water, and open at each end to the entrance of the tides. This phaenomenon is unique, and we can account for their formation in no other way than by supposing them to have been once true rivers, having their source in the mountains of New Guinea, and reduced to their present condition by the subsidence of the intervening land.

This view of the origin of the Aru fauna is further confirmed by considering what it is not, as well as what it is; its deficiencies teach as much as what it possesses. There are certain families of birds highly characteristic of the Indian Archipelago in its western and better-known portion. In the Peninsula of Malacca, Sumatra, Java, Borneo and the Philippine Islands, the following families are abundant in species and in individuals. They are everywhere *common* birds. They are the Buceridae, Picidae, Bucconidae, Trogonidae, Meropidae, and Eurylaimidae; but not one species of all these families is found in Aru, nor, with two doubtful exceptions, in New Guinea. The whole are also absent from Australia. To complete our view of the subject, it is necessary also to consider the Mammalia, which present peculiarities and deficiencies even yet more striking. Not one species found in the great islands westward inhabits Aru or New Guinea. With the exception of only of pigs and bats, not a genus, not a family, not even an order of mammals is found in common. No Quadrumana, no Sciuridae, no Carnivora, Rodentia, or Ungulata inhabit these depopulated forests. With the two exceptions above mentioned, all the mammalia are *Marsupials*; in the great western islands there is not a single marsupial! A kangaroo inhabits Aru (and several New Guinea), and this, with three or four species of *Cuscus*, two or three little rat-like marsupials, a wild pig and several bats, are all the mammalia I have been able either to obtain or hear of.

It is to the full development of such interesting details that the collector and the systematist contribute so largely. In this point of view the discovery of every new species is important, and their correct description and accurate identification absolutely necessary. The most obscure and minute species are for this purpose of equal value with the largest and most brilliant, and a correct knowledge of the distribution and variations of a beetle or a butterfly as important as those of the eagle or the elephant. It is to the elucidation of these apparent anomalies that the efforts of the philosophic naturalist are directed; and we think, that if this highest branch of our science were more frequently alluded to by writers on natural history, its connexion with geography and geology discussed, and the various interesting problems thence arising explained, the too prevalent idea—that Natural History is at best but an amusement, a trivial and aimless pursuit, a useless accumulating of barren facts,—would give place to more correct views of a study, which presents problems as vast, as intricate, and as interesting as any to which the human mind can be directed, whose objects are as infinite as the stars of heaven and infinitely diversified, and whose field of research extends over the whole earth, not only as it now exists, but also during the countless changes it has undergone from the earliest geological epochs.

Let us now examine if the theories of modern naturalists will explain the phaenomena of the Aru and New Guinea fauna. We know (with a degree of knowledge approaching to certainty) that at a comparatively recent geological period, not one single species of the present organic world was in existence; while all the Vertebrata now existing have had their origin still more recently. How do we account for the places where they came into existence? Why are not the same species found in the same climates all over the world? The general explanation given is, that as the ancient species became extinct, new ones were created in each country or district, adapted to the physical conditions of that district. Sir C. Lyell, who has written more fully, and with more ability, on this subject than most naturalists, adopts this view. He illustrates it by speculating on the vast physical changes that might be effected in North Africa by the upheaval of a chain of mountains in the Sahara. "Then," he says, "the animals and plants of Northern Africa would disappear, and the region would gradually become fitted for the reception of a population of species *perfectly dissimilar in their forms, habits, and organization*." Now this theory implies, that we shall find a general similarity in the productions of countries which resemble each other in climate and general aspect, while there shall be a complete dissimilarity between those which are totally opposed in these respects. And if this is the general law which has determined the distribution of the existing organic world, there must be no exceptions, no striking contradictions. Now we have seen how totally the productions of New Guinea differ from those of the Western Islands of the Archipelago, say Borneo, as the type of the rest, and as almost exactly equal in area to New Guinea. This difference, it must be well remarked, is not one of species, but of genera, families, and whole orders. Yet it would be difficult to point out two countries more exactly resembling each other in climate and physical features. In neither is there any marked dry season, rain falling more or less all the year round; both are near the equator, both subject to the east and west monsoons, both everywhere covered with lofty forest; both have a great extent of flat, swampy coast and a mountainous interior; both are rich in Palms and Pandanaceae. If, on the other hand, we compare Australia with New Guinea, we can scarcely find a stronger contrast than in their physical conditions: the one near the equator, the other near and beyond the tropics; the one enjoying perpetual moisture, the other with alternations of excessive drought; the one a vast ever-verdant forest, the other dry open woods, downs, or deserts. Yet the faunas of the two, though mostly distinct in species, are strikingly similar in character. Every family of birds (except Menuridae) found in Australia also inhabits New Guinea, while all those striking deficiencies of the latter exist equally in the former. But a considerable proportion of the characteristic Australian *genera* are also found in New Guinea, and, when that country is better known, it is to be supposed that the number will be increased. In the Mammalia it is the same. Marsupials are almost the only quadrupeds in the one as in the other. If kangaroos are especially adapted to the dry plains and open woods of Australia, there must be some other reason for their introduction into the dense damp forests of New Guinea, and we can hardly imagine that the great variety of monkeys, of squirrels, of Insectivora, and of Felidae, were created in Borneo because the country was adapted to them, and not one single species given to another country exactly similar, and at no great distance. If there is any reason in the hardness of the woods or the scarcity of wood-boring insects, why woodpeckers should be absent from Australia, there is none why they should not swarm in the forests of New Guinea as well as in those of Borneo and Malacca. We can hardly help concluding, therefore, that some other law has regulated the distribution of existing species than the physical conditions of the countries in which they are found, or we should not see countries the most opposite in character with similar productions, while others almost exactly alike as respects climate and general aspect, yet differ totally in their forms of organic life.

In a former Number of this periodical we endeavoured to show that the simple law, of every new creation being closely allied to some species already existing in the same country, would explain all these anomalies, if taken in conjunction with the changes of surface and the gradual extinction and introduction of species, which are facts proved by geology. At the period when New Guinea and North Australia were united, it is probable that their physical features and climate were more similar, and that a considerable proportion of the species inhabiting each portion of the country were found over the whole. After the separation took place, we can easily understand how the climate of both might be considerably modified, and this might perhaps lead to the extinction of certain species. During the period that has since elapsed, new species have been gradually introduced into each, but in each closely allied to the pre-existing species, many of which were at first common to the two countries. This process would evidently produce the present condition of the two faunas, in which there are many allied species,—few identical. The great well-marked groups absent from the one would necessarily be so from the other also, for however much they might be *adapted* to the country, the law of close affinity would not allow of their appearance, except by a long succession of steps occupying an immense geological interval. The species which at the time of separation were found only in one country, would, by the gradual introduction of species allied to them, give rise to groups peculiar to that country. This separation of New Guinea from Australia no doubt took place while Aru yet formed part of the former island. Its separation must have occurred at a very recent period, the number of species common to the two showing that scarcely any extinctions have since taken place, and probably as few introductions of new species.

If we now suppose the Aru Islands to remain undisturbed during a period equal to about one division of the Tertiary epoch of geologists, we have reason to believe that the change of species of Vertebrata will become complete, an entirely new race having gradually been introduced, but all more or less closely allied to those now existing. During the same period a new fauna will also have arisen in New Guinea, and then the two will present the same comparative features that North Australia and New Guinea do now. Let the process of gradual change still go on for another period regulated by the same laws. Some species will then have become extinct in the one country, and unreplaced, while in the other a numerous series of modified species may have been introduced. Then the faunas will come to differ not in species only, but in generic groups. There would be then the resemblance between them that there is between the West India Islands and Mexico. During another geological period, let us suppose Aru to be elevated, and become a mountainous country, and extended by alluvial plains, while New Guinea was depressed, reduced in area, and thus many of its species perhaps extinguished. New species might then be more rapidly introduced into the modified and enlarged country; some groups, which had been early extinct in the other, might thus become very rich in species, and then we should have an exact counterpart of what we see now in Madagascar, where the families and some of the genera are African, but where there are many extensive groups of species forming peculiar genera, or even families, but still with a general resemblance to African forms. In this manner, it is believed, we may account for the facts of the present distribution of animals, without supposing any changes but what we know have been constantly going on. It is quite unnecessary to suppose that new species have ever been created "perfectly dissimilar in forms, habits, and organization" from those which have preceded them; neither do "centres of creation," which have been advocated by some, appear either necessary or accordant with facts, unless we suppose a "centre" in every island and in every district which possesses a peculiar species.

It is evident that, for the complete elucidation of the present state of the fauna of each island and each country, we require a knowledge of its geological history, its elevations and subsidences, and all the changes it has undergone since it last rose above the ocean. This can very seldom be obtained; but a knowledge of the fauna and its relation to that of the neighbouring countries will often throw great light upon the geology, and enable us to trace out with tolerable certainty its past history. A consideration of the birds of Aru has led us at some length into this subject, both on account of the interest attached to it, and because we are not aware of any attempt to explain in detail how the existing distribution of species has arisen, or strictly to connect it with those changes of surface which all countries have undergone . . .

* * * * *

Two years later Wallace produced a larger-scale analysis of faunal patterns based on the same general principles. This was 'On the zoological geography of the Malay Archipelago,' communicated by Charles Darwin to the Linnean Society in late 1859. This work—a classic of its genre—contains a detailed description of the faunal discontinuity that now bears his name: 'Wallace's Line.'

On the zoological geography of the Malay Archipelago (ref 53)

In Mr. [P. L.] Sclater's paper on the Geographical Distribution of Birds, read before the Linnean Society, and published in the *Proceedings* for February 1858, he has pointed out that the western islands of the Archipelago belong to the Indian, and the eastern to the Australian region of Ornithology. My researches in these countries lead me to believe that the same division will hold good in every branch of Zoology; and the object of my present communication is to mark out the precise limits of each region, and to call attention to some inferences of great general importance as regards the study of the laws of organic distribution.

The Australian and Indian regions of Zoology are very strongly contrasted. In one the Marsupial order constitutes the great mass of the mammalia,—in the other not a solitary marsupial animal exists. Marsupials of at least two genera (*Cuscus* and *Belideus*) are found all over the Moluccas and in Celebes; but none have been detected in the adjacent islands of Java and Borneo. Of all the varied forms of Quadrumana, Carnivora, Insectivora, and Ruminantia which abound in the western half of the Archipelago, the only genera found in the Moluccas are *Paradoxurus* and *Cervus*. The Sciuridae, so numerous in the western islands, are represented in Celebes by only two or three species, while not one is found further east. Birds furnish equally remarkable illustrations. The Australian region is the richest in the world in Parrots; the Asiatic is (of tropical regions) the poorest. Three entire families of the Psittacine order are peculiar to the former region, and two of them, the Cockatoos and the Lories, extend up to its extreme limits, without a solitary species passing into the Indian islands of the Archipelago. The genus *Palaeornis* is, on the other hand, confined with equal strictness to the Indian region. In the Rasorial order, the Phasianidae are Indian, the Megapodiidae Australian; but in this case one species of each family just passes the limits into the adjacent region. The genus *Tropidorhynchus*, highly characteristic of the Australian region, and everywhere abundant as well in the Moluccas and New Guinea as in Australia, is quite unknown in Java and Borneo. On the other hand, the entire families of Bucconidae, Trogonidae and Phyllornithidae, and the genera *Pericrocotus*, *Picnonotus*, *Trichophorus*, *Ixos*, in fact, almost all the vast family of Thrushes and a host of other genera, cease abruptly at the eastern side of Borneo, Java, and Bali. All these groups are *common birds* in the great Indian islands; they abound everywhere; they are the characteristic features of the ornithology; and it is most striking to a naturalist, on passing the narrow straits of Macassar and Lombock, suddenly to miss them entirely, together with the Quadrumana and Felidae, the Insectivora and Rodentia, whose varied species people the forests of Sumatra, Java, and Borneo.

To define exactly the limits of the two regions where they are (geographically) most intimately connected, I may mention that during a few days' stay in the island of Bali I found birds of the genera *Copsychus*, *Megalaima*, *Tiga*, *Ploceus*, and *Sturnopastor*, all characteristic of the Indian region and abundant in Malacca, Java, and Borneo; while on crossing over to Lombock, during three months collecting there, not one of them was ever seen; neither have they occurred in Celebes nor in any of the more eastern islands I have visited. Taking this in connexion with the fact of *Cacatua*, *Tropidorhynchus*, and *Megapodius* having their western limit in Lombock, we may consider it established that the Strait of Lombock (only 15 miles wide) marks the limits and abruptly separates two of the great Zoological regions of the globe. The Philippine Islands are in some respects of doubtful location, resembling and differing from both regions. They are deficient in the varied Mammals of Borneo, but they contain no Marsupials. The Psittaci are scarce, as in the Indian region; the Lories are altogether absent, but there is one representative of the Cockatoos. Woodpeckers, Trogons, and the genera *Ixos*, *Copsychus*, and *Ploceus* are highly characteristic of India. *Tanysiptera* and *Megapodius*, again, are Australian forms, but these seem represented by only solitary species. The islands possess also a few peculiar genera. We must on the whole place the Philippine Islands in the Indian region, but with the remark that they are deficient in some of its most striking features. They possess several isolated forms of the Australian region, but by no means sufficient to constitute a real transition thereto.

Leaving the Philippines out of the question for the present, the western and eastern islands of the Archipelago, as here divided, belong to regions more distinct and contrasted than any other of the great zoological divisions of the globe. South America and Africa, separated by the Atlantic, do not differ so widely as Asia and Australia: Asia with its abundance and variety of large Mammals and no Marsupials, and Australia with scarcely anything but Marsupials; Asia with its gorgeous Phasianidae, Australia with its dull-coloured Megapodiidae; Asia the poorest tropical region in Parrots, Australia the richest: and all these striking characteristics are almost unimpaired at the very limits of their respective districts; so that in a few hours we may experience an amount of zoological difference which only weeks or even months of travel will give us in any other part of the world!

Moreover there is nothing in the aspect or physical character of the islands to lead us to expect such a difference; their physical and geological differences do not coincide with the zoological differences. There is a striking homogeneity in the two *halves* of the Archipelago. The great volcanic chain runs through both parts; Borneo is the counterpart of New Guinea; the Philippines closely resemble the equally fertile and equally volcanic Moluccas; while in eastern Java begins to be felt the more arid climate of Timor and Australia. But these resemblances are accompanied by an extreme zoological diversity, the Asiatic and Australian regions finding in Borneo and New Guinea respectively their highest development.

But it may be said: "The separation between these two regions is not so absolute. There *is* some transition. There *are* species and genera common to the eastern and western islands." This is true, yet (in my opinion) proves no transition in the proper sense of the word; and the nature and amount of the resemblance only shows more strongly the absolute and original distinctness of the two divisions. The exception here clearly proves the rule.

Let us investigate these cases of supposed transition. In the western islands almost the only instance of a group peculiar to Australia and the eastern islands is the *Megapodius* in North-west Borneo. Not one of the Australian forms of Mammalia passes the limits of the region. On the other hand, Quadrumana occur in Celebes, Batchian, Lombock, and perhaps Timor; Deer have reached Celebes, Timor, Buru, Ceram, and Gilolo, but not New Guinea; Pigs have extended to New Guinea, probably the true eastern limit of the genus *Sus*; Squirrels are found in Celebes, Lombock, and Sumbawa: among birds, *Gallus* occurs in Celebes and Sumbawa, Woodpeckers reach Celebes, and Hornbills extend to the North-west of New Guinea. These cases of identity or resemblance in the animals of the two regions we may group into three classes; 1st, identical species; 2nd, closely allied or representative species; and 3rd, species of peculiar and isolated genera. The common Grey Monkey (*Macacus cynomolgus*) has reached Lombock, and perhaps Timor, but not Celebes. The Deer of the Moluccas seems to be a variety of the *Cervus rufus* of Java and Borneo. The Jungle Cock of Celebes and Lombock is a Javanese species. *Hirundo javanica*, *Zosterops flavus*, *Halcyon collaris*, *Eurystomus gularis*, *Macropygia phasianella*, *Merops javanicus*, *Anthreptes lepida*, *Ptilonopus melanocephala*, and some other birds appear the same in the adjacent islands of the eastern and western divisions, and some of them range over the whole Archipelago. But after reading Lyell on the various modes of dispersion of animals, and looking at the proximity of the islands, we shall feel astonished, not at such an amount of interchange of species (most of which are birds of great powers of flight), but rather that in the course of ages a much greater and almost complete fusion has not taken place. Were the Atlantic gradually to narrow till only a strait of twenty miles separated Africa from South America, can we help believing that many birds and insects and some few mammals would soon be interchanged? But such interchange would be a fortuitous mixture of faunas essentially and absolutely dissimilar, not a natural and regular transition from one to the other. In like manner the cases of identical species in the eastern and western islands of the Archipelago are due to the gradual and accidental commingling of originally absolutely distinct faunas.

In our second class (representative species) we must place the Wild Pigs, which seem to be of distinct but closely allied species in each island; the Squirrels also of Celebes are of peculiar species, as are the Woodpeckers and Hornbills, and two Celebes birds of the Asiatic genera *Phaenicophaeus* and *Acridotheres*. Now these and a few more of like character are closely allied to other species inhabiting Java, Borneo, or the Philippines. We have only therefore to suppose that the species of the western passed over to the eastern islands at so remote a period as on one side or the other to have become extinct, and to have been replaced by an allied form, and we shall have produced exactly the state of things now existing. Such extinction and such replacement we know has been continually going on. Such has been the regular course of nature for countless ages in every part of the earth of which we have geological records; and unless we are prepared to show that the Indo-Australian Archipelago was an altogether exceptional region, such must have been the course of nature here also. If these islands have existed in their present form only during one of the later divisions of the Tertiary period, and if interchange of species at very rare and distant intervals has occurred, then the fact of some identical and other closely allied species is a necessary result, even if the two regions in question had been originally peopled by absolutely distinct creations of organic beings, and there had never been any closer connexion between them than now exists. The occurrence of a limited number of representative species in the two divisions of the Archipelago does not therefore prove any true transition from one to the other.

The examples of our third class—of peculiar genera having little or no affinity with those of the adjacent islands—are almost entirely confined to Celebes, and render that island a district *per se*, in the highest degree interesting. *Cynopithecus*, a genus of Baboons, the extraordinary Babirusa and the singular ruminant *Anoa depressicornis* have nothing in common with Asiatic mammals, but seem more allied to those of Africa. A quadrumanous animal of the same genus (perhaps identical) occurs in the little island of Batchian, which forms the extreme eastern limit of the highest order of mammalia. An allied species is also said to exist in the Philippines. Now this occurrence of quadrumana in the Australian region proves nothing whatever as regards a transition to the western islands, which, among their numerous monkeys and apes, have nothing at all resembling them. The species of Celebes and Batchian have the high superorbital ridge, the long nasal bone, the dog-like figure, the minute erect tail, the predaceous habits and the fearless disposition of the true Baboons, and find their allies nowhere nearer than in tropical Africa. The *Anoa* seems also to point towards the same region, so rich in varied forms of Antelopes.

In the class of birds, Celebes possesses a peculiar genus of Parrots (*Prioniturus*), said to occur also in the Philippines; *Meropogon*, intermediate between an Indian and an African form of Bee-eaters; and the anomalous *Scissirostrum*, which Prince Bonaparte places next to a Madagascar bird, and forms a distinct subfamily for the reception of the two. Celebes also contains a species of *Coracias*, which is here quite out of its normal area, the genus being otherwise confined to Africa and continental India, not occurring in any other part of the Archipelago. The Celebes bird is placed, in Bonaparte's *Conspectus*, between two African species, to which therefore I presume it is more nearly allied than to those of India. Having just received Mr. Smith's Catalogue of the Hymenoptera collected during my first residence in Celebes, I find in it some facts of an equally singular nature. Of 103 species, only 16 are known to inhabit any of the western islands of the Archipelago, while 18 are identical with species of continental India, China, and the Philippine Islands, two are stated to be identical with insects hitherto known only from tropical Africa, and another is said to be most closely allied to one from the Cape.

These phenomena of distribution are, I believe, the most anomalous yet known, and in fact altogether unique. I am aware of no other spot upon the earth which contains a number of species, in several distinct classes of animals, the nearest allies to which do not exist in any of the countries which on every side surround it, but which are to be found only in another primary division of the globe, separated from them all by a vast expanse of ocean. In no other case are the species of a genus or the genera of a family distributed in *two* distinct areas separated by countries in which they do not exist; so that it has come to be considered a law in geographical distribution, "that both species and groups inhabit continuous areas."

Facts such as these can only be explained by a bold acceptance of vast changes in the surface of the earth. They teach us that this island of Celebes is more ancient than most of the islands now surrounding it, and obtained some part of its fauna before they came into existence. They point to the time when a great continent occupied a portion at least of what is now the Indian Ocean, of which the islands of Mauritius, Bourbon [Réunion], etc. may be fragments, while the Chagos Bank and the Keeling Atolls indicate its former extension eastward to the vicinity of what is now the Malayan Archipelago. The Celebes group remains the last eastern fragment of this now submerged land, or of some of its adjacent islands, indicating its peculiar origin by its zoological isolation, and by still retaining a marked affinity with the African fauna.

The great Pacific continent, of which Australia and New Guinea are no doubt fragments, probably existed at a much earlier period, and extended as far westward as the Moluccas. The extension of Asia as far to the south and east as the Straits of Macassar and Lombock must have occurred subsequent to the submergence of both these great southern continents; and the breaking up and separation of the islands of Sumatra, Java, and Borneo has been the last great geological change these regions have undergone. That this has really taken place as here indicated, we think is proved by the following considerations. Not more than twenty (probably a smaller number) out of about one hundred land birds of Celebes at present known are found in Java or Borneo, and only one or two of twelve or fifteen Mammalia. Of the Mammalia and birds of Borneo, however, at least three-fourths, probably five-sixths, inhabit also Java, Sumatra, or the peninsula of Malacca. Now, looking at the direction of the Macassar Straits running nearly north and south, and remembering we are in the district of the monsoons, a steady south-east and north-west wind blowing alternately for about six months each, we shall at once see that Celebes is more favourably situated than any other island to receive stray passengers from Borneo, whether drifted across the sea or wafted through the air. The distance too is less than between any of the other large islands; there are no violent currents to neutralize the action of the winds; and numerous islets in mid-channel offer stations which might rescue many of the wanderers, and admit, after repose, of fresh migrations. Between Java and Borneo the width of sea is much greater, the intermediate islands are fewer, and the direction of the monsoons *along* and not *across* the Java sea, accompanied by alternating currents in the same direction, must render accidental communication between the two islands exceedingly difficult; so that where the facilities for intercommunication are greatest, the number of species common to the two countries is least, and *vice versa*. But again, the mass of the species of Borneo, Java, etc., even when not *identical* are *congeneric*, which, as before explained, indicates *identity* at an earlier epoch; whereas the great mass of the fauna of Celebes is widely different from that of the western islands, consisting mostly of genera, and even of entire families, altogether foreign to them. This clearly points to a former total diversity of forms and species,—existing similarities being the result of intermixture, the extreme facilities for which we have pointed out. In the case of the great western islands a former more complete identity is indicated, the present differences having arisen from their isolation during a considerable period, allowing time for that partial extinction and introduction of species which is the regular course of nature. If the very small number of western species in Celebes is all that the most favourable conditions for transmission could bring about, the complete similarity of the faunas of the western islands could never (with far less favourable conditions) have been produced by the same means. And what other means can we conceive but the former connexion of those islands with each other and with the continent of Asia?

In striking confirmation of this view we have physical evidence of a very interesting nature. These countries are in fact *still connected*, and that so completely that an elevation of only 300 feet would nearly double the extent of tropical Asia. Over the whole of the Java Sea, the Straits of Malacca, the Gulf of Siam, and the southern part of the China Sea, ships can anchor in less than fifty fathoms. A vast submarine plain unites together the apparently disjointed parts of the Indian zoological region, and abruptly terminates, exactly at its limits, in an unfathomable ocean. The deep sea of the Moluccas comes up to the very coasts of Northern Borneo, to the Strait of Lombock in the south, and to near the middle of the Strait of Macassar. May we not therefore from these facts very fairly conclude that, according to the system of alternate bands of elevation and depression that seems very generally to prevail, the last great rising movement of the volcanic range of Java and Sumatra was accompanied by the depression that now separates them from Borneo and from the continent?

It is worthy of remark that the various islands of the Moluccas, though generally divided by a less extent of sea, have fewer species in common; but the separating seas are in almost every case of immense depth, indicating that the separation took place at a much earlier period. The same principle is well illustrated by the distribution of the genus *Paradisea*, two species of which (the common Birds of Paradise) are found only in New Guinea and the islands of Aru, Mysol, Waigiou, and Jobie, all of which are connected with New Guinea by banks of soundings, while they do not extend to Ceram or the Ké Islands, which are no further from New Guinea, but are separated from it by deep sea. Again, the chain of small volcanic islands to the west of Gilolo, though divided by channels of only ten or fifteen miles wide, possess many distinct representative species of insects, and even, in some cases, of birds also. The Baboons of Batchian have not passed to Gilolo, a much larger island, only separated from it by a channel ten miles wide, and in one part almost blocked up with small islands.

Now looking at these phenomena of distribution, and especially at those presented by the fauna of Celebes, it appears to me that a much exaggerated effect, in producing the present distribution of animals, has been imputed to the accidental transmission of individuals across intervening seas; for we have here as it were a test or standard by which we may measure the possible effect due to these causes, and we find that, under conditions perhaps the most favourable that exist on the globe, the percentage of species derived from this source is extremely small. When my researches in the Archipelago are completed, I hope to be able to determine with some accuracy this numerical proportion in several cases; but in the mean time we will consider 20 per cent. as the probable maximum for birds and mammals which in Celebes have been derived from Borneo or Java.

Let us now apply this standard to the case of Great Britain and the Continent, in which the width of dividing sea and the extent of opposing coasts are nearly the same, but in which the species are almost all identical,—or to Ireland, more than 90 per cent. of whose species are British,—and we shall at once see that no theory of transmission across the present Straits is admissible, and shall be compelled to resort to the idea of a very recent separation (long since admitted), to account for these zoological phenomena.

It is, however, to the oceanic islands that we consider the application of this test of the most importance. Let any one try to realize the comparative facilities for the transmission of organized beings across the Strait of Macassar from Borneo to Celebes, and from South Europe or North Africa to the island of Madeira, at least four times the distance, and a mere point in the ocean, and he would probably consider that in a given period a hundred cases of transmission would be more likely to occur in the former case than one in the latter. Yet of the comparatively rich insect-fauna of Madeira, 40 per cent. are continental species; and of the flowering plants more than 60 per cent. The Canary Islands offer nearly similar results. Nothing but a former connexion with the Continent will explain such an amount of specific identity (the weight of which will be very much increased if we take into account the representative species); and the direction of the Atlas range towards Teneriffe, and of the Sierra Nevada towards Madeira, are material indications of such a connexion.

The Galapagos are no further from South America than Madeira is from Europe, and, being of greater extent, are far more liable to receive chance immigrants; yet they have hardly a species identical with any inhabiting the American continent. These islands therefore may well have originated in mid-ocean; or if they ever were connected with the mainland, it was at so distant a period that the natural extinction and renewal of species has left not one in common. The character of their fauna, however, is more what we should expect to arise from the chance introduction of a very few species at distant intervals; it is very poor; it contains but few genera, and those scattered among unconnected families; its genera often contain several closely allied species, indicating a single antitype.

The fauna and flora of Madeira and of the Canaries, on the other hand, have none of this chance character. They are comparatively rich in genera and species; most of the principal groups and families are more or less represented; and, in fact, these islands do not differ materially, as to the general character of their animal and vegetable productions, from any isolated mountain in Europe or North Africa of about equal extent.

On exactly the same principles, the very large number of species of plants, insects, and birds, in Europe and North America, either absolutely identical or represented by very closely allied species, most assuredly indicates that some means of land communication in temperate or sub-arctic latitudes existed at no very distant geological epoch; and though many naturalists are inclined to regard all such views as vague and unprofitable speculations, we are convinced they will soon take their place among the legitimate deductions of science.

Geology can detect but a portion of the changes the surface of the earth has undergone. It can reveal the past history and mutations of what is now dry land; but the ocean tells nothing of her bygone history. Zoology and botany here come to the aid of their sister science, and by means of the humble weeds and despised insects inhabiting its now distant shores, can discover some of those past changes which the ocean itself refuses to reveal. They can indicate, approximately at least, where and at what period former continents must have existed, from what countries islands must have been separated, and at how distant an epoch the rupture took place. By the invaluable indications which Mr. Darwin has deduced from the structure of coral reefs, by the surveys of the ocean-bed now in progress, and by a more extensive and detailed knowledge of the geographical distribution of animals and plants, the naturalist may soon hope to obtain some idea of the continents which have now disappeared beneath the ocean, and of the general distribution of land and sea at former geological epochs.

Most writers on geographical distribution have completely overlooked its connexion with well-established geological facts, and have thereby created difficulties where none exist. The peculiar and apparently endemic faunae and florae of the oceanic islands (such as the Galapagos and St. Helena) have been dwelt upon as something anomalous and inexplicable. It has been imagined that the more simple condition of such islands would be to have their productions identical with those of the nearest land, and that their actual condition is an incomprehensible mystery. The very reverse of this is however the case. We really require no speculative hypothesis, no new theory, to explain these phenomena; they are the logical results of well-known laws of nature. The regular and unceasing extinction of species, and their replacement by allied forms, is now no hypothesis, but an established fact; and it necessarily produces such peculiar faunae and florae in all but recently formed or newly disrupted islands, subject of course to more or less modification according to the facilities for the transmission of fresh species from adjacent continents. Such phenomena therefore are far from uncommon. Madagascar, Mauritius, the Moluccas, New Zealand, New Caledonia, the Pacific Islands, Juan Fernandez, the West India Islands, and many others, all present such peculiarities in greater or less development. It is the instances of identity of species in distant countries that presents the real difficulty. What was supposed to be the more normal state of things is really exceptional, and requires some hypothesis for its explanation. The phenomena of distribution in the Malay Archipelago, to which I have here called attention, teach us that, however narrow may be the strait separating an island from its continent, it is still an impassable barrier against the passage of any considerable number and variety of land animals; and that in all cases in which such islands possess a tolerably rich and varied fauna of species mostly identical, or closely allied with those of the adjacent country, we are forced to the conclusion that a geologically recent disruption has taken place. Great Britain, Ireland, Sicily, Sumatra, Java and Borneo, the Aru Islands, the Canaries and Madeira, are cases to which the reasoning is fully applicable.

In his introductory Essay on the Flora of New Zealand, Dr. [Joseph] Hooker has most convincingly applied this principle to show the former connexion of New Zealand and other southern islands with the southern extremity of America; and I will take this opportunity of calling the attention of zoologists to the very satisfactory manner in which this view clears away many difficulties in the distribution of animals. The most obvious of these is the occurrence of Marsupials in America only, beyond the Australian region. They evidently entered by the same route as the plants of New Zealand and Tasmania which occur in South temperate America, but having greater powers of dispersion, a greater plasticity of organization, have extended themselves over the whole continent though with so few modifications of form and structure as to point to a unity of origin at a comparatively recent period. It is among insects, however, that the resemblances approach in number and degree to those exhibited by plants. Among Butterflies the beautiful Heliconidae are strictly confined to South America, with the exception of a single genus (*Hamadryas*) found in the Australian region from New Zealand to New Guinea. In Coleoptera many families and genera are characteristic of the two countries; such are Pseudomorphidae among the Geodephaga, Lamprimidae and Syndesidae among the Lucani, Anoplognathidae among the Lamellicornes, Stigmoderidae among the Buprestes, *Natalis* among the Cleridae, besides a great number of representative genera. This peculiar distribution has hitherto only excited astonishment, and has confounded all ideas of unity in the distribution of organic beings; but we now see that they are in exact accordance with the phenomena presented by the flora of the same regions, as developed in the greatest detail by the researches of Dr. Hooker.

It is somewhat singular, however, that not one *identical species* of insect should yet have been discovered, while no less than 89 species of flowering plants are found both in New Zealand and South America. The relations of the animals and of the plants of these countries must necessarily depend on the same physical changes which the Southern hemisphere has undergone; and we are therefore led to conclude that insects are much less persistent in their specific forms than flowering plants, while among Mammalia and land birds (in which no genus even is common to the countries in question) species must die and be replaced much more rapidly than in either. And this is exactly in accordance with the fact (well established by geology) that at a time when the shells of the European seas were almost all identical with species now living, the European Mammalia were almost all different. The duration of life of species would seem to be in an inverse proportion to their complexity of organization and vital activity.

In the brief sketch I have now given of this interesting subject, such obvious and striking facts alone have been adduced as a traveller's note-book can supply. The argument must therefore lose much of its weight from the absence of detail and accumulated examples. There is, however, such a very general accordance in the phenomena of distribution as separately deduced from the various classes or kingdoms of the organic world, that whenever one class of animals or plants exhibits in a clearly marked manner certain relations between two countries, the other classes will certainly show similar ones, though it may be in a greater or a less degree. Birds and insects will teach us some truths; and even animals and plants, though existing under such different conditions, and multiplied and dispersed by such a generally distinct process, will never give conflicting testimony, however much they may differ as regards the amount of relationship between distant regions indicated by them, and consequently notwithstanding the greater or less weight either may have in the determining of questions of this nature.

This is my apology for offering to the Linnean Society the present imperfect outline in anticipation of the more detailed proofs and illustrations, which I hope to bring forward on a future occasion.

* * * * *

By 1864, Wallace had fully integrated natural selection into his approach to historical biogeography, and went on to identify many important connections between 'the survival of the fittest' and the nature of animal and plant distribution, in so doing contributing significantly to each field of study. Typical of his productions was a short letter concerning the causes of bird migration printed in Nature *in 1874:*

Migration of birds (ref 244)

The subject to which Prof. [Alfred] Newton has called attention is one of great interest to all naturalists, and requires to be studied systematically; for I can hardly think that the solution is so "simple in the extreme" as Mr. Newton thinks it may be.

It appears to me probable that here, as in so many other cases, "survival of the fittest" will be found to have had a powerful influence. Let us suppose that in any species of migratory bird, breeding can as a rule be only safely accomplished in a given area; and further, that during a great part of the rest of the year sufficient food cannot be obtained in that area. It will follow that those birds which do not leave the breeding area at the proper season will suffer, and ultimately become extinct; which will also be the fate of those which do not leave the feeding area at the proper time. Now, if we suppose that the two areas were (for some remote ancestor of the existing species) coincident, but by geological and climatic changes gradually diverged from each other, we can easily understand how the habit of incipient and partial migration at the proper seasons would at last become hereditary, and so fixed as to be what we term an instinct. It will probably be found, that every gradation still exists in various parts of the world, from a complete coincidence to a complete separation of the breeding and the subsistence areas; and when the natural history of a sufficient number of species in all parts of the world is thoroughly worked out, we may find every link between species which never leave a restricted area in which they breed and live the whole year round, to those other cases in which the two areas are absolutely separated. The actual causes that determine the exact time, year by year, at which certain species migrate, will of course be difficult to ascertain. I would suggest, however, that they will be found to depend on those climatal changes which most affect the particular species. The change of colour, or the fall, of certain leaves; the change to the pupa state of certain insects; prevalent winds or rains; or even the decreased temperature of the earth and water, may all have their influence. Ample materials must exist, in the case of European birds, for an instructive work on this subject. The two areas should be carefully determined for a number of migratory birds; the times of their movements should be compared with a variety of natural phenomena likely to influence them; the past changes of surface, of climate, and of vegetation should be taken account of; and there seems no reason to doubt that such a mode of research would throw much light on, if it did not completely solve, the problem . . .

* * * * *

In an essay contributed to the journal Nineteenth Century *in 1879, Wallace considered how the evolutionary process leads to the development of 'disjunct' distribution patterns:*

Animals and their native countries (ref 302)

In the December [1878] number of this Review, Mr. P. L. Sclater called attention to the subject of the geographical distribution of animals in its bearing on the theory of evolution, and gave numerous special cases in which the actual distribution of particular species and groups is very difficult to explain on that theory without making assumptions which, in his opinion, the evidence at our disposal does not warrant. Difficulties of this nature are so numerous, and many of them seem to him so weighty, that, in order to explain them, he is led to question, what is almost an axiom with evolutionists, that *identity of structure is, without exception, an indication of descent from a common parent*. Similar doubts, though not stated in exactly the same terms, have been felt by Professor Mivart (*Genesis of Species*, chap. iii); and it therefore becomes a matter of interest to examine a little more closely into the alleged difficulties, in order to see whether they are not really explicable on the principle of descent with modification, only calling to our aid such general assumptions as are fully warranted by what we actually know of the migrations and extinctions of living things, and of the past changes in the physical condition of the earth and its inhabitants.

As Mr. Sclater's article gives an excellent summary of the nature and meaning of zoological distribution, and of the main general conclusions arrived at by naturalists, our purpose will be best attained by proceeding at once to consider his special cases of difficulty; and in doing so we shall have occasion to discuss, as fully as may be required, the general principles and particular illustrations needed to elucidate them.

We have first the case of the *Little Blue Magpie of Spain*, which has a very close ally in the extreme parts of Eastern Asia and Japan, while there is nothing closely allied to these in all the intervening regions or in any other part of the world. This is said to be an infringement of the canon as to the continuity of specific areas, and as such to require explanation. Before proceeding further, it will be well to inquire into the value of this canon of continuity, and whether it is so clear and well established that infringements of it are altogether exceptions to the usual course of nature. So far from this being the case, I believe it will be found that, between the complete continuity of the area occupied by a species or a genus and such wide discontinuity as occurs in the present example, there is every possible gradation; and further, that the instances of discontinuity are very numerous, while those of complete continuity are far less generally the rule than appears at first sight.

In order to understand the bearing of this class of phenomena on the theory of derivation, let me briefly indicate the probable course of a *genus* of animals from its birth or origin to its final extinction.

Genera are groups of species which agree among themselves, and differ from all other groups in the same family order, by the possession of some *structural* peculiarities. We must therefore suppose a genus to have had its origin in some variation of *structure* which was useful to its possessors—such as a modification of the bill, feet, or wings of a bird, or of the teeth, claws, or horns of a mammal. According to the theory of natural selection, the possessors of such a useful peculiarity would increase at the expense of their close allies who did not possess it, and would soon form a distinct group of individuals breeding together and constituting a species—the first species of the new genus. This species having in time supplanted the parent species, and being better adapted than it for the battle of life, would almost certainly cover a wider area, and thus come into competition with several of the allied species of the old genus, some of which it would also probably supplant, and occupy the areas they formerly occupied. But as they had been modified into distinct *species* (differing, perhaps, slightly in colour or habits in accordance with the varying physical conditions), so the now widespread species of the new genus would vary, and become modified in a somewhat analogous manner, forming a *genus* consisting of several *species*. Now, if the generic form thus produced was one of great inherent vigour and adaptability, and if the peculiarity of structure it possessed was of considerable importance, it would become what Mr. Darwin terms a dominant group: that is, it would spread widely over the earth under various modified forms suited to the various conditions it became subject to. At last it would reach its maximum of development, and cease to spread further, either owing to its inability to adapt itself to further changes of climate, etc., or, what is more likely, from its coming into competition with other dominant groups which had in like manner spread from some other centres.

Now, during all this time, which may be termed the period of growth of the genus, its area will have been almost necessarily continuous, and the areas occupied by its several species (also continuous) will probably overlap each other. But now commences its period of decay. Other groups of the same or allied families have given rise to varieties which also become dominant species and genera, which, under the somewhat changed physical conditions that in time have come about, beat it in the battle of life, and force it to retire step by step from the vast area it had overrun. First one species and then another will dwindle away and become finally extinct, and by so doing will necessarily leave gaps in its area of distribution. This process going steadily on, the time will at last come when two or three species only will remain, most likely in widely separated parts of its former area; their position being determined either by the competition being there somewhat less severe, or by some speciality of conditions which are exceptionally favourable to the dying-out group. Then one and then another of these species will die out, and the once extensive genus will only be represented by a single species inhabiting a very restricted locality. This will become rarer and rarer, the necessary preliminary to that final extinction which we know to be the fate, sooner or later, of every group of living things.

Most working naturalists (and none better than Mr. Sclater) are acquainted with genera whose distribution will illustrate all the successive phases of this hypothetical history; while palaeontology furnishes us with some actual examples of the progress of a group from its rise to its decay, though, owing to the extreme imperfection of the geological record (and its total absence for important epochs in many parts of the globe), we can never trace the complete history of such a group. A little consideration will show us, however, why it is that continuity of generic and specific areas appears to be the rule, discontinuity the exception. There can be no doubt that the development of an extensive genus is a slow process, while its decay and final extinction need not be slow, and may conceivably be extremely rapid. Geological and geographical changes may be long in preparation, but finally very abrupt. Land may sink a thousand feet without producing any very important effect except diminution of area, but the next hundred feet of depression may cut it off from a continent, and may alter the direction of ocean currents, thus producing a greater organic and physical change than had been brought about by the previous subsidence occupying ten times as long. Again, such a change as that which admitted the highly organised Miocene mammalia of Europe into Tropical and South Africa must have led at once to the extermination of many of the indigenous species, and have restricted the area of many more. It is also important to remember that the dominant or growing species and genera, which are those having continuous areas, will be necessarily more prominent, more numerous in species and individuals, and therefore far better known; while those in process of extinction, and for that very reason having discontinuous areas, will be less numerous, far less common, and in fact often very rare, and therefore much less known. In many cases, too, it will happen that the discontinuity is not great as regards distance, and it will then not be noticed, or will be imputed to want of knowledge, although it may be quite as real as when half a continent lies between the two species.

It appears, therefore, that the discontinuity of many genera and higher groups, so far from being difficult of explanation, is really one of the inevitable results of the process of extinction which is always going on. The peculiarity of the particular case we are considering is that it is somewhat extreme in the fact of two species only being left, occupying limited areas situated at the opposite extremities of the immense Palaearctic region. But this is not very extraordinary, because there are in Western Europe and Japan a number of pairs of closely allied species whose extinction in the intervening areas would lead to an exactly similar phenomenon to that we are considering. Such are the European and Japanese jays, bullfinches, gold-crest warblers, and wrens, all of which are closely allied to each other, while they are separated by a wide area in Central Asia often occupied by species which differ considerably from both. Should either of these groups die out, we might expect that the species inhabiting the comparatively desert and inhospitable regions of Central Asia would succumb first, while those living in the milder and more equable climates of Western Europe and Japan would probably linger on, the last of their race. It is very interesting to note that in most cases of such widely separated but closely allied species or groups there is a decided similarity in the general physical conditions of the countries they inhabit. The ally of the Spanish blue magpie is found in North-eastern Asia from Shanghai to Pekin and the Amoor, as well as in Japan; and these countries reproduce the hot summers and the cold winters, the rugged mountains and the sheltered valleys of Spain, while both areas are subject to the influence of the vicinity of the ocean in an almost equal degree.

Before going further we must guard against a misconception as to the progressive rise and decay of species, genera, and higher groups. It is not maintained that this will always take place uninterruptedly or continuously. On the contrary, it is certain that the decay of a group may run its course for a time, and then, owing to changed conditions, may be checked, and even be changed into a new growth and development. Hence arise those isolated groups, which yet, by their abundance in species and the considerable area they occupy, show that they are in a flourishing condition—of which the tanagers in South America, the broadbills (Euryloemidae) of Asia, and the colies of Africa, may serve as examples. When an extensive group is in process of extinction, it may become broken up into many isolated portions, sometimes in juxtaposition to each other, sometimes separated in remote parts of the globe. Changes of conditions, whether physical or organic, may favour first one, then another, of these portions, or the same portion may be subject to alternate phases of progression and decay several times repeated. How can we wonder that the final result of such complex processes, whose general nature we can understand, but whose details it is impossible for us to trace, should often lead to anomalies in geographical distribution? And when we consider that these processes have been often intensified and further complicated by geographical mutations, and by those forced migrations induced by the climatal changes which culminated in the glacial epoch, the wonder rather is that we can account for so much, than that there should be matters of detail which we cannot explain.

It is clear, then, that the case of the *Little Blue Magpie* of Spain and its close ally in Eastern Asia is simply an example of a dying-out group, of which two species only remain isolated in countries favourable to their existence; and further, that the supposition of their common ancestors having once occupied the intervening region, so far from being unwarranted, is supported by the analogy of several other groups of birds in the same area.

Mr. Sclater's next two cases may be sufficiently explained by the application of the general considerations and examples already adduced. We have two allied species of *Oxyrhamphus* isolated in South-East Brazil and Central America—countries, it may be remarked, about equally removed from the equator and enjoying very similar climates; while in the cuckoos of the genus *Neomorpha* we have a similar phenomenon in nearly the same two areas, with the addition of three species in the intervening districts which are not closely allied to the two others . . .

Mr. Sclater's third problem, that of *Pitta angolensis*, is very interesting, and will afford us an opportunity of discussing some of the most curious phenomena of distribution, and of bringing forward some considerations which I believe will go far towards the removal of most of the difficulties they present. The case is that of an extensive and very beautiful group of birds, ranging from the Himalayas to Eastern Australia, but having one species isolated in Western Africa. Mr. Sclater has himself mentioned a few parallel cases, but there are others equally interesting, a consideration of which may aid us in our attempted explanation. We have first the gorilla and chimpanzee, corresponding to the orangs of Borneo and Sumatra; and among the quadrumana the genus *Cercocebus* allied to the Eastern macaques is wholly West African, while *Colobus*, closely allied to the Asiatic *Semnopithecus*, is found in Abyssinia as well as in West Africa. Among birds we have *Alethe*, a genus of babblers, and *Pholidornis*, one of the Dicoeidae, allied to Asiatic forms; while the genus of crested hornbills, *Berenicornis*, has one species in Sumatra and the only other in West Africa. The fruit-thrushes of the genus *Criniger*, so abundant in Asia and Malaya, are also found in West and in South Africa, while the beautiful eastern parroquets of the genus *Palaeornis* inhabit West Africa and Abyssinia. Among reptiles and amphibia we have three families which follow the same rule. Lizards of the family Acontiadae are confined to the Moluccas, Ceylon, and West and South Africa; toads of the family Engystomidae have nearly the same range, but are more widely spread in Asia; while snakes of the family Homalopsidae are abundant in tropical Asia and America, and are even found in Europe, while in Africa they are confined to the western districts.

These numerous cases of the occurrence of what are otherwise Eastern groups in West Africa, undoubtedly suggest some correspondence of physical conditions which renders this portion of the continent alone suitable to them. The further question, how they got there at all, is elucidated by what we know of the past history of Africa and Europe. It is now generally admitted that, before the Miocene period, Africa was cut off from the great continent of the northern hemisphere by a wide arm of the sea. It was then in fact an island, or perhaps a group of large islands, and probably contained only some of the lower forms of mammalian life, among which the lemurs and the insectivora were conspicuous. When, during the Miocene period, it became united to Europe and Asia, it was at once overrun by a number of the large mammalia of that continent, such as elephants, rhinoceroses, lions, giraffes, antelopes, hippopotami, apes, and many other forms whose remains are found in abundance in the Miocene deposits of France, Germany, Hungary, Italy, and Greece. South Europe must have then possessed a sub-tropical, if not a tropical, climate, the gradual deterioration of which led to the extinction of most of these animals, while in Africa they have survived and greatly multiplied. Among the European fossils of this period we find some of especial interest. Such are, the genus *Hyomoschus*, mentioned by Mr. Sclater as one of the Indian group of chevrotains still living isolated in West Africa; a monkey closely allied to *Colobus* and *Semnopithecus*, and perhaps an ancestral form of these two groups now separated in Asia and West Africa; and among birds the snake-eating secretary bird; *Necrornis*, supposed to be allied to the peculiarly African plantain-eaters; and a parrot allied to the well-known grey parrots of Africa. Here, then, we have traced the origin of the resemblances between certain animal forms of tropical Asia and West Africa to their having been derived from a common source in the great northern continent, and we are justified in supposing that the *Pitta angolensis* is also a descendant of one of these Miocene forms. A species of *Pitta* even now inhabits Japan, so that we may well suppose the genus to have originated in Europe or Western Asia in the warm Miocene period. It was probably once more abundant in Africa, but, along with the *Hyomoschus* and the crested hornbill, is now in process of extinction in the one continent, while its allies continue to flourish in the other.

Leaving for the present the next three cases of difficulty adduced by Mr. Sclater, we will pass on to the subject of allied forms occurring in the tropics of both hemispheres. Besides the barbets, which occur in the tropics of Asia, Africa, and America in almost equal abundance, we have the trogons, abundant in Asia and America, but with a single peculiar genus of two species in Africa, and the tapirs confined to the Malay islands and tropical America, while equally remarkable are two genera of snakes, *Dryiophis* and *Dipsadoboa*, confined to West Africa and tropical America. Towards an explanation of these curious anomalies we have the very interesting fact, that tapirs closely resembling those now living abounded in Europe during the Miocene period, and continued to live in the Pliocene period, both in France and England, as well as in North America. This suggests that a tropical climate is not essential to these animals, and that their present restricted range is due to other than climatal causes. We may also be sure that if they could live so far north as our island in the Pliocene period, they might have ranged considerably further north during the earlier and warmer Miocene. The only difficulty is, how did these Miocene tapirs reach America? And if we can find any reasonable answer to this question we may consider that it will equally apply to all the other cases which have been mentioned.

The close relation between many of the extinct mammalia of North America and Europe at successive periods, while in other cases entire groups have always been restricted to one continent only, renders it certain that there existed at several distinct epochs some land connection sufficient to enable terrestrial animals to pass between them. The sea at Behring Strait is so shallow that we may safely conclude that the continents of Asia and America have here been recently connected, while the shallow Okhotsk, Japan, and Yellow seas indicate a large extension of the lowlands of Eastern Asia; but the very deep Atlantic comes up to beyond 55 degrees N. latitude on the east side of Kamschatka, so that this part of the connecting land would probably always have had a temperate rather than a tropical climate. On the European side we find between the west of Ireland and Newfoundland a maximum depth of about 12,000 feet, but with large areas between 5,000 and 10,000 feet deep; and although this implies an immense subsidence, it is not very improbable that all the area from this line northward to Greenland and Iceland was dry land during some part of the Miocene period. In support of this view it may be noted that the Alps, the Pyrenees, the Rocky Mountains, and even the Himalayas, were all in early Miocene times many thousand feet lower than they are now. This is proved by the fact of Eocene and Miocene marine deposits of great thickness, which must have been formed in rather deep water, being found elevated from ten to sixteen thousand feet above the sea-level. As an exmple we may mention the Dent du Midi in Switzerland, where marine shells of early Miocene or late Eocene type are found at an elevation of 10,940 feet; and, as this mountain must have suffered enormous denudation, these figures can only represent a portion of the rise of the land, most of which has occurred during the Miocene period. To balance this rise over extensive areas on both sides of the Atlantic, there must have been corresponding areas of subsidence, and we may fairly locate these where the indications of palaeontology and geography concur in rendering them probable. We have already seen that the migrations of mammalia between Europe and America have been such as to render some land route necessary (for particulars of these migrations see the writer's *Geographical Distribution of Animals*, vol. i, pp. 140, 153), while the broken-up character of the coasts of Ireland and Newfoundland, Labrador, Greenland, and Iceland, with the extensive bank of the Azores, all point to a certain amount of recent sinking of land on the outskirts of this area of great depression.

To Mr. Sclater's question—Where did the tropical land exist which afforded the passage of the tropicopolitan forms from one continent to the other?—it may therefore be answered: It existed in the north temperate zone during some part of the Miocene period, at the time probably when a rich temperate flora covered what are now the icy wastes of Greenland and Spitzbergen . . . It has been necessary [here] to treat the question broadly, and to omit many details which require fuller elucidation. I can only now call attention to the obvious fact that the geological age of the remains of any animal type in a given area cannot be held to denote the period of its earliest appearance in that area by migration or otherwise, because, till it became somewhat abundant, there would be little chance of its remains being preserved or discovered. This will apply to the case of the tapirs which are supposed to have migrated to North America in the Miocene period, but whose fossil remains are not found in any deposits earlier than the Pliocene.

We will now return to Mr. Sclater's fourth problem, that of the occurrence of the curious insectivorous mammal, *Solenodon*, in the Antilles, while its nearest allies are to be found in Madagascar. By the help of the conclusions we have already arrived at, much of the marvel and difficulty of this curious case of geographical distribution vanishes. It is simply an extreme instance of a family of animals which has been long dying out, but which maintains a lingering existence in two remote island groups where it is comparatively free from the competition of higher types, and where the general physical conditions are favourable. The fact that the family Centetidae consists of six very distinct genera (five in Madagascar and one in the Antilles) is a sufficient indication that it was once an extensive group. In the Lower Miocene of Auvergne, the fossil remains of a small animal has been found, which is provisionally classed in this very family; and both in Europe and America a considerable number of the remains of Insectivora of peculiar genera have been discovered, indicating that this order of mammals is a very ancient one, which probably long ago arrived at its maximum of development, and has been diminishing in proportion as the larger and more perfectly organised forms have been increasing. It is interesting to note that the two localities where the Centetidae still linger have many remarkable similarities and correspondences. Both are insular groups of the first rank; both are separated from their adjacent continents by very deep sea; both are situated just within the line of the tropic; both are subject to hurricanes; both are very mountainous; in both all the higher mammalia are very deficient; and the differences of their forms of life from those of adjacent lands are such as to indicate that they have both remained insulated for a considerable period geologically. There can be little doubt that these resemblances have something to do with the continued existence in both of isolated members of a once widespread group of mammals, of a comparatively low type of organisation, and unable to bear the competition to which they have been exposed in continental areas. The same principles will, of course, explain the presence in Madagascar of a mouse allied to an American group, of three American genera of colubrine snakes, and of lizards belonging to the peculiar American family Iguanidae, as well as of the beautiful green diurnal moths of the genus *Urania*, and several beetles of decidedly South American affinities. In some other cases we have, as it were, a relic of the former wide extension of now restricted groups. Thus, one genus of snakes, *Ahaetulla*, is found in Africa and South America as well as in Madagascar; while a genus of geckoes, *Phyllodactylus*, inhabits also America and Australia, and there are many similar cases among insects.

After what has been now advanced, the distribution of the lemurs (which forms Mr. Sclater's fifth case) will offer little difficulty. Every indication points to this being a group of great antiquity, and to its having been once very widely spread. Its still existing remnants are scattered from Sierra Leone to Celebes, and from Natal to Eastern Bengal and South China; and they are so varied that they require to be classed in three distinct families and thirteen genera. Still more important is the proof of their extreme antiquity afforded by the recent discovery, in the Eocene deposits of the South of France, of a skull of an unmistakable lemur, allied to one of the still living forms of West Africa known as the "Potto;" while several other fossils of the same age are also believed to belong to the lemurine group. In North America, too, abundant remains have been found in the Lower Eocene deposits, which are believed to be intermediate between lemurs and the South American marmosets. This clear evidence both as to the antiquity and the wide range of the lemurs renders it quite unnecessary to postulate any special changes of sea and land to account for their actual distribution. Inhabiting Europe in Eocene times, they were probably spread over the whole northern continent, and would as easily migrate southward into their present habitats as the hedgehogs, the civets, the chevrotains, or the porcupines, which have all a somewhat similar, but far more extensive distribution. Like the Centetidae, the lemurs find Madagascar best suited to them, more no doubt from the absence of competitive forms than from any peculiar physical conditions. On the great continents they are usually scarce, and are protected by their nocturnal habits and by frequenting dense forests. They thus continue to survive in the midst of creatures of a higher type and more recent origin than themselves, and, together with the opossums of America and some of the smaller marsupials of Australia, seem to have handed down to us a sample of the forms of life which flourished in the earliest tertiary or even in mesozoic times.

The last case of anomalous distribution—that of the giant land-tortoises of the Mascarene and Galapagos islands—offers perhaps less real difficulty than any of the preceding, on account of the existing wide range and the extreme antiquity of the genus *Testudo* to which they all belong. This genus occurs in the Miocene deposits of Europe and India, and in the Eocene of North America; and living species are abundant in Africa and Asia, and are also found in South Europe and North America. It has evidently, therefore, been a dominant group during a large portion of the tertiary epoch, and it still maintains a vigorous existence. There does not seem to be any evidence that these giant species of the two hemispheres are more closely allied to each other than the smaller forms of remote regions; for though the Galapagos tortoises and the extinct species of the Mascarene islands both belong to a flat-headed type, they may have differed in important external characters. Their gigantic size is evidently due to their seclusion for countless generations in islands where they were entirely free from the attacks of enemies, and where they could procure abundance of food; both natural and sexual selection giving advantage to the larger and stronger individuals. The only difficulty is how they reached the Galapagos. But as we may go back to the middle of the tertiary epoch for this event, it is not an improbable supposition that the coast of South America then extended considerably westward, while the islands themselves may have been more extensive, thus reducing the dividing strait to a width across which either the adult animals or their eggs might be floated by currents or surface-drifts. Their entrance to the Mascarene islands from

Africa might have been effected in a similar manner. This is the solution suggested by Dr. Günther himself (*Nature*, vol. xii, p. 297), and it is one which perfectly harmonises all the known facts.

The only other difficulty suggested by Mr. Sclater rather applies to the theory of natural selection itself than to geographical distribution. He asks how we are to account for closely allied forms so frequently inhabiting the same area, while in so many other cases allied species are strictly limited to distinct areas, to the diverging physical and organic conditions of which they are supposed to be adapted. We think that Mr. Sclater has himself furnished a clue to one mode of solution, in his statement that the willow-warbler and the chiff-chaff, though so alike externally, are yet quite distinct in mode of life and habits. It is evident, that a variation in the habits of a portion of the individuals of a species would lead to their mutual association and at the same time to their separation from the parent form, and would thus obviate that tendency to the intercrossing of the different varieties which would undoubtedly occur if the variation were of colour or form only. Is it not probable, therefore, that where two or more closely allied species inhabit the same area, they have arisen *at first* by means of useful variations of instincts or habits; while those which inhabit adjacent but separate areas may have arisen by favourable variations of colour, form, or constitution only? In support of this view it may be noted that the coal and marsh titmice, which are very closely allied, differ considerably in habits; while the great and blue titmice, which are very different in external characters, agree closely in habits, and are often seen together.

Returning to the general question of zoological distribution and its anomalies, it has been shown, I trust, that the only mode of explaining the existing distribution of living things is by a constant reference to those comparatively slight but often important changes of sea and land, which the most recent researches show to be alone probable; and, what is still more important, by recognising the undoubted fact that every group of animals whose distribution is discontinuous is now more or less in a fragmentary condition, and has, in all probability, once had a much more extensive range, to which its present distribution may offer no clue whatever. Who would ever have imagined, for example, that the horse tribe, now confined to Africa and Asia, formerly ranged over the entire American continent, north and south, in great abundance and variety; or that the camel tribe, now confined to Central Asia and the Andean region of South America, formerly abounded in North America, whence in fact our existing camels were almost certainly derived? How easy it is to imagine that analogous causes to those which have so recently exterminated the horses of America and Europe might have acted in a somewhat different direction, and have led to the survival of horses in South America and Africa, and their extermination elsewhere. Had this been the case, how strong would have been the argument for a former union of these two continents; yet we now know that these widely separated species would be merely the relics of a once dominant group which had occupied and become extinct in all the northern continents.

Discoveries of extinct forms remote from the countries they now inhabit, are continually furnishing us with new proofs that the great northern continents of the two hemispheres were really the birthplace of almost if not quite all the chief forms of animal life upon the globe; while change of climate, culminating in the glacial epoch, seems to have been the motive power which has driven many of these forms into the tropical lands where they now alone exist.

If we give full weight to these various considerations, and at the same time bear constantly in mind the extreme imperfection of our knowledge of extinct land animals, we shall, I believe, have no difficulty in explaining most of the apparent anomalies in zoological distribution, and in imagining a possible and even probable solution for those extreme cases of difficulty which the facts at our command do not yet permit us to explain in detail.

Let us now briefly summarise the general principles on which the solution of problems in zoological distribution depends.

During the evolution of existing forms of animal life, we may picture to ourselves the production of successive types, each in turn increasing in variety of species and genera, spreading over more or less extensive regions of the earth's surface, and then, after arriving at a maximum of development, passing through various stages of decay, dwindling to a single genus or a single species, and finally becoming extinct. While the forms of life are thus, each in turn, moving on from birth to maturity and from maturity to decay and death, the earth's surface will be undergoing important physical changes, which will sometimes unite and sometimes separate contiguous continents or islands, leading now to the intermingling, now to the isolation, of the progressing or diminishing groups of animals. Again, we know that climates have often changed over a considerable portion of the earth, so that what was at one time an almost tropical region has become at another time temperate, and then even arctic; and these changes have, it is believed, been many times repeated, leading each time to important changes, migrations, and extinctions of animal and vegetable life.

It is by the combined effect of these three distinct sets of causes, acting and reacting on each other in various complex ways, that have been produced those curious examples of erratic distribution of species and genera which have been so long a puzzle to the naturalist, but which have now, it is believed, been shown to be the natural and inevitable results of the process of animal development, combined with constant changes in the geography and in the climate of the earth.

* * * * *

The circumstances of distribution of island species have always interested biogeographers and evolutionists alike, and Wallace was no exception. Many observers have considered Island life *(ref 721), published in 1880, his finest scientific study. In late 1879, Wallace delivered a public lecture on islands (ref 316) which highlighted some of the biogeographical (and physical geography) material that would appear in the larger work:*

Islands, as illustrating the laws of the geographical distribution of animals

. . . there is no branch of Natural History on which the theory of evolution has thrown so much light as on the geographical distribution of animals and plants. So long as every species was believed to have been independently created the place where they were found was of little importance. The idea prevailed that each animal had been formed in exact adaptation to its native country, and to that only, and that the chief or perhaps the only reason why the animals of different countries were not the same was that their climates, soils, or other features, were correspondingly different. On this theory it would be said that America had no horses, Australia no rabbits, and New Zealand no pigs, because these animals were not *adapted* to them. We now know, however, that they are pre-eminently adapted to them; for when introduced they all flourish and increase, often to such an extent as to become a perfect nuisance, and to exterminate or drive away native productions. We therefore conclude that the reason these animals are not natives of Australia and New Zealand is because they could not reach these countries till man carried them there. This, however, is not a universal explanation, for innumerable fossil horses are found in America (both North and South), and some of these have only recently become extinct. Here, then, we are driven to conclude that some changes took place in America, either in climate or in its animal or vegetable life, which were so inimical to horses as to cause their extinction; that these unfavourable conditions afterwards passed away, but that, the entire race of horses having been extinguished, there was no means of renewing it except by immigration from the Old World. This was prevented by the isolation of America. But when it was effected by man, horses ran wild, and increased both on the prairies of the northern and the pampas of the southern continent. Hence we have to modify our first statement, and to say, that the absence of any particular animal or group of animals from any country depends either on its not having been able to reach the country, or on its having been exterminated, owing to unfavourable conditions, after having reached it or after having developed in it. Thus modified, our statement embodies the fundamental principle of the modern theory of the geographical distribution of animals and plants . . .

In order to bring this subject before you in its simplest and most intelligible form, I propose to confine myself in the present lecture, to the phenomena presented by *Islands*, giving you a somewhat detailed account of the facts in a few typical examples.

Islands differ much in their *origin*, and this difference of origin determines the kinds of animals and plants which inhabit them. They may be divided for our purpose into three classes—1. Recent Continental Islands. 2. Ancient Continental Islands. 3. Oceanic Islands.

By *Continental Islands* we mean those which have been separated from a continent, either recently (in a geological sense) or at some very remote epoch. *Oceanic Islands*, on the contrary, are such as have never formed part of a continent, but have been produced in mid-ocean. *Continental Islands* geologically resemble continents, having a variety of formations usually comprising stratified rocks of different ages. *Oceanic Islands* are always either volcanic or coralline, the only known exceptions being the Seychelles Islands in the Indian Ocean.

Recent Continental Islands are always surrounded by a shallow sea (almost always less than 100 fathoms deep), so that they are connected with the continent by a submarine bank, and an elevation of 600 feet would again join them to it. They are generally situated *near* to continents, but sometimes at a distance of several hundred miles, as in the case of Borneo, about 600 miles from Cambodia, the intervening sea being everywhere less than 100 fathoms deep.

Ancient Continental Islands are always separated from the adjacent continent by a *deep sea*, though they are usually no further off than the more recent islands. Thus, Madagascar is separated from Africa by a strait which is about 2,000 fathoms deep, and there is more than 1,000 fathoms between Hayti and the coast of South America.

Oceanic Islands are usually situated in mid-ocean and surrounded by very deep seas, examples of which are the Azores, St. Helena, and the Sandwich Islands. But there are some exceptions. Thus, the Cape de Verde Islands, the Canaries, and Madeira, are all within 400 miles of the coast of Africa, while some of the Canary Islands are only 50 miles distant from it; but even these are separated from the mainland by a depth of more than 1,000 fathoms.

Such are the physical characteristics of the three classes of islands. Their zoological features are equally well marked. 1. *Recent Continental Islands* always possess the same kinds of animals as the mainland—mammalia, birds, and reptiles especially—all the more important orders and families being represented, generally by the same, but sometimes by distinct though closely-allied species. 2. *Ancient Continental Islands* possess animals which are on the whole different from those of the mainland. They always possess mammalia, birds, and reptiles; but many of the families and orders of the continent are wanting, and some which are present in the island are not represented on the continent. 3. *Oceanic Islands* have *no land mammalia*, and of reptiles and amphibia either none or very few. Birds are more or less numerous, and more or less the same as those of the continent according as the facilities for their passage are many or few.

We will now take in order an example or two of each class of islands, pointing out their peculiarities and the conclusions that may be drawn from them.

There is no more perfect and beautiful example on the globe of a recent continental island than that in which we live—Great Britain. This island, as all know who know anything of geology, contains, as it were, a little epitome of Europe. Almost every formation, from the oldest to the newest, is found in our country. It is also surrounded by an exceedingly shallow sea, which stretches out so as to include the whole of the islands up to the 100-fathom line . . . Outside that line the sea bottom suddenly sinks down, so that a few miles further you get to 500 and then 1,000 fathoms, whereas between that and the shores of Europe the whole sea is less than 100 fathoms deep. That is one striking characteristic of recent continental islands.

In the case of our own country we know a great deal of its past history. We know that at a comparatively recent geological period this country formed part of the Continent. We have proofs of this, not only in the community of animals which could not have got to us in any other way, but we have some physical proofs of a very interesting kind. For instance, around a great part of the coast of England there are found submarine forests—that is, down at low-water mark, and stretching below, are found the stems of trees growing in the mud, with their roots spreading out just as they originally grew. It is clear that trees could not grow in the sea in the place where they stand, and it is therefore perfectly clear that the land must have been higher or the sea have been lower.

. . . There are also in Glamorganshire, on the north side of the Bristol Channel, some caves, now opening into perpendicular sea cliffs, which are washed at their base by the sea; yet these caves contain thousands, and sometimes tens of thousands, of the antlers of deer, showing that they must have been surrounded by forest country inhabited by the deer, and that hyenas and bears, whose bones are also found in them, dragged the deer into the caves and devoured them. It is certain, therefore, that a considerable part of the Bristol Channel must once have been dry land covered with forest. But there is a still more striking proof of the former elevation of our island in Scotland, and no doubt there are plenty of other examples in England, namely, old river channels buried hundreds of feet deep in the earth . . . Some of these old river channels are 260 feet below the present level of the sea. Not only do we know that the land was once higher, but we know that the change took place at a comparatively recent period, because these old river channels are filled up with glacial drift . . . Well, that height of 260 feet is more than sufficient to join our island to the Continent and make it a part of Europe.

Another interesting proof is that dead shells, such as now live in shallow water, have been dredged up off the Shetland Islands at a depth of 90 fathoms. These are arctic shells, now inhabiting cold regions, and they always live either between the tide marks or just below the lowest tides, so that these shells found in 90 fathoms of water were once living on the sea-shore.

These facts render it perfectly certain—even if we had no land animals to tell us so—that our country was once connected with the Continent. Therefore we are not surprised to find that our animals are practically identical with those of the Continent. We have in Great Britain twenty-eight different kinds of wild mammalia, without counting the bats, which can fly. We must add to these four which we know lived in our country during the historic period, and which have been destroyed by man—the wolf, the bear, the wild ox, and the wild boar. These animals were all hunted in olden time, and have become extinct . . . When we go further back, into the time when prehistoric men, who made flint weapons, lived in our country, then we have a still larger number of animals, including the mammoth, the rhinoceros, bears, hyenas, lions, horses, bisons, and reindeer, all of which were natives of this country.

These and many smaller animals were found at the same time in England and on the Continent, so that when Britain and Ireland formed part of the mainland of Europe they contained at least 50 species of mammalia.

The reason we have so few now is, that besides the elevations of land joining England with the Continent there have been subsidences, which reduced our country to an archipelago of very small islands, formed by the mountains of Scotland, Wales, and Ireland. The last of these subsidences took place during the glacial epoch, as proved by the occurrence of gravel with marine arctic shells, at an elevation of more than 1,000 feet, in Wales and Scotland. At this period nearly all the mammalia must have perished, and what we now possess must have entered the country during the latest period of elevation, which coincided with the era of the stone palaeolithic weapons and the cave men who manufactured and used them. But the emigration of mammalia into a new and barren country recently raised above the sea will be a slow process, and only the larger, more abundant, and hardier species will at first pass over or succeed in establishing themselves. This is shown by the fact that our deficiency is chiefly in the smaller species, for while Germany has 32 rodents . . . we have only 13, but we have a total of 32 mammalia against 67 in Germany, which is a considerably larger proportion. Ireland has fewer than we have (20 species), because they had a still longer journey to take to reach that country. This is still better seen in the case of the reptiles and amphibia, including our snakes, lizards, frogs, toads, and newts. Of these the small kingdom of Belgium has 22 species, Britain has 13, and Ireland 4—so few that it has led to the common saying that Ireland has no reptiles, which is not quite true.

We have thus, I think, sufficiently accounted for what animals our country has, and also for what it has not. There remains, however, to be noticed a small but decided amount of peculiarity in the animals of our country. We possess no peculiar mammal, all being exactly similar to those of the Continent; but we do possess one peculiar bird, the red grouse, which inhabits England, Scotland, and Ireland, but is known in no other country. Its nearest ally is the willow grouse of Norway, which turns white in winter, whereas our bird becomes darker in winter. Besides this, two of our little garden favourites, the coal-tit and the long-tailed-tit, are so differently coloured from their Continental allies that some naturalists class them as distinct and peculiar species.

Among our fresh-water fishes we have some very remarkable cases of peculiarity. It has been long known to fishermen that many of the mountain lakes of Scotland, Wales, and Ireland produced peculiar kinds of trout, but few of these had been carefully examined till Dr. Günther, of the British Museum, obtained specimens from all our chief lakes, and compared them with those of the Continent, and he has come to the conclusion that we possess no less than eleven different kinds of trout which are altogether peculiar to our islands. One is peculiar to the lakes of the Orkney Islands, one to Loch Leven, and another to Loch Killin, in Scotland; one to the Llanberris lakes in Wales, three to the lakes of Ireland, while some are found both in Wales and Scotland.

Let us see how these curious facts can be explained. Our red grouse is probably a descendant of the willow grouse of Norway, which entered our country when it was united to the Continent during the latter part of the glacial period. After the separation it remained and spread over all the healthy parts of the country, and as the climate became milder, and there was less snow in winter, it gradually lost the property of changing white at that season, because it was better protected and concealed by remaining the colour of the heather among which it lives . . . The slight changes in the tits would be brought about in a similar way, owing to some slight peculiarity of tint in our woods rendering it advantageous for the colour of the birds to be changed accordingly; and it is to be remarked that the change has occurred in birds which do not migrate to the Continent, and which therefore must become adapted to the conditions of the country they inhabit, or cease to exist.

The case of the peculiar lake fishes is more difficult, because we know less of their mode of life, and especially of how they pass from one part of the country to another. Yet that their eggs are somehow conveyed for long distances is certain, because we have the very same species inhabiting remote mountain lakes in both Wales and Scotland. Mr. Darwin remarks that small living fish have been carried through the air and dropped by whirlwinds, while the ova may be carried in the same manner, or by becoming attached to the feet of aquatic birds. It is also possible that some of the fishes may have survived the glacial period, and during the time that the whole country was buried in ice some of them may have been carried in the pools which often exist on the surface of glaciers to great distances.

It may be said that these causes are very improbable, and could only have acted very rarely. But this is just what we want; for if the eggs of fishes were carried *frequently* from one lake to another then species peculiar to a lake could never have arisen. It is because a fish, once brought to a lake by a rare accident, has remained undisturbed there for perhaps thousands of years that it has become modified. All that time it has been subjected to peculiar conditions. The temperature, the colour, and the composition of the water are peculiar, as are the food it can obtain and the dangers it has to avoid. In time it becomes adapted to all these special conditions, and in doing so becomes slightly modified in colour, form, and habits, so as to form a new species. we see then that the animals which inhabit our country are, both in their peculiarities and deficiencies, sufficiently accounted for by what we know of the country's past history.

We will now take one more example of a *Recent Continental Island*, but one which is considerably less recent than Britain, namely, the island of Formosa, which, instead of being in the temperate zone, is situated close upon the tropics, about a hundred miles from the coast of China . . . The island . . . is about half the size of Ireland. In the centre it has a magnificent chain of mountains, rising to a height of 12,000 feet, so that, although near the tropics, their summits are covered with snow during a great part of the year. The lower parts of these mountains and valleys are covered with magnificent forests, so that in every respect this island is well adapted to support a great number and variety of animals. That this is a continental island is proved by the characteristic feature that the sea around it is less than 100 fathoms deep; and though the island is comparatively small it contains the large number of 30 species of true land mammals, not counting bats. This is a very large number, for in the adjacent coast districts of all tropical China there are only 34 or 35. This is very different from what is found in our country, where we have a much smaller number than the Continent. But the richness of Formosa in animal life is no doubt due to the variety of its beautiful mountains and forests, and to the probable fact that it has not been depopulated, as our country has, by sinking beneath the sea and rising up again. Among these 30 species of mammals, 13 are peculiar, namely, a monkey, a mole, a civet, a pig, two kinds of deer, a goat-antelope, several mice, and three flying squirrels. Birds are also very numerous, 128 species of land birds being known, of which 42 are peculiar to the island. This great amount of peculiarity indicates a long period of isolation, although the island, like our own, stands on a submarine bank less than 100 fathoms below the surface.

But the comparative antiquity of the island is still more clearly shown by the peculiar affinities of several of its mammalia and birds, which in several cases are either identical with or allied to species not found in China, but inhabiting distant countries. Thus the clouded tiger and Asiatic wild cat inhabit Formosa and the Malay Archipelago, but are unknown in China. One of the deer and the goat-antelope are also allied to Malay species, as are two of the flying squirrels; while the spotted deer, the wild pig, and the fruit-bat of Formosa are allied to peculiar Japanese species. This all proves that there must have been a great change in the distribution of animals since the time when Formosa was separated from the mainland, and as such changes are slow it proves a great lapse of time.

Birds exhibit the same phenomenon even more strikingly; for more than half the peculiar birds of Formosa have their allies, not in China, but in such remote regions as the Himalayas, the Nilgherries in Southern India, the Malay Islands, or Japan; and many of them are totally unlike anything found in China. The lapse of time which has allowed such changes of distribution to take place has also been sufficient for so many of the species to become changed by long isolation under peculiar conditions. But there are some few species which have resisted this process of change, although their distribution has been greatly altered. Four birds—a red-breasted fly-catcher, a handsome wood-pigeon, a spotted hawk-eagle, and a brown wood-owl, are found only in Formosa and the Himalaya Mountains, while four other Formosan birds inhabit different parts of India and the Malay Islands, but not China.

Here then we have an island which in every respect answers to our definition of a recent continental island, but has yet been separated from the continent sufficiently long to allow of considerable change in the species of animals, and also of considerable change in their distribution. Now as our own island dates back to the close of the glacial period, and exhibits hardly any such changes, we may be sure that Formosa dates very much further back, and must have been an island during some considerable portion of the Pliocene period.

We will now go on to examine an *Ancient Continental Island*, and there is no better example than Madagascar. This great island is separated from the

African continent by a channel about 250 miles wide, and more than 10,000 feet deep. It is larger than the whole of the British Isles. It has a great central plateau and lofty mountains, a varied geological structure, a luxuriant vegetation, and a tropical climate. It possesses about 60 different kinds of land mammalia, besides bats, and has therefore certainly been connected with some continent.

But these mammalia are very peculiar, and do not at all resemble those which now chiefly inhabit Africa, which, as you know, is remarkable for its numerous monkeys and baboons, its lions, leopards, and hyenas, its zebras, rhinoceroses, elephants, giraffes, buffaloes, and antelopes. Not one of these animals, and nothing like them, is found in Madagascar, and thus the first impression would be that it cannot possibly have derived its animals from Africa; but a closer examination may lead us to alter our opinion.

Considerably more than half the . . . mammals of Madagascar are lemurs. These are singular animals, of low organisation, usually considered to be the lowest of the Quadrumana, or monkey tribe, but by some naturalists classed as a distinct order. Lemurs are found in Africa, and also in tropical Asia and the Malay Islands, but those of Madagascar are very peculiar, all forming distinct genera, and one a distinct family. Here then we have no indications of a specially African origin. The next important group are the Insectivora, to which our mole and hedgehog belong. In Madagascar there are about a dozen species, but strange to say all but one belong to a family only found elsewhere in the West India Islands! and that one is a shrew mouse, such as are found all over the world. We then come to the Carnivora, which consist chiefly of eight species of civets, all of peculiar genera, but many of them most nearly allied to African genera. The only other animal of importance is a river hog belonging to an African genus; but as these animals are good swimmers it may have reached the island without any actual land union; and the same may be said of the hippopotamus, whose remains have been found in a semi-fossil condition.

The birds are equally remarkable. There are more than a hundred peculiar species of land birds, and only twelve found elsewhere—some in Africa and some in Asia. The plantain-eaters, the barbets, the honey-guides, the glossy-starlings, the ground-hornbills, and other peculiarly African families, are quite unknown; and there are instead of these thirty-five peculiar genera, most of which have no near allies in any part of the world. Of the peculiar species, however, sixteen belong to African genera, while only about five belong to Asiatic genera. Among the African allies are a guinea-fowl, a parrot, and a cuckoo, all birds which cannot cross over wide arms of the sea.

Turning for some further light to the reptiles, we find that many of the African groups of snakes are absent, and that there are instead three peculiarly American genera. The lizards are more African, but there is one group which is only found elsewhere in America and Australia . . .

The extraordinary affinity of some of the Madagascar Insectivora and snakes for those of the West Indies and South America has greatly puzzled naturalists, and it has even been suggested that there was once a continent running quite across from Madagascar to the West Indies; that this continent completely disappeared, leaving a fragment only in Madagascar; and that subsequently Africa was formed. But when we consider the depths of the intervening oceans, and especially that, after all, these affinities are exceptional, we shall find that no such enormous changes are necessary, especially as we have many facts which point the way to the true solution of the problem. All the large and characteristic African animals, whose absence from Madagascar is the first great fact in its natural history, inhabited Europe and Western Asia as far as Northern India during the Miocene period. About the same time we know that a sea cut off tropical Africa from the northern continent, and the natural inference is that it was only when that sea-bed was raised and Africa joined to Europe and Asia, that all these animal forms entered Africa, and subsequently, as the climate became unsuitable, became extinct in Europe.

But in the Eocene and Early Miocene periods we know that the distribution of animals was very different from what it is now. Lemurs were then found in Europe and America, and opossums, now confined to America, also inhabited Europe. We believe then that the types now common to Madagascar and remote parts of the world were then widespread over Europe and Asia, and reached Madagascar during its early connection with the continent. The birds which are of Eastern affinities probably reached it by means of islands in the Indian Ocean, of which the Laccadive and Maldive coral reefs, the Chagos Islands, and the extensive submarine bank of the Seychelles, are the relics.

Madagascar therefore is an island which dates back certainly to the Miocene, probably to the Eocene, period, and has preserved to us in greater abundance than elsewhere some of the animal forms of that ancient world.

We will now pass to the third class of islands of which I spoke—the *Oceanic Islands*; and to illustrate these I will take two examples, the Galapagos Islands, which are 600 miles from the west coast of South America, and the Azores, in the middle of the North Atlantic Ocean. The Galapagos are a group of considerable extent, separated from South America by a sea 10,000 feet deep. Like almost all oceanic islands they are entirely volcanic, and possess no indigenous mammalia. There are rats and mice, but these were conveyed thither by ships. They possess reptiles—and it is a curious question how they got there—living tortoises of two different kinds, and one species extinct. The living tortoises are of enormous size, equal to that in the Zoological Gardens, which is six feet long, and able to carry two men standing on its back . . . There are also in these islands five species of lizards, all of American families or genera. One of them is a marine lizard, which is found nowhere else, and lives in the sea like the old lizards of the Lias period. Though it is marine it belongs to a family abundant in America, and which inhabits trees. These lizards undoubtedly originally came from America, and have become modified by living in this volcanic island with little vegetation. How their ancestors got to the islands is a problem not yet solved. Perhaps their eggs were floated there by the sea. There is no island in the Pacific where you do not find lizards, though there may be no mammalia, frogs, toads, or snakes. In the Galapagos there are two species of snakes, almost identical with those found on the mainland, and these may have been conveyed in ships, because snakes often conceal themselves amongst the cargo.

The birds are exceedingly interesting, there being 57 species, and more than half of them belong to peculiar genera; but they are all allied to American birds, so that it is certain their ancestors came from America. There are 31 species of land birds, and of these 30 are peculiar to the Galapagos. It is found that those that have changed most are forest birds that do not migrate, and those that have changed least are birds that wander about more and do migrate. The one bird that has not changed at all is perhaps the most interesting bird on the continent of America. It is known as the rice-bird, or "bob-o-link," its scientific name being *Dolichonyx oryzivorus*. It breeds in Canada, and migrates to the Bermudas, and is found all over the United States, and right away to Paraguay in South America, so that it is the greatest wanderer of any land bird in America and therefore it is not so remarkable that it should have found its way to these islands. The chief reason why the birds of the Galapagos have become so much altered is their being isolated. If fresh emigrants keep coming to an island there is little chance of the race becoming modified, because the breed is kept pure. The migrating rice-bird keeps the race pure; but the other birds, which come more rarely, have been subjected to, and modified by, the altered conditions of the islands. We shall see the reason why very few birds migrate to these islands when we have examined the next group—the Azores.

The Azores or Western Islands form the best contrast to the Galapagos, for though equally oceanic they are temperate instead of tropical, and are situated in the stormy North Atlantic instead of the calm equatorial Pacific Ocean; and this difference of position has resulted in a great difference in their animal inhabitants. The Azores have no mammalia and no reptiles. There are 22 different kinds of land birds; but instead of being all but one peculiar, as in the Galapagos, there is only one peculiar, all the rest being identical with European birds. This is a most remarkable fact, and is at first very puzzling. It has, however, been clearly ascertained by residents on the islands that the reason of this identity of the species with those of Europe is simply this—the migration of these birds is still going on. The Azores are situated in a very stormy sea, and are subject to violent storms, which blow from all points of the compass, especially in spring and autumn. The regular winds are westerly, and if these winds brought the birds, then they ought to be American, but there is not a single American bird. The interesting fact is this, that after every great storm in spring and autumn the people invariably find some new birds in the islands which they never saw before. These birds are generally watched for and shot. Sometimes there are only one or two fresh birds, and they very rarely maintain themselves, not being in sufficient numbers or the islands are not adapted to them. But some of the 18 permanently resident kinds also came in these storms, and the original migration which took place thousands of years ago is still kept up by birds coming every few years from the same country, and thus the breed is kept pure. The result is that the birds of the Azores may be said to be every one identical with those of Europe, except the bullfinch, which is a bird that does not take long flights, and it is presumed got blown there once a long time ago and has seldom come since. The result is that the peculiar conditions of the climate, food, etc., have acted upon it so as to modify it into a distinct species. An interesting proof that this is the mode in which these islands have been peopled is shown by the number of species in the different sub-groups. The Azores consist of about 20 islands, and they are separated into three groups—Eastern, Central, and Western. The central group is the largest. If, as some people maintain, these islands once formed part of the European continent, we should expect to find the largest number of animals and birds in the central group. On the contrary, the eastern group, which is the smallest, has the most birds, showing that the eastern group, which is nearest to the Continent, first receives the majority of the emigrants, and that many go no farther; for the central group has less, while the western group has the fewest. This fact agrees with the theory that these birds are simply stragglers from Europe.

The insects show the same phenomena, being mainly identical with those in Europe. The plants, too, are decidedly European, with about ten per cent of peculiar species. Thus we see that the same cause has operated in every case. We learn this important fact, too, that it is not the currents of the ocean that help chiefly to stock islands with plants, insects, and birds, and it is not the ordinary winds that help them there; but it is the violent and occasional storms. By these storms birds get caught up against their will, and are carried out to sea, and then are obliged to fly with the storm. No doubt for every one that alights on these islands hundreds perish; but the few that arrive serve as the parents of a fresh stock. Now we see the reason why the Galapagos have such an immense number of peculiar species . . . There is hardly such a thing as a storm known there, and comparatively light winds. The consequence is, that there is no regular migration of birds to these islands, and the few that do reach them come at such remote intervals that they do little towards keeping up the purity of the original stock . . .

I have now been able to give you an outline of only a small portion of a very large subject; for every island is a special study of itself, no two being alike in their past history or present relations. . . . the islands of the globe offer materials of intense interest to the naturalist who studies not only the present but the past history of the animal world; and I trust that I have succeeded in making clear to you the nature of the zoological problems they present, and have also enabled you to understand the principles by which the solution of such problems may be successfully attempted.

* * * * *

Wallace's contributions to systematic regional biogeography—including the earlier-mentioned 'Wallace's Line'—are still relatively well known. His first commentary on this subject took the form of an 1859 letter to ornithologist P. L. Sclater that was published in the first volume of the Sclater's journal Ibis*:*

Letter from Mr. Wallace concerning the geographical distribution of birds (ref 52)

My Dear Mr. Sclater,—Your paper on 'The Geographical Distribution of Birds' (*Journ. Proc. Linn. Soc.* ii, p. 130) has particularly interested me, and I hope that a few remarks and criticisms thereon may not be unacceptable to you. With your division of the earth into six grand zoological provinces I perfectly agree, and believe they will be confirmed by every other department of zoology as well as by botany. On the number of species you give to each province I can say nothing, you being in so much better a position than myself for arriving at a correct conclusion; but in the *areas* of the several provinces I believe you have made some very important errors, which of course affect materially the proportionate richness in species of the several provinces. These I will take the liberty of pointing out, as well as of defining, as closely as possible, the limits of each division.

1. The Palaearctic Region I would extend over all Africa north of the great Desert, for I think none of the peculiar forms of Tropical Africa are found there. For its southern limits further eastwards I take the parallel of 30 degrees N. lat. as an average from Suez to the east boundary of Afghanistan, then turning north, to exclude the Punjaub, follow the Himalayas to Birmah, then bend a little south, and then north again to where the 30th parallel crosses the Yung-ling Mountains on the west of China, and follow the same parallel across China to Chusan. This is, I think, as fair a give-and-take as can be chosen with our imperfect knowledge. Now for the area:—

Continental Europe and Asia	14,850,000 (Humboldt.)
Deduct Continental part of "Regio Indica" (Arabia–North Africa)	2,630,000
	12,220,000
Add for islands	280,000
Total area of Palaearctic Region	12,500,000

—instead of 14,000,000 given by you.

2. Æthiopian Region. In this you make a very large area, which must certainly be a mistake. You give 12,000,000 square miles. Now Balbi gives only 8,500,000 for all Africa and Madagascar; and, if we estimate the part of Arabia taken from Asia as equal to the part of Africa given to Europe, this will be the true area. But I would suggest that such an anomalous tract as the Sahara coming between two regions should be given to neither; it should in fact be considered as a *Sea*. It is certainly quite as unproductive of animal life as the sea, perhaps more so; and it gives quite an erroneous idea of the productiveness of Tropical Africa to add this immense desert to it. I take, therefore, Africa south of the Sahara, and after reaching the Nile as far north as the 1st Region, following a diagonal across Arabia from Mount Sinai to the

eastern extremity. A careful measurement gives me the area of this with Madagascar as 6,500,000 square miles.

3. Indian Region. Of this we have already defined the north limit, and I would add a tract of Arabia on the western shores of the Persian Gulf. Its south-eastern limits I draw between the islands of Bali and Lombok, and between Celebes and Borneo, and the Moluccas and the Philippines. Barbets reach Bali, but not Lombok; *Cacatua* and *Tropidorhynchus* reach Lombok, but not Bali: this I think settles that point. *Cacatua*, *Trichoglossus*, and *Scythrops* in Celebes, and not in Borneo, settle the other. No doubt many Indian forms reach Celebes; but we must remember the proximity, and in the course of ages the only wonder is there has not been more intercommunication. A careful estimate of these islands, with Formosa, etc. added to the continental portions, gives an area of only 3,100,000 square miles.

4. Australian Region. You have rather over-estimated this: with the Moluccas, etc., North Guinea, New Zealand, and the Pacific Islands (except the Sandwich, which I think should go with America), I cannot make an area of more than 2,600,000 square miles. In the island of Batchian I have found *Podargus*, *Coriphilus*, and *Paradisea*, which shows that the several subdivisions of this region are very closely connected.

5 and 6. North and South American Regions. I put the limit between these at 22 degrees N.; on the coast it may be further north, on the table-land further south, but this will be near the mean. This gives for the North American region 5,500,000 square miles, South American region with West Indies 5,600,000 square miles, making a total which agrees with Balbi.

Now, taking my corrected areas and your number of species, the proportionate richness or square miles to each species is as follows:—

Zoological Regions	Proportionate richness	
	Wallace	Sclater
1. Palaearctic, area 12,500,000, sp. 650	1/19,200	1/21,000
2. Æthiopian, area 6,500,000, sp. 1250	1/ 5,200	1/ 9,600
3. Indian, area 3,100,000, sp. 1500	1/ 2,050	1/ 2,600
4. Australian, area 2,600,000, sp. 1000	1/ 2,600	1/ 3,000
5. North American, area 5,500,000, sp. 660	1/ 8,300	1/ 9,000
6. South American, area 5,600,000, sp. 2250	1/ 2,500	1/ 2,400

From the above estimate it results that the Indian region is in proportion to its area the richest in species of birds, and, when we consider how much less known it is than South America, we may expect that the proportion will increase. In Birmah, Cochin China, and China, the ornithology of extensive regions is absolutely unknown; while in South America there is scarcely a spot which has not been repeatedly and thoroughly explored. Though no one is more profoundly impressed than myself with the vast and inexhaustible riches of South America in every department of natural history, and pre-eminently in birds, yet I am also convinced that a considerable portion of its great apparent superiority to all other countries is due to the universal spread of the Spanish and Portuguese races over every part of its vast interior, which is thus absolutely as free as Europe to the researches of naturalists, who have not been slow to take advantage of it. No other part of the Tropical world is in this condition. In three-fourths of the African and Indian regions, the naturalist only penetrates at the risk of his life; and, even where this is safe, I can speak from personal experience of the great difference between these regions and those of South America as to the inducements to a traveller to prolong his stay. In these countries we have absolutely no community of ideas, feelings, or wants with the population; whereas, even in the most remote districts of South America, one is always in contact with men of European race and feelings, and with at least a remanant of the usages and wants of European civilization. South America too possesses a physical superiority to every other region, which would lead us to expect a vast richness in its natural productions. Nowhere else in the Tropics are such vast and fertile alluvial plains, such mighty forests, such gigantic rivers, such an extensive and lofty range of mountains. There is no such compact mass of intertropical land as South America. There is no mountain range but the Andes, both sides of which are in the Tropics. This long succession of temperate plateaux, together with the southern extremity of South America, adds immensely to the diversity of its fauna, combining in fact all the varied physical features and stations of the Old World in a space of barely one-fifth its extent. It thus happens that in many cases the natural productions of South America will bear comparison not only with any *one* of the other regions, but with all the rest of the world, especially if we leave out Australia as an altogether peculiar region, having no more connexion with the old than with the new continent.

There is perhaps no fact connected with geographical distribution more extraordinary, and at first sight inexplicable, than the division of such an apparently homogeneous tract as the Indian Archipelago between two provinces which have less in common than any other two upon the earth. To the geographer and geologist, there is absolutely nothing to mark the division between the two regions.

Borneo differs more from Java than does the former from Celebes, or the latter from Timor. The Phillippines strikingly resemble the Moluccas in their physical features; Borneo has much resemblance to New Guinea, yet their zoological productions differ greatly. Between the Indian and Australian zoological regions, as above defined, I believe there is absolutely no true transition—that is, no *species* are common to the two, which we cannot easily account for by the various accidents and migrations which in the course of ages must have tended to mingle the productions of islands so close to each other. I believe that these two regions are as absolutely distinct as South America and Africa, and it is only because they are separated by straits of from 20 to 100 miles wide, instead of the Atlantic, that they have become slightly connected by the interchange of a few species and genera . . .

* * * * *

Wallace's involvement with systematic regional biogeography culminated in his massive two-volume study The geographical distribution of animals *(ref 718) in*

1876. Through this, the six-region classification scheme became a fixture. Nonetheless, it was often challenged, and on several occasions Wallace took up the debate. In 1883 Nature *provided the stage for one such discussion (ref 360):*

On the value of the "Neoarctic" as one of the primary zoological regions

In the *Proceedings of the Academy of Natural Sciences of Philadelphia* (December, 1882) Prof. Angelo Heilprin has an article under the above title, in which he seeks to show that the Neoarctic and Palaearctic should form one region, for which he proposes the somewhat awkward name "Triarctic Region," or the region of the three northern continents. The reasons for this proposal are, that in the chief vertebrate classes the proportion of peculiar forms is less in both the Nearctic and Palaearctic than in any of the other regions; while, if these two regions are combined, they will, together, have an amount of peculiarity greater than some of the tropical regions.

This may be quite true without leading to the conclusion argued for. The best division of the earth into zoological regions is a question not to be settled by looking at it from one point of view alone; and Prof. Heilprin entirely omits two considerations—peculiarity due to the absence of widespread groups, and geographical individuality. The absence of the families of hedgehogs, swine, and dormice, and of the genera *Meles*, *Equus*, *Bos*, *Gazella*, *Mus*, *Cricetus*, *Meriones*, *Dipus*, and *Hystrix*, among mammals; and of the important families of flycatchers and starlings, the extreme rarity of larks, the scarcity of warblers, and the absence of such widespread genera as *Acrocephalus*, *Hypolais*, *Ruticilla*, *Saxicola*, *Accentor*, *Garulus*, *Fringilla*, *Emberiza*, *Motacilla*, *Yunx*, *Cuculus*, *Caprimulgus*, *Perdix*, *Coturnix*, and all true pheasants, among birds, many of which are groups which may almost be said to characterise the Old World as compared with the New, must surely be allowed to have great weight in determining this question.

The geographical individuality of the two regions is of no less importance, and if we once quit these well-marked and most natural primary divisions we shall, I believe, open up questions as regards the remaining regions which it will not be easy to set at rest. There runs through Prof. Heilprin's paper a tacit assumption that there should be an equivalence, if not an absolute equality, in the zoological characteristics and peculiarities of all the regions. But even after these two are united, there will remain discrepancies of almost equal amount among the rest, since in some groups the Neotropical, in others the Australian, far exceed all other regions in their speciality. The temperate and cold parts of the globe are necessarily less marked by highly peculiar groups than the tropical areas, because they have been recently subjected to great extremes of climate, and have thus not been able to preserve so many ancient and specialised forms as the more uniformly warm areas. But, taking this fact into account, it seems to me that the individuality of the Nearctic and Palaearctic regions is very well marked, and much greater than could have been anticipated; and I do not think that naturalists in general will be induced to give them up by any such arguments as are here brought forward.

* * * * *

In 1894, in an address to the Cambridge Natural Science Club, Wallace produced his definitive analysis of Sclaterian regionalism:

What are zoological regions? (ref 494)

The subject which I now propose to discuss, is the purport and use, and therefore the essential nature, of what are termed zoological regions. This seems necessary because, although such regions have been more or less generally adopted for more than thirty years, there has of late grown up a conception as to their nature and purport which seems to me to be altogether erroneous, and which, if generally adopted, is calculated to lead to confusion, and to minimise, if not to destroy, whatever advantages may be derived from their use.

The time has therefore come when this question must be discussed and, if possible, settled, and, in the hope of leading to such a settlement, I propose to point out what seem to me to be the essential characteristics of such regions, considering them to be established for the purpose of facilitating the study of geographical distribution as one phase of the problem of evolution. In order to come at once to the question at issue, I will first summarise the statements or assumptions which seem to me to imply a misconception of the nature and uses of zoological regions. These fall under two heads:—

(1) It is asserted that the same regions will not answer to show the distribution of all groups of land animals. Some of the classes, or orders, or sometimes even the families, require us to establish different sets of regions—regions which may differ both as to their number and their limits—in order to represent and study the distribution of such groups.

(2) As a guide to what constitutes a region, it is laid down that areas which have few peculiarities in the higher groups—such as families, even though of continental extent, rich and varied in genera and species, and having a large number of peculiar types, are not of regional status. The criterion of a region is said to be the exclusive possession of peculiar groups of higher rank than genera; and this without any regard to proportionate area, or to the poverty and monotony of the fauna as a whole.

Now the first of these assumptions . . . raises the whole question of the nature and practical utility of zoological regions, and is a proposition which the chief purpose of this article is to disprove: it must therefore be considered in some detail.

In the first place, it implies that the students of any particular group—reptiles, beetles, butterflies, land shells, etc.—should each mark out the globe into regions exhibiting the chief features of the distribution of its families, genera, and species, and that any other division, arrived at by the study of other groups, will be of little or no use to them. But if this is true, it must be carried further; for not only do the various classes and orders of animals differ considerably in their distribution, but many of the tribes and families. To take the case of the mammalia, which, for distributional study, has always been treated as a whole, how different is the distribution of the Edentata from that of the Ungulata. In the former group South America is so rich that it is of more importance than all the rest of the globe, while in the latter it is so poor that even when joined with North America it would hardly equal either of the other continental regions in importance. But if we constructed a set of regions to correspond with the distribution of each of these orders, we should not bring out the facts more clearly than can be done by means of the regions most usually adopted for the whole class, while we should lose the advantage of easy comparison with each other, and with the remaining orders of the class, as well as with other classes of animals. But *comparative* distribution is the one essential feature of our study, without facilities for which the bare facts are uninstructive and of hardly any scientific value.

This point has been very clearly brought out in the case of birds, in a work specially devoted to the geographical distribution of one family—the plovers. These birds are, as a whole, cosmopolitan, so much so that Mr. [Henry] Seebohm tells us "they have not even a remote connection" with the usually adopted zoological regions; and he adds: "These birds only recognise three regions—Arctic, Temperate, and Tropical." Again, after describing the distribution of the chief genera during the breeding season, he says: "The inevitable conclusion is that the Charadriadae *do* pay considerable attention to the climatic or isothermal regions, but appear practically to ignore the Sclaterian regions."

These very positive statements would lead a reader to conclude that here, at all events, the regions established by Dr. Sclater for birds as a whole are of no use. Yet we find that in the great work above referred to—the *Geographical Distribution of the Charadriadae*—Mr. Seebohm rarely uses these three climatal regions, but throughout the book gives the distribution of the species of each genus in terms of the six Sclaterian regions. And if we consider the habits of these birds, so many of which get their food on sea-shores and tidal estuaries, while all of them have great powers of flight, and many of them migrate along the coasts of all the continents, it is really surprising to find so many of the genera and species which are nevertheless strictly limited to certain of the Sclaterian regions. Owing, no doubt, to the peculiarities of habit just referred to, about half the genera are cosmopolitan, being found in all the six regions during some part of the year; but, even of these, certain groups of species are often confined to one or two regions.

When we turn to the non-cosmopolitan genera, however, we find some very instructive facts, which well serve to illustrate my main contention as to the sufficiency of one set of regions . . . [table omitted] . . . Now we have here to notice two points:

(1) That in most of these genera not only are they absent from one or more of the Sclaterian regions and present in others, but in the regions where they do occur they are usually widely dispersed, thus showing that their range is defined and limited by the very same barriers which so well mark out the general range of land birds.

(2) We also find that the use of the old-established and widely accepted six

regions of Dr. Sclater, enables us very clearly and concisely to describe or to tabulate the comparative distribution of the genera and species of this great family of wading birds, which have been thought to be such erratic wanderers . . .

Now this case of the plovers is perhaps as strong as any that can be brought to prove that different groups require different sets of regions; and it at once brings us to the question at issue, which is, whether anything would be gained by establishing a set of Charadriine regions. The climatic regions—which Mr. Seebohm suggests as more natural in this case—would not bring out such facts as the absence of *Ædicnemus* from the Nearctic and of *Limosa* from the Ethiopian regions; the limitation of *Glareola* to the eastern hemisphere, and of *Phegornis* to the Australian and Neotropical regions, unless the Sclaterian regions were also used as sub-regions—thus introducing complication in place of simplicity, and gaining, so far as I can see, no advantage whatever.

But further, if the plovers are to have their own regions and sub-regions, there are probably 50 or 100 of the orders and families of the animal kingdom which would equally require to be so treated; and as in all these cases the new regions must have separate names, it is clear that by far the larger part of them would remain for ever unknown, except to their inventors.

A little consideration will, I think, convince us that this plan, of practically unlimited distinct sets of regions, would be a positive hindrance to any intelligent study of the distribution of animals, a study which derives its chief interest and importance from its relation to the theory of organic evolution, and which must therefore include the *comparative distribution* of the various classes, orders, and minor groups. But how will it be possible to make the necessary comparison if the distribution of the groups to be compared is given in terms of as many distinct sets of regions all differing in their names and in their boundaries? It would be like comparing the structures of different animals as described in the works of a number of anatomists each of whom had a different classification and a different set of technical terms, so that before a single comparison could be made the terms used in one description would have to be translated into the terms used in the other. In the study of geographical distribution, should this system prevail, the student would find it necessary to adopt some one set of regions for his own use, and then endeavour to translate the facts given by each specialist into terms of that set before he could obtain any clear conception or accurate knowledge of their comparative distribution.

An idea seems to be prevalent among biologists that there is some *law* of distribution, that may differ for different groups, and that may require different regions to exhibit it or to conform to it. This, however, is a mere supposition; but, if it is a correct one, we shall certainly not be likely to discover the "law" by recording the facts of distribution in such a way as to render a *comparative* study of them as difficult as possible. Laws of distribution can only be arrived at by comparative study of the different groups of animals, and for this study we require a common system of regions and a common nomenclature.

It appears to me, however, that the "law," or at all events the general principles on which the diversities of distribution among land animals depend, is already fairly well understood. What we require is to be able to work out the details in the different groups, and thus explain certain difficulties or anomalies. To detect anomalies it is essential to compare the distribution of the different groups by means of a common system of regions. If we construct regions to fit each group, the study of each separate group will be apt to forget that it presents any anomalies which require explanation.

Before leaving this part of the subject it may be well to give a short account of the reasons which led to the original establishment of the six Sclaterian regions for the purpose of facilitating the study of the geographical distribution of animals; in order to show that they are not arbitrary divisions, but are founded on a large body of observations. It is evident, in the first place, that many of the ordinary divisions of the geographer serve well to define the areas characterised by special groups of animals or plants. The South European, the Malayan, the Brazilian, or the South African faunas and floras, are constantly referred to, because those districts are really characterised by distinct assemblages of animals and plants, and this undoubtedly depends partly on their possessing peculiarities of climate resulting in peculiarities of vegetation—as forest, prairie, desert, or woodland; partly in their being limited by more or less effective barriers, climatic or geographical; and partly on their past geological history and on the more recent changes of physical geography they have undergone. But such areas as these are too small and too numerous to enable us to express the broader features of the distribution of animals, and the larger or primary geographical divisions—which were those used by the older naturalists—are often unsuitable and mis-leading, because they are *not* coincident in their boundaries with those more permanent natural barriers which have mainly determined the zoological specialities of different parts of the globe. Yet some of the divisions of the geographer are such well-defined and ancient areas that they *do* nearly coincide with characteristic assemblages of animals; and thus the geographical units, Europe, Asia, Africa, North America, South America, and Australia, can be easily modified into six zoological regions, which do represent with considerable accuracy the broad features of animal distribution. These modifications may be briefly enumerated in order to show how the limits of the regions have been arrived at.

Beginning with Europe, we see at once that it is zoologically homogeneous, since a large proportion of the species and all the larger genera range over the whole of it. But the same genera, in the case of the higher animals, at all events, prevail in North Africa, mingled only with a few desert types, and we therefore, for zoological purposes, add this area to Europe. It is interesting to note that we have a clear explanation of this identity, in the proofs that quite recently—that is, during the Pleistocene period—Europe and North Africa were connected both at Gibraltar, and from Sicily and Malta to Tripoli, as indicated both by submarine banks which still unite them, and by the fossil hippopotami and elephants of the Maltese, Sicilian, and Gibraltan caverns. But further, if we go eastward from Europe into Siberia and Central Asia, we find the same genera and many of the same species of mammals and birds ranging all the way to the shores of the Pacific. To such an extent is this the case that about fifty-six species of British passerine birds range to Central and North-East Asia, while no less than fifty-three species (or representative sub-species) of land-birds are common to Great Britain and Japan. Europe and North Asia are therefore parts of one zoological region, the reason being that there is not, nor has been in recent geological times, any effective barrier between them, while in climate they are sufficiently alike.

Here, then, we have roughly marked out our first great zoological region—the Palaearctic, or northern old-world region. Southward its limits are undefined, and where there are no well-marked barriers, such as the Himalayas or the desert, there will always be a greater or less width of border-land between two conterminous regions.

Now this Palaearctic region is, fortunately, the only one that differs very largely from the ordinary geographical quarters or divisions of the globe. For when we go to Africa we find that, leaving out the northern portion, which we have seen to be essentially European, the remainder constitutes a very distinct and compact area zoologically—the land of giraffes, zebras, hippopotami, baboons, and antelopes—which has been termed the Ethiopian region. Then we have southern or tropical Asia, together with the larger Malay Islands, which we know must recently have formed a part of it, constituting the Indian or Oriental region, and corresponding almost exactly with tropical Asia. Then we come to Australia, which forms the nucleus of another well-marked region, including with it, however, most of the Pacific Islands, with New Guinea and the Moluccas. Turning now to the western hemisphere, we have South America and North America, which, with slight modifications, form two well-defined regions—the Neotropical, including all South America with the tropical portion of North America and the West Indian Islands; the Nearctic, comprising the remainder of North America . . .

Now the only real interest of the study of geographical distribution lies in its giving us a clue to the causes which have brought the very divergent and often conflicting distribution of the various species, genera and higher groups, and by thus being able to explain most of the anomalies of distribution. These causes we can trace, in many cases, either to geographical or climatal changes in the past, which temporarily removed the barriers that now exist or interposed others that are now absent; or, on the other hand, to the recent extinction of groups in certain regions where they formerly abounded; or, again, to the very different powers of dispersal possessed by different organisms, which enable some groups to spread easily where others are stopped by an insurmountable barrier. Now it is usually this last phenomenon, of varying powers of dispersal, that has led the students of certain groups to urge that the old-established regions do not serve their purpose. But when a group can more or less easily traverse the barrier between two regions, however permanent that barrier may be, the fact enables us to explain the exceptional distribution of that group, but it does not render the established regions less natural, or require a fresh set of regions, which would certainly not be natural in any broad sense, to explain them.

Again, it will usually, perhaps always, be found that even in the groups appealed to as requiring a new set of regions, a portion of the species, and even of the genera, are limited to the older regions. I have already shown that this is the case with the almost cosmopolitan plovers, and it also occurs in another instance where it has been very strongly urged that the Sclaterian regions will not apply. I allude to the distribution of insects in the Oriental and Australian portions of the Malay Archipelago. Here, in the case of birds and mammals, there is a most abrupt and striking change on passing from Borneo and Java to the Moluccas and New Guinea; but in insects this is not conspicuously the case, and it has been said that the whole Archipelago, from Sumatra to New Guinea, and even to the Solomon Islands, is characterised by one uniform insect-fauna. This, however, is by no means a correct statement. There are undoubtedly many genera common to the whole Archipelago, as might be expected from the great similarity of climate and the uniformly forest-clad nature of the islands, together with the power of crossing narrow seas possessed by all winged insects. Yet, both among butterflies and beetles, especially the latter, there are a considerable number of genera confined respectively to the Indo-Malayan and Austro-Malayan divisions of the Archipelago, giving to the fauna of each a characteristic facies.

Now if we adopt for insects, as has been proposed, a single Malayan region including the whole of the Archipelago, we should be apt to lose sight of the two distinct elements it contains, the one due to an ancestral diversity corresponding to that which still exists in all the higher animals, the other dependent on a comparatively recent process of intermigration between the two portions of what are fundamentally distinct insect faunas; while it is not clear what corresponding advantage would be obtained by the student of geographical distribution.

From the point of view I have now endeavoured to set forth, we may, I think, draw the conclusion that the six Sclaterian regions are natural zoological divisions, because they are separated by barriers of considerable antiquity and permanence, which have led to their being characterised each by well-marked assemblages of the higher animals. Further, when groups of organisms which from their exceptional powers of dispersal, or from any other cause, have been able to extend themselves beyond these barriers, that is no reason whatever for establishing new regions—which would *not* be marked out by equally important barriers—since the divergencies in the distribution of the various classes or orders, as exhibited by means of a common series of regions, is one of those interesting problems of distribution which can only be solved by comparative study. Not only, therefore, is one set of regions all that is required to exhibit the distribution of the various terrestrial organisms; but, for all purposes of comparative study it is immeasurably superior to the establishment of numerous sets of special regions, constructed so as to accord with the distribution of special animal groups.

We now come to the second objection—the supposed inequality of the six Sclaterian regions. Some of them are said to be really only sub-regions, while others are said to be so diverse as to be rendered more equal if divided into two regions.

This question of equality is decided almost exclusively by one characteristic, and one that seems to me to be not the most important for the purpose we have in view. This character is the possession of peculiar groups of the rank of family or order, taking no account either of the richness and variety of life-development, or of the geographical extent of the area in question. From this point of view Australia is sometimes said to be equal to all the rest of the world, both on account of its rich development of the marsupial order, but especially because in the duck-bill and spiny ant-eater it possesses a distinct sub-class of mammals. From another point of view, however, Australia, Africa, and South America are united in one primary region, because they alone possess one of the sub-classes of birds—the Ratitae.

New Zealand and Madagascar have each been proposed as regions, the first on account of its *Apteryx* and moas, with its isolated lizard-like *Hatteria* [tuatara]; the second for its peculiar families of Lemurs and Insectivora, and equally peculiar families and genera of birds. In contrast with these, we have the proposal to unite the rich and extensive Palaearctic and Nearctic regions to form one region only, because they do not possess a sufficient number of peculiar families and genera of mammals and birds, although the new region thus constituted is perhaps twenty times as rich as New Zealand in varied forms of life.

Those who adopt these views appear to me to attach a very exaggerated importance to the possession by a limited area of some remnant of an otherwise extinct group, which has been preserved owing to its long-continued isolation in a district where it has been secure from the competition of higher forms—almost always, therefore, in an island. Such survivals are exceedingly interesting; but I cannot see what they have to do with the division of the whole land-area of the globe into zoological regions, whose sole purport and use is to facilitate the study of the geographical distribution of all land animals.

The conception of zoological regions expressed in the views I am now combating seems to me to be altogether erroneous, and to lead to results which are neither useful nor instructive, and far less natural than that which takes account of a variety of characters as the best guides to an approximate equality. I urge, therefore, that zoological regions, to be at once natural and useful in the highest degree, must be founded on a combination of essential features, as follows:—

(1) They should be founded upon, and approximate to, the great primary geographical divisions of the earth, which there is reason to believe have been permanent during considerable geological periods.

(2) They should be rich and varied in *all* the main types of animal life.

(3) They should possess great individuality; whether exhibited by the *possession* of numerous peculiar species, genera, or families, or by the entire *absence* of genera or families which are abundant and widespread in some of the adjacent regions.

Tested by these conditions the six Sclaterian regions seem all that can be desired—subject of course to modification in details. If we make some allowance for the inevitable poverty of the temperate as compared with the tropical regions—due both to present and to past conditions of climate—they present a greater amount of equality than might be expected. The Neotropical region is somewhat the richest—very much the richest in birds and insects—and this may be traced to its possessing so enormous an area of tropical forest-clad land, together with the greatest of the mountain ranges situated wholly within the tropics—the Andes, and two other isolated mountain groups of great extent and antiquity in Brazil and Guiana; while the Nearctic is the poorest—due perhaps to its rather limited area, its large extent of arid lands, but more especially to its extreme climate, a severe winter prevailing to considerably south of the parallel of 40 degrees N. Latitude.

The subdivisions of the primary regions is far less important; and with the same facts before them, naturalists arrive at different conclusions. I would suggest, therefore, that for the present, at all events, no definite named subdivisions should be attempted, but that the continental portion of each region be subdivided by the use of the terms north, south, east, west, and central, with their combinations where required. By the use of these terms the range of a genus or species within the regions may be defined with sufficient accuracy, and in a manner at once intelligible to every student.

The conclusions to which this discussion has led us may now be briefly summarised as follows: Zoological regions are those primary divisions of the earth's surface of approximately continental extent, which are characterised by distinct assemblages of animal types. Though strictly natural, in the sense already pointed out, they have no absolute character as equal independent existences, since they may have been different in past ages, but are more or less conventional, being established solely for the purpose of facilitating the study of the existing geographical distribution of animals in its bearing on the theory of evolution. There is thus, in my opinion, no question of who is *right* and who is *wrong* in the naming and grouping of these regions, or of determining what are the *true* primary regions. All proposed regions are, from some points of view, natural, but the whole question of their grouping and nomenclature is one of convenience and of utility in relation to the object aimed at.

It is because I think that the future progress of a branch of biological study in which I take great interest will depend on our arriving at some uniformity of view as to this question of zoological regions, that I have devoted so much space to its discussion.

* * * * *

As a natural extension of his interest in regional systematics, Wallace took up the subject of the equitable measurement of faunal affinities. In 1877 he presented one of the first suggestions for a method of statistical treatment of unequal-sized biotas:

The comparative richness of faunas and floras tested numerically (ref 280)

In his letter in *Nature*, vol. xvii, p. 9, Prof. [Alfred] Newton has strongly brought out the absurdity of comparing districts of *very different areas* by the proportionate number of species to area in each. On this principle he shows that to be equally rich with the small island of Rodriguez, Madagascar ought to possess four times as many species of birds as exist throughout the whole world. It does not, however, by any means follow that the method thus exposed may not be of value in comparing regions of approximately equal area, as is the case with several of the primary regions, to determine the comparative richness of which Mr. Sclater first applied it. I have not Mr. Sclater's paper at hand, but it is my impression that he made no attempt to show—"that the proper mode of comparing the wealth or poverty of one fauna with another was to state the proportion which the number of species composing it bears to the area over which they range"—as Prof. Newton implies that he did, but that he merely adopted this method as the only one readily available for the comparison of his regions. Although I took the opportunity of making some corrections in the figures, I never committed myself to the principle; and I very soon afterwards found that it was not to be trusted. As, however, several later writers have made use of it without remark, it will be interesting to consider where the exact point of the fallacy lies, and with what modifications the method can be trusted to give useful and consistent results.

If we compare two islands of almost exactly equal areas, such as Ceylon and Tasmania, and find that the one has twice or three times as many species of mammals or birds as the other, it will be generally admitted that we express the fact correctly when we say that, as regards such a group of animals, the one is twice or thrice as rich as the other; and the same may be said of two countries or two continents of identical areas. For on the supposition that there is a general correspondence between the numbers of rare and common, of local and of wide-spread species in the two areas compared (and this seems probable), then the average number of distinct species to be met with on one spot, or to be seen during a journey of equal length, will be proportionate to the total number of species in the two areas. But now let us divide one of the two continents or islands which we are comparing into two or more parts. We know, as a matter of fact, that one-half the area will always contain much more than half the total number of species, while one-tenth of the area will contain immensely more than one-tenth of the species. To take an example: the county of Sussex is about one-eightieth part the area of the British Isles, yet it actually contains full two-thirds of the total number of flowering plants, both being estimated by the same flora (Babington's *Manual*, fifth edition, British Isles 1,536 species, Sussex 1,059 species). If we now compare either Britain or Sussex with an *equal area* on the continent of Europe or North America, we may obtain an instructive estimate of the comparative richness of their respective floras; but if we compare *unequal areas*, and then endeavour to equalise them by getting the proportions of species to area, we shall obtain erroneous results, which will become literally absurd when the areas compared are very unequal.

The problem remains, how to compare unequal areas of which we possess the zoological or botanical statistics. We can only do so by equalising them, and this may not be so difficult as at first sight appears. For example, let us take the Palaearctic and North American regions, in which the species of birds are nearly equal in number, but the areas are as about seven to three. The number of the Palaearctic species have, however, been proportionately increased of late years, and if we take the western half of the Palaearctic region so as to include North Africa and Persia we shall have an area about equal to the Nearctic region, and a number of species perhaps one-sixth or one-eighth less, which will thus represent the comparative richness of these two areas. The eastern half of the region, including Japan and North China, is probably as rich as the western; while the intermediate portion is poorer in species. Combining these three portions, and taking the average, we should perhaps find the Palaearctic region about four-fifths or five-sixths as rich as the Nearctic, instead of less than one-half, as shown by the method of proportionate areas.

Whenever we know how many *peculiar* species any district contains, we can deduct its area from the total area of the region to be compared, and this number of peculiar species, from the fauna of the region; and by this means we may reduce two unequal regions to comparative equality. Again, all detached portions or islands should be omitted in estimating the comparative richness of regions, because they affect these regions very unequally. By adding Britain to Europe you increase the area without adding to the fauna, and thus make the region seem poorer; while by adding Madagascar to Africa, or New Zealand to Australia, you add to the fauna in a greater proportion than you increase the area, and thus make the region seem richer. For a fair comparison continents should be compared with continents, and islands with islands, and these should in every case be brought to an approximate equality of area by lopping off outlying portions with their peculiar species. We shall then get results which will be instructive, and which will afford us a fair estimate of the comparative richness of different countries in the several classes of animals and plants.

* * * * *

One of Wallace's late-period works in biogeography, the now all but forgotten 'The birds of paradise in the Arabian nights,' was published in 1904 (ref 615). This essay is too long for full reproduction here, but merits a short excerpt, at least, for the fascinating way it integrates the studies of animal distribution, literature, myth, physical and economic geography, and ethnology. At a point in the essay (pp. 564–5) in which he is describing the travels of the hero of one of the Arabian nights *tales, Wallace remarks:*

. . . when we leave this "land of camphor" [an area Wallace places as being in southern China] and Hasan's last journey to the actual Islands of Wák-Wák [i.e., the Aru Islands] is described, we seem to leave reality altogether, and to be involved in all kinds of contradictions and impossibilities. And the reason of this is, that we have here got beyond the limits actually visited by any of the travelling merchants of Persia or Arabia, who could only tell what they had heard from the Malay, Javan, or Bugis traders, who then carried on the trade between the Malay Peninsula and the remotest islands of the Eastern

Archipelago. Even thus, it is probable that the Arab merchants who visited Malay ports, then the centre for much of the trade of the Far East, had whatever information they picked up at second or third hand; since the Bugis of Celebes would probably bring the spices, beche-de-mer, mother-of-pearl shells, sandal-wood, and other products, as well as the skins of birds-of-paradise, to the port of Macassar, whence the more western Malay and Javanese traders would convey these and various articles of commerce to the chief ports of Java, Sumatra, and the Malay Peninsula. From the latter ports they were distributed, as they are now from Singapore, to India and Persia on the one side, and to Siam and China on the other. We must also remember that the whole voyage from Singapore to the Aru Islands is through, perhaps, the calmest sea in the world for the entire distance (2500 miles); that the coasts of large or small islands are everywhere in sight, rendering navigation of the easiest; and that the monsoons blow with great regularity, the somewhat variable west monsoon from December to June, and the much steadier and stronger east monsoon from July to November. On the other hand, the sea between Southern China and the Aru Islands, by way of the Philippines and Northern Moluccas, or by Celebes, is immensely more dangerous, typhoons and violent storms being frequent; so that we may be quite sure that, in the time of the authors of the *Arabian Nights*, even more exclusively than now, the products of New Guinea, the Aru Islands, and the Moluccas reached the Western world by way of Java, Sumatra, and Malaya only. These facts will enable us to some extent to unravel the extraordinary misconceptions of these old story-tellers, as to the position and relative importance of the more remote and inaccessible regions they described or referred to, while at the same time showing that there was a substratum of fact, and often of very accurate general observation, even in their wildest stories . . .

* * * * *

Biogeography—especially zoogeography—was a field of study to which Wallace's intellectual talents were ideally suited. To date he remains its historically most significant practitioner. But his long-term domination of the subject is now viewed by some as not having been an entirely constructive influence on its progress. The entire story is beyond the scope of present discussion; suffice it to say that his dogmatic support of Sclater's regional faunas classification and accompanying position on the fixed nature of continents and ocean basins (see Section Seven) may have had the effect of slowing progress toward ultimately more sophisticated appreciations of related problems.

This is not to say, however, that there remains little reason for referring back to Wallace's opinions on the subject of biogeography. His support of currently untenable positions was in good part a function of the lack of critical factual data at that time (but see Fichman 1977); there can be little doubt, for example, that he would have been one of the first to grasp the biogeographic significance of the theory of plate tectonics had he been alive when it emerged in the early 1960s. More importantly, Wallace's contribution to biogeography was fully as much as could be expected from any "father" of a discipline: he convincingly argued for its recognition as a subject deserving of inquiry, showed how it could be treated systematically, and introduced and illustrated lasting methods for its study.

Related writings

on invertebrate distribution and faunas: refs 10, 13, 25, 40, 53, 74, 81, 96, 109, 127, 135, 181, 193, 205, 269, 284, 290, 488, 714, 715, 716, 718, 719, 720, 721, 725, 727

on avian distribution and faunas: refs 5, 15, 25, 31, 32, 37, 38, 41, 52, 53, 57, 58, 62, 63, 64, 66, 67, 71, 72, 79, 84, 85, 89, 101, 102, 114, 141, 244, 269, 274, 279, 290, 302, 469, 500, 615, 709, 714, 715, 718, 719, 720, 721, 725, 727, 729

on mammalian distribution and faunas: refs 8, 22, 26, 53, 81, 97, 269, 290, 302, 323, 350, 500, 714, 715, 718, 719, 720, 721, 725, 727, 729

on plant collecting, distribution, and systematics: refs 19, 22, 27, 146, 160, 248, 257, 272, 289, 292, 294, 313, 384, 393, 441, 503, 668, 669, 689, 710, 713, 714, 715, 718, 719, 720, 721, 725, 727, 729, 731, 732

on systematic regional classification: refs 52, 53, 78, 81, 269, 275, 286, 289, 290, 360, 441, 494, 500, 575, 715, 718, 719, 721, 725, 727, 729, 732

on the statistical measurement of biotas: refs 52, 85, 280, 500, 718, 727, 732

on analytical and theoretical biogeography: refs 20, 38, 40, 41, 53, 74, 78, 81, 82, 85, 89, 96, 131, 181, 188, 193, 244, 269, 275, 286, 302, 316, 441, 469, 488, 714, 715, 716, 718, 719, 720, 721, 724, 725, 727, 729

on island biogeography: refs 38, 40, 53, 78, 81, 85, 181, 188, 193, 257, 316, 340, 393, 469, 689, 715, 719, 720, 721, 727

on biogeography–physical geography relations: refs 8, 20, 38, 41, 53, 78, 81, 82, 85, 89, 96, 97, 102, 131, 181, 188, 244, 257, 269, 286, 292, 302, 316, 340, 393, 469, 547, 615, 714, 715, 716, 718, 719, 720, 721, 726, 727, 728, 729, 730, 732

on the relations between biogeography and evolutionary theory: see Section Six, p. 364

on cultural biogeography: refs 10, 27, 248, 289, 292, 384, 441, 615, 713, 714, 715, 718, 719, 720, 721, 725, 727, 728, 732

on miscellaneous topics: refs 248, 256, 262, 286, 334, 413, 448, 711, 718, 729

additional related topics: see Section Eight, p. 471

book reviews and notices: refs 181, 193, 205, 229, 250, 274, 384

secondary literature: Beddall (1969), Brooks (1984), Browne (1983), Fichman (1977, 1981), George (1964, 1981), McKinney (1972), Nelson (1978), C. H. Smith (1989)

ALFRED R. WALLACE. 1853.

§ 6 · EVOLUTION

... none of my differences of opinion from Darwin imply any real divergence as to the overwhelming importance of the great principle of natural selection, while in several directions I believe that I have extended and strengthened it. The principle of "utility," which is one of its chief foundation-stones, I have always advocated unreservedly; while in extending this principle to almost every kind and degree of coloration, and in maintaining the power of natural selection to increase the infertility of hybrid unions, I have considerably extended its range. Hence it is that some of my critics declare that I am more Darwinian than Darwin himself, and in this, I admit, they are not far wrong . . .

(*My life*, 1905: ref 729, Vol 2, p. 22)

MUCH of what has been written on Wallace revolves around his independent discovery of the principle of natural selection and subsequent deferral to Darwin's lead on the subject. About as much as deserves to be said about the personal aspects of this relationship has in one place or another already been said; here, we will concern ourselves entirely with Wallace's point of view and how this both paralleled Darwin's and contrasted with it.

Wallace notwithstanding, it was Darwin's highly uniformitarian perspective and meticulously assembled evidence that assured the success of the natural selection idea. It is not without reason that he is considered one of the great figures in the history of science: evolutionary concepts, prone as they are to distortion by special interest thinking, present the ultimate challenge for reduction to objective forms of representation. Darwin pointedly avoided assumptions that seemed to resist unequivocal test, and therefore had little sympathy for arguments invoking final causes. In his opinion, evolution was to be treated most dispassionately as a matter of history in the making; within this framework natural selection was the main process through which new forms came into being.

Wallace, on the other hand, not only admitted the possibility that final causes might be operating, but believed right from the beginning that the overall process of evolution could not be contextualized otherwise. It has usually been thought that Wallace's cosmology became more and more fixated on final causes as he aged, but this is a half-truth only. Initially (i.e., in 1845), he looked on evolution as a process implicit in the uniformitarian operation of a few universal laws of physical–biological interaction (one of which was gravitational attraction), and leading eventually to the development of advanced beings (and social organization). This fundamentally deterministic position allowed for—that is, *assumed*—the existence of innumerable and more or less ungeneralizable *immediate* causes of change. His recognition of the principle of natural selection in 1858: (1) made him realize that the immediate causes of biological change *could* be generalized, (2) caused him to change his opinion on how the immediate and final causes of evolution were to be related, and (3) *automatically* led him to seek proximate causes, paralleling natural selection, that could explain those phenomena for which 'the survival of the fittest' seemed unable to account. In short, in contrast with Darwin, Wallace was always working from *both* the 'bottom, up' *and* the 'top, down.' It is all the more remarkable that the two men conceived models of natural selection that turned out to be as similar as they were.

The critical elements of Darwin's natural selection model are the concepts 'variation' and 'utility'. He argued that within every population every trait (i. e., character or adaptation) varies in one respect or another in its expression from individual to individual, and that, in response to the varying contingencies of (biological and physical) environment, the utility value of these must also vary. Accordingly, those individuals whose assemblage of traits best meets the requirements for survival demanded by the environment at the time differentially survive and pass these assemblages on to their progeny.

Wallace's model of natural selection differs little from Darwin's on the subject of the relation of variation to utility. But Wallace did not feel (as Darwin did) that the higher human attributes could have evolved for reasons of their biological utility value alone. In the mid- or early 1850s (or perhaps even the 1840s) he was already beginning to wonder whether natural forces more subtle than those of a rote physical or biological kind might also be influencing the direction of evolution. Wallace was impressed with the apparent discontinuities of cause and effect that we observe in nature. The elements carbon, hydrogen, oxygen, nitrogen, etc., for example, seemed to combine in ways producing structures that transcended physicalistic modes of description. Living things were composed of atoms; but no known properties of these fundamental units could account for the mysteries of growth, behaviour, and other seemingly irreducible phenomena assigned to the organizational level 'life'. None the less, the actions of physical processes were obviously being modified by biological ones, and vice versa. Perhaps this two-directional pathway incorporated causal influences originating in yet higher levels of organization—levels the implications of whose operation we could but dimly perceive, if at all. Thus, the causal discontinuities might be apparent only, and just as a physical environment modified by biological activity was necessary to produce the most diverse and organized biological communities, so too might a human biological capacity 'enriched' by the operation of influences transcending our immediate appreciation of things be necessary to continue a universally 'progressive' evolution. Granting this scenario, a provisional *raison d'etre* could be recognized for those human characters that seemed to have no purely biological utility function: simply, their 'use' lay in their relation to such biology-transcending forces of organization.

By about 1865 Wallace had concluded that spiritualist writings described a natural realm whose properties gave evolutionary meaning to the higher human faculties. Wallace already believed that species diversification was facilitated by the interplay of various physical and biological forces—climate and climatic change, the 'survival of the fittest', and so forth. He now accepted the spiritualists' argument that moral/intellectual evolution—'spiritual diversification'—was in like fashion facilitated by the individual's considered exploitation of his earthly opportunities (see Section Two).

Very few workers, of course, have been willing to pursue studies grounded in such divergent thinking. Depending on how one approaches the question, however, Wallace's views on biological organization *per se* may or may not have been less conservative than Darwin's. Both before 1865 and after that date, Wallace regarded natural selection as a *law of interaction*—much like gravity—'ruling' one level of the natural hierarchy. As a result, he found it necessary (quite rightly, it seems) to distinguish between its immediate characteristics of operation and its longer-term results. Darwin accepted that structural continuities—the continuing *result* of adapted organisms—evidenced the causal continuity of the *process* of natural selection. Wallace never thought this line of reasoning to be quite trustworthy. Gravity's inability to regulate the cosmos in a manner making life a *necessary* consequence of its operation (alone) was paralleled, he felt, by natural selection's inability to sustain organization of a kind evidenced by the higher intellectual and moral attributes. Wallace eventually attributed his posited hierarchy of organizational levels to the operation of a causally universal will. But again, no 'first causes' or 'special effects' were implied: the entire hierarchy—all of its level-by-level and between-level organization and interrelations—operated upon 'laws of nature' as that term is generally understood. Otherwise put, 'reality' was the full and orderly self-expression of that will.

One of the implications of Wallace's cosmology is the idea that outside a certain domain of operation, natural selection should be powerless. Conversely, within that domain it was a catalyst for change nearly as universal as gravity was within the purely physical realm. Thus it was that Wallace came to believe (as stated in the opening quotation in this section) that he was in some ways 'more Darwinian than Darwin'. Natural selection's importance to the diversification of life he did not doubt, but he strongly distinguished between those characteristics it served to maintain in a population from those it could merely work alongside of, and whose maintenance had to be assigned to causes originating beyond the biological level of organization. Still, by rationally excluding from consideration what he felt to be results stemming from the operation of such extra-biological forces, he was able to continue to assign the causes of most adaptive structures to immediate (ecological) influences.

In a section of the essay 'Evolution' (ref 589), solicited by the New York Sun *in 1900, Wallace commented on:*

The nature and limits of evolution

So many of the objections which are still made to the theory of evolution, and especially to that branch of it which deals with living organisms, rest upon a misconception of what it professes to explain, and even of what any theory can possibly explain, that a few words on its nature and limits seem to be necessary.

Evolution, as a general principle, implies that all things in the universe as we see them have arisen from other things which preceded them by a process of modification, under the action of those all-pervading but mysterious agencies known to us as "natural forces," or more generally the "laws of nature." More particularly the term evolution implies that the process is an "unrolling" or "unfolding," derived probably from the way in which leaves and flowers are usually rolled up or crumpled up in the bud and grow into their perfect form by unrolling or unfolding. Insects in the pupa and vertebrates in the embryo exhibit a somewhat similar condition of folding and the word is therefore very applicable to an extensive range of phenomena; but it must not be taken as universally applicable, since in the material world there are other modes of orderly change under natural laws to which the terms development or evolution are equally applicable. The "continuity" of physical phenomena as illustrated by the late Sir William Grove in 1866 has the same general meaning, but evolution implies more than mere continuity or succession—something like growth or definite change from form to form under the action of unchangeable laws.

The point especially to be noted here is, that evolution, even if it is essentially a true and complete theory of the universe, can only explain the existing conditions of nature by showing that it has been derived from some pre-existing condition through the action of known forces and laws. It may also show the high probability of a similar derivation from a still earlier condition; but the farther back we go the more uncertain must be our conclusions, while we can never make any real approach to the absolute beginnings of things. Herbert Spencer, and many other thinkers before him, have shown that if we try to realize the absolute nature of the simplest phenomena, we are inevitably landed either in a contradiction or in some unthinkable proposition. Thus, suppose we ask—is matter infinitely divisible or is it not? If we say it is, we cannot think it out, since all infinity, however it may be stated in words, is really unthinkable.

If we say there is a limit—the ultimate atom—then, as all size is comparative, we can imagine a being to whom this atom seems as large as an apple or even a house does to us: and we then find it quite unthinkable that this mass of matter should be in its nature absolutely indivisible even by an infinite force. It follows that all explanations of phenomena can only be partial explanations. They can inform us of the last change or the last series of changes which brought about the actual conditions now existing, and they can often enable us to predict future changes to a limited extent; but both the infinite past and the remote future are alike beyond our powers. Yet the explanations that the theory of evolution gives us are none the less real and none the less important, especially when we compare its teachings with the wild guesses or the total ignorance of the thinkers of earlier ages.

* * * * *

Thirty years before the Sun *article, on pages 358–360 and 364–371 of the last essay (ref 165) in* Contributions to the theory of natural selection *(ref 716) Wallace explained his views on the nature of the greater reality into which natural selection fits:*

. . . those faculties which enable us to transcend time and space, and to realize the wonderful conceptions of mathematics and philosophy, or which give us an intense yearning for abstract truth (all of which were occasionally manifested at such an early period of human history as to be far in advance of any of the few practical applications which have since grown out of them), are evidently essential to the perfect development of man as a spiritual being, but are utterly inconceivable as having been produced through the action of a law which looks only, and can look only, to the immediate material welfare of the individual or the race.

The inference I would draw from this class of phenomena is, that a superior intelligence has guided the development of man in a definite direction, and for a special purpose, just as man guides the development of many animal and vegetable forms. The laws of evolution alone would, perhaps, never have produced a grain so well adapted to man's use as wheat and maize; such fruits as the seedless banana and bread-fruit; or such animals as the Guernsey milch cow, or the London dray-horse. Yet these so closely resemble the unaided productions of nature, that we may well imagine a being who had mastered the laws of development of organic forms through past ages, refusing to believe that any new power had been concerned in their production, and scornfully rejecting the theory (as my theory will be rejected by many who agree with me on other points), that in these few cases a controlling intelligence had directed the action of the laws of variation, multiplication, and survival, for his own purposes. We know, however, that this has been done; and we must therefore admit the possibility that, if we are not the highest intelligences in the universe, some higher intelligence may have directed the process by which the human race was developed, by means of more subtle agencies than we are acquainted with. At the same time I must confess, that this theory has the disadvantage of requiring the intervention of some distinct individual intelligence, to aid in the production of what we can hardly avoid considering as the ultimate aim and outcome of all organized existence—intellectual, ever-advancing, spiritual man. It therefore implies, that the great laws which govern the material universe were insufficient for his production, unless we consider (as we may fairly do) that the controlling action of such higher intelligences is a necessary part of those laws, just as the action of all surrounding organisms is one of the agencies in organic development. But even if my particular view should not be the true one, the difficulties I have put forward remain, and I think prove, that some more general and more fundamental law underlies that of "natural selection." The law of "unconscious intelligence" pervading all organic nature, put forth by Dr. Laycock and adopted by Mr. Murphy, is such a law; but to my mind it has the double disadvantage of being both unintelligible and incapable of any kind of proof. It is more probable, that the true law lies too deep for us to discover it; but there seems to me, to be ample indications that such a law does exist, and is probably connected with the absolute origin of life and organization . . . [pp. 358–60]

* * * * *

. . . Those organic compounds of which organized beings are built up, consist, as is well known, of matter of an extreme complexity and great instability; whence result the changes of form to which it is continually subject. This view enables us to comprehend the *possibility*, of the phenomena of vegetative life being due to an almost infinite complexity of molecular combinations, subject to definite changes under the stimuli of heat, moisture, light, electricity, and probably some unknown forces. But this greater and greater complexity, even if carried to an infinite extent, cannot, of itself, have the slightest tendency to originate consciousness in such molecules or groups of molecules. If a material element, or a combination of a thousand material elements in a molecule, are alike unconscious, it is impossible for us to believe, that the mere addition of one, two, or a thousand other material elements to form a more complex molecule, could in any way tend to produce a self-conscious existence. The things are radically distinct. To say that mind is a product or function of protoplasm, or of its molecular changes, is to use words to which we can attach no clear conception. You cannot have, in the whole, what does not exist in any of the parts; and those who argue thus should put forth a definite conception of matter, with clearly enunciated properties, and show, that the necessary result of a certain complex arrangement of the elements or atoms of that matter, will be the production of self-consciousness. There is no escape from this dilemma,—either all matter is conscious, or consciousness is something distinct from matter, and in the latter case, its presence in material forms is a proof of the existence of conscious beings, outside of, and independent of, what we term matter.

Matter is Force.—The foregoing considerations lead us to the very important conclusion, that matter is essentially force, and nothing but force; that matter, as popularly understood, does not exist, and is, in fact, philosophically inconceivable. When we touch matter, we only really experience sensations of resistance, implying repulsive force; and no other sense can give us such apparently solid proofs of the reality of matter, as touch does. This conclusion, if kept constantly present in the mind, will be found to have a most important bearing on almost every high scientific and philosophical problem, and especially on such as relate to our own conscious existence.

All Force is probably Will-Force.—If we are satisfied that force or forces are all that exist in the material universe, we are next led to enquire what is force? We are acquainted with two radically distinct or apparently distinct kinds of force—the first consists of the primary forces of nature, such as gravitation, cohesion, repulsion, heat, electricity, etc.; the second is our own will-force.

Many persons will at once deny that the latter exists. It will be said, that it is a mere transformation of the primary forces before alluded to; that the correlation of forces includes those of animal life, and that *will* itself is but the result of molecular change in the brain. I think, however, that it can be shown, that this latter assertion has neither been proved, nor even been proved to be possible; and that in making it, a great leap in the dark has been taken from the known to the unknown. It may be at once admitted that the *muscular force* of animals and men, is merely the transformed energy derived from the primary forces of nature. So much has been, if not rigidly proved, yet rendered highly probable, and it is in perfect accordance with all our knowledge of natural forces and natural laws. But it cannot be contended that the physiological balance-sheet has ever been so accurately struck, that we are entitled to say, not one-thousandth part of a grain more of force has been exerted by any organized body or in any part of it, than has been derived from the known primary forces of the material world. If that were so, it would absolutely negative the existence of will; for if will is anything, it is a power that *directs* the action of the forces stored up in the body, and it is not conceivable that this *direction* can take place, without the exercise of some force in some part of the organism. However delicately a machine may be constructed, with the most exquisitely contrived detents to release a weight or spring by the exertion of the smallest possible amount of force, *some* external force will always be required; so, in the animal machine, however minute may be the changes required in the cells or fibres of the brain, to set in motion the nerve currents which loosen or excite the pent up forces of certain muscles, *some force* must be required to effect those changes. If it is said, "those changes are automatic, and are set in motion by external causes," then one essential part of our consciousness, a certain amount of freedom in willing, is annihilated; and it is inconceivable how or why there should have arisen any consciousness or any apparent will, in such purely automatic organisms. If this were so, our apparent WILL would be a delusion, and Professor Huxley's belief—"that our volition counts for something as a condition of the course of events," would be fallacious, since our volition would then be but one link in the chain of events, counting for neither more nor less than any other link whatever.

If, therefore, we have traced one force, however minute, to an origin in our own WILL, while we have no knowledge of any other primary cause of force, it does not seem an improbable conclusion that all force may be will-force; and thus, that the whole universe, is not merely dependent on, but actually *is*, the WILL of higher intelligences or of one Supreme Intelligence . . .

These speculations are usually held to be far beyond the bounds of science; but they appear to me to be more legitimate deductions from the facts of science, than those which consist in reducing the whole universe, not merely to matter, but to matter conceived and defined so as to be philosophically inconceivable. It is surely a great step in advance, to get rid of the notion that *matter* is a thing of itself, which can exist *per se*, and must have been eternal, since it is supposed to be indestructible and uncreated,—that force, or the forces of nature, are another thing, given or added to matter, or else its necessary properties,—and that mind is yet another thing, either a product of this matter and its supposed inherent forces, or distinct from and co-existent with it;—and to be able to substitute for this complicated theory, which leads to endless dilemmas and contradictions, the far simpler and more consistent belief, that matter, as an entity distinct from force, does not exist; and that FORCE is a product of MIND. Philosophy had long demonstrated our incapacity to prove the existence of matter, as usually conceived; while it admitted the demonstration to each of us of our own self-conscious, ideal existence. Science has now worked its way up to the same result, and this agreement between them should give us some confidence in their combined teaching.

The view we have now arrived at seems to me more grand and sublime, as well as far simpler, than any other. It exhibits the universe, as a universe of intelligence and will-power; and by enabling us to rid ourselves of the impossibility of thinking of mind, but as connected with our old notions of matter, opens up infinite possibilities of existence, connected with infinitely varied manifestations of force, totally distinct from, yet as real as, what we term matter.

The grand law of continuity which we see pervading our universe, would lead us to infer infinite gradations of existence, and to people all space with intelligence and will-power; and, if so, we have no difficulty in believing that for so noble a purpose as the progressive development of higher and higher intelligences, those primal and general will-forces, which have sufficed for the production of the lower animals, should have been guided into new channels and made to converge in definite directions. And if, as seems to me probable, this has been done, I cannot admit that it in any degree affects the truth or generality of Mr. Darwin's great discovery. It merely shows, that the laws of organic development have been occasionally used for a special end, just as man uses them for his special ends; and, I do not see that the law of "natural selection" can be said to be disproved, if it can be shown that man does not owe his entire physical and mental development to its unaided action, any more than it is disproved by the existence of the poodle or the pouter pigeon, the production of which may have been equally beyond its undirected power.

The objections which in this essay I have taken, to the view,—that the same law which appears to have sufficed for the development of animals, has been alone the cause of man's superior physical and mental nature,—will, I have no doubt, be over-ruled and explained away. But I venture to think they will nevertheless maintain their ground, and that they can only be met by the discovery of new facts or new laws, of a nature very different from any yet known to us . . . [pp. 364–371]

* * * * *

Wallace's pre- 'On the tendency . . .' deliberations over the nature of the evolutionary process had channelled themselves into an internally unresolvable position. As to evolution's general causes, he had recognized a relationship between the products of nature (especially in their diversity and distributional aspects) and evolution starting around 1845, when he read The vestiges of the natural history of Creation. *For some ten years, Wallace cogitated on the relationship between evolution and geographical/ecological distribution. By 1855 he had realized a statable position; the result was 'On the law which has regulated the introduction of new species' (ref 20) (see pages 220–231). But even at that point the relationship of* adaptation *to organic change still mystified him. The basic problem was adaptation's connection to utility. He could see that many organisms had peculiarities of adaptation linked to their style of life, but was quite unwilling to argue on the basis of this correlation (as Lamarck and Chambers had earlier) that these were* necessary *to their existence: to do so was apparently to embrace Creationism. Moreover, he was already thinking that many peculiarly human characteristics were not necessary to survival* per se.

Considering the selection from 'On the Habits of the Orang-utan of Borneo' (ref 26) reproduced in Section One, the relationship of adaptation to utility was still baffling Wallace as of 1856. In this work, it will be remembered, he quite explicitly states that there are adaptive characteristics which have no function with respect to (biological) survival: 'Yes, we reply, we do mean to assert that many animals are provided with organs and appendages which serve no material or physical purpose'.

At last, in 1858, it occurred to Wallace that utility arguments need not be grounded in Creationist concepts. This revelation was written up as the now famous 'Ternate essay':

On the tendency of varieties to depart indefinitely from the original type (ref 43)

One of the strongest arguments which have been adduced to prove the original and permanent distinctness of species is, that *varieties* produced in a state of domesticity are more or less unstable, and often have a tendency, if left to themselves, to return to the normal form of the parent species; and this instability is considered to be a distinctive peculiarity of all varieties, even of those occurring among wild animals in a state of nature, and to constitute a provision for preserving unchanged the originally created distinct species.

In the absence or scarcity of facts and observations as to *varieties* occurring among wild animals, this argument has had great weight with naturalists, and has led to a very general and somewhat prejudiced belief in the stability of species. Equally general, however, is the belief in what are called "permanent or true varieties,"—races of animals which continually propagate their like, but which differ so slightly (although constantly) from some other race, that the one is considered to be a *variety* of the other. Which is a *variety* and which the original *species*, there is generally no means of determining, except in those rare cases in which the one race has been known to produce an offspring unlike itself and resembling the other. This, however, would seem quite incompatible with the "permanent invariability of species," but the difficulty is overcome by assuming that such varieties have strict limits, and can never again vary further from the original type, although they may return to it, which, from the analogy of the domesticated animals, is considered to be highly probable, if not certainly proved.

It will be observed that this argument rests entirely on the assumption, that *varieties* occurring in a state of nature are in all respects analogous to or even identical with those of domestic animals, and are governed by the same laws as regards their permanence or further variation. But it is the object of the present paper to show that this assumption is altogether false, that there is a general principle in nature which will cause many *varieties* to survive the parent species, and to give rise to successive variations departing further and further from the original type, and which also produces, in domesticated animals, the tendency of varieties to return to the parent form.

The life of wild animals is a struggle for existence. The full exertion of all their faculties and all their energies is required to preserve their own existence and provide for that of their infant offspring. The possibility of procuring food during the least favourable seasons, and of escaping the attacks of their most dangerous enemies, are the primary conditions which determine the existence both of individuals and of entire species. These conditions will also determine the population of a species; and by a careful consideration of all the circumstances we may be enabled to comprehend, and in some degree to explain, what at first sight appears so inexplicable—the excessive abundance of some species, while others closely allied to them are very rare.

The general proportion that must obtain between certain groups of animals is readily seen. Large animals cannot be so abundant as small ones; the carnivora must be less numerous than the herbivora; eagles and lions can never be so plentiful as pigeons and antelopes; the wild asses of the Tartarian deserts cannot equal in numbers the horses of the more luxuriant prairies and pampas of America. The greater or less fecundity of an animal is often considered to be one of the chief causes of its abundance or scarcity; but a consideration of the facts will show us that it really has little or nothing to do with the matter. Even the least prolific of animals would increase rapidly if unchecked, whereas it is evident that the animal population of the globe must be stationary, or perhaps, through the influence of man, decreasing. Fluctuations there may be; but permanent increase, except in restricted localities, is almost impossible. For example, our own observation must convince us that birds do not go on increasing every year in a geometrical ratio, as they would do, were there not some powerful check to their natural increase. Very few birds produce less than two young ones each year, while many have six, eight, or ten; four will certainly be below the average; and if we suppose that each pair produce young only four times in their life, that will also be below the average, supposing them not to die either by violence or want of food. Yet at this rate how tremendous would be the increase in a few years from a single pair! A simple calculation will show that in fifteen years each pair of birds would have increased to nearly ten millions! whereas we have no reason to believe that the number of the birds of any country increases at all in fifteen or in one hundred and fifty years. With such powers of increase the population must have reached its limits, and have become stationary, in a very few years after the origin of each species. It is evident, therefore, that each year an immense number of birds must perish—as many in fact as are born; and as on the lowest calculation the progeny are each year twice as numerous as their parents, it follows that, whatever be the average number of individuals existing in any given country, *twice that number must perish annually*,—a striking result, but one which seems at least highly probable, and is perhaps under rather than over the truth. It would therefore appear that, as far as the continuance of the species and the keeping up the average number of individuals are concerned, large broods are superfluous. On the average all above *one* become food for hawks and kites, wild cats and weasels, or perish of cold and hunger as winter comes on. This is strikingly proved by the case of particular species; for we find that their abundance in individuals bears no relation whatever to their fertility in producing offspring. Perhaps the most remarkable instance of an immense bird population is that of the passenger pigeon of the United States, which lays only one, or at most two eggs, and is said to rear generally but one young one. Why is this bird so extraordinarily abundant, while others producing two or three times as many young are much less plentiful? The explanation is not difficult. The food most congenial to this species, and on which it thrives best, is abundantly distributed over a very extensive region, offering such differences of soil and climate, that in one part or another of the area the supply never fails. The bird is capable of a very rapid and long-continued flight, so that it can pass without fatigue over the whole of the district it inhabits, and as soon as the supply of food begins to fail in one place is able to discover a fresh feeding-ground. This example strikingly shows us that the procuring a constant supply of wholesome food is almost the sole condition requisite for ensuring the rapid increase of a given species, since neither the limited fecundity, nor the unrestrained attacks of birds of prey and of man are here sufficient to check it. In no other birds are these peculiar circumstances so strikingly combined. Either their food is more liable to failure, or they have not sufficient power of wing to search for it over an extensive area, or during some season of the year it becomes very scarce, and less wholesome substitutes have to be found; and thus, though more fertile in offspring, they can never increase beyond the supply of food in the least favourable seasons. Many birds can only exist by migrating, when their food becomes scarce, to regions possessing a milder, or at least a different climate, though, as these migrating birds are seldom excessively abundant, it is evident that the countries they visit are still deficient in a constant and abundant supply of wholesome food. Those whose organization does not permit them to migrate when their food becomes periodically scarce, can never attain a large population. This is probably the reason why woodpeckers are scarce with us, while in the tropics they are among the most abundant of solitary birds. Thus the house sparrow is more abundant than the redbreast, because its food is more constant and plentiful,—seeds of grasses being preserved during the winter, and our farm-yards and stubble-fields furnishing an almost inexhaustible supply. Why, as a general rule, are aquatic, and especially sea birds, very numerous in individuals? Not because they are more prolific than others, generally the contrary; but because their food never fails, the sea-shores and river-banks daily swarming with a fresh supply of small mollusca and crustacea. Exactly the same laws will apply to mammals. Wild cats are prolific and have few enemies; why then are they never as abundant as rabbits? The only intelligible answer is, that their supply of food is more precarious. It appears evident, therefore, that so long as a country remains physically unchanged, the numbers of its animal population cannot materially increase. If one species does so, some other requiring the same kind of food must diminish in proportion. The numbers that die annually must be immense; and as the individual existence of each animal depends upon itself, those that die must be the weakest—the very young, the aged, and the diseased,—while those that prolong their existence can only be the most perfect in health and vigour—those who are best able to obtain food regularly, and avoid their numerous enemies. It is, as we commenced by remarking, "a struggle for existence," in which the weakest and least perfectly organized must always succumb.

Now it is clear that what takes place among the individuals of a species must also occur among the several allied species of a group,—viz. that those which are best adapted to obtain a regular supply of food, and to defend themselves against the attacks of their enemies and the vicissitudes of the seasons, must necessarily obtain and preserve a superiority in population; while those species which from some defect of power or organization are the least capable of counteracting the vicissitudes of food, supply, etc., must diminish in numbers, and, in extreme cases, become altogether extinct. Between these extremes the species will present various degrees of capacity for ensuring the means of preserving life; and it is thus we account for the abundance or rarity of species. Our ignorance will generally prevent us from accurately tracing the effects to their causes; but could we become perfectly acquainted with the organization and habits of the various species of animals, and could we measure the capacity of each for performing the different acts necessary to its safety and existence under all the varying circumstances by which it is surrounded, we might be able even to calculate the proportionate abundance of individuals which is the necessary result.

If now we have succeeded in establishing these two points—1st, *that the animal population of a country is generally stationary, being kept down by a periodical deficiency of food, and other checks*; and, 2nd, *that the comparative abundance or scarcity of the individuals of the several species is entirely due to their organization and resulting habits, which, rendering it more difficult to procure a regular supply of food and to provide for their personal safety in some cases than in others, can only be balanced by a difference in the populations which have to exist in a given area*—we shall be in a condition to proceed to the consideration of *varieties*, to which the preceding remarks have a direct and very important application.

Most or perhaps all the variations from the typical form of a species must have some definite effect, however slight, on the habits or capacities of the individuals. Even a change of colour might, by rendering them more or less distinguishable, affect their safety; a greater or less development of hair might modify their habits. More important changes, such as an increase in the power or dimensions of the limbs or any of the external organs, would more or less affect their mode of procuring food or the range of country which they inhabit.

It is also evident that most changes would affect, either favourably or adversely, the powers of prolonging existence. An antelope with shorter or weaker legs must necessarily suffer more from the attacks of the feline carnivora; the passenger pigeon with less powerful wings would sooner or later be affected in its powers of procuring a regular supply of food; and in both cases the result must necessarily be a diminution of the population of the modified species. If, on the other hand, any species should produce a variety having slightly increased powers of preserving existence, that variety must inevitably in time acquire a superiority in numbers. These results must follow as surely as old age, intemperance, or scarcity of food produce an increased mortality. In both cases there may be many individual exceptions; but on the average the rule will invariably be found to hold good. All varieties will therefore fall into two classes—those which under the same conditions would never reach the population of the parent species, and those which would in time obtain and keep a numerical superiority. Now, let some alteration of physical conditions occur in the district—a long period of drought, a destruction of vegetation by locusts, the irruption of some new carnivorous animal seeking "pastures new"—any change in fact tending to render existence more difficult to the species in question, and tasking its utmost powers to avoid complete extermination; it is evident that, of all the individuals composing the species, those forming the least numerous and most feebly organized variety would suffer first, and, were the pressure severe, must soon become extinct. The same causes continuing in action, the parent species would next suffer, would gradually diminish in numbers, and with a recurrence of similar unfavourable conditions might also become extinct. The superior variety would then alone remain, and on a return to favourable circumstances would rapidly increase in numbers and occupy the place of the extinct species and variety.

The *variety* would now have replaced the *species*, of which it would be a more perfectly developed and more highly organized form. It would be in all respects better adapted to secure its safety, and to prolong its individual existence and that of the race. Such a variety *could not* return to the original form; for that form is an inferior one, and could never compete with it for existence. Granted, therefore, a "tendency" to reproduce the original type of the species, still the variety must ever remain preponderant in numbers, and under adverse physical conditions *again alone survive*. But this new, improved, and populous race might itself, in course of time, give rise to new varieties, exhibiting several diverging modifications of form, any of which, tending to increase the facilities for preserving existence, must, by the same general law, in their turn become predominant. Here, then, we have *progression and continued divergence* deduced from the general laws which regulate the existence of animals in a state of nature, and from the undisputed fact that varieties do frequently occur. It is not, however, contended that this result would be invariable; a change of physical conditions in the district might at times materially modify it, rendering the race which had been the most capable of supporting existence under the former conditions now the least so, and even causing the extinction of the newer and, for a time, superior race, while the old or parent species and its first inferior varieties continued to flourish. Variations in unimportant parts might also occur, having no perceptible effect on the life-preserving powers; and the varieties so furnished might run a course parallel with the parent species, either giving rise to further variations or returning to the former type. All we argue for is, that certain varieties have a tendency to maintain their existence longer than the original species, and this tendency must make itself felt; for though the doctrine of chances or averages can never be trusted to on a limited scale, yet, if applied to high numbers, the results come nearer to what theory demands, and, as we approach to an infinity of examples, become strictly accurate. Now the scale on which nature works is so vast—the numbers of individuals and periods of time with which she deals approach so near to infinity, that any cause, however slight, and however liable to be veiled and counteracted by accidental circumstances, must in the end produce its full legitimate results.

Let us now turn to domesticated animals, and inquire how varieties produced among them are affected by the principles here enunciated. The essential difference in the condition of wild and domestic animals is this,—that among the former, their well-being and very existence depend upon the full exercise and healthy condition of all their senses and physical powers, whereas, among the latter, these are only partially exercised, and in some cases are absolutely unused. A wild animal has to search, and often to labour, for every mouthful of food—to exercise sight, hearing, and smell in seeking it, and in avoiding dangers, in procuring shelter from the inclemency of the seasons, and in providing for the subsistence and safety of its offspring. There is no muscle of its body that is not called into daily and hourly activity; there is no sense or faculty that is not strengthened by continual exercise. The domestic animal, on the other hand, has food provided for it, is sheltered, and often confined, to guard it against the vicissitudes of the seasons, is carefully secured from the attacks of its natural enemies, and seldom even rears its young without human assistance. Half of its senses and faculties are quite useless; and the other half are but occasionally called into feeble exercise, while even its muscular system is only irregularly called into action.

Now when a variety of such an animal occurs, having increased power or capacity in any organ or sense, such increase is totally useless, is never called into action, and may even exist without the animal ever becoming aware of it. In the wild animal, on the contrary, all its faculties and powers being brought into full action for the necessities of existence, any increase becomes immediately available, is strengthened by exercise, and must even slightly modify the food, the habits, and the whole economy of the race. It creates as it were a new animal, one of superior powers, and which will necessarily increase in numbers and outlive those inferior to it.

Again, in the domesticated animal all variations have an equal chance of continuance; and those which would decidedly render a wild animal unable to compete with its fellows and continue its existence are no disadvantage whatever in a state of domesticity. Our quickly fattening pigs, short-legged sheep, pouter pigeons, and poodle dogs could never have come into existence in a state of nature, because the very first step towards such inferior forms would have led to the rapid extinction of the race; still less could they now exist in competition with their wild allies. The great speed but slight endurance of the race horse, the unwieldly strength of the ploughman's team, would be useless in a state of nature. If turned wild on the pampas, such animals would probably soon become extinct, or under favourable circumstances might each lose those extreme qualities which would never be called into action, and in a few generations would revert to a common type, which must be that in which the various powers and faculties are so proportioned to each other as to be best adapted to procure food and secure safety,—that in which by the full exercise of every part of his organization the animal can alone continue to live. Domestic varieties, when turned wild, *must* return to something near the type of the original wild stock, *or become altogether extinct.*

We see, then, that no inferences as to varieties in a state of nature can be deduced from the observation of those occurring among domestic animals. The two are so much opposed to each other in every circumstance of their existence, that what applies to the one is almost sure not to apply to the other. Domestic animals are abnormal, irregular, artificial; they are subject to varieties which never occur and never can occur in a state of nature: their very existence depends altogether on human care; so far are many of them removed from that just proportion of faculties, that true balance of organization, by means of which alone an animal left to its own resources can preserve its existence and continue its race.

The hypothesis of Lamarck—that progressive changes in species have been produced by the attempts of animals to increase the development of their own organs, and thus modify their structure and habits—has been repeatedly and easily refuted by all writers on the subject of varieties and species, and it seems to have been considered that when this was done the whole question has been finally settled; but the view here developed renders such an hypothesis quite unnecessary, by showing that similar results must be produced by the action of principles constantly at work in nature. The powerful retractile talons of the falcon- and the cat-tribes have not been produced or increased by the volition of those animals; but among the different varieties which occurred in the earlier and less highly organized forms of these groups, *those always survived longest which had the greatest facilities for seizing their prey.* Neither did the giraffe acquire its long neck by desiring to reach the foliage of the more lofty shrubs, and constantly stretching its neck for the purpose, but because any varieties which occurred among its antitypes with a longer neck than usual *at once secured a fresh range of pasture over the same ground as their shorter-necked companions, and on the first scarcity of food were thereby enabled to outlive them.* Even the peculiar colours of many animals, especially insects, so closely resembling the soil or the leaves or the trunks on which they habitually reside, are explained on this princple; for though in the course of ages varieties of many tints may have occurred, *yet those races having colours best adapted to concealment from their enemies would inevitably survive the longest.* We have also here an acting cause to account for the balance so often observed in nature,—a deficiency in one set of organs always being compensated by an increased development of some others—powerful wings accompanying weak feet, or great velocity making up for the absence of defensive weapons; for it has been shown that all varieties in which an unbalanced deficiency occurred could not long continue their existence. The action of this principle is exactly like that of the centrifugal governor of the steam engine, which checks and corrects any irregularities almost before they become evident; and in like manner no unbalanced deficiency in the animal kingdom can ever reach any conspicuous magnitude, because it would make itself felt at the very first step, by rendering existence difficult and extinction almost sure soon to follow. An origin such as is here advocated will also agree with the peculiar character of the modifications of form and structure which obtain in organized beings—the many lines of divergence from a central type, the increasing efficiency and power of a particular organ through a succession of allied species, and the remarkable persistence of unimportant parts such as colour, texture of plumage and hair, form of horns or crests, through a series of species differing considerably in more essential characters. It also furnishes us with a reason for that "more specialized structure" which Professor Owen states to be a characteristic of recent compared with extinct forms, and which would evidently be the result of the progressive modification of any organ applied to a special purpose in the animal economy.

We believe we have now shown that there is a tendency in nature to the continued progression of certain classes of *varieties* further and further from the original type—a progression to which there appears no reason to assign any definite limits—and that the same principle which produces this result in a state of nature will also explain why domestic varieties have a tendency to revert to the original type. This progression, by minute steps, in various directions, but always checked and balanced by the necessary conditions, subject to which alone existence can be preserved, may, it is believed, be followed out so as to agree with all the phenomena presented by organized beings, their extinction and succession in past ages, and all the extraordinary modifications of form, instinct, and habits which they exhibit.

* * * * *

The preceding answer to the mystery of the origin of species, logical as it seemed (and still seems), must have come as quite a shock to Wallace. Before its conception he could not have foreseen that a model depending *on the notion of utility of specific characters both followed from and explained the relationships of geographical/geological distribution he had been puzzling over. Unhappily, however, he still found himself left with the problem of what to do about the higher intellectual and moral faculties of man, which yet appeared unnecessary to survival* per se.

The solution to this dilemma, in the form of a hypothesized third realm of nature, was worked out over the years 1858 to 1865. During that period, Wallace occupied himself with exploring the logical implications of the Darwin–Wallace theory of species origins—to determine: (1) how internally consistent it was within its domain of application, and, just as importantly, (2) how far that domain extended. It proved to be highly consistent, and, so far as Wallace could make out, no more helpful in explaining the blossoming of man's higher faculties than he had undoubtedly originally suspected (i.e., in 1858). Accordingly, he added to his model what he felt to be the examinable proposition that a biology-influencing 'spirit realm' existed. The result was a synthesis in which biological origins *could be related to the operation of natural selection, and biological* functions *could be assigned, in the last analysis, to a forward-leading development into wholly spiritual beings. It is central to an appreciation of Wallace's work to recognize when he was speaking strictly in terms of 'how things originated' as opposed to 'where things were going'. This difference of perspective is sometimes alluded to in his writings, as in the following selection from 'The origin of species and genera,' (ref 322, pp. 94–96), in which he distinguishes between the concepts 'descent with modification' (biological evolution in general) and 'the origin of species by means of natural selection':*

. . . At first the opponents of Darwinism opposed evolution as well; but of late years the opposition is directed wholly to the adequacy of the causes which Mr. Darwin maintains are sufficient to explain the origin of each species from some pre-existing species, and therefore the origin of all existing species from some one or more ancestral forms. It is maintained that there are other laws at work besides natural selection, and Mr. Darwin has himself admitted that there probably are such. Most of the opponents of Darwinism argue in favour of some guiding or organising power, either internal or external, as absolutely necessary for the development of the various complex organs and special adaptations which characterise each important class of animals. Others go still further, and maintain that "natural selection" is powerless to produce new species in any case, its function being to keep those which are produced in a state of health and perfection by weeding out all that are imperfect; or they argue that, so long as the "cause of variation" is unknown, the power that preserves those variations when they have arisen plays a very subordinate part. These last writers maintain that the causes, whatever they are, which produce certain variations in certain species at certain times, are the true and only causes of the origin of species.

Now all these objections, in so far as they refer to the origin of the different *species* of one *genus* from a common ancestral species, or even of all the *species* and *genera* of one *family* from some still more remote ancestor, may, I think, be shown to be invalid; because we have direct evidence, almost amounting to demonstration, that changes to this extent are producible by the known laws of variation and the admitted action of natural selection. But when we go further back, and propose to account for the origin of distinct *families*, *orders*, and *classes* of animals by the same process, the evidence becomes far less clear and decisive. We find groups with organs of which no rudiment exists in other groups; we find classes differing radically in structure from other classes; and we have no direct evidence that changes of this nature are now in process, as we have that the lesser changes resulting in new *species* and new *genera* are in progress.

Yet the evidence that these deeper and more important changes in the structure of organised beings have taken place by gradual steps through the ordinary processes of generation is overwhelming. The numerous intermediate links that have been discovered both among living and extinct animals, and especially the wonderful community perceptible in the embryological development of the most diverse living types, force upon us the conclusion that the entire animal and vegetable kingdoms owe the wonderfully diversified forms they now exhibit to one unbroken process of "descent with modification" from a few primeval types. It is indeed generally assumed that if we go so far, we must admit *one* original type of living organism; but this does not seem necessary. By means of whatever laws we suppose living things first to have originated, why should not the primeval germs have appeared many times over, and in forms determined or modified by the infinitely varied chemical and physical conditions to be found in the crust of the earth? The identity of ultimate structure and wonderful similarities of development of all organisms may be due to the unity of the laws by which organic life was first produced; the diversity of the great types of animal and vegetable forms may be due to the operation of those laws at different places, acting on different combinations of elements, which are subject to unlike physical conditions.

The point here insisted upon is, that the origin of all organisms, living and extinct, by "descent with modification," is not necessarily the same thing, and is not included in, "the origin of species by means of natural selection." The latter we not only know has occurred, but we can follow the process step by step by means of known facts and known laws; the former, we are almost equally certain, has occurred, but we cannot trace its steps, and there may have been facts and laws involved of which we have no certain knowledge. The terms "laws of growth," "laws of development," "laws of inheritance," "laws of variation," "laws of correlation," "direct action of the environment," "laws of habit and instinct," with some others, are used to express the action of causes of which we are almost wholly ignorant, as we are of the nature of life itself. Now Mr. Darwin has himself admitted that there are these unknown causes at work, and that "natural selection is the most important but not the exclusive means of modification." There may be some question as to the term "most important," if, as is not improbable, the most radical differences in animals and their most important organs could not have been produced by it alone in the same way as the specific modifications of a genus or family may be produced. This, however, is a fair matter for discussion and research, and will probably continue to be so for many generations; and even if it should be ever proved that higher laws than "natural selection" have brought about the more fundamental divergences of the animal and vegetable kingdoms, this will not be held to detract in any way from the greatness and the value of Mr. Darwin's work, any more than it will be held to detract from the greatness of Newton, if it should some day be demonstrated that the law of gravitation as expressed by him is not absolutely true, but that (as some physicists now suppose) it should be found to be subject to a higher law for remote stellar distances.

No thoughtful person can contemplate without amazement the phenomena presented by the development of animals. We see the most diverse forms—a mollusc, a frog, and a mammal—arising from apparently identical primitive cells, and progressing for a time by very similar initial changes, but thereafter each pursuing its highly complex and often circuitous course of development, with unerring certainty, by means of laws and forces of which we are totally ignorant. It is surely a not improbable supposition that the unknown power which determines and regulates this marvellous process may also determine the initiation of those more important changes of structure and those developments of new parts and organs which characterise the successive stages of the evolution of animal forms. In so far as Mr. Darwin denies the necessity of any such power, and maintains that the origin of all the diverse forms and types and all the complex structures of the organic world are due to identically the same laws and processes as are adequate to produce the different species of *Rubus* or of *Canis*, from some ancestral bramble or dog respectively, his opponents have undoubtedly a case well worthy of being argued out in the courts of science. They should, however, remember that no final judgment has been given or can be given while the evidence on both sides is not only circumstantial but imperfect and contradictory; and it would be well not to declare too confidently that Mr. Darwin's theory has hopelessly broken down, since a majority both of naturalists and geologists, whose evidence as experts will undoubtedly have great weight with the educated public, are at present altogether in his favour . . .

* * * * *

Once one is alerted to the way Wallace distinguished between evolutionary 'origins' and 'destinations', the directions and variety of his work after 1858 fall neatly into place. His natural selection studies deal entirely with proximate causes. There was, simply, no need to relate them to the matter of final causes (at least so far as an understanding of the immediate workings of the process of diversification was concerned). The same is true of all his studies in physical geography, geology and biogeography. His manner of treating problems pertaining to the future *evolution of the world, however, was quite different. Especially as related to man, this subject could only be addressed by taking into consideration the expanding "progression" involved. If man were to evolve (i.e., spiritually, as Wallace believed it was along these lines that evolution would primarily continue) in the right direction, he would have to come to recognize what was important, and to strive for it. Therein, the teachings of spiritualism entered, upholding Wallace's various recommendations for social change.*

In sum, it seems both unrealistic and unproductive to try to restrict Wallace's 'theory of evolution' to natural selection and its applications. Most of the content of the other sections of this volume concerns the way Wallace put his 'evolutionary theory' to work. His world-view in many respects approximates to the more recently conceived 'Gaia hypothesis', the idea that the earth has evolved in a synergistic manner, continually 'preparing itself' for the production of more and more complex beings. But Wallace's view is both more highly structured and more comprehensive than the Gaia hypothesis, to this degree more closely resembling the evolutionary model projected through theosophy (interestingly, however, Wallace ended up feeling that the teachings of theosophy themselves were 'too inaccessible').

We are now in a position to survey Wallace's contributions to the theory of biological change—taken in its purely 'material' sense, especially as interpreted through natural selection. His writings on this subject are eclipsed only by Darwin's, and it was Wallace and Thomas Huxley who figured most prominently in popularizing the idea and defending it—and Darwin—against both real and imagined criticism.

It is now seldom remembered that Wallace was the leading figure in the development of the theory of colour as related to evolutionary change. One early dividend of this work was his introduction of the concept of 'polymorphism' as part of a long presentation to the Linnean Society in 1864 (ref 96). Wallace later composed an article-length abstract of the paper for publication in Reader; *from this source is excerpted the following summary of the polymorphism discussion:*

. . . No less than 120 species of the family [Papilionidae] inhabit the Malay Archipelago, distributed throughout the islands in varying proportions, Borneo containing the largest number (twenty-nine species), while many of the smaller islands have only from six to ten species. A careful study of large series of these insects, collected in about thirty islands from Sumatra to New Guinea, has brought to light a number of curious facts in variation and geographical distribution which it was the chief object of the paper to illustrate. The author considered that variation was by no means one simple fact, but included distinct phenomena which have been very often confounded. He would class the phenomena of variation under the heads of—1st, Simple Variability; 2nd, Dimorphism, or Polymorphism; 3rd, Local Forms; 4th, Co-existing Varieties; 5th, Races, or Sub-species; and, 6th, True Species.

Simple variability includes all cases of great instability of specific form, or those in which the offspring differ more or less from the parents, but differ irregularly, and, as it were, accidentally. In such cases, so great is the tendency to vary that it is difficult to find two individuals exactly alike. One of the Malayan species (*Papilio severus*) exhibits this in a remarkable degree, and several others less strikingly. It may be called irregular variability, and is of the same nature as that so characteristic of domestic breeds.

Polymorphism, or dimorphism, differs from simple variability in this—that the offspring differ from the parents in a considerable degree, and in a manner more or less constant and regular; so that, of the offspring of a single pair, some will resemble their parents, while others will differ from them; but the differences will be tolerably fixed and definite, and intermediate varieties will never occur. A large and handsome Malayan butterfly, *Papilio memnon*, is a good case of dimorphism. The male is nearly uniform bluish black, with rounded hind wings, and never varies. One portion of the females resemble the males in shape, but are coloured brown or ashy, and with more or less white markings on the hind wings. Another set of females are found, however, which differ remarkably in the shape of the wings, the under ones being lengthened behind into a large spoon-shaped tail; and they have also white lines radiating from the base of the wing. Intermediates in form or colour between these two kinds of females never occur. These very distinct kinds of female insect do not produce young like themselves only; for, from eggs laid by one of them, both kinds of butterflies are produced, as well as the male, which is different from either. It is just as if a blue-eyed flaxen-haired white man had two wives, one a straight-haired red-skinned Indian squaw, the other a woolly-headed coal-black negress; and all the boys from either mother were real white boys just like their father . . . but the Indian mother should sometimes have negro daughters, and the negro mother Indian daughters. Such a thing seems absurd and contrary to nature; yet this is exactly what takes place in the butterfly called *Papilio memnon*—the males are always exactly like the male parent, while the female offspring of each mother are partly like and partly unlike herself.

The butterfly called *Papilio pammon*, inhabiting all parts of India, is another case. It is nearly black, with a band of white spots along the margin of the fore wings and across the middle of the hind wings. Some of the females are exactly like the male, having only a small additional red spot behind. The most abundant females are, however, quite different, having a large white and brick-red patch on the hind wings and a row of red spots. This was long supposed to be a different species named *Papilio polytes*; but, from the eggs laid by it, *Papilio pammon* was produced as well as others resembling itself; and a further proof is that no male of *P. polytes* has ever yet been found, although the species is very common. There is, however, in India yet another butterfly, named *Papilio romulus*, of which no male has ever been found, although female specimens exist in every collection; and, from a careful examination of this insect, the author was convinced that it was really a third form of the female of *Papilio pammon*. To parallel this case, we must suppose that our white man has a third wife, white like himself, and that the children of both white, black, and red wives, if boys, are alike white; but, if girls, may be white, black, or red, and often without any regard to the colour of their mothers. Further east, in the Moluccas and New Guinea, there is a butterfly named *Papilio ormenus*, which has also three kinds of females, but in this case all are different from the male; just as if our supposed polygamist had an oblique-eyed yellow Chinese wife in addition to his black and red partners. The more exact parallel would be, however, to suppose an island inhabited only by white men, with black, red, and yellow women, and that, after many generations had passed away, the men all remained pure white, and the women of their respective colours and races equally pure. This would exactly represent the phenomena of *polymorphism*, which is the ordinary course of nature in several species of insects in various parts of the world, and will sufficiently show that it is quite a different thing from simple variation, with which it has hitherto been confounded.

There are, however, other species which show how this state of things may have been brought about. One inhabiting the Philippine Islands (*Papilio alphenor*) has females which vary very much while the males are very different and constant. Now Mr. Darwin has shown that very small differences in external appearance are often accompanied by a very important difference in constitution, so that one variety, being better adapted to surrounding circumstances than another, may increase and flourish, while its companion diminishes, and ultimately dies out. But, in the case of an insect producing a very variable female, it may come to pass that the extreme varieties will each possess some qualities which render them better adapted to exist than the intermediate forms, which will therefore die out and leave only the extremes. The difficulty remains, however, of accounting for the fact that these intermediate varieties are not continually reproduced; and this can only be explained by the tendency of natural selection being to preserve continually more and more of the offspring of those individuals which produce only one or both of the favourable forms, and few or none of the unfavourable ones; and also from the fact that among the offspring of those which produce the largest proportion of the extreme favourable varieties will be found those most perfect and highly-developed individuals which, when adverse circumstances threaten the extermination of the whole race, will alone maintain their existence and transmit to the whole succeeding race the physiological as well as external peculiarities of their parents.

The cases of different broods of the same species having very distinct characters, as in many European butterflies, was mentioned as a somewhat different though allied phenomenon, which may be called seasonal or alternate dimorphism. Ants exhibit another case in three or four differently-formed sets of individuals being produced from the same parents; but here each form was of use to the whole community, and Mr. Darwin has shown how they could have been produced by "natural selection;" this might be called structural or economical dimorphism. Yet another is that shown to exist in many plants (as *Linum* and *Primula*) by Mr. Darwin, where the reproductive organs are modified so that each form bears a definite functional relation to the other. This may be called sexual or reproductive dimorphism. All these different cases, however, have this character in common, that the union of the distinct forms does not produce intermediate varieties, but reproduces the two or more parent forms altogether pure; whereas, when simple varieties or races intermingle, the offspring seldom resemble either parent exactly, but are more or less intermediate between them . . .

* * * * *

Wallace's work on protective coloration and mimicry was his most significant contribution to the study of colour in nature. He practically invented the first subject single-handedly, originating the theories of 'warning colours,' 'disruptive colours,' 'alluring colours,' 'recognition colours' (and marks), and 'deflexion colours.' In 1879 Wallace prepared an essay on protective coloration for inclusion in the collection Science for all:

The protective colours of animals (ref 318)

. . . It is an almost universal rule that each animal either has enemies which seek to feed upon it, or that it seeks itself to feed upon other animals. In the first case, it has to escape its enemies or it cannot long continue to live. This it does either by its swiftness of flight, by its watchfulness, or by hiding itself from view. Some species come abroad only at night, some burrow under ground, many hide themselves among leaves, or bark, or stones, and thus escape destruction. Their enemies, however are as swift and as watchful as they are themselves, and they can in most cases only escape them by avoiding observation. To do this, they must not be too conspicuous; and thus any kind of colouring that renders them hardly visible while seeking their food or attending to their young, actually tends to preserve their lives, and often alone enables them to secure the safety of their offspring. But the enemy who is in pursuit of them is in just the same predicament. He, too, must be concealed by his colour, or he will be seen afar off and his prey will seek a secure concealment. In that case he will simply starve to death, and his race will cease to exist. It thus appears that almost every kind of animal requires concealment; and it might therefore be thought that colour must always be injurious, and ought never to exist. And as colour not only exists, but abounds among the various classes of animals, it may be thought that we have a *reductio ad absurdum*, and that protective colouring cannot be of much importance. Further examination, however, shows us that even gay colours are very often protective, because the earth and the sky, the leaves and the flowers, themselves glow with pure and vivid hues. In other cases conspicuous colouring is useful to an animal, as when it is protected by the possession of a deadly sting or a nauseous taste, and the bright or unusual colour warns its would-be enemies to avoid it. There are also a great number of animals who appear to be sufficiently able to take care of themselves without resorting to concealment, and with these the tendency to the production of colour, which seems to be inherent in organic beings, exhibits itself unchecked. Taking all these facts into consideration, we find that there is an ample field for the development of bright and conspicuous colour on the one hand, and for the display of an infinite variety of protective tints on the other, dependent on the structure, the habits, and the instincts of the different kinds of animals.

Let us now consider a few familiar examples of protective colouring. Owing to the mildness of the winter of 1877, and the dampness of the following spring, my garden was overrun with slugs, and I had to wage continual war against them. On every damp evening I would go round the borders, examining the choicest plants, and, taking the slugs off with a knife, deposit them in a jar of strong brine. While doing this, many of them, on being touched, would contract and drop to the ground, and though they fell close under my eyes, I often had some trouble to find them again, owing to their close resemblance to the small pebbles with which the soil abounded . . . It may be said that this was only an accidental resemblance, and at first it did not strike me as being anything else; but [eventually] . . . the conviction forced itself upon me that here was a case of true protection, and that what deceived me would also probably sometimes deceive the birds and other animals that feed upon slugs. In the tropical forests I had often . . . to resort to the sense of touch to supplement that of sight, in distinguishing between the Phasmidae or "stick insects" and real pieces of stick; and as in this case it is universally admitted that the resemblance is a protection to the insects, since it saves them from the attacks of the numerous tropical insectivorous birds, we may well believe that our familiar slugs are similarly protected . . .

We will now consider some other cases of protection by colour among animals of our own country, before proceeding to those more wonderful developments which occur chiefly in tropical lands. Every collector of beetles must have observed how many of our Curculionidae or weevils are brown or speckled, and also that they have the habit, on being touched or alarmed, of falling down on the ground, drawing in their legs and antennae, and there becoming undistinguishable from small lumps of earth or stones. Others, however, which are found constantly on nettles and herbage, are beautifully green, and these usually run or fly away when alarmed. A curious little beetle, *Onthophilus sulcatus*, is brown and furrowed, so as exactly to resemble the seed of some umbelliferous plant . . . The Lappet-moth (*Gastropacha querci*), when at rest, so disposes its rich brown wings as to resemble, both in shape and colour, a dead leaf; while the Buff-tip moth (*Pygaera bucephala*) so contracts its wings that it looks exactly like a thick piece of broken stick, the yellow patch at the extremity of the wings giving the appearance of the freshly-broken end. This is a case which well illustrates how impossible it is to decide from the appearance of a specimen in a cabinet whether the colours of an animal are or are not protective, for no one would imagine that this handsome and conspicuously-coloured moth could ever deceptively resemble a bit of dead stick, and so obtain protection from its enemies. It is a very common thing in the tropics to find beetles and moths which resemble bird's droppings, and the same occurs in this country . . . It has also been noticed that the general tints of the moths which are on the wing in autumn and winter correspond to the prevailing hues of nature at those seasons. The Rev. Joseph Greene states that the great majority of the autumnal moths are of various shades of yellow and brown, like those of the autumnal foliage; while the winter moths of the genera *Cheimatobia* and *Hybernia* are of grey and silvery tints.

It is among the caterpillars, however, that protective colouring is the most general and conspicuous. An immense number of these creatures are green, corresponding with the tints of the leaves on which they feed, or brown when they rest on bark or twigs; while a large number of the larvae of the Geometridae or Loopers have the habit of sticking themselves out rigidly like sticks, which they exactly resemble in shape as well as in colour. Every one knows, however, that there are a number of very brightly-coloured caterpillars, and it may be asked how these are protected, or why the others need protection if these can do without it. The answer to this question is most instructive, and affords the most conclusive proof that various examples of protective tints in nature really have the effect we impute to them. It has been found by repeated observation and experiment that every green and brown caterpillar, without exception, is greedily eaten by birds, and even by frogs, lizards, and spiders, and that they endeavour to conceal themselves from these numerous enemies by feeding usually at night, while during the day they remain motionless upon leaves, twigs, or bark, of the same colour as themselves. The brightly-coloured caterpillars, on the other hand, were found to be universally rejected by birds when offered to them, and even by lizards, frogs, and spiders . . . The same rule was found to apply to all the hairy or spiny caterpillars; and, what is very interesting, the habits of these creatures are correspondingly different from those of the green and brown eatable species. They all feed during the day; they do not conceal themselves, but feed openly, as if courting observation, and secure in the knowledge of their safety from all enemies . . .

This connection of gay colours and bold habits with non-edibility, throws light on many other cases of bright colouring which might otherwise be adduced as opposed to the theory of protection. Thus, among our beetles we have such conspicuous creatures as the lady-birds (Coccinellidae) and the "soldiers and sailors" among the Malacoderms, which are all conspicuous and defenceless insects, never hiding themselves, or seeking concealment, or feigning death, as do so many other beetles. The reason is now found to be that, like gaudy caterpillars, they are generally unfit for food. The same explanation may be given of the conspicuous whiteness of certain moths. One of these, *Spilosoma menthrasti*, is very common, but when given by Mr. Stainton to a brood of young turkeys among hundreds of other worthless moths after a night's "sugaring," it was always rejected, each bird in succession picking it up and then throwing it down again, as if too nasty to eat. The same thing has been observed with the showy butterflies forming the family Danaidae. Insect-eating birds were observed by Mr. Belt in South America, catching butterflies which they brought to their nest to feed their young; yet during half an hour they never brought one of the Danaidae, which were flying lazily about in great numbers.

But there are other modes of protection, besides a nauseous taste which renders concealment unnecessary. Either weapons or armour have the same effect, if they are sufficiently perfect of their kind to render it useless or dangerous for their enemies to attack them. The best example of armed insects are the bees and wasps, and among these conspicuous colours are the rule, while they usually fly about and seek their food without any attempt at concealment. Other insects have so hard a covering, or such awkward spines, as to be practically uneatable, and among tropical insects many of these are conspicuously or gaudily coloured . . . Others, again, obtain protection by extreme rapidity of flight, and by concealing themselves in holes or among flowers when at rest, and these are often brilliantly coloured, as in the case of the common Rosechafer. These few examples are merely intended to show that it is no argument against the use of protective colours in some animals, that many others have brilliant and clearly non-protective hues. In those cases, the creatures have certainly some substitute which enables them to live and continue their race. What this substitute is we can in some cases find out, but in many others we are too ignorant of the habits and surroundings of the species to determine whether its peculiar colours are or are not protective, or, if they are not, to determine what are the peculiar conditions which enable it to dispense with this particular kind of safeguard . . .

Leaving now these familiar examples, to be found everywhere around us, let us cast a glance over a wider field, and see how the general conditions of existence, affecting many different groups of animals at once, influence their coloration for protective purposes. And first let us transport ourselves to the great deserts of the earth, and inquire what kind of animal life we find there. Canon Tristram has travelled much in the Sahara, and he thus describes the characteristic colours of its animal life:

> In the desert, where neither trees, brushwood, nor even undulations of the surface, afford the slightest protection against its foes, a modification of colour which shall assimilate an animal to that of the surrounding country is absolutely necessary. Hence, without exception, the upper plumage of every bird, whether lark, chat, sylvian, or sand-grouse, and also the fur of all the smaller mammals, and the skin of all the snakes and lizards, is of one uniform isabelline or sand colour.

This is not a characteristic of one desert, but of all. In a recent account of the Steppe of Erivan in Asia Minor, it is said that "a remarkable feature of the animal inhabitants of the Steppe, insects and reptiles, and especially of the lizards, is the most perfect coincidence of their colouring with the colouring of the Steppe." More prominent examples of this prevalent tint are such animals as the camel and the lion, which are exactly of the usual tints of sand and sandy rock.

Let us go now to the arctic regions, and we find these reddish-yellow tints entirely wanting, and instead of them pure white, or in a few cases dark-brown or black, where conspicuousness seems of more importance than concealment. All the bears of the globe are brown or black, except the polar bear, which is white. The polar hare, the snow-bunting, the snowy-owl and the jer-falcon, are also white or nearly so; while the arctic fox, the ermine, and the Alpine hare, change white in winter, as does our own Highland ptarmigan. This last bird is a fine example of protective colouring, for its summer plumage so exactly harmonises with the lichen-covered stones among which it delights to sit, that a person may walk through a flock of them without seeing a single bird; and when it changes to white in winter it is equally protected amid the snow which covers the mountains. A striking exception to the usual white covering of arctic animals is the Musk-sheep, or Musk-ox as it is often erroneously called. This animal is of a dark-brown colour, easily seen among the snow and the ice, but the reason of this is not difficult to explain. The Musk-sheep is gregarious, and derives its protection from this habit. A solitary strayed animal would soon become the prey of the polar bears or even of the arctic foxes; it is therefore of more importance that it should see its comrades at a distance, and so be able to rejoin them, than that it should be concealed from its few enemies. Another case is that of the sable, which retains its rich brown fur throughout the severity of a Siberian winter, but at that season it frequents trees, feeding on fruits and berries, and is so active that it catches birds among the branches. Again, the common raven is found in the extreme arctic regions, but is always black; and this is probably because it has no enemies, while, as it feeds on carrion, it does not need to be concealed from its prey. These three cases are exceedingly valuable from a theoretical point of view, for they prove the incorrectness of a common notion that animals may change to white in the arctic regions either from the direct effect of cold, or from some influence of the white reflections from the snow; and they teach us that only those animals become white to whom that colour is useful, while those which either do not require protection or to whom dark colours are actually beneficial, remain totally unaffected. The cause of change must therefore be sought, not in the direct action of external conditions, but in the same general laws of variation and selection which have modified all the other characters of animals in the way most beneficial to them.

Nocturnal animals offer equally good examples of protective colouring. Mice, rats, bats, and moles, are all of dusky or blackish hues, and are therefore very difficult to be seen at night; when alone they move about, while during the day they conceal themselves in holes or underground. When concealment by day as well as by night is required, as in the case of owls and goatsuckers, we find dusky mottled tints, assimilating with bark or earth during the day, and not very conspicuous at night. In some few cases nocturnal animals are conspicuous, a striking example of which is the North American skunk, which has much white about it and a large white tail which it carries erect in the most conspicuous manner possible. But the horrible odour emitted by this animal makes it universally dreaded, and its conspicuous tail is thus a signal-flag to all carnivorous animals not to attack it—a parallel case, in fact, to the white moth, which we have already seen was rejected by birds which eat so many other moths.

Equally striking as a proof that colour is largely protective is the fact, that nowhere but among the evergreen forests of the tropical and sub-tropical zones do we meet with birds the ground-colour of whose plumage is green. Parrots, which are confined to such countries, are generally green, with small patches of vivid colours. In the Eastern tropical islands many pigeons are as green as parrots, and there are numbers of other groups which are of the same colour. Such are the barbets, a family of fruit-eating birds, especially abundant in tropical Asia; the green bulbuls (Phyllornithidae); the Bee-eaters; the Turacos of tropical Africa, the little White-eyes (*Zosterops*) of the eastern tropics; and many other groups. These all frequent thick foliage, with which their colours so exactly harmonise that it is most difficult to detect them.

Contrast these with the ordinary colouring of the birds of the region of deciduous trees, of which our own country is a fair example. Here anything approaching a pure green is unknown, while brown or olive is the almost universal body-colour of the plumage. This is the tint which is least conspicuous among the leafless trees and bushes, which prevail for so large a part of the year, and when the need of protection is greatest.

Among reptiles these protective tints are very apparent. Our lizards and snakes are all more or less brown or olive tinged, while in the tropics alone they are often of a vivid green, exactly corresponding with the vegetation they dwell among. The curious geckos—flat lizards with dilated toes, which cling to the trunks of trees or to rocks—are often finely marbled with green and grey, so as exactly to resemble the lichen-covered surface on which they cling. Some arboreal snakes of the genus *Dipsas* are, however, nocturnal; and these, like all other nocturnal animals which require to be concealed, are of dusky colours, being of various shades of black, brown, and olive.

Many fishes even, present clear examples of protective colouring. Such as rest on the bottom, like the flounder, skate, sole, or Miller's Thumb, are invariably of the colour of the bottom, and often singularly speckled, so as to resemble sand or gravel. Such as swim near the surface of the water are almost always dark-bluish or greenish above, and white beneath, colours which evidently tend to their concealment from enemies in the air above them or in the water below. The brilliantly-coloured fishes from warm seas are many of them well concealed when surrounded by the brilliant sea-weeds, corals, sea-anenomes, and other marine animals, which make the sea-bottom sometimes resemble a fantastic flower-garden. The pipefish and sea-horses (*Hippocampus*) are excellent examples of this style of colouring. Some of them are greenish, resembling floating sea-weed; but in Australia there is a large species which is covered with curious leafy appendages, and all of a brilliant red colour, and this lives among red sea-weed, and is then perfectly concealed.

It is, however, among tropical insects that the most perfect and wonderful cases of protection by colour and marking are to be found, and a very few examples of these must now be given. The best known and most celebrated are the leaf-insects of the genus *Phyllium*—curious large insects, whose wings and wing-covers are broad and flat, shaped and veined exactly like leaves, while their legs, head, and thorax have all flat dilatations, like the stipules of many plants; and the whole being of the exact green tint of the foliage of the plant they live on, it is actually impossible to detect them when they are not in motion. The walking-stick insects, or spectres, are equally curious. These are long cylindrical insects, often nearly a foot long, and of the exact colour of pieces of greenish or brown sticks. If they have wings, these fold up closely, and are concealed under wing-covers of the same stick-like appearance; while the head and legs are so shaped and jointed as either to fit closely on to the stick-like body, or to appear like branched twigs. These creatures hang about shrubs in the forests, and can seldom be distinguished from small twigs and branches which have fallen from the trees overhead. They remain quite motionless during the day, and feed at night, and they hang anyhow across the foliage, holding on by two or three of their legs only, and they thus give themselves that unsymmetrical appearance which belongs to accidentally-broken twigs. A few of the species are still further protected by curious green, leafy excrescences all over the body, so as to look exactly like a piece of dead twig overgrown with a delicate moss . . .

Even among butterflies, whose gay colours seem only adapted to render them conspicuous, there are equally wonderful examples of protective marking. It was first pointed out by Mr. T. W. Wood . . . that our beautiful little Orange-tip butterfly (*Anthocharis cardamines*), although so conspicuous when on the wing, is perfectly concealed when resting in the evening in its favourite position among the flower-heads of the wood parsley (*Anthriscus sylvestris*). Its under surface is beautifully mottled with white and green, which strikingly assimilate with the white and green flower-heads of this plant. Much more wonderful, however, and perhaps the most wonderful of all imitative insects, is the leaf-butterfly of India (*Kallima inachis*). This is a rather large and handsome butterfly, of a deep bluish colour, with a broad orange band across the wings. It is thus sufficiently conspicuous; but it flies very quickly, and in a zigzag manner, so as to be caught with great difficulty. It is when at rest that it requires protection, and this it obtains by its colour and markings on the under surface, and by its peculiar habits. The upper wings have an acute lengthened apex, which is exactly the shape of the tip of the leaf of many tropical trees and shrubs; while the hind wings are produced into a short narrow tail, which well represents the stalk of a leaf. Between these points runs a dark curved line, representing the mid-rib, and from this radiate a few oblique markings for the veins of the leaf. The colour of the under side of the wings closely imitates that of dead leaves, but it varies almost infinitely through shades of bright yellow, reddish, ochre, brown, and ashy, just as leaves vary in their different stages of drying and decay. Even more remarkable is the manner in which the diseases and decay of leaves are represented by powdered dots and blotches, often gathered into little groups, so as to imitate in a most marvellous way the various fungi which attack decaying leaves. But to render the disguise effective, it is necessary that the insect should assume the position of a leaf, and this it does most perfectly. It always settles on an upright twig or branch, holding on by its fore legs, while its body (concealed between the margins of the wings) rests against the stem which the extremity of the tail, representing the stalk, just touches. The head and antennae are concealed between the front margins of the wings, and thus nothing is seen at a little distance but what appears to be a dead leaf still attached to the branch . . .

In this case, and in that of the moss-covered stick-insect, we see the extreme perfection of imitative colouring; and we can only understand how this has been produced, by always keeping in mind the very much more numerous cases of slight or partial protection by colour or marking. We can only now briefly indicate some of the steps by which such protection is brought about.

None of the characters of animals are more variable than their colours, though this may appear doubtful when we look at the constant tints and markings of so many animals in a state of nature. There is, however, good reason to believe that even, in cases, these variations are constantly occurring, but, owing to the fact that the tint of each animal is useful to the species, all important deviations from it soon die out. Certain it is that almost every domesticated animal varies in colour, and these varieties, not being hurtful as in a state of nature, are increased and multiplied without end. Now, if we suppose an animal to suffer from being too conspicuous, any variation of colour or marking tending to make it less conspicuous will give it a better chance of life; and as offspring tend to be like their parents, these less conspicuous varieties will often leave successors similarly endowed; but these again varying, some among them will be still more protected; and thus the protective tints will tend to become more and more perfect in each succeeding generation, till their enemies, finding the pursuit too difficult, will confine their attention chiefly to other species. Then there will be no more change till some new enemy appears, when a further advance may take place till the protection becomes sufficiently perfect to place our supposed animal in a slightly better position than its neighbours.

It has been a difficulty to many persons to understand how such variations could explain the curious cases of the Alpine hare, the ptarmigan, and many other animals which become white in winter only, when the ground is covered with snow and that colour serves as a protection. It has, however, been observed, that a slight seasonal change takes place in many animals. Thus, in Siberia, the wolf, the horse, the cow, the roe, elk, reindeer, and two kinds of antelope, all become paler in colour during the winter. Now, if either of these species migrated northward, till it came to inhabit a country where the winter snow remained on the ground for half the year, varieties in which the seasonal change was more and more pronounced would have an advantage, and thus, in the course of many generations, an animal might be produced which changed colour as completely as do the arctic fox or the ptarmigan.

We must now conclude this very brief outline of one of the most curious chapters in natural history. We have shown how varied and how widespread are protective colours among animals; and, if we add to these the cases in which conspicuous colours are useful, sometimes to warn enemies from such as are distasteful or are possessed of dangerous weapons, at other times to aid wandering species to recognise their companions or to find their mates, we shall become satisfied that we have a clue to much of the varied coloration and singular markings throughout the animal kingdom, which at first sight seem to have no purpose but variety and beauty.

* * * * *

Though his friend and colleague Bates actually originated the theory of mimicry, Wallace wrote a good deal on the phenomenon and emerged as an equally important figure in its early study. In the same abstract of the paper on Malayan Papilionidae (ref 96) containing his discussion of polymorphism, Wallace explained mimicry as follows:

. . . a number of the Malayan and Indian *Papilios* resemble very closely species of Danaidae inhabiting the same districts. These Danaidae are almost a nuisance to the collecting entomologist from their abundance and ubiquity. Every garden, every roadside, the suburbs of every village are full of them—indicating very clearly that their existence is an easy one, and that they are free from persecution by the foes that keep down the population of less favoured races. Their strong and peculiar odour is believed to be the cause of their safety, and they are for this reason habitually passed over by insectivorous creatures. When, therefore, another insect which has not this peculiar scent resembles one of the Danaidae of the same district, it also is passed over by mistake, and thus gains an advantage. When this resemblance is general and slight, the benefit will be correspondingly small. Such a slight general resemblance may, however, evidently occur accidentally between very differently constructed animals; and what is important to observe is, that when once this happens the resemblance and the corresponding advantage will necessarily go on increasing by the action of "natural selection."

For example, let a *Papilio* resemble a *Danais* so slightly that it is only mistaken for it by very unobservant birds, etc., or at a considerable distance. Even this will be some advantage to it, for many individuals that would otherwise have been devoured will now live and leave offspring. But every animal varies more or less, sooner or later. Among the varieties of this *Papilio* some must be more like, some less like the *Danais*. The former will escape persecution more than the latter—will increase, therefore, while the latter will diminish. In each succeeding generation this preservation of the more like and the destruction of the less like will go on, which must slowly, but surely, produce a gradually increasing likeness, till the one insect can hardly be distinguished from the other.

Such cases occur everywhere. Some of the more remarkable have been pointed out by Mr. Bates as occurring in South America, where every streak and spot, as well as the exact outline, has been copied by insects of a quite different structure. In tropical Africa cases quite as extraordinary occur, two or more insects of widely different structure coming outwardly to resemble a third equally different from both; and, in every case whether there are two or three, one of them is a *Danais*, or belongs to one or two other groups which have also some special protection. In every case, too, the insects which resemble each other inhabit the same country, and, in cases where their habits are known, are found always to frequent the same places. This takes the phenomenon altogether out of the region of accidental resemblance, for it is evident that the chances of one insect resembling another are much greater when we compare those of all parts of the world than when we confine ourselves to those inhabiting the same country, the same locality, and even the same station. Yet cases of this very close resemblance between species of distinct families and genera are very rare in the former case—just as rare, in fact, as accidental resemblances should be; whereas, in the latter case, where, according to the law of chances, they should be infinitely more rare, they are so frequently to be met with in every part of the world that they may almost be called common.

There must be, therefore, in the cases pointed out by Mr. Bates and Mr. Wallace, a true relation between the species so resembling each other; and, if the explanation given by them be objected to, it should be shown that it is inadequate to account for the observed facts. Instead of this, however, the objection is made that there ought to be more cases of mimicry if this explanation is true; that almost all Pieridae should imitate *Heliconias*, and almost all *Papilios* should have come to resemble Danaidae. To this the answer appears sufficient, that it is only when, among the countless variations of insects, a remote resemblance accidentally occurs between the distinct groups, that "natural selection" can begin to act to increase that resemblance for the creature's benefit. The remaining species of the group will be preserved in a different manner. Some will acquire increased powers of flight, others facilities for concealment, others, again, increased reproductive powers, which will enable them to resist the persecution of their enemies. And it will depend, at some critical period of the existence of the species, on the proportionate amount and rapidity of variation in these different directions to determine the special means by which its perpetuation shall be secured. The comparative isolation of cases of mimicry is therefore no argument either against its serving as a means of protection, or against its having been produced in the manner here indicated. No less than fourteen Indian and Malayan *Papilios* mimic species of distinct groups, in several cases so closely that they have been placed by entomologists in the same species, although really having no close affinity; and the writer was constantly deceived by them on the wing. We cannot wonder, therefore, that birds and insects also confound them . . .

* * * * *

The following letter, printed in the Athenaeum *of 1 December 1866, was another of Wallace's early treatments of mimicry:*

Natural selection (ref 123)

At the last meeting of the Entomological Society, the subject of "Mimicry," or protective resemblances, was brought forward by Prof. Westwood and myself, and during the discussion that ensued some objections were made to the explanation of the phenomena first given by Mr. Bates, and adopted by me as the only sufficient one. Time did not permit of an answer being made to those objections, and as they are at first sight very plausible, and were brought forward not by one of Mr. Darwin's opponents, but by a gentleman who fully admits the great principles of evolution and development in organic nature, they may probably have weight with some persons. Believing, however, that they are entirely unsound, may I beg a little of your space to give my reasons for rejecting them.

Mr. [David] Sharp stated that four different causes might be sufficient to produce the phenomena of "mimicry" more or less completely, viz., first, accidental resemblances; secondly, similar conditions of life; thirdly, heredity, or reversion to a common ancestral type; and, fourthly, the perservation of useful variations.

To the first, or accidental resemblances, it was admitted that very few, if any, of the cases adduced by Mr. Bates or myself could be due. The last is the one we adopt. The second and third only remain, and these, Mr. Sharp argued, would account for most, or perhaps all, of our cases without the agency of natural selection at all. Now, all I can admit is, that in some cases of very closely allied species of the same or of closely allied genera, an accurate external resemblance, such as we term "mimicry," might possibly be produced either by "heredity," or by the action of like conditions. But in all the cases in which the insects resembling each other belong to distinct *orders*, or distinct *families*, or to *genera* not intimately allied, or even to well-marked *sections* of extensive genera, I entirely deny that either or both of these causes could have produced the whole series of phenomena presented by *mimicking* insects, and for the following reasons, which appear to me sufficiently conclusive:—

1. In all cases of mimicry, the resemblance of the one species to another in a different group is entirely superficial, and is always strictly confined to those characters which cause the one to *look like the other.* The structure, the habits, the form of inconspicuous parts, the colour of inconspicuous parts, the nature of the food, or the character of the larva and pupa, are not, as far as we know, ever modified in a similar manner. But if such general causes as "heredity" or "similar conditions" produced resemblances, these resemblances should affect various parts of the organization, not those conspicuous to the eye only. The effect being limited with strict reference to external resemblance, seems to me a fatal objection to referring it to any cause or causes of a general nature.

2. There are no grounds for believing that minute details of colouration and marking are due to climatal conditions at all, still less that they can be produced so identically alike in species of groups widely differing in organization; neither is there any evidence that such details are ever continued by heredity to one species only in each of two distinct family groups which contain hundreds of other variously-coloured species.

3. It is only a very few groups of insects which are the subjects of imitation by many other groups. But "heredity" should affect nearly all groups not too remotely allied; and "common conditions" should affect all species inhabiting the same forests with some approach to an average frequency. The fact that there is no such miscellaneous character in the resemblances (the group of Danaioid butterflies being the mimicked in the great majority of cases) tells us plainly that no causes affecting all insects alike can be at the bottom of this curious phenomenon.

4. Protective resemblance to a species of a distinct order sometimes occurs, as in the curious *Orthopterous* insect adduced by Prof. Westwood, which had been always taken for a *Coleopterous* insect that inhabits the same country (*Tricondyla* sp.). Neither "heredity" nor "like conditions" can be called in here; yet the phenomenon is so similar to that of the mimicking butterflies, that the idea of a similar cause in both instances is irresistibly forced upon us.

5. Resemblances of the most perfect kind occur between insects and inanimate objects. Phasmidae imitate sticks, leaves, or moss most wonderfully. The larvae of Geometrae also imitate sticks. Thousands of tropical Coleoptera imitate bark (and it is always those that cling to bark); others that sit motionless on leaves cannot be distinguished from the dung of birds dropped on a leaf. These are most clearly *protective imitations*, and they can none of them possibly be produced by "heredity" or "similarity of conditions," but, if produced at all by natural causes, seem clearly due to the continued preservation of useful variations. The mimicry of other insects is equally protective, and there is every probability that both were produced in a similar manner.

6. This is rendered still more certain by the fact that in both classes of resemblance it is the female only that is most frequently protected, for reasons which I have already explained, but cannot now enter into. It is only the female "leaf-insect" that is so wonderfully like a leaf; and in many species of *Pieris* and

Diadema it is the females only that mimic *Heliconias* and *Euplaeas*. This fact alone renders it in the highest degree improbable that the two groups of phenomena should have been due to distinct causes, even if the preceding arguments had not shown us how impossible it was to explain any of the main features of "mimicry" by such causes as "heredity" or "the action of like external conditions."

For these reasons it appears to me indisputable that "natural selection," or the continued survival of variations useful to the possessor, is the only theory yet before us which is capable of explaining the whole of the facts presented by "mimicking insects."

* * * * *

Four years later, as part of ref 175, a response to an article by Alfred W. Bennett entitled 'The theory of natural selection from a mathematical point of view,' Wallace produced another analysis concerning the mimetic process:

. . . Mr. Bennett next discusses the phenomena of "mimicry," and proposes to show, by mathematical calculations, that the effects could not be produced by natural selection. But, at the very outset, he makes an important error, which seriously affects his subsequent reasonings; for he leads his readers to understand that there is only one completely mimicking species of *Leptalis*, while the majority are of the normal white-butterfly type. The fact is, however, that but few species of *Leptalis* retain the simple colouring of their allies the Pieridae, while the great majority are either coloured like the Heliconidae, or show a considerable amount of colour or marking in that direction. He is also apparently unaware that some Heliconidae (*Ithomia eurimediae*, for example) approximate in colour to the normal white and yellow species of *Leptalis*, and this renders it much less difficult to understand how a sufficient amount of variation in colour might occur at a first step, to produce a resemblance which, viewed at some considerable distance, would be deceptive, and therefore useful.

We next come to the demonstration by means of figures, and we here find still more serious errors. Mr. Bennett says, that supposing a *Leptalis* may vary in twenty different ways, one only being the direction required,

> the chance of any individual producing a descendant which will take its place in the succeeding generation varying in the required direction, is 1/20; the chance of this operation being repeated in the second generation is $1/20^2 = 1/400$; the chance of this occurring for ten successive generations is $1/20^{10}$, or about one in ten billions,

whence it is concluded that there are overwhelming chances against any progressive variation in the right direction ever taking place. But first, I do not admit the assumption that only one variation out of twenty would be in the right direction; when it is remembered how great is the variety of the Heliconidae, both in colour and marking. It seems more likely one-fourth or one-third at least would help to approximate to some of them, and thus be useful. Taking, however, Mr. Bennett's own figures, there are three great oversights in this one short sentence. The first, is, that each *Leptalis* produces, not one only, but perhaps twenty or fifty offspring; the second is, that the right variation has, by the hypothesis, a greater chance of surviving than the rest; and the third, that at each succeeding generation the influence of heredity becomes more and more powerful, causing the chance of the right variation being reproduced to become greater and greater.

Now with these three modifications the weight of the argument is entirely destroyed; for, allowing the *Leptalis* to produce only twenty offspring (a small number for a butterfly), the chances become even that one out of the twenty varies in the right direction. But nineteen out of the twenty, on the average, are soon killed off by the various causes that keep down the population of the species, and the chances are very much in favour of that one surviving which, by the hypothesis, has varied in the right direction. It is not pretended that this one would survive always, or even on the average, but in a large number of cases it would certainly do so; and taking Mr. Bennett's own estimate of a million individuals as the population of a rare species, we may fairly estimate that in a quarter, or say even in a tenth part of these, the surviving offspring would possess the favourable variation. But now a new factor enters into the problem, of which Mr. Bennett takes no account. Those that have already varied tend to leave offspring varying in the same direction as themselves; and as these will all have an advantage, the offspring of the one-tenth will increase at the expense of those of the nine-tenths; and this tendency being still more powerful in the third generation, with the additional advantage as the numbers increase of the chance of both parents being favourable varieties, we may fairly expect the favourable to have completely exterminated the unfavourable variations, and to have firmly established themselves as a well-marked race. The enormous possible rapidity of multiplication, enabling a pair of individuals to produce millions in a few generations; the survival of the fittest, giving to favourable variations—not their bare numerical chance, as Mr. Bennett supposes, but—a certainty in the long run of living at the expense of the rest; and the powerful influence of heredity, which actually increases the *tendency* to produce the favourable variations with each succeeding generation,—are three of the main foundation-stones of the theory of natural selection, yet all three are ignored in this attempted mathematical demonstration of its insufficiency.

There is one other point in the theory of the origin of "mimicry" that deserves notice. It is, that the modifications leading to it are much more easy to explain than those leading to new genera and families, because the changes effected are wholly superficial and are almost entirely confined to colour. Now colour is both more variable than any other character, and is less intimately correlated with structure, so that great changes of colour may rapidly occur without in any other way affecting the individual, as we see in almost all our domestic animals. Experiments in breeding show that very large spontaneous variations of colour are frequent in insects; and thus the number of steps to produce a required amount of change may be much fewer than in cases of structural modification, in which every other part of the organism has to be coordinated to work harmoniously with the modified organ.

I may here take the opportunity of denying that I have argued, as Mr. Bennett says I have, that "an infinitesimal and inappreciable distinction may make the difference of a slightly longer span of life being allowed to the butterfly to lay its eggs in safety;" and I cannot imagine how he could have imputed to me anything so absurd. What I have maintained is, that for natural selection to act, either in producing "mimicry," or structural changes, no large or special variations are required, because the usual amount of *variability* which occurs in every part of the organism is sufficient (*Contributions*, pp. 287–291). But so far from supposing this to be "infinitesimal" or "inappreciable," I show that it is so palpable and so readily appreciated by horticulturists and breeders as to have enabled them to produce all the wonderful variety in our domestic animals and cultivated plants. And every entomologist knows that similar variability exists in insects, and that the constantly occurring variations of colour are especially great . . .

* * * * *

In 1882, Wallace discussed new contributions by Fritz Müller to the theory of mimicry:

Dr. Fritz Müller on some difficult cases of mimicry (ref 353)

In his original explanation of the cause of mimicry, Mr. Bates referred to the occurrence of many cases in which species of different genera of Heliconidae resemble each other quite as closely as the mimicking Leptalides and Papilios resemble species of Ithomia and other Heliconoid butterflies. In these cases both the imitating and the imitated species are protected by distastefulness, and it was not therefore clear how the one could derive any benefit by resembling the other. Accordingly, Mr. Bates did not consider these to be true cases of mimicry, but to be due, either to identical parallel variations of externally similar form, or "to the similar adaptation of all to the same local, probably inorganic, conditions."

Examples of this close resemblance of species of different genera of protected groups have now become very numerous, and they often extend to three or more distinct genera, some species of which imitate each other in most parts of tropical America, each changing in a corresponding manner as we pass from one district to another.

In my Address to the Biological Section of the British Association at Glasgow, in 1876 [ref 257] . . . , I connected these cases with a number of others in which peculiarities of colour or of form appear together in several groups not closely allied, but always among those inhabiting the same locality and as frequently among unprotected (that is, eatable) as among protected groups of butterflies; and I concluded, generally, as Mr. Bates had done, that these curious phenomena were due to "unknown local causes."

Thus the matter rested, till, in 1879, Dr. Fritz Müller published in *Kosmos* a paper on 'Ituna and Thyridia; a remarkable case of Mimicry in Butterflies', and in 1881 a second paper, entitled 'Remarkable cases of acquired resemblance among Butterflies,' in which he gives a solution of the problem as really a case of mimicry. The first of these papers was translated by Mr. R. Meldola, and communicated to the Entomological Society of London in May, 1879, and the same gentleman has kindly furnished me with a translation of the second paper . . . which devotes much space to a criticism of my suggested "unknown local causes" as a sufficient explanation of the phenomena. I may at once say that I admit this criticism to be sound; and that Dr. F. Müller's theory appears to me to afford a clue (with some slight modifications) to most of the cases of close individual resemblance of not-nearly-related species of butterflies yet observed. I therefore wish to state, as briefly as possible, the exact nature of the explanation now afforded us, and this is the more necessary because Dr. Müller's theory did not receive much support when brought before the Entomological Society, nor did it then satisfy Mr. Bates, the discoverer of the true meaning and importance of the phenomena of mimicry as interpreted by the doctrine of Natural Selection.

The explanation depends on the assumption, that some at least, if not all, young insectivorous birds learn by experience that the Heliconoid butterflies are distasteful, and in so doing sacrifice a certain number of individuals of each distinct species. But if two species, both equally distasteful, closely resemble each other, then the number of individuals sacrificed is divided between them in the proportion of the square of their respective numbers; so that if one species (*a*) is twice as numerous as the other (*b*), then *b* will only lose one-fourth as many individuals as it would do if it were quite unlike *a*; and if it is only one tenth as numerous then it will benefit in the proportion of 100 to 1. It is an undoubted fact that the species of protected butterflies, like those of other groups, differ greatly in abundance of individuals, some being very rare while others are among the commonest of all butterflies. The proportion of 100 to 1, therefore, is far below the amount of benefit an uncommon species might derive by resembling a common one. The benefit to be derived is thus clear, if the protected species are subject to the danger of attacks by young birds before they learn that such species are uneatable. I agree with Dr. Müller that they are exposed to this danger; and when we consider the great number and variety of insectivorous birds in South America the danger must be considerable, and quite sufficient to render it important for a numerically weak species to reduce it to a minimum, although to a species abounding in individuals it may be of little importance. It has been suggested that young birds have an hereditary instinct enabling them to distinguish uneatable butterflies antecedent to experience; but this seems in the highest degree improbable. It has no doubt been shown by Mr. Darwin that monkeys in captivity have a dread of snakes, and Mr. Jenner Weir believes that birds have an instinctive knowledge of uneatable caterpillars. But even admitting that in these two cases there is an instinctive hereditary aversion, it does not follow that the same will occur with regard to protected butterflies. Snakes form one well-marked group, and it is not alleged that monkeys distinguish between poisonous and harmless snakes; and caterpillars can also be readily divided into the two classes of edible and inedible by their green or brown (protective) colours on the one hand, and their gaudy or conspicuous colouration or hairy bodies on the other. But the protected butterflies have no such general mark of inedibility. Their colours and forms vary greatly, and cannot as a group be readily differentiated from those of other butterflies; and it is not to be accepted without actual proof that a young bird knows instinctively every Heliconoid or Danaioid butterfly in its district, as well as the protected Papilios and moths, almost infinitely varied as they are in colour and marking, among the equally numerous and equally varied butterflies of other groups. It therefore seems clear to me that we have here a *vera causa* for the acquisition of true protective mimicry by the less abundant species of inedible butterflies.

There is however yet another cause which may have led to mimicry in these cases, and one which does not appear to have been discussed by Dr. Müller. The fact that the majority of butterflies are edible and are actually eaten by birds and other insectivorous creatures, while a considerable minority are distasteful and are thus protected, renders it pretty certain, *a priori*, that there exist many degrees of distastefulness. Certain species appear to be rejected by all insectivorous creatures, while some, though not eaten by birds, may be devoured by lizards, dragon-flies, or spiders. Some, too, may be eaten by some birds and rejected by others, and no ornithologist will think it strange or improbable that a trogon should have somewhat different tastes from a tyrant-shrike or a swallow. Again, in some species the distastefulness may extend to all the stages of egg, larva, pupa, and perfect insect, while in others it may be confined to one or more of these stages; or special dangers may exist for one species which are absent in the case of another. But it is evident, that, if these differences exist, it will be advantageous for the less protected to mimic the more completely protected species, and the fact of the affinity between the different genera, with perhaps some tendency to revert to a common style of colouration or marking, will afford facilities for the development of this class of mimicry even greater than occur in the case of the distinct and often remote families of completely unprotected butterflies. We need not, therefore, be surprised to find whole series of species of distinct genera of Heliconoid butterflies apparently mimicking each other; for such mimicry is antecedently probable on account of the greater need of protection of some of these species than others, arising either from some species being less distasteful to certain enemies, or less numerous, and therefore likely to suffer to a serious extent by the attacks of inexperienced birds. When these two conditions are combined, as they often would be, we have everything necessary for the production of mimicry.

The explanation now given, so far as it refers to the various degrees of protection, may be extended to explain those cases in which various groups of Nymphalidae or other families appear to mimic each other; such as Catagramma, Callithea, and Agrias in one series, and Apatura with Heterochroa in another. In my *Tropical Nature* (p. 257) I have remarked—"Here, again, neither genus is protected, and the similarity must be due to unknown local causes"; but this is more than we know, and I now think that some of these groups—perhaps Catagramma and Heterochroa—are partially protected, and the advantage of sharing in this partial protection has led species of altogether unprotected and much persecuted groups to gain some protection by mimicking them, whenever their general form, habits, and style of coloration offered a suitable groundwork for variation to act upon.

If these views are correct we shall have the satisfaction of knowing that all cases of mimicry are explicable by one general principle; and it seems strange to me now that I should not have seen how readily the principle is applicable to these abnormal cases. The merit of the discovery is however wholly due to Dr. Fritz Müller; and it is to be hoped that he will complete his work by obtaining, if possible, evidence of its correctness. The chief thing required is an experimental proof of various degrees of inedibility in butterflies, during the different stages of their life-history; and also some observations as to the comparative abundance of the species of protected butterflies which mimic each other. If to this can be added the proof that such groups as Catagramma, which seem to be the objects of mimicry, are partially protected by inedibility, the chief remaining difficulty in the application of the theory of natural selection to all known cases of protective imitation will have been cleared up.

*　*　*　*　*

In 1883, Wallace sent another communication to Nature *concerning Müller's work:*

Difficult cases of mimicry (ref 359)

I have received from Mr. Thos. Blakiston, of Tokio, Japan, a communication to the *Japan Mail* by himself and Prof. Alexander, commenting on my article in *Nature*, vol. xxvi, p. 86, and pointing out some errors as to the estimated advantage derived by the mimicking butterflies. On referring to my article, I find that I have, by an oversight, misstated the mathematical solution of the problem as given by Dr. Fritz Müller and confirmed by Mr. Meldola, and have thus given rise to some confusion to persons who have not the original article in the *Proceedings of the Entomological Society* to refer to. Your readers will remember that the question at issue was the advantage gained by a distasteful, and therefore protected, species of butterfly, which resembled another distasteful species, owing to a certain number being annually destroyed by young insectivorous birds in gaining experience of their distastefulness. Dr. Müller says:

> If both species are equally common, then both will derive the same benefit from their resemblance—each will save half the number of victims which it has to furnish to the inexperience of its foes. But if one species is commoner than the other, then the benefit is unequally divided, and the *proportional advantage* for each of the two species which arises from their resemblance is as *the square* of their relative numbers.

This is undoubtedly correct, but in my article I stated it in other words, and incorrectly . . .

[My original] statement is shown by Messrs. Blakiston and Alexander to be untrue; but as some of your readers may not quite see how, if so, Dr. Müller's statement can be correct, it will be well to give some illustrative cases. Using small and easy figures, let us first suppose one species to be twice as numerous as the other, *a* having 2000 and *b* 1000 individuals, while the number required to be sacrificed to the birds is 30. Then, if *b* were unlike *a* it would lose 30 out of 1000, but when they become so like each other as to be mistaken, they would lose only 30 between them, *a* losing 20, and *b* 10. Thus *b* would be 20 better off than before, and *a* only 10 better off; but the 20 gained by *b* is a gain on 1000, equal to a gain of 40 on 2000, or four times as much *in proportion* as the gain of *a*. In another case let us suppose *c* to consist of 10,000 individuals, *d* of 1000 only, and the number required to be sacrificed in order to teach the young birds to be 110 for each species. Then, when both became alike, they would lose 110 between them, *c* losing 100, *d* only 10. Thus *c* will gain only 10 on its total of 10,000, while *d* will gain 100 on its total of 1000, equal to 1000 on 10,000, or 100 times as much *proportional gain* as *c*. Thus, while the gain in actual numbers is inversely proportional to the numbers of the two species, the *proportional* gain of each is inversely as the *square* of the two numbers.

I am, however, not quite sure this way of estimating the *proportionate* gain has any bearing on the problem. When the numbers are very unequal, the species having the smaller number of individuals will presumably be less flourishing, and perhaps on the road to extinction. By coming to be mistaken for a flourishing species it will gain an amount of advantage which may long preserve it as a species; but the advantage will be measured solely by the fraction of *its own numbers* saved from destruction, not by the proportion this saving bears to that of the other species. I am inclined to think, therefore, that the benefit derived by a species resembling another more numerous in individuals is really in inverse proportion to their respective numbers, and that the proportion of the squares adduced by Dr. Müller, although it undoubtedly exists, has no bearing on the difficulty to be explained.

* * * * *

Other than the 'man' issue, the most significant disagreement between Wallace and Darwin over the proper application of the principle of natural selection occurred as Darwin worked out his theory of sexual selection. Darwin argued that sexual differences—especially in colouration and ornamentation—were largely a result of the operation of female choice. Wallace felt, however, that environmental agents of selection were more generally at work, and defended this notion in a paper presented at the British Association for the Advancement of Science meetings in 1867:

Birds' nests and plumage (ref 139)

Underlying the various structural peculiarities of birds' nests there is a functional difference, which enables us to divide them into two classes, the first comprising all those nests in which the sitting bird is either fully exposed to view, or but imperfectly concealed by the situation of the nest among foliage, etc.; the second, in which the nest is either completely roofed in, or is placed in a dark hole or cavity, so that the eggs and young, as well as the sitting bird, are effectually hidden. Our thrushes, warblers, and finches, as well as all raptorial birds and pigeons, and the chatterers and tanagers of the tropics, are examples of the former style of nest; while the kingfishers, which build underground; the parrots and woodpeckers, which build in holes of trees; the Icteridae of America, which build hanging covered nests; and our own wren, which makes a domed nest, are examples of the latter mode of nidification.

Turning now to the colours and markings of birds, we shall find that, from this point of view also, we can divide them into groups, which are probably functional, since the division depends, not upon the colour or mark of the whole species, but on the difference or identity of colour in the two sexes. The striking difference of colour in the sexes of many birds has attracted much attention, but has not hitherto been bought under any general law. In some groups, as the toucans for example, although the colours are very varied and very brilliant, they suffer no diminution in the female, which is always equally adorned with her mate. In the majority of birds this is not the case, but the female is almost always less brilliant than the male; and in many cases, as in the gorgeous chatterers and Paradise birds, the hen bird is so dull and dusky in colour that it can rarely be recognised as belonging to the same species.

Now, I believe I have discovered a causal connection between these two groups of facts, for with but very few exceptions I find that, whenever both sexes are of bright, gay, or conspicuous colours, the nest is of the second class, or such as to conceal the sitting bird; while, whenever there is a striking contrast of colours, the male being gay or conspicuous, the female dull and obscure, the nest is open, and the sitting bird fully exposed to view. The following are the chief facts on which this generalization is founded:—The entire families of kingfishers, motmots, and Bucconinae, or American barbets, build their nests in holes deep in the earth. They are almost all gaily coloured or at least conspicuous birds, and in every case the female is either of exactly the same colour and markings as the male, or when different is scarcely less conspicuous. The Capitonidae, the toucans, the plantaineaters, the hoopoes, the trogons, the woodpeckers, and the parrots, all build their nests in holes of trees, and in all these birds (comprising 500 or 600 species), which are almost universally gay-coloured or conspicuous, the females are either coloured exactly like the males, or, when different, are little less conspicuous. The Eurylaemidae, another group of strikingly handsome eastern birds, make a covered nest, and there is no distinction of sexes whatever. The beautiful little group of Paridae or tits differ from most of their allies in always building a nest in which the female is concealed, for while some of the species build a roofed nest, others place it in a hole. The genera *Sitta*, *Sittella*, and *Climacteris* either build their nests in holes, or conceal them carefully among twigs and dense foliage. They are all pretty and conspicuous birds, and the males are always either coloured like the females, or even sometimes more gaily. The fine metallic starlings of the East (*Eulabes* and *Calornis*) have the sexes exactly alike, and the former builds its nest in holes, while the latter forms a roofed hanging nest. The ground-cuckoos (*Centropus*) are showy birds, which build a domed nest, and the females are not different from the males. The Icteridae, or hangnests of America, are almost all showy birds, generally adorned with bright red or yellow patches on their glossy black plumage. They all build roofed or purse-shaped nests, generally pensile and in all the females are as gay as their mates.

These facts strikingly contrast with the number of birds of showy plumage which build open nests, for in almost every case the female birds are totally devoid of the gay plumage of their partners. The superb blue and crimson chatterers, the exquisitely varied tanagers, the gay little manakins, with their orange, blue, scarlet, or white heads, have all females of the dullest brown, or olive-green tints, generally without a vestige of colour over their whole bodies. Hundreds of species of tropical flycatchers, shrikes, thrushes, and warblers are exceedingly handsome birds, adorned with the most brilliant hues, but they build open nests, and in every case the females are much less conspicuous than the males, being especially deficient in those bright patches of colour on the head and upper parts which would render them visible while sitting on their eggs, and in a large proportion of instances being of uniform earthy brown or olive-green tints best adapted for concealment.

We learn from these curious facts that, in harmony with innumerable facts of a similar nature among insects, the female bird being exposed to danger during the important epoch of incubation, and it being of the highest necessity for the preservation of the species that she should be protected during that epoch, all gay colours which would betray her presence to carnivorous birds, beasts, or reptiles are denied here, except in those cases when the structure or situation of the nest itself effectually hides her; then, as if plainly to teach us the reason of this difference, we find female birds adorned with all the gorgeous hues which, in other cases, are confined to the male sex.

A most striking proof that this is the true explanation of these phenomena is found in some very curious and anomalous facts in the natural history of birds. There are a few cases in which the male birds either chiefly or wholly perform the duties of sitting on the eggs; and when this is the case, and the bird is sufficiently defenceless to need protection, the usual external characters of the sexes are also reversed, and the female becomes the larger and the finer-coloured bird. The sooty phalarope, the dotterel, and the small Indian quails of the genus *Turnix*, are examples.

The origin of this curious and unexpected connection between the mode of abird's nesting, and the colour of the female plumage, will be differently explained according as we adopt the theory of "special creations" of every slightly varied form, instinct, or habit, or believe that these are all due to the action of the laws of variation and hereditary transmission, and the cumulative effect of "natural selection" or survival of the fittest. It appears to me that the views of Mr. Darwin best enable us to understand these facts, and that they well illustrate how large a part the need of protection has played in producing many of the most striking peculiarities and anomalies in the animal kingdom. It is true there are some exceptions to the law here laid down, but they are few in number, and can generally be explained by the peculiar habits of the species. On any theory of evolution or development of the forms of life, we must admit that the present condition of nature is the resultant of all past conditions. The long-continued actions and reactions of all known and unknown causes, must have been intricate in the highest degree, and their effects must be proportionately complex. Anomalies and exceptions must therefore be expected to occur, for how can we explain all the phenomena of nature so long as we know only an infinitesimal fraction of its past history? Yet there are existing facts in nature which enable us to trace some of the more important influences which have been mainly instrumental in modifying and remoulding the forms and colours and habits of all living things.

* * * * *

Wallace escalated his attack on sexual selection when Darwin expanded the theory into half of the book The descent of man and selection in relation to sex *in 1871. Wallace criticized the theory directly in a review of that work published the same year (ref 186), later refining his arguments in* Tropical nature and other essays *(ref 719) in 1878 and* Darwinism *(ref 724) in 1889. Related commentary appeared in an overwhelmingly aggressive review (ref 304) of Grant Allen's* The colour sense *printed in* Nature *in 1879. The following excerpts from that notice also convey some sense of how thoroughly Wallace had immersed himself in the study of colour:*

. . . The next chapter, on the direct action of the colour sense upon the animal integuments, deals with the theory of sexual selection as advanced by Mr. Darwin, and endeavours to support it by a variety of general considerations. Many of these arguments are very weak, and are often founded on insufficient or erroneous facts, some of which I shall endeavour to point out. The great aim of this chapter is to prove that the colours of animals are intimately associated with the colours of the objects they feed upon. Butterflies and moths being the most beautifully coloured of all insects and feeding on flowers, is held to be the first great fact in support of this view; and this is backed up by the remark that "the colours of caterpillars are mostly protective, being due to natural selection alone, while those of butterflies are mostly attractive, being largely due to sexual selection." To this we must altogether demur, as slurring over what is really a stupendous difficulty in the way of the theory. So far from the colours of caterpillars being "mostly protective" every entomologist knows that a large number of caterpillars in every part of the world are conspicuously coloured, and what is more to the point that their colours are as brilliant and varied as those of butterflies themselves, if we take into account the nature of their integument, the small amount of surface, and the uniform cylindrical form of their bodies. The caterpillar of *Papilio dissimilis*, for instance, on a bluish green ground has a series of broad irregular longitudinal bands of the richest orange yellow, and between these there are a number of round red spots; while those of many of the *Euplaeas* are adorned with exquisite pink and yellow markings, and with a number of long fleshy processes of equally brilliant colours. Owing to caterpillars being so difficult to preserve, and being rarely collected and figured in their native countries, comparatively few of them are known, but it is certain that they often exhibit the most brilliant hues and the most exquisite patterns; and as they may be said to feed invariably on green leaves, while sexual selection cannot affect them, the natural inference is that the same general laws which produce colour in them are quite sufficient for the production of even more varied hues in the perfect insects, whose expanded wing surfaces, ever varying in size, form, and neuration, offer a field so much better fitted for its development.

In beetles the appearance of colour is also attempted to be correlated with their flower-haunting habits by means of equally doubtful facts. The magnificent Buprestidae and Longicorns are, as far as my experience goes, almost wholly wood-feeders, frequenting the bark of dead trees, and very rarely found on flowers; the Cleridae and Silphidae feeding on dead animal matter, are often brilliantly coloured; and generally in beetles, the absence of colour may be traced to the need of concealment and protection, while whenever a special mode of protection exists, whether by nauseous secretions, hard integuments, rapid flight, or facilities for concealment, then colour appears in infinitely various phases; and this law generally prevails throughout the whole insect-world. In his argument in favour of bright hues being attractive to the opposite sexes of insects, Mr. Allen seems always to forget that it is the male that is attracted to the female, and not *vice versa*; and when he says (p. 158) that he "cannot see why Mr. Wallace, who allows the attractive nature of colouring in flowers, should deny its attractive nature in the question of sex," I reply, that in flowers colour enables the insect to recognise the species, but no one has ever asserted that insects improve and alter the colour of flowers by their preference for certain varieties of colour irrespective of the honey or pollen produced; and in like manner I maintain that the colour of an insect is a guide to easy recognition by its mate, but that there is not one single particle of evidence to show that minute differences in the colour of the same species are observed by insects, still less that such differences are so important to them as to lead to the rejection of a healthy and well-organised mate; yet unless this is the case, the whole theory of sexual selection falls to the ground.

Again, the general connection between coloured flowers and coloured insects is by no means so general and constant as Mr. Allen supposes. Perhaps the richest displays of gay flowers in the world are to be found in temperate Australia, in South Africa, and in the South European Alps, yet in all these countries the butterflies are very inferior to those of tropical forests, where flowers are comparatively rare. In the forests of Pará, for instance, gay flowers are very scarce, as noticed by Mr. Bates as well as by myself, yet the butterflies are endless in their variety of lovely hues. Of course there *are* bright flowers in the tropics, and as travellers notice these whenever they see them and also notice the handsome butterflies, it is easy to infer, as is here done, that the two invariably go together. We may also remark that the sexual allurement of a peculiar odour given out by special patches of scales on butterflies' wings has been discovered by Fritz Müller in the genera *Mechanitis*, *Dircenna*, and *Thecla*, all very brilliantly coloured groups, a clear indication that colour is not a sexual allurement, or we should find it most developed, not in conjunction with, but in the absence of, the attraction of odour.

We must now pass on to the vertebrates, and we here find very good evidence adduced of the existence of a colour-sense in fishes, reptiles, and birds, as we should expect from the known structure of their eyes; while in the case of mammals it is far less decisive. The attempt to associate the brilliant colours of these animals with their food and surroundings, acting through sexual selection, is, however, what we have now to consider; and though many alleged facts are adduced in support of it, several of them are as doubtful and inconclusive as in the case of insects. We shall confine our attention to the birds, which are the stronghold of the theory, and are so much more completely known than the less highly-organised fishes and reptiles. Mr. Allen claims the parrots as fruit-eaters, but they are really seed-eaters, their bills being specially formed to crack the shells and extract and grind up the kernels of nuts and other fruits. They do not therefore aid in the disposal of seeds, as they feed on brown nuts or unripe green fruits from which they extract the seeds, much more frequently than on coloured ripe fruits. The general green colour of parrots is undoubtedly protective, and this green colour is lost, and vivid tints appear just in proportion as, owing to various conditions, the need of concealment diminishes. This is especially the case in countries where mammals are few and a low type of organisation prevails, as in the Australian region, in Madagascar, and in South America; while in Africa and Asia, where a higher type of organisation prevails, the colours of parrots are more sober and protective. A little further on we find the Australian honey-suckers noted for their magnificent coloration; the fact being that they are decidedly a dull-coloured group, hardly superior to our thrushes, and not equal to our finches. Yet they are as universally flower-feeders as the humming-birds themselves; and the total absence of brilliant colour from these birds, which are *the* characteristic family of Australia, and have been developed in correlation with the brilliant Australian flora, absolutely negatives the idea of colour in birds being dependent on the amount of colour in the food and surroundings of certain groups. Again, the ground-feeding pheasant family are passed over as containing only one brilliant bird, the peacock, whereas it abounds in species of the most gorgeous colour. Such are the Impeyan pheasant of the Himalayas, whose metallic plumage is that of a gigantic sun-bird; the golden pheasant, the silver pheasant, and Reeves' pheasant of China, all unsurpassed for gay and conspicuous colouring; the glorious crimson and white-spotted tragopans, the elegant peacock-pheasants, and the intensely brilliant fire-backed pheasants of the Malay countries—together composing a group of birds whose colours are unsurpassed for beauty and splendour, and thus are directly opposed to the general gloom and absence of colour in their habitual surroundings.

In treating of mammals we find an equal want of discrimination in estimating comparative colour and conspicuousness. The tigers, the zebras, the beautifully marked antelopes, and the spotted deer and giraffes, which are really among the most brightly-coloured of mammals, are passed over as less beautifully coloured than the squirrels and monkeys, in order to support the theory that arboreal mammals feeding on fruits should be (but unfortunately for the theory they are not) the most brightly coloured. Monkeys, as a rule, are very dingy brown or black, about one or two per cent. of the species have patches of bright colour on the bare skin of various parts of their bodies, while the nut-eating squirrels as a whole are certainly not superior to the grazing antelopes. In the summary of facts given at pages 184 and 185 there are many errors. *Scissirostrum pagei* does not "belong to a family generally dull," while it is itself decidedly dull-coloured; the "pretty cigana" is a very plain coloured bird; Santarem, of which it is said "the pastures are destitute of flowers, and also of animal life, with the exception of a few small plain-coloured birds," is one of the richest localities for flowering shrubs in South America, and one of the few places where I remember the conspicuously coloured fruits on many of these shrubs, while the butterflies in the adjacent forests are gorgeous in the extreme; and lastly, the "gay-coloured squirrel," for which I myself am made responsible, is one of the dullest of the group, pretty indeed as are all squirrels, owing to its brown and yellowish ringed tail, but in no sense whatever "gay," while I certainly say not a word about its feeding on "bright-coloured fruits."

Such mistakes as these pervade this portion of the work, and are made the foundation for repeated argument and illustration; and they serve to show how impossible it is even for the most earnest and enthusiastic student to make a few months' labour suffice for a correct appreciation of the bearing of the overwhelming mass of facts presented by the countless species of the animal and vegetable world. I have marked a number of other passages to which I altogether demur, but many of them involve arguments which would extend far beyond the limits of an article . . .

* * * * *

Occasional conflicts of opinion notwithstanding, Darwin and Wallace were on very friendly terms both personally and professionally, and Wallace quickly established himself as one of Darwin's most steadfast champions. Wallace's many defences of 'Darwinism' are typified by a classic analysis that appeared in The Annals and Magazine of Natural History *of October 1863:*

Remarks on the Rev. S. Haughton's paper on the bee's cell, and on the origin of species (ref 83)

. . . Mr. Haughton combats the views not only of those who believe that the regular structure of the Bee's comb can be accounted for through the agency of "natural selection" and variation, but also of the opposite school, who impute to the Bee a supernatural or divinely inspired instinct, by which it is enabled to construct its cells on true mechanical and mathematical principles, so as to combine the requisite accommodation for rearing its brood and storing its honey, with the greatest amount of strength and the utmost economy of material. In his opinion of this last school I quite agree with him, but think he has not pointed out its weakest points. If we consider the cell as adapted to the size of the grub and young bee, and in its relations to the cells immediately surrounding it, there can be no doubt that the form of the cell itself, with its pyramidal base and arrangement in double tiers, gives the greatest economy of space and material possible. But if we look at the whole comb suspended vertically by its upper side only, we shall immediately perceive that the strain upon its uppermost rows of cells is many times greater than that upon its lower ones; so that, if economy of material was the main object of this beautiful structure, and the attainment of such economy was secured by unerring wisdom, the walls of the cells should regularly decrease in thickness from the upper to the lower part of the comb. The same mathematical knowledge that enables us to see the beauty and economy of the form of the individual cells, as surely points out the great waste of material in building the upper and lower portions of the comb of the same thickness and strength. We have here, I think, a conclusive argument against the notion that the bees are guided by any supernatural impulse to construct their cells on the best mathematical principles, so as to economize, in the highest degree, labour, space, and material.

When Mr. Haughton attempts to overthrow the theory of Mr. Darwin on this subject, we are compelled to demur to many of his statements, which, indeed, are often so deficient in clearness as to suggest the idea that *The Origin of Species* has been but superficially studied by him. In his first paragraph, for example, he speaks of a class of writers by whom "the geometrical properties of the cells are alleged as a sufficient cause for the production of the insects that make them, from the advantage which these forms of cells are supposed to possess over other forms—advantages said to be so important as to decide the battle of life in favour of the insects that adopt the geometrical plan of making their cells." This is surely a most unfair statement of the doctrine that simultaneous favourable variations in structure and habits, accumulated by natural selection, may act and react on each other, and thus ultimately lead to such a modification of the insect as may better adapt it for constructing the most advantageous form of cell. Mr. Haughton's statement of the case is, that the cell made by the bee is a sufficient cause for the production of the bee; and he would have his readers believe that this absurdity is maintained by the writers he alludes to.

The author then describes the following three forms of cells which he has observed, but does not always express his meaning with sufficient accuracy:—1. Hexagonal cells, somewhat pyramidal, with a rounded extremity. The British tree-wasp and the genus *Polistes* make cells of this form, in small groups, and often of a very fragile papery material. 2. "Hexagonal cells formed of adjoining prismatic figures, with rectilinear axes, terminated by a truncated plane, at right angles to the axes of the prisms." I have quoted this elaborate description literally, because I am quite unable to understand what the author means by a "truncated plane," which renders his meaning somewhat obscure. The cells of this form are said to occur in wasps' nests from the West Indies and South Africa. 3. The bee's hexagonal cell terminated by three faces of a rhombic dodecahedron, each of which forms one-third of the base of one of the cells of the opposite layer. It is not stated, but may be inferred, that the first two forms of cells are in a single layer only; and all these varieties of cells, it is said, may be accounted for "simply by the mechanical pressure of the insects against each other during the formation of the cell." Again, at page 428, "The true cause of the shape of the cell is the crowding together of the bees at work, as was first shown by Buffon. From this crowding together they cannot help making cells with the dihedral angles of 120 degrees of the rhombic dodecahedron; and the economy of wax has nothing to do with the origin of the cell, but is a geometrical property of the figure named." There are, however, several important objections to this pressure-theory. Many exotic tree-wasps construct little groups of three or four hexagonal cells, only one or two insects working at them together. Here is no crowding, yet they are hexagonal. Again, a Mexican bee (*Melipona domestica*) makes a comb of cylindrical cells, only partially hexagonal; and in the Malay Islands there is a domesticated bee which makes oval cells, and though the insects are kept in hollow bamboos for hives, yet the crowding together does not make their cells hexagonal. The wild bee of Borneo, on the other hand, suspends its comb from the arms of lofty trees in the free air; and if crowding had *all* and economy *nothing* to do with it, one would think that here the cells should retain their normal cylindrical form; instead of which, they are as beautifully geometrical as those of our own hive-bee. But, what is still more important, Mr. Darwin states (*Origin of Species*, ed. 3, p. 251) that our bees build the cell-wall at first rough and ten times as thick as it is to remain when finished, it being afterwards gnawed down to the proper thickness. Here is a complete proof of economy of wax rather than economy of labour, and a complete disproof of the theory of circular walls pressed into hexagons by the crowds of struggling bees, which is given us as a new theory of the formation of the bee's cell, unsupported by a single original observation.

To finish this subject of bees, we will now pass to page 427, where Mr. Haughton produces his most crushing argument. He seems to suppose that it is necessary to the theory of Mr. Darwin that there should have been a number of species of bees, now extinct, filling up the gap between the single round cell of the humble-bee and the perfect geometrical structure of the hive-bee, each of them using a little less wax than the preceding one, and that, to effect this, it is necessary that there should have been a bee building a *triangular* cell, and after that, one building a *square* cell, before arriving at the *hexagonal* cell of the hive-bee. But in this view there is a misconception of the conditions of the problem. It is true that, to fill up a given space with cells of a given *area* and walls of equal thickness, the triangle will be more economical of material than the circle (with solid intervals), and the square more economical than the triangle. The primary use of the cell, however, is not the storing of honey—but the accommodation of the larva and pupa; for this it must have a certain *diameter*, and the triangular cell must therefore circumscribe the circular one, and will then be found to require more materials even than the circular cell with solid intervals, without taking into account the fact that the sides of the triangular cells, being without support in their whole length, would have to be thicker than those of any other form, if of equal strength. The same argument will apply in a less degree to the walls of a square cell.

A still more serious error exists, however, in supposing any such extravagantly shaped cells requisite to form the gradual passage from the circle to the hexagon, in order that every step of the process may give its proportionate saving of material. Let the reader draw a number of equal circles in contact, and he will at once perceive how very simple it is (considering that the bees build the cell-wall of a uniform thickness, and reduce it to the smallest serviceable dimensions by gnawing down the growing walls) to suppose them, when material was scanty, to gnaw out a little of the solid triangles left between the circles. The amount of intelligence perceptible in the habits of most insects renders such an act by no means beyond their capacities; and as every step in this direction would tend to the well-being of the community, what was at first done under the pressure of necessity would at length become a regular practice, and finally settle into that class of hereditary habits which we call instinct. Some of these steps do actually occur in the *Melipona domestica* and other bees; and the immense quantity of honey consumed by the hive-bee to make a small quantity of wax, as well as its curious habit of cutting down the walls of its cells to a uniform thickness, are certainly very strong arguments in favour of this view.

Exactly the same arguments will apply to the origin, step by step, of the lozenge-formed planes forming the pyramidal base of the cell as to the hexagonal form of its wall; for these planes are the simple result of gnawing away the superfluous wax in the angles between the alternate spherical bases of the opposite layer of cells; and when this wax is so much gnawed away as to reduce all the walls of the cells to an equal thickness, the true geometrical figure which we see is the necessary result. (*Origin of Species*, p. 247) It is evident, therefore, that all the minute calculations of geometers respecting the amount of saving in this pyramidal base over a *flat* base to the cell is altogether beside the question, because a flat base could not arise out of spherical alternate bases in contact, by any such simple successive steps as are shown to result in the existing form.

On the question of the "origin of species" Mr. Haughton enlarges considerably; but his chief arguments are reduced to the setting-up of "three unwarrantable assumptions," which he imputes to the Lamarckians and Darwinians, and then, to use his own words, "brings to the ground like a child's house of cards." The first of these is "*the indefinite variation of species continuously in the one direction*." Now this is certainly never assumed by Mr. Darwin, whose argument is mainly grounded on the fact that variations occur in *every direction*. This is so obvious that it hardly needs insisting on. In every large family there is almost always one child taller, one darker, one thinner than the rest; one will have a larger nose, another a larger eye: they vary morally as well; some are more poetical, others more morose; one has a genius for numbers, another for painting. It is the same in animals: the puppies, or kittens, or rabbits of one litter differ in many ways from each other—in colour, in size, in disposition; so that, though they do not "*vary continuously in one direction*," they do vary continuously in many directions; and thus there is always material for natural selection to act upon in *some* direction that may be advantageous.

In his remarks upon this "unwarrantable assumption" (which is altogether his own), Mr Haughton has the following passage:—"In the writings of Darwin there is this singular inconsistency, that, while he shows the utmost effects of human breeding on domestic animals to be capable of production in ten or twenty years, he denies the right of his adversaries to appeal to the unaltered condition of the ass, the ostrich, and the cat for 3000 years," etc. The first part of this sentence is so completely out of the pale of grammatical construction, that I must conclude Mr. Haughton writes a very bad hand, and did not correct the proofs. But, so far from Mr. Darwin denying his opponents the use of the facts above alluded to, he himself offers them far stronger ones, in the many species of shells which have lived unchanged since the middle tertiary epochs, and of mammals whose remains are found in beds which testify that they have survived important changes of the earth's surface. No one who understands the theory of natural selection will imagine that these facts are in any way opposed to it.

The second supposed "unwarrantable assumption" is, "*That the causes of variation*, viz. natural advantage in the struggle for existence (Darwin), *are sufficient to account for the effects asserted to be produced*." There certainly never was a more unwarrantable assertion made, than that Darwin assigned "natural advantage in the struggle for existence" as "the cause of variation." Darwin over and over again declares that the *cause* of variation is unknown (*Origin of Species*, pp. 8, 38), though the *fact* is certain and undeniable. Natural selection, acting through advantage in the struggle for existence, *accumulates* favourable variations, but in no sense *causes* them. This is the very foundation of Mr. Darwin's theory; yet even this is misunderstood or misrepresented by Mr. Haughton.

The third "unwarrantable assumption" charged upon Mr. Darwin is, "*That succession implies causation*," "that the Palaeozoic Cephalopoda produced the Red-Sandstone fishes," "that these in turn gave birth to the Liassic reptiles," etc., etc. Now those who have read the *Origin of Species* know that such absurd doctrines as these are nowhere taught there; and I can only say to those who have not read it that I challenge them or Mr. Haughton to produce any passages which will bear such a meaning.

In conclusion, it is asserted "that naturalists who have accepted by multitudes the new theory of the origin of species are, as a class, untrained in the use of the logical faculties, which, however, they may be charitably supposed to possess in common with other men." This is the judgment of the Rev. S. Haughton on such men as Lyell, Hooker, Lubbock, Huxley, and Asa Gray. A perusal of his paper, with the remarks I have now made upon it, will enable any one to judge how far Mr. Haughton himself possesses those "logical faculties" which he is half inclined to deny to the mass of British naturalists. There are several other minor points in his paper which might be alluded to; but it has already occupied as much space as it deserves, and I will only, in conclusion, quote from it a short paragraph which contains an important truth, but which may very fairly be applied in other quarters than those for which the author intended it:—"No progress in natural science is possible as long as men will take their rude guesses at truth for facts, and substitute the fancies of their imagination for the sober rules of reasoning."

* * * * *

In October 1867 Wallace mounted another defence of Darwin's work in a long notice (ref 140) of the Duke of Argyll's book The reign of law. *The following discussion appeared on pages 474–7 of that critique:*

. . . "It is curious," says the Duke of Argyll, "to observe the language which this most advanced disciple of pure naturalism [i.e., Darwin] instinctively uses, when he has to describe the complicated structure of this curious order of plants (the Orchids). 'Caution in ascribing intentions to nature,' does not seem to occur to him as possible. Intention is the one thing which he does see, and which when he does not see, he seeks for diligently until he finds it. He exhausts every form of words and of illustration by which intention or mental purpose can be described. 'Contrivance'—'curious contrivance'—'beautiful contrivance,'—these are expressions which occur over and over again. Here is one sentence describing the parts of a particular species; 'the Labellum is developed into a long nectary, *in order* to attract Lepidoptera, and we shall presently give reason for suspecting that the nectar is *purposely* so lodged, that it can be sucked only slowly *in order* to give time for the curious chemical quality of the viscid matter setting hard and dry.' " Many other examples of similar expressions are quoted by the Duke, who maintains that no explanation of these "contrivances" has been or can be given, except on the supposition of a personal contriver, specially arranging the details of each case, although causing them to be produced by the ordinary processes of growth and reproduction.

Now there is a difficulty in this view of the origin of the structure of Orchids which the Duke does not allude to. The majority of flowering plants are fertilized, either without the agency of insects or, when insects are required, without any very important modification of the structure of the flower. It is evident, therefore, that flowers might have been formed as varied, fantastic, and beautiful as Orchids and yet have been fertilized by insects, in the same manner as Violets, or Clover, or Primroses or a thousand other flowers. The strange springs and traps and pitfalls found in the flowers of Orchids cannot be necessary *per se*, since exactly the same end is gained in ten thousand other flowers which do not possess them. Is it not then an extraordinary idea to imagine the Creator of the Universe *contriving* the various complicated parts of these flowers as a mechanic might contrive an ingenious toy or a difficult puzzle? Is it not a more worthy conception that they are some of the results of those general laws which were so co-ordinated at the first introduction of life upon the earth as to result necessarily in the utmost possible development of varied forms?

But let us take one of the simpler cases adduced and see if our general laws are unable to account for it.

> There is a Madagascar Orchis—the *Angraecum sesquipedale*—with an immensely long and deep nectary. How did such an extraordinary organ come to be developed? Mr. Darwin's explanation is this. The pollen of this flower can only be removed by the proboscis of some very large moths trying to get at the nectar at the bottom of the vessel. The moths with the longest proboscis would do this most effectually; they would be rewarded for their long noses by getting the most nectar; whilst on the other hand, the flowers with the deepest nectaries would be the best fertilized by the largest moths preferring them. Consequently, the deepest nectaried Orchids and the longest nosed moths would each confer on the other a great advantage in the 'battle of life.' This would tend to their respective perpetuation and to the constant lengthening of nectar and noses.

The Duke of Argyll then quotes Darwin's diffident statement "that we can thus *partially* understand how this astonishing nectary was produced," and says it is indeed but a *partial* understanding,—but he does not show what point the explanation given fails to meet. I maintain, on the contrary, that the laws of multiplication, variation, and survival of the fittest . . . would under certain conditions *necessarily* lead to the production of this extraordinary nectary. Let it be remembered that what we have to account for is only the unusual length of this organ. A nectary is found in many orders of plants and is especially common in the Orchids, but in this case only is it more than a foot long. How did this arise? We begin with the fact, proved experimentally by Mr. Darwin, that moths do visit Orchids, do thrust their spiral trunks into the nectaries, and do fertilize them by carrying the pollinia of one flower to the stigma of another. He has further explained the exact mechanism by which this is effected, and the Duke of Argyll admits the accuracy of his observations. In our British species, such as *Orchis pyramidalis*, it is not necessary that there should be any exact adjustment between the length of the nectary and that of the proboscis of the insect, and thus a number of insects of various sizes are found to carry away the pollinia and aid in the fertilization. In the *Angraecum sesquipedale*, however, it is necessary that the proboscis should be forced down into a particular part of the flower, and this would only be done by a large moth straining to drain the nectar from the bottom of the long tube. (It is a peculiarity of this species that the nectar only occupies a depth of one or two inches at the bottom of the nectary.) Now let us start from the time when the nectary was only half its present length or about six inches, and was chiefly fertilized by a species of moth which appeared at the time of the plant's flowering, and whose proboscis was of the same length. Among the millions of flowers of the *Angraecum* produced every year some would always be shorter than the average, some longer. The former, owing to the structure of the flower, would not get fertilized, because the moths could get all the nectar without forcing their trunks down to the very base. The latter would be well fertilized, and the longest would on the average be best fertilized of all. By this process alone the average length of the nectary would annually increase, because, the short ones being sterile and the long ones having abundant offspring, exactly the same effect would be produced as if a gardener destroyed the short ones and sowed the seed of the long ones only; and this we know by experience would produce a regular increase of length, since it is this very process which has increased the size and changed the form of our cultivated fruits and flowers.

But this would lead in time to such an increased length of the nectary that many of the moths could only just reach the surface of the nectar, and only the few with exceptionally long trunks be able to suck up a considerable portion.

This would cause many moths to neglect these flowers because they could not get a satisfying supply of nectar, and if these were the only moths in the country the flowers would undoubtedly suffer and the further growth of the nectary be checked by exactly the same process which had led to its increase. But there are an immense variety of moths of various lengths of proboscis, and as the nectary became longer other and larger species would become the fertilizers, and would carry on the process till the largest moths became the sole agents. Now, if not before, the moth would also be affected, for those with the longest probosces would get most food, would be the strongest and most vigorous, would visit and fertilize the greatest number of flowers, and would leave the largest number of descendants. The flowers most completely fertilized by these moths being those which had the longest nectaries, there would in each generation be on the average an increase in the length of the nectaries, and also an average increase in the length of the proboscis of the moths, and this would be a *necessary result* from the fact that nature ever fluctuates about a mean, or that in every generation there would be flowers with longer and shorter nectaries, and moths with longer and shorter probosces than the average. No doubt there are a hundred causes that might have checked this process before it had reached the point of development at which we find it. If, for instance, the variation in the quantity of nectar had been at any stage greater than the variation in the length of the nectary, then smaller moths could have reached it and have effected the fertilization. Or if the growth of the probosces of the moths had from other causes increased quicker than that of the nectary, or if the increased length of proboscis had been injurious to them in any way, or if the species of moth with the longest proboscis had become much diminished by some enemy or other unfavourable conditions, then in any of these cases the shorter nectaried flowers which would have attracted and could have been fertilized by the smaller kinds of moths would have had the advantage. And checks of a similar nature to these no doubt have acted in other parts of the world, and have prevented such an extraordinary development of nectary as has been produced by favourable conditions in Madagascar only and in one single species of Orchid . . .

Now, instead of this beautiful self-acting adjustment, the Duke of Argyll's theory is, that the Creator of the Universe by a direct act of his Almighty power so disposed the natural forces influencing the growth of this one species of plant as to cause its nectary to increase to this enormous length, and at the same time by an equally special act determined the flow of nourishment in the organization of the moth so as to cause its proboscis to increase in exactly the same proportion, having previously so constructed the *Angraecum* that it could only be maintained in existence by the agency of this moth. But what proof is given or suggested that this was the mode by which the adjustment took place?

None whatever, except a feeling that there is an adjustment of a delicate kind and an inability to see how known causes could have produced such an adjustment. I believe I have shown, however, that such an adjustment is not only possible but inevitable, unless at some point or other we deny the action of those simple laws which we have already admitted to be expressions of existing facts . . .

* * * * *

On a number of occasions Wallace wrote on that most difficult of topics, instinct. The conservatism of his 'no first causes' approach to the subject dominates a short essay that appeared in the collection Contributions to the theory of natural selection *(ref 716) in 1870:*

On instinct in man and animals (ref 164)

The most perfect and most striking examples of what is termed instinct, those in which reason or observation appear to have the least influence, and which seem to imply the possession of faculties farthest removed from our own, are to be found among insects. The marvellous constructive powers of bees and wasps, the social economy of ants, the careful provision for the safety of a progeny they are never to see manifested by many beetles and flies, and the curious preparations for the pupa state by the larvae of butterflies and moths, are typical examples of this faculty, and are supposed to be conclusive as to the existence of some power or intelligence, very different from that which we derive from our senses or from our reason.

How Instinct may be best Studied. Whatever we may define instinct to be, it is evidently some form of mental manifestation, and as we can only judge of mind by the analogy of our own mental functions and by observation of the results of mental action in other men and in animals, it is incumbent on us, first, to study and endeavour to comprehend the minds of infants, of savage men, and of animals not very far removed from ourselves, before we pronounce positively as to the nature of the mental operations in creatures so radically different from us as insects. We have not yet even been able to ascertain what are the senses they possess, or what relation their powers of seeing, hearing, and feeling have to ours. Their sight may far exceed ours both in delicacy and in range, and may possibly give them knowledge of the internal constitution of bodies analogous to that which we obtain by the spectroscope; and that their visual organs do possess some powers which ours do not, is indicated by the extraordinary crystalline rods radiating from the optic ganglion to the facets of the compound eye, which rods vary in form and thickness in different parts of their length, and possess distinctive characters in each group of insects. This complex apparatus, so different from anything in the eyes of vertebrates, may subserve some function quite inconceivable by us, as well as that which we know as vision. There is reason to believe that insects appreciate sounds of extreme delicacy, and it is supposed that certain minute organs, plentifully supplied with nerves, and situated in the subcostal vein of the wing in most insects, are organs of hearing. But besides these, the Orthoptera (such as grasshoppers, etc.) have what are supposed to be ears on their fore legs, and Mr. Lowne believes that the little stalked balls, which are the sole remnants of the hind wings in flies, are also organs of hearing or of some analogous sense. In flies, too, the third joint of the antennae contains thousands of nerve-fibres, which terminate in small open cells, and this Mr. Lowne believes to be the organ of smell, or of some other, perhaps new, sense. It is quite evident, therefore, that insects may possess senses which give them a knowledge of that which we can never perceive, and enable them to perform acts which to us are incomprehensible. In the midst of this complete ignorance of their faculties and inner nature, is it wise for us to judge so boldly of their powers by a comparison with our own? How can we pretend to fathom the profound mystery of their mental nature, and decide what, and how much, they can perceive or remember, reason or reflect! To leap at one bound from our own consciousness to that of an insect's, is as unreasonable and absurd as if, with a pretty good knowledge of the multiplication table, we were to go straight to the study of the calculus of functions, or as if our comparative anatomists should pass from the study of man's bony structure to that of the fish, and, without any knowledge of the numerous intermediate forms, were to attempt to determine the homologies between these distant types of vertebrata. In such a case would not error be inevitable, and would not continued study in the same direction only render the erroneous conclusions more ingrained and more irremovable?

Definition of Instinct. Before going further into this subject, we must determine what we mean by the term instinct. It has been variously defined as—"disposition operating without the aid of instruction or experience," "a mental power totally independent of organization," or "a power enabling an animal to do that which, in those things man can do, results from a chain of reasoning, and in things which man cannot do, is not to be explained by any efforts of the intellectual faculties." We find, too, that the word instinct is very frequently applied to acts which are evidently the result either of organization or of habit. The colt or calf is said to walk instinctively, almost as soon as it is born; but this is solely due to its organization, which renders walking both possible and pleasurable to it. So we are said instinctively to hold out our hands to save ourselves from falling, but this is an acquired habit, which the infant does not possess. It appears to me that instinct should be defined as—"the performance by an animal of complex acts, absolutely without instruction or previously-acquired knowledge." Thus, acts are said to be performed by birds in building their nests, by bees in constructing their cells, and by many insects in providing for the future wants of themselves or their progeny, without ever having seen such acts performed by others, and without any knowledge of why they perform them themselves. This is expressed by the very common term "blind instinct." But we have here a number of assertions of matters of fact, which, strange to say, have never been proved to be facts at all. They are thought to be so self-evident that they may be taken for granted. No one has ever yet obtained the eggs of some bird which builds an elaborate nest, hatched these eggs by steam or under a quite distinct parent, placed them afterwards in an extensive aviary or covered garden, where the situation and the materials of a nest similar to that of the parent birds may be found, and then seen what kind of nest these birds would build. If under these rigorous conditions they choose the same materials, the same situation, and construct the nest in the same way and as perfectly as their parents did, instinct would be proved in their case; now it is only assumed, and assumed, as I shall show further on, without any sufficient reason. So, no one has ever carefully taken the pupae of a hive of bees out of the comb, removed them from the presence of other bees, and loosed them in a large conservatory with plenty of flowers and food, and observed what kind of cells they would construct. But till this is done, no one can say that bees build without instruction, no one can say that with every new swarm there are no bees older than those of the same year, who may be the teachers in forming the new comb. Now, in a scientific inquiry, a point which can be proved should not be assumed, and a totally unknown power should not be brought in to explain facts, when known powers may be sufficient. For both these reasons I decline to accept the theory of instinct in any case where all other possible modes of explanation have not been exhausted.

Does Man possess Instincts? Many of the upholders of the instinctive theory maintain, that man has instincts exactly of the same nature as those of animals, but more or less liable to be obscured by his reasoning powers; and as this is a case more open to our observation than any other, I will devote a few pages to its consideration. Infants are said to suck by instinct, and afterwards to walk by the same power, while in adult man the most prominent case of instinct is supposed to be, the powers possessed by savage races to find their way across a trackless and previously unknown wilderness. Let us take first the case of the infant's sucking. It is sometimes absurdly stated that the new-born infant "seeks the breast," and this is held to be a wonderful proof of instinct. No doubt it would be if true, but unfortunately for the theory it is totally false, as every nurse and medical man can testify. Still, the child undoubtedly sucks without teaching, but this is one of those *simple* acts dependent upon organization, which cannot properly be termed instinct, any more than breathing or muscular motion. Any object of suitable size in the mouth of an infant excites the nerves and muscles so as to produce the act of suction, and when at a little later period, the will comes into play, the pleasurable sensations consequent on the act lead to its continuance. So, walking is evidently dependent on the arrangement of the bones and joints, and the pleasurable exertion of the muscles, which lead to the vertical posture becoming gradually the most agreeable one; and there can be little doubt that an infant would learn of itself to walk, even if suckled by a wild beast.

How Indians travel through unknown and trackless Forests. Let us now consider the fact, of Indians finding their way through forests they have never traversed before. This is much misunderstood, for I believe it is only performed under such special conditions, as at once to show that instinct has nothing to do with it. A savage, it is true, can find his way through his native forests in a direction in which he has never traversed them before; but this is because from infancy he has been used to wander in them, and to find his way by indications which he has observed himself or learnt from others. Savages make long journeys in many directions, and, their whole faculties being directed to the subject, they gain a wide and accurate knowledge of the topography, not only of their own district, but of all the regions round about. Every one who has travelled in a new direction communicates his knowledge to those who have travelled less, and descriptions of routes and localities, and minute incidents of travel, form one of the main staples of conversation round the evening fire. Every wanderer or captive from another tribe adds to the store of information, and as the very existence of individuals and of whole families and tribes, depends upon the completeness of this knowledge, all the acute perceptive faculties of the adult savage are devoted to acquiring and perfecting it. The good hunter or warrior thus comes to know the bearing of every hill and mountain range, the directions and junctions of all the streams, the situation of each tract characterized by peculiar vegetation, not only within the area he has himself traversed, but for perhaps a hundred miles around it. His acute observation enables him to detect the slightest undulations of the surface, the various changes of subsoil and alterations in the character of the vegetation, that would be quite imperceptible to a stranger. His eye is always open to the direction in which he is going; the mossy side of trees, the presence of certain plants under the shade of rocks, the morning and evening flight of birds, are to him indications of direction, almost as sure as the sun in the heavens. Now, if such a savage is required to find his way across this country in a direction in which he has never been before, he is quite equal to the task. By however circuitous a route he has come to the point he is to start from, he has observed all the bearings and distances so well, that he knows pretty nearly where he is, the direction of his own home and that of the place he is required to go to. He starts towards it, and knows that by a certain time he must cross an upland or a river, that the streams should flow in a certain direction, and that he should cross some of them at a certain distance from their sources. The nature of the soil throughout the whole region is known to him, as well as all the great features of the vegetation. As he approaches any tract of country he has been in or near before, many minute indications guide him, but he observes them so cautiously that his white companions cannot perceive by what he has directed his course. Every now and then he slightly changes his direction, but he is never confused, never loses himself, for he always feels at home; till at last he arrives at a well-known country, and directs his course so as to reach the exact spot desired. To the Europeans whom he guides, he seems to have come without trouble, without any special observation, and in a nearly straight unchanging course. They are astonished, and ask if he has ever been the same route before, and when he answers "No," conclude that some unerring instinct could alone have guided him. But take this same man into another country very similar to his own, but with other streams and hills, another kind of soil, with a somewhat different vegetation and animal life; and after bringing him by a circuitous route to a given point, ask him to return to his starting place, by a straight line of fifty miles through the forest, and he will certainly decline to attempt it, or, attempting it, will more or less completely fail. His supposed instinct does not act out of his own country.

A savage, even in a new country, has, however, undoubted advantages, from his familiarity with forest life, his entire fearlessness of being lost, his accurate perception of direction and of distance, and he is thus able very soon to acquire a knowledge of the district that seems marvellous to a civilized man; but my own observation of savages in forest countries has convinced me, that they find their way by the use of no other faculties than those which we ourselves possess. It appears to me, therefore, that to call in the aid of a new and mysterious power to account for savages being able to do that which, under similar conditions, we could almost all of us perform, although perhaps less perfectly, is almost ludicrously unnecessary . . .

* * * * *

In a letter printed in the 20 February 1873 issue of Nature, *Wallace discussed one particular question related to instinct in a way consistent with the position he had set out three years earlier:*

Inherited feeling (ref 222)

The remarkable case of an inherited feeling of dislike for a special class of persons, communicated by Mr. Darwin, appears to me to support a view I have long held (but not yet published) as to the explanation of another class of so-called instincts. The three separate instances given in which the dogs showed a violent antipathy to butchers, either without seeing them or when they were dressed as gentlemen, clearly indicates that it was through the sense of smell that the painful sensation was experienced; and this is quite in accordance with the wonderful delicacy and importance of this sense in most animals, and especially in dogs. It is natural to suppose that some ancestor of these dogs was systematically and cruelly ill-treated by several butchers, perhaps from some thievish propensity or other bad habit which required frequent punishment, so that the smell of a butcher came to be invariably associated with pain and a desire for revenge. But the most important fact to observe is, that there must be some peculiar odour developed in human beings by constant contact with flesh, which a dog can recognise apart from individual peculiarities and in spite of perfect disguise. Now the power many animals possess to find their way back over a road they have travelled blindfolded (shut up in a basket inside a coach for example) has generally been considered to be an undoubted case of true instinct. But it seems to me that an animal so circumstanced will have its attention necessarily active, owing to its desire to get out of its confinement, and that by means of its most acute and only available sense it will take note of the successive odours of the way, which will leave on its mind a series of images as distinct and prominent as those we should receive by the sense of sight. The recurrence of these odours in their proper inverse order—every house, ditch, field, and village having its own well-marked individuality—would make it an easy matter for the animal in question to follow the identical route back, however many turnings and cross-roads it may have followed. This explanation appears to me to cover almost all the well-authenticated cases of this kind.

* * * * *

Wallace's views on evolution and natural selection are most completely set out in his monograph Darwinism, *published in 1889. But he continued to contribute after that date, serving most effectively in the role of 'establishment' critic. 'The method of organic evolution' (ref 510), for example, an important review paper, appeared in* Fortnightly Review *in 1895. In another review paper, published the next year, Wallace re-examined one of the most fundamental elements of Darwinian/Wallaceian theory:*

The problem of utility: are specific characters always or generally useful? (ref 527)

The above stated question is discussed at great length in the second part of the late Mr. [George] Romanes' work on *Darwin and After Darwin*, fully half of the volume being devoted to it; and in the preface the author states his belief that his arguments are so conclusive that he has "broken to fragments" the doctrine of utility, and that he has "made a full end thereof." A careful perusal of the volume, and a full consideration of all the facts and arguments adduced therein, seem to me to leave the problem just where it was before; but the variety of subjects discussed, the great mass of details referred to, and the ingenuity of some of the arguments in support of the author's view, lead me to think that I have not hitherto set forth the facts and arguments in favour of the utility-theory with sufficient completeness, while I am indebted to the lamented author for pointing out one or two weak points in my discussion of the question, and for a number of useful references to Darwin's statements on the points at issue, some of which I had overlooked. Although Mr. Romanes' discussion of the question is so lengthy, the problem itself is in its essence a comparatively simple one, and is I believe capable of being solved by a reference to well-known facts and admitted principles. The reason why Mr. Romanes is able to support his views by so many quotations from Darwin's works, is due to the fact that Darwin was firmly convinced of the heredity of acquired characters, and especially of the influence of food and climate and the effects of use and disuse; and this belief must be borne in mind whenever he speaks of specific characters being due to other causes than natural selection. It must also be remembered that Darwin was not acquainted with the evidence we now possess as to the extreme frequency of variation everywhere in nature, its large amount, and its universality in every organ and every character that can be measured or otherwise estimated. Had he known what we now know on this subject, he would not so frequently have made the proviso—"if they vary, for without variation natural selection can do nothing," or have alluded to the possibility of variations of the same kind occurring "perhaps after a long interval of time." We now know that variations of almost every conceivable kind occur, in all the more abundant species, in every generation, and that the material for natural selection to work upon is never wanting. Accepting, then, these facts of variation, and always keeping in mind the severity of the struggle for existence, nine tenths at least of the progeny of the higher animals perishing annually before reaching maturity, thus leading to a systematic and continual weeding out of the less fit—let us endeavour to realize the process of the formation of new species and the nature of the characters which distinguish allied species from each other.

In my article on 'Mimicry and other Protective Resemblances among Animals' [ref 134], first published in 1867, I laid down the principle of utility, perhaps a little too absolutely, in the following passage:—

> Perhaps no principle has ever been announced so fertile in results as that which Mr. Darwin so earnestly impresses upon us, and which is indeed a necessary deduction from the theory of Natural Selection, namely—that none of the definite facts of organic nature, no special organ, no characteristic form or marking, no peculiarities of instinct or of habit, no relations between species or between groups of species, can exist but which must now be or once have been *useful* to the individuals or races which possess them . . .

Now these characters, in virtue of which the variety has become a species, are in fact its "specific characters," and they alone will absolutely differentiate it from all other species. We need not trouble ourselves about the cases of doubtful species, in which the distinctive characters are either so minute or so unstable that we cannot invariably determine them. On the theory of evolution by natural selection there must be such cases. They are species in the making and not quite completed. But in the great majority of species definite characters do exist by which any single individual can be recognized and the species to which it belongs be determined; and the question is, whether or not the characters, or combination of characters, which thus differentiate it are now useful or were useful at the time of its origination. * In order to answer this question, we must briefly summarize both the facts and the admitted principles or theories which bear upon it.

Every extensive area contains a number of large and dominant species which appear to be, and probably are for considerable periods, stable, both in average population and in the extent of the area they occupy. Taking any one of these species—say of bird or mammal—so long as the whole conditions of its environment remain unchanged or very little changed it will, theoretically, continue to maintain itself, as we know many species have maintained themselves during the whole period since the glacial epoch, and some very much longer. The species, however, is not absolutely homogeneous. It varies in every generation, not minutely or infinitesimally as was formerly supposed, but very considerably, the variations being easily seen and measured by any one who looks for them; and they extend, so far as we know, to every part of

* To this should be added—"or were correlated with some useful characters." I have referred to such correlations in my *Natural Selection and Tropical Nature* [ref 725], pp. 172 and 175; and as to apparently useless characters being in some cases correlated with those which are useful, in my *Darwinism*, p. 140; but it is cumbersome to restate this part of the theory whenever it is stated that all specific characters are useful.

the organism, external and internal, since no part has yet been found to be invariable when a large number of individuals have been compared. The species is therefore composed of a fluctuating mass of variable units which yet maintain the same general average of characters, and this it can only do by a constant or intermittent weeding out of the extremes in every direction. Such a weeding out on a large scale takes place annually, because, although the annual increase by birth is very large, the population of adults remains approximately fixed. The species is maintained in harmony with its environment by the survival of the fittest.

But now let some important change occur, either in climate, in abundance of food, or by the irruption of some new and hitherto unknown enemies, a change which at first injuriously affects the species. It must, therefore, undergo some amount of modification, either structural or functional, in order to succeed under the new conditions; and the constant variations of every part around its mean furnish the materials for adapting the organism to these new conditions. If a new enemy is the danger to be guarded against, this adaptation may be effected in several ways. Swiftness in running or flying, habits of concealment, or seeking new kinds of food in places inaccessible to the enemy, may each lead to the survival of those individuals which were sufficiently intelligent to adopt them or sufficiently favoured by rapid variation in the desired direction. Survival of the fittest in these respects, going on year by year, might lead to the formation of two or more diverging races each able to maintain itself in the presence of the new enemy, while the former average type of the species rapidly became extinct. We should thus have two or three incipient new species; but they would not become well differentiated species till they had acquired certain definite and important characteristics. These are (1) some amount of infertility when crossed with the parent form or with each other; and (2) some distinct and conspicuous external characters by means of which the new varieties could readily distinguish their own kind even when at considerable distances or when partially concealed; or, in the case of flowering plants, be distinguished by the insects which fertilize them.

The greatest danger to a species under new and adverse conditions is, that it should not be able to adapt itself to them with sufficient rapidity. It is for this reason that, as Darwin concludes, new species arise, mainly, from those which have a large population, which occupy a wide area, and which present much variation—a combination rarely found except in continental areas. But this danger is evidently much increased if crossing with the parent form is not at first checked and soon afterwards completely prevented, except as a quite exceptional occurrence. The means of preventing this intercrossing are, for animals, either infertility, external distinctions leading to the preferential mating of similar forms, or physical isolation. The latter I believe, with Darwin, to be of comparatively little importance and to have very rarely been the chief agent in modification. In the great majority of cases a new species must arise amidst the population of an existing species; and while its adaptation is progressing any intercrossing with the parent form will be injurious. I have endeavoured to show, and can still find no flaw in my reasoning, that mutual infertility would be usually brought about by natural selection wherever the two forms were in contact, and also that the early occurrence of well-marked external differences would assist greatly in the rapidity of adaptation (*Darwinism*, pp. 174–180). This view will explain the curious fact of the well-marked differences of colour or form which almost invariably characterize allied species. These "recognition marks," as I have termed them, are of great use even to existing well-defined spcies, but they must have been of still greater use during the earlier stages of differentiation, when the very existence of the new form must have largely depended on them.

I may here remark that it is because these external differences of colour or marking are quite as constantly present in peculiar insular species as in those inhabiting a continent, that I do not believe in local isolation as of any importance in species-formation. Insular species may have been produced in two ways. Either a portion of a declining species may have reached the island, where it survived through the more favourable conditions while it became extinct on the continent; or, a few individuals of a dominant species reached the island, where, owing to the absence of competition, they rapidly increased till the island became fully stocked with the unchanged species. Then (and only then) survival of the fittest would begin to act, and the differences of food and climate, with the different kinds of enemies, would render some modifications of structure, form, or colour advantageous, and thus a new species would be formed by adaptation from the old one in almost exactly the same way as on the continent. In both these cases recognition-characters, to aid in the prevention of intercrossing, would be produced by natural selection. But if insular species have usually been formed by a few individuals somewhat different from the type having first reached the island and thereafter preserved their peculiarities, there is no reason why any distinctive and stable form of coloration or marking should have been developed, since there would be no similar species from which it would need to be differentiated. Neither is the small amount of divergence that usually prevails between the mean of a few individuals taken at random, such as might have accidentally reached an island, and the average type of the species, at all comparable with the well-marked characters that usually distinguish insular forms, and there is nothing in mere isolation without selection which can increase the difference. As examples we may refer to the many peculiar species of butterflies and birds found in the various islands of the West Indian and Malayan Archipelagoes, which are quite as distinct from each other as are allied continental species, and which exhibit all the characteristics of forms which have been fully differentiated by natural selection.

The sketch now given of the usual mode of formation of new species under natural selection leads to the conclusion that every species (of the higher animals at all events) will usually possess at least three peculiarities: in the first place, it must exhibit some difference of structure or function adapting it to new conditions; secondly, some distinction of colour, form, or peculiar ornament serving as distinctive recognition marks; and, thirdly, the physiological peculiarity of some amount of infertility when crossed with allied species. The first two constitute its "specific characters." But if we consider that every species in the long line of its ancestry must have had similar specific characters, adapting it to the peculiar conditions of its environment and distinguishing it from its nearest allies; that some of these characters, when generally useful, have persisted, and now constitute generic or family characters; that others have been again and again modified so as to adapt them to new and sometimes quite different conditions; and that others again, becoming useless, persist when quite harmless or remain in a more or less rudimentary condition; and when we further consider that many genera and families extend far back into geological time and must have originated in the midst of a physical and biological environment very different from that which now prevails, we shall dimly understand how complex are the forces and processes which have led to the assemblage of characters now presented by each organism, and how difficult it must be to determine positively that any one of these characters is not, *nor ever has been*, useful to its possessor. Yet this is what is done by those writers who maintain, as did the late Mr. Romanes, that the majority of specific characters are not and never have been useful, but have arisen through definite variation under the influence of definite causes, and, when neither useful nor hurtful, persist and constitute the main external differences which we observe between species and species. This theory, which, although to some extent held by Darwin himself, I consider to be wholly erroneous, we will now proceed to discuss.

It may be well first to dispose of a point, made much of by Mr. Romanes, that I do not urge utility as a characteristic either of varieties or of genera and higher groups, and that it is therefore illogical to claim it for species. But this is a misapprehension, since I *do* claim that when varieties are constant, are hereditary, and occupy a definite area, and are therefore what Darwin termed "incipient species," the characteristics which distinguish them from the parent species are, to some extent, adaptive and useful, and will become fully so when the variety becomes a fully differentiated species. And as to genera and families, it is obvious that every one of their distinguishing characters was once a specific character, since genera are merely groups of species, all of which were derived from one parent species, and which have become more or less isolated by the extinction of intermediate forms. Families are, in the same way, derived from a single genus and ultimately from a single species, and the same reasoning applies to them. The reason why my argument on this question has been limited to species is, because the whole problem is included in that of species: it is in them that the process and laws of development can be best studied free from many of those complexities of modification and survival or disused and partially aborted parts and organs which often constitute generic or family characters. If every one of the new characters or new combinations of characters which arise when a new species becomes differentiated from its parent-form,—if every one of these is adaptive and utilitarian, then no higher groups can possess characters other than those which were once adaptive, since genera and families can never acquire new characters except through every one of their component species acquiring those characters. The problem as exhibited in species includes therefore the problem in all higher groups.

I have already set forth in some detail the argument for utility founded on the fact of the continuous progresss of the discovery of utilities with the continuous growth of our knowledge of the life-histories and inter-relations of plants and animals (*Darwinism*, pp. 131–142). I will therefore now devote more special attention to the fundamental argument, that whereas every modification of a species which arises under the influence of natural selection must, from the very nature of its origin, be useful to the new form, no other agency has been shown to exist capable of producing non-utilitarian characters *in every individual constituting a species, neither more nor less*. Now the general cause which is adduced as being able to do this is stated by Darwin in the following passages, which are quoted by Mr. Romanes as expressing his own views:—

> There must be some efficient cause for each slight individual difference, as well as for more strongly marked variations which occasionally arise; and if the unknown cause were to act persistently, it is almost certain that *all* the individuals of the species would be similarly modified (*Origin of Species*, p. 171).

Again, after referring to cleistogamic flowers and degraded parasitic animals, he says:—

> We are ignorant of the exciting cause of the above specified modifications; but if the unknown cause were to act almost uniformly for a length of time, we may infer that the result would be almost uniform; and in this case all the individuals of the species would be modified in the same manner. *

Now these passages, merely as stating a possibility or a probability, appear to me to be wanting both as regards logic and in the absence of any appeal to the actual facts of variation. For the argument is, briefly, that the same causes will always produce the same or closely similar results. But this is only true when the same causes *act upon identical materials and under identical conditions*. But the very foundation of the Darwinian theory is, that the materials—the individuals of a species—are *not* identical, but that they vary indefinitely and in many directions even under closely similar conditions. How then can any external or internal causes produce an identical result—a definite new variation—in *all* the individuals of a species, born as they are of varying parents, of different ages, and subject to ever fluctuating conditions? It seems to me, therefore, that the *à priori* probabilities are all against Darwin's supposition.

Now let us see how far the *facts* of variation give any support to the theory of useless specific characters. If there is one thing better established than another it is that the individual variations which are constantly occurring in all common species are indefinite in their character and very unequal in their amount. Some species are much more variable than others, and Darwin has shown reasons for believing that any change of conditions induces *variability*, but not that it causes definite *variations*. The two things are radically distinct. So far as I am aware, no evidence has been adduced of any special conditions which have produced a definite variation in the whole offspring of all the individuals subjected to it. But it must do more than this. For it must produce a variation

* *Origin*, p. 175. In my *Darwinism*, p. 141, I have stated my opinion that Darwin did not believe in the production of useless characters in *all* the individuals of a species. I had overlooked the passages quoted by Mr. Romanes and given above, which certainly show that he did believe it.

so exceptionally stable that it constantly recurs in all the offspring of successive generations, even though those offspring are subjected to considerable change of conditions, as are the individuals of all species except the rarest or the most local. Only with such constancy and stability of inheritance could a useless character become fixed in every individual of a species, which it must be to be a "specific" character. It must, therefore, from the very first have been invariable. But this feature of invariability without selection has not been found to characterize any variation, whether occurring among wild or domesticated organisms. Such an occurrence would necessarily have forced itself upon the attention of breeders and horticulturists. For if the theory is true that the majority of specific characters are of this useless kind, their occurrence as permanent and unchangeable variations must be a common phenomenon, and we ought to find that foreign plants when first cultivated very often present new characters, not sporadically but appearing in every individual, and which cannot be got rid of, since they do not vary and selection would therefore be powerless to eliminate them. Has any indication of a phenomenon of this kind ever been noted?

Let us come now to the actual causes said to produce useless specific characters. According to Mr. Romanes they are five in number: Climate, Food, Sexual Selection, Isolation, and Laws of Growth. Let us consider how these are known to act or are alleged to act. Climate and Food undoubtedly produce modification in the individual, but it has not yet been proved that these modifications are hereditary. If this could be proved the whole discussion on the heredity of acquired characters would be settled in the affirmative. The supposed proof that these causes produce definite changes which are hereditary is derived from the fact that there is often a simultaneous change in the colours of many animals, or in the form or texture of the foliage of many plants, in different parts of the area they occupy which are characterized by differences of climate. But in every case these changes can be interpreted as adaptations for protection in the case of animals, and as either adaptations or individual non-hereditary modifications in the case of the plants. The firm belief that such individual characters were usually, if not always, inherited led to some looseness in Darwin's reasoning on this point, and still more so in that of most modern upholders of the theory.

The next alleged cause, Sexual Selection, whether we limit it as I do, to the struggles of the males, leading to the development of weapons and defensive armour, or with Darwin extend it to the choice by the females of the more ornamental males, thus leading to the development of decorative plumes, etc., is really a form of natural selection, and sexual characters are therefore useful characters. It is true that, from my point of view, male distinctive colour and ornament have not this particular use; and Mr. Romanes makes a good point against me when he says that in imputing their origin and development to the surplus vitality and energy of the male I give away my case, since I admit that useless specific characters may be developed independently of natural selection. This is owing to my having omitted to lay special stress on the *specific* part of each ornament being really a "recognition mark," and therefore essential both to the first production and subsequent well-being of every species. In the summary of my argument (*Darwinism*, p. 298) I have adduced the need of recognition as the cause of specific specialization of colour, but in the body of my discussion as to sexual ornaments I have not referred to it, and this omission greatly weakens my argument. I should have said that the accessory plumes and other ornaments *originate* at points of great nervous and muscular excitation, and are *developed* through surplus energy; and that, from their first appearance, they were *utilized* for purposes of recognition, which explains both their comparative stability in each species and their distinctness in allied forms. (Since writing this paper I have carefully studied Professor Weismann's new theory of "Germinal Selection," which seems to me to have a high degree of probability, and which, if true, enables us to explain two phenomena which have not hitherto been fully explicable. These are (1) the complete or almost complete disappearance of many characters which have become useless; and (2) the development of secondary sexual characters far beyond the point of utility as recognition marks, and, apparently, up to the extreme point of incipient hurtfulness. It thus furnishes one link necessary in the chain of argument proving that these secondary sexual characters are explicable without calling in the very problematical agency of female choice.)

The next alleged cause, Isolation, I do not admit to be a *vera causa* at all, for reasons already given. It is, at most, an aid to the differentiation of new species by natural selection.

The last alleged cause, the Laws of Growth, can never, of itself, account for specific characters, but only for those structural and histological peculiarities of organisms which characterize the higher groups such as classes and sometimes perhaps orders and families; and even these must always, when they first originated, have had a utilitarian character, since it is almost impossible to conceive that the details of structure of the various tissues or organs produced under the action of these laws were absolutely indifferent to the well-being of the organism.

If, then, we admit, as I do admit, that certain growths, appendages, or markings, which are of no use to the organism, do occasionally appear, no agency has been adduced which could, first, cause these useless characters to appear in every individual or a species, and then totally cease to appear whenever any portion of this species is selected and slightly modified so as to occupy a new place in nature or to save itself from extinction by some new enemy. Whenever useless characters are said to be "specific," it seems to be forgotten that one species has always passed continuously into another by a process of normal individual variation and survival of the fittest. There is no chasm in such a process, no sudden transition from one creature to another of a different nature. The transition is by a purely normal and almost imperceptible process of adaptation to new conditions, and in itself furnishes no reason whatever why any useless character, if it had constantly reappeared in the countless millions of individuals during all the millions of generations of the duration of the species, should at once disappear, or be replaced by some new character equally universal, equally invariable, and equally useless.

I strongly urge, therefore, that the general causes suggested by Darwin as possibly leading to the production of useless specific characters, as well as the more special causes enumerated by Mr. Romanes, do not apply to the actual facts of variation and and heredity so far as they are yet known to us; and further, that no attempt has been made to show, even hypothetically, how, through the action of known causes, such characters, when they do arise, can become first extended to every individual of a species, and then be totally obliterated as regards any portion of the species which may become modified so as to constitute a new species. Useful characters thus strictly limited are the necessary and logical results of modification through survival of the fittest. No agency has been shown to exist capable of producing useless characters similarly limited. And as it is beyond the powers of human reason to know absolutely that any characters so limited as to be really specific are and always have been useless, it is both unscientific and illogical to postulate such characters as being present in all or many species, and therefore as constituting an essential characteristic feature of specific forms.

The preceding discussion may, I hope, be considered sufficient to show that useless specific characters, if they exist, can only be the result of some comparatively rare and exceptional conditions, and that they certainly are not, as has been alleged, a general characteristic of species; but it may be as well to notice . . . [two] . . . special cases which have been adduced by Mr. Romanes and others as examples of their existence or as illustrating their formation.

The Niata cattle of South America, which have strangely upturned jaws, are said to breed very true and to form a definite well-marked race which, if the character were not injurious but simply indifferent, might lead to the formation of a species defined by this useless specific character. The short-legged Ancon sheep, and the six-toed cats, are other examples of such remarkable abnormalities or sports which have the curious property of being strongly hereditary, and yet, apparently, of never leading to the formation of new species. Almost all students of evolution now admit that "sports" or large and sudden divergencies from the specific type are *not* the materials from which new species have been formed, the reason being that they are extremely rare occurrences; and when any such "sport" appeared in a species, the individual presenting it would either be avoided by its fellows and leave no offspring, or by repeated crossings with the normal type the sport would disappear. We may, no doubt, imagine conditions under which a sport of this kind, once appearing in both sexes, might lead to the formation of a breed and ultimately of a species; but the combination of conditions requisite to bring this about is so improbable that we can only look upon it as a bare possibility. But the question we are discussing is not whether, under certain very rare and exceptional conditions, a few species may possibly be formed which are distinguished only by altogether useless characters, but whether such characters are common in the majority of species and, to use Mr. Romanes' words, exist in "enormous numbers." The case of abnormal sports or monstrosities such as those here referred to can certainly not be adduced as giving any support to this view.

The next case, that of the Porto Santo rabbits, is held by Mr. Romanes to prove that the constant characters which distinguished them from common rabbits were only results of the action of peculiar conditions on individuals, and were not produced by natural selection. He arrives at this conclusion from the fact that one of the two which died at the Zoological Gardens after four years' captivity was sent to Darwin, who found that the special colouring that distinguished the breed—the absence of black on the tail and ear-tips and the reddish colour on the back—had almost disappeared, and that the whole colouring was very little different from that of the common wild rabbit. Hence Mr. Romanes concludes that other wild species may be really only climatal forms, and their peculiar characters be non-adaptive. But no mention is made of the remarkably small size of these rabbits, which were only about half the weight of the common wild species and which looked no larger than average rats. If this also were a result of the action on the individuals of scanty food or a peculiar climate, it would have rapidly disappeared with ample food at the Zoological Gardens; and neither in this point nor in the peculiar form of the posterior end of the skull and interparietal bone, which was so distinct that Darwin figured it (see *Animals and Plants under Domestication*, i, p. 118), did he note any difference in the dead animal. It seems probable, therefore, that the colour-peculiarities of the Porto Santo rabbits were due to a change of tint of the longer hairs which may have been lost during the illness which led to the animal's death. And as we have no information as to the supposed change having been progressive during the four years of confinement, or that it affected the second specimen, no such conclusion as that drawn by Mr. Romanes can be held to be established . . .

Several biological friends with whom I have discussed this question, while agreeing that the majority of specific characters are useful, have suggested that useless characters may have been produced in some such manner as the following. If some useless character appears as a variation in some individuals of exceptional vigour, it may increase by interbreeding, and its repeated production being perhaps favoured by some local conditions, it may come to form a marked local variety. Now, if the conditions become unfavourable to the species in the area occupied by the type, this may in course of time become extinct, and the variety distinguished by the altogether useless character will remain as the only representative of the species. It may be admitted that such a mode of origin of a non-utilitarian specific character is conceivable, but whether it ever actually occurs in nature may be doubted; while if it does occur, it must be owing to so rare a combination of circumstances that it can produce no such general prevalence of useless specific characters as is claimed by the advocates of that theory. (If, however, the variation is preserved because it occurs in exceptionally vigorous individuals, it is correlated with a character which is useful.)

In order to ascertain whether the immediate antecedent to such a mode of species-formation as is suggested is at all common, and thinking that British flowering plants offer the best materials for its detection, I put the case to two experienced British botanists as follows:—Are there any examples within your knowledge of well-marked varieties (not mere individual states due to local conditions) which occupy a considerable area to the exclusion of the parent species, and which do not occupy any area, or only a very small one, with the type? Each of them suggested several species which seemed to answer to the conditions, but on further consideration it appeared that they did not do so, and we were finally reduced to a single case, that of one of the species of *Rubus*, a genus which most botanists will regard as a very unsafe one to draw any any conclusions from. *Rubus radula* Weihe, is said to be abundant in the Midland parts of England, but in the Southern and South-western counties to be replaced by the variety *anglicanus* of W. M. Rogers, the type never having been found in the area occupied by this variety. If this is the case, and the two forms, said to be easily recognizable, really occupy distinct areas and nowhere overlap, or very slightly so, then we have the condition precedent to the formation of a species by the extinction of the type, thus leaving the variety to represent the species. Of course in this case we do not know that the characters which distinguish the variety are useless; but if they are so, and if the variety should possess some superior vigour of constitution or other useful peculiarity which enables it to survive when the type dies out, we should have an illustration of one mode in which useless specific characters may possibly have arisen.

The enquiry is interesting, however, because it brings to light the rather unexpected fact, that fixed varieties of plants occupying considerable areas to the exclusion of the type are not common, and, perhaps, in our island do not exist. And should they be found to occur more frequently in other countries—as varieties of birds, mammals, and reptiles do occur in separate areas in North America—they may be usually explained as adaptations to very different climatic conditions, in which case the distinguishing characters will be utilitarian, and the local varieties will be really incipient species.

The preceding enquiry leads us to certain very definite conclusions. In the first place, we see that *species*, which have been differentiated as such by the laws of variation and survival of the fittest, must be characterized by certain peculiarities whereby they have obtained an advantage in the struggle with their fellows. These peculiarities constitute their "specific characters," and these *must* be useful. As this applies also to every species in the direct line of descent, the characters which are sectional or generic must also, at the time of their origin, have been useful.

In the second place, although non-utilitarian characters do undoubtedly appear in the normal course of variation, no agency has yet been detected adequate to the extension of these useless peculiarities to all the individuals which constitute a species, and, further, to prevent their extension to any of the varieties which are destined to become new species. Unless the power in question can have this twofold effect it cannot lead, except by accident, to the production of useless specific characters.

Under conceivable conditions, however, it is possible that certain useless characteristics may become limited to the individuals of a single species. But what we know of the modes of variation and the distribution of varieties indicates that, if at any time so produced, they must be altogether exceptional and of the nature of chance products; and that they cannot possibly constitute such a general characteristic of species as has been suggested.

Our final conclusion is that, whether we can discover their use or no, there is an overwhelming probability in favour of the statement that every truly *specific* character is or has been useful, or, if not itself useful, is strictly correlated with such a character.

* * * * *

In 1908—at the age of 85—Wallace attempted to reaffirm the dominant place of Darwinism in evolutionary biology in yet another review article entitled 'The present position of Darwinism'. This featured critiques of three studies that seemingly threatened the theory: 'Neo-Lamarckianism,' 'Mutationism,' and 'Mendelism':

The present position of Darwinism (ref 660)

The general public are being told to-day that Darwinism is played out; that, as a means of explaining the origin of *species* and the general development of the organic world, it is entirely superseded by newer and more scientific views. Of course the public, ever ready to accept new things in science, believes these statements, which are put forward with so much confidence and, apparently, on such good authority; while the theologians are especially glad to seize upon this weapon against what they have long considered to be their most formidable enemy.

As an example of this latter phase of opinion, we have the work of Dr. Rudolf Otto, Professor of Theology in the University of Göttingen, said to be in the front rank of the literature to which it belongs. According to a reviewer of this book in the *Inquirer* (of April 6th, 1907), the author, after a fair statement of the Darwinan theory, goes on to say:

> The modern prevailing view in contrast is that these chance individual variations play but a minor part in the production of new species. New species spring from old species by a disturbance of general vital equilibrium, from which a new state results immediately . . . The struggle for existence is an unfavourable, not an advantageous factor, since it operates to prevent new developments, and the new arises, not where the struggle is severe, but where it is weak. The theory of the building up of sporadic variations is thus giving place to the theory of the development of inherent organic tendencies and characters, which are neither produced by nor dependent upon environment, but often assert themselves against it.

This is the reviewer's summary, and he then gives us Dr. Otto's conclusion, as being, that Darwinism is "an unsuccessful hypothesis."

To turn from a German theologist to an English pressman, we find in the *Sheffield Daily Independent* of July 3rd a lengthy article on the Darwinian Jubilee of the Linnean Society, the larger portion of which consists of "Objections," among which we find such statements as the following: "The comparatively recent science of 'Heredity' has raised insuperable objections to the Darwinian assumption regarding the alleged transmission of variations;" and then, after explaining the phenomena of "Mendelism," he asks: "Where does natural selection come in then? Nobody knows exactly where . . . but the whole question of the origin of species is as much a mystery to-day as when Sir John Herschel called it 'the mystery of mysteries'." I may state in passing that the first of the above statements is simply untrue, since the "transmission" of the variations on which Darwinism rests is not an assumption, but a universally admitted fact.

Such statements as these have been going the round of the Press for the last few years, and . . . readers are naturally impressed with the apparent weight of these criticisms; and as they are quite unable to detect the mis-statements or the misinterpretation of facts, and are quite unaware of the ever-growing mass of evidence and of the more weighty scientific opinion on the other side, even well-educated people are acquiring very erroneous ideas on this subject. I therefore propose now to give a short outline sketch of the theories which are claimed to be, in whole or in part, a substitute for Darwin's explanation of organic evolution by means of Natural Selection. They are commonly known as those of the Neo-Lamarckists, the Mutationists, and the Mendelians.

The great French zoologist, Lamarck, was the first who, a century ago, proposed any detailed scientific theory of the origin of the various species of organisms by a natural process of modification; but his views were never widely adopted by naturalists, because it was clear that they would not account for all the facts, especially those of adaptation to the infinitely varied conditions of existence. This theory was, briefly, that every organ or part used in satisfying a creature's wants was increased in strength or size, or otherwise modified by use and effort, that the modifications thus produced were transmitted to their offspring, and thus led in the course of time to the production of the diverse forms we see everywhere in nature. In the case of plants, it was more especially the direct influence of the environment, heat or cold, calms or storms, a moist or dry atmosphere, a rocky or an alluvial soil, that caused the modifications of form and structure, the transmission of which in time produced new species. The author tells us that otters, beavers, water-fowl, frogs and turtles were not made web-footed in order that they might swim; but, their wants having attracted them to the water in search of food, they stretched out the toes of their feet to strike the water and more rapidly move along its surface; and thus, in course of time, the broad membranes that now connect their toes were produced.

But it was soon observed by zoologists that many of the characters of animals could not possibly have been produced by use or disuse. How, for instance, could hair or wool be so produced, how could hair change into the prickles of the hedgehog, the erectile spines of the porcupine, the nose-horn of the rhinoceros? How could the hard wing-covers of beetles be modified in a thousand ways into polished or rugged surfaces, often with knobs or even spines so closely resembling those of the bark on which they rest as to serve obviously for concealment? How could any possible *use* of the wing-coverts of the peacock, the wings of the Argus-pheasant, or the scull-cap of the crown-pigeon or the umbrella-bird produce the wonderful developments of feathers, quite useless for either flight or covering, which we find in those birds? And, as a final example, how could the infinitely varied colours and patterns which we see upon the wings of butterflies or the feathers of birds have been possibly caused by the mere *use* of those organs; yet these colours are now known, in a vast number of cases, probably in all, to serve purposes of concealment, or warning, or of recognition, which are of real, often of supreme utility, but which could not possibly have been caused or even fixed in each species by any conceivable *direct* action, either of the organism itself or of its environment.

In the case of plants, the botanist found that the direct action of the causes adduced by Lamarck could not possibly have produced the infinite variety of structural diversities in leaf, in flower, and especially in fruit, that are everywhere to be met with.

Yet though Lamarckism had been dropped as inadequate for more than half a century, at a later period, after Darwin had shown how every difficulty is presented was obviated by his own theory of natural selection, the discarded view was revived, largely by the influence of the palaeontologist [Edward D.] Cope in America. This was, in part, due to the circumstance that one of Lamarck's fundamental assumptions—the inheritance of whatever changes were produced in the individual either by the use of its own organs or by the direct agency of the environment—was believed to be true, and was actually accepted by Darwin, though he always maintained that it had very little or no effect in producing modification of species. Cope supported Lamarckism for nearly a quarter of a century, and in his last work, *The Primary Factors of Organic Evolution* [see ref 524], published in 1896, he tells us that "the stimuli of chemical and physical forces, and also molar motion, or use or its absence, are abundantly sufficient to produce variations of all kinds in organic beings." But as he deals solely with extinct animals, and carefully avoids any attempt to account for the phenomena of living things already referred to, the accuracy of the above statement can hardly be admitted.

It is a remarkable thing that so many American biologists had more or less adopted Cope's theory, to the extent at least of minimising the influence of natural selection, while none of them seem to have recognised the very small portion of the phenomena it can possibly be made to explain. But since Cope's death in 1897 a decided change of opinion has taken place, and some very valuable experiments and observations, showing the universal action of natural selection, have been carried out . . .

When we further consider that in every one of the numerous cases in which the Lamarckian theory wholly fails to account for the facts, that of Darwin fully explains them, and usually in their minutest details, we shall not be surprised that every close observer who makes himself acquainted with these facts by direct observation becomes absolutely certain of the complete breakdown of the former theory as elucidating the origin of species.

We have next to consider the more modern theory of Mutation, founded in 1901 by Dr. Hugo de Vries, of Amsterdam, in his work, entitled *Die Mutations Theorie*. This consists of two large volumes devoted to the elaborate study of the various forms of a species of evening primrose (*Œnothera Lamarckiana*), which has run wild in Holland, as the common species *Œ. biennis* has in England and elsewhere. By growing this plant from seed in large quantities for several years, a few individuals were found of such different appearance in foliage, mode of growth, size, etc., as to appear like distinct species. These are termed "mutations," and are said to come as true from seed as the parent plant usually does. De Vries, therefore, maintains that they *are* new species which he has actually produced; and from this experiment with a single species of plant, he comes to the tremendous conclusion that it is in this way that new species are produced, *per saltum*, not by the slower process of variation and selection as maintained by Darwin. Yet all his efforts to find a wild European plant behaving in the same way have been so far in vain.

It must be remembered that the genus *Œnothera* is almost peculiar to America, and is altogether absent from Europe and temperate Asia. Moreover, the plant named *Œ. Lamarckiana* is totally unknown in America (or anywhere else) in a wild state, and is said to have originated in the Jardin des Plantes at Paris; it is, therefore, thought to be a hybrid between two or more species cultivated there, and it is known that hybrids and mongrels not unfrequently produce new and eccentric forms. There is, therefore, no proof whatever that in a state of nature such mutations are produced, except, perhaps, very rarely; while the assumption that they have been, and are, produced so frequently as to constitute the mode by which *all* existing species have come into existence is a most illogical conclusion to draw from the phenomena presented by *one* species of plant of totally unknown parentage!

To show what claims are made for the theory thus founded, I will refer to a recent article by Prof. A. A. W. Hubrecht, in the *Popular Science Monthly*. He maintains that De Vries' work proves that in nature new species never arise through fluctuating variability, but are exclusively due to "mutations." He tells us that, forty years after Darwin, "the birth of a new species *has actually been observed by him*"—referring only to the abnormal "sports" of the Œnotherae of unknown parentage, as already described. He declares that "as long as the mutation has not appeared there can be no origin of a new species; the species is constant, and only subject to fluctuating variability, which never leads to the formation of a species." He concludes with the statement that De Vries "has been the first to show us the sharp distinction that exists between chance variation and fluctuating variation, and to prove that it is not the latter, but the former, that calls forth in Nature the origin of species." These positive assertions as to what has occurred throughout the whole realm of organic nature in the whole course of its development rest wholly on experiments with one plant, although those experiments are rendered comparatively valueless owing to its not being itself a known wild species, but probably a hybrid. Was there ever such a mountain of theory reared upon such an almost infinitesimal basis of fact!

Let us now see what a few biologists of eminence think of these amazing claims. There is probably no greater living authority on the subject of the variations among wild and cultivated plants than Sir W. T. Thiselton Dyer, the late Director of Kew Gardens, and at the end of a careful discussion of this subject last year in *Nature*, he concludes with the following weighty remarks:

> That mutations inevitably appear sooner or later under cultural conditions is not an assumption but a fact. If, as Mr. Lock seems to argue, there is an equal chance of their *occurrence* in either case, then their *appearance* should be more frequent in nature than in cultivation, because the former has a larger population to work with. But it is not so. I therefore conclude with Darwin that cultivation introduces some provocative condition which is lacking (or latent) in nature.

In a previous letter he had shown that such mutations are unable to hold their own in nature, and are thus constitutionally unfitted to become new species. And this is just what we should expect. For these large accidental variations or sports, occurring rarely, would have enormous chances against their being in exact adaptation to the whole inorganic and organic environment at the time and in the place where they happen to appear. They would thus necessarily die out.

Turning to the animal world, Mr. Power found several examples of natural "mutants" during his long investigation of the genus *Leptinotarsa*, and he studied them in the same complete way as all other variations. His conclusion is

> that all inheritable variations behave alike, and in no case is there any evidence of a fundamental difference between "mutants" and any other heritable variations . . . There is then no necessary incongruity between gradual small variation and rapid large variation in the origin of species, but the two are the extremes of the same process.

This entirely cuts away the basis of the mutation theory. For as the fluctuating variations are enormously frequent and in every direction, while mutations are admittedly very rare, and in only one or very few directions in each case, it is clear that the former present materials for adaptation to changed conditions which the latter do not afford.

But perhaps the most complete proof of the absolute incompetence of the mutation theory is that adduced by Professor E. B. Poulton in the Introduction to his volume of *Essays on Evolution*, just published. This is afforded by the innumerable examples of Protective Resemblance and Mimicry, which form such a striking feature of the insects of every part of the world, but especially of the tropics. In many cases these resemblances in outward form, and especially in the exact patterns and colours of the wings of butterflies and moths, which are very different in structure, and often belong to widely separated groups—moths sometimes resembling butterflies, at other times wasps—are so accurate to the smallest detail that, as he remarks, it is as impossible that they can be produced by sudden mutation as that a piece of iron could be struck blindfold with a chisel and hammer so as to produce a key that would open an elaborate safety lock . . .

We now come to the third modern theory that is so generally supposed to have replaced Darwinism in the opinion of the majority of Evolutionists, and which is known as Mendelism. For the information of the general reader, it may be well here to state, as briefly and simply as possible, what Mendelism is . . .

[Mendel's] chief experiments were with common garden peas, which, as most people know, are now usually classed in two distinct forms or varieties, those whose ripe seeds are yellow and those in which they are bluish-green. The former comprise the split-peas of grocers and many of the older and hardier sorts of garden peas, while the latter are known as marrow-fat peas, from their superior eating qualities. There are, however, many other distinctive characters, both in seed, in pod, in foliage, stature, etc. Each of these varieties, as is well known, when sown comes true to name.

But when two different varieties were crossed, Mendel found that in each case the hybrids were of one kind only. In the case of yellow and blue peas, for instance, the hybrids were all yellow. Hence in this pair he termed the yellow the "dominant" or (as Darwin would have termed it) the "prepotent" form. But the special discovery of Mendel was that when these yellow peas, the produce of the first cross, were grown by themselves, instead of producing all yellow peas, the result was both kinds in the proportion, approximately, of three yellow to one green, both colours occurring in the same pod. Then, in the third generation, another surprise occurred, for the "greens" bred true for several successive generations; but the yellows produced only one yellow which continued to breed true, while the other two broke up into yellows and greens in the same proportions as before. The final result of a large number of generations is found to be that yellows and greens, each breeding true, occur in equal proportions; whence it follows that what are termed "dominant" characters are not permanently so, but constitute a purely temporary feature which, after a large number of generations, has only a minute numerical superiority. The ultimate result of the Mendelian law seems to be *a tendency to perpetuate certain pairs of characters without intermixture.*

This fixity of character may at first sight be supposed to favour the evolution of a new species, as strongly urged by many of the Mendelians, but this will entirely depend on what is the nature of these characters. It is clear that if they are in the slightest degree, and in any way, injurious, they will interfere with adaptation, and therefore by the constant effect of natural selection will rapidly die out. Professor J. A. Thomson [in *Heredity*] tells us that Mendelian phenomena have been observed in peas, mice, rabbits, poultry, snails, and several other plants and animals. But in every case one of the parents at least is either a sport which has occurred under domestication or a doubtful natural variety or "mutation." One of the mice is an albino; of the rabbits, albino or long-haired; of the poultry, a breed with abnormal combs; of the cattle, hornless, etc.; while of two European nettles (*Urtica pilulifera* and *U. Dodartii*) the latter is an almost entire-leaved form, apparently of unknown origin, and not admitted as a species either by Hooker or Nyman . . .

It is evident that we have here to do with abnormal forms that rarely or never occur in nature as permanent self-sustaining species. The curious phenomena that present themselves when these are crossed with allied forms either domesticated or natural, though they may be of considerable interest as furnishing materials for the study of the theory of heredity, have absolutely nothing whatever to do with the origin or modification of species.

It must not be forgotten also that the results of Mendelian experiments are not constant, some observers obtaining different results with the same or very similar forms, of which Prof. Thomson gives many examples; and much ingenious reasoning is devoted to explaining (or explaining away) these diverse results.

Neither is the whole conception quite so novel as usually stated, since most of the facts were known to Darwin. He states that so long ago as 1729 it was observed that blue and white peas planted near each other mutually crossed, and in the Autumn peas of both colours were found in one pod, while none were of an intermediate colour. Wiegmann, Gärtner, and the Rev. J. M. Berkeley all repeated this experiment, and found the fact to be correct. (*Animals and Plants under Domestication*, vol. I, p. 397) The fundamental fact of Mendelism, that certain characters in a few special cases do not blend when crossed, was not only well known to Darwin, but carefully discussed by him. He has an interesting section headed: *On certain characters not blending*, which begins thus:

> When two breeds are crossed their characters usually become intimately fused together; but some characters refuse to blend, and are transmitted in an unmodified state, either from both parents or from one. When grey and white mice are paired the young are not piebald nor of an intermediate tint, but are pure white or of the ordinary grey colour; so it is when white and common turtledoves are paired. In game fowls, a great authority, Mr. J. Douglas, remarks: 'I may here state a strange fact: that if you cross a black with a white game, you get birds of both breeds of the clearest colour.' Sir R. Heron crossed during many years white, black, brown and fawn-coloured Angora rabbits, and never once got these colours mingled in the same animal, but often all four colours in the same litter. Additional cases could be given, but this form of inheritance is very far from being universal even with respect to the most distinct colours. When turnspit dogs and Ancon sheep, both of which have dwarfed limbs, are crossed with common breeds, the offspring are not intermediate in structure, but take after either parent. When tailless or hornless animals are crossed with perfect individuals it frequently, but by no means invariably, happens that the offspring are either perfectly furnished with these organs or are quite destitute of them.

He then gives cases of partially intermediate forms when Dorking fowls, hairless dogs, or solid-hoofed pigs are crossed with normal types.

He then discusses similar phenomena in plants in which still more curious facts had been observed. Major Trevor Clarke crossed a small glabrous-leaved annual stock with pollen of a large, red-flowered, rough-leaved biennial stock, the seeds of which produced half glabrous and half rough-leaved forms, but none intermediate.

> In the succeeding generations raised from rough-leaved crossed seedlings some glabrous plants appeared, showing that the glabrous character, though incapable of blending with or modifying the rough leaves, was all the time latent in this

> family of plants. The numerous plants, formerly referred to, which I raised from reciprocal crosses between the peloric and common Antirrhinum, offer a nearly parallel case; for in the first generation all the plants resembled the common form, and in the next generation, out of one hundred and thirty-seven plants, two alone were in an intermediate condition, the others perfectly resembling either the peloric or the common form.

In these last two cases we have the Mendelian law clearly brought out that the product of a first cross, presenting the characters of one of the parents only, produced seeds which, in the next generation, gave plants with the characters of the two original forms. If, therefore, Mendel had never made his experiments, or had never published them, here was the sufficient basis for what is now known as Mendelism, published by Darwin in 1868, and remaining practically unnoticed by the host of workers, some of whom declare Mendel's paper to be "the most important contribution of its size ever made to biological science."

The reason why Darwin did not prosecute the research further, so as to detect the numerical law of successive generations, is sufficiently shown in his closing remarks on the subject. In the first place, he was quite satisfied, from the large mass of facts he had accumulated during more than twenty years of research, that hybridisation or the intercrossing of very distinct forms had no place whatever in the natural process of species-formation . . .

It appears to me that in . . . [his own writings] . . . Darwin has given us a better idea of the real nature and bearing of Mendelian inheritance, as regards any possible influence on the problem of the origin of species, than any amount of study of the complex diagrams and tabular statements which the Mendelians are for ever putting before us with great flourish of trumpets and reiterated assertions of their importance.

As playing any essential part in the scheme of organic development, the phenomena seem to me to be of the very slightest importance. They arise out of what are essentially abnormalities, whether called varieties, "mutations," or sports. These abnormalities are very rare in a state of nature, as compared with the ever-present individual variability ample in amount and affecting every part and organ which furnishes the material for both man's and for nature's selection. The very fixity of these abnormalities and the small number of the characters affected by them, as well as their rarity, all show them to be the refuse material of nature's workshop, as proved by the fact that *none of them ever maintain themselves in a state of nature.* It seems to me that the most probable interpretation of the phenomena is, that both the fixity and the rarity of such characters are beneficial to the species in which they appear, because it renders their extinction under natural conditions more certain and more rapid, thus preventing the injurious effects that might result from their competing with the normal form while undergoing slow adaptive modification, or from checking that adaptation by the continued production of less adapted forms through intercrossing. It is, I think, clear that any species which gave birth to a large number of such abnormal and unchangeable individuals would be so hampered by them whenever adaptive modification became necessary that the whole species might be in danger of extinction. Hence, among allied species, those which gave rise to fewest of such abnormal forms would be best adapted to become the parents of new species, whenever the need for adaptation to new conditions arose.

But the absolute denial of any part in the process of organic evolution for these abnormal outcasts of nature does not imply that their study may not have a certain value for a comprehension of the mysterious phenomena of inheritance. The study of disease in all its strange forms, together with that of the numerous deformities to which the human body is subject, may in many cases have thrown light on obscure physiological processes and upon the conditions essential to health; but we should hardly claim them as being essential to the processes of a healthy and normal growth.

To anyone who has devoted a considerable portion of his life to the study of nature, both in field and in cabinet, both at home and in distant regions, the vast complex of phenomena presented by the organic world, with its endless specific forms, their myriad relations and adaptations, the laws of their development in the past and their distribution in the present, is almost overwhelming in its grandeur and its beauty. Almost all such loving students of nature have found in the theory of Darwin, in his many stimulating works and in those of his friends and followers, the only intelligible clue to the mighty labyrinth of nature. To such students of nature the claims of the Mutationists and the Mendelians, as made by many of their ill-informed supporters, are ludicrous in their exaggeration and total misapprehension of the problem they profess to have solved. To set upon a pinnacle this mere side-issue of biological research, as if it comprised within itself all the phenomena and problems presented by the organic cosmos, is calculated to bring ridicule upon what, in its place, may be an interesting and perhaps useful line of study. To myself these monstrous claims suggest a comparison with those of the perhaps equally enthusiastic and equally ill-informed admirers of the immortal Pickwick, who believed his "Speculations on the Source of the Hampstead ponds with some Observations on the Theory of Tittlebats," to have been a most important contribution to the science of that period . . .

* * * * *

Though some observers (for example Gould 1980) have charged Wallace with promoting a teleological 'law' of natural selection that is untestable as stated, this may be a premature judgment. Again, it must be emphasized that Wallaceian natural selection is a purely ecological concept; it describes a state of interaction between population and environment, not, as in Darwin's setting of the principle, a process. As a result, Wallace's ideas on the subject cannot be analysed in strictly Darwinian terms, and until they are recognized for what they are and subjected to forms of test that are consistent with their own logic, there is little point in launching off-target criticisms of them (see C. H. Smith 1989). Of course, it is fully reasonable to complain that Wallace's evolutionary cosmology *as a whole falls short of being an exploitable scientific structure, but once again this is not because it is demonstrably incorrect. It is, instead, and simply, too vague. Were the entire concept couched in terms more formal than Wallace himself was able to provide for it, it might even yield some very new kinds of answers to some very old and vexing questions. This is a big 'if'; on the other hand, existing responses to Wallace's reasoning fall a good deal short of showing that such an 'upgrading' is not possible.*

In any case, from the standpoint of sheer inventiveness, Wallace has been approached by but few evolutionary thinkers. Others—including Darwin, of course—have contributed more, to the extent that their writings have been more widely read, accepted, and cited, but one is hard put to identify any figure who attempted to cope, systematically, with a wider range of evolutionary questions. Although in this regard his vision can hardly be credited as unerring (that he underestimated the importance of mutation studies and Mendelian genetics, for example, is plain enough), it yet touched on a variety of subjects that have proven to be of enduring interest. Despite this fact—and the related one that a number of the more unorthodox positions he defended remain unanswered—some workers still reject Wallace as a careless purveyor of pseudo-scientific objectives. The dust, however, is still settling.

Related writings

on the utility hypothesis: refs 26, 57, 134, 153, 165, 303, 314, 389, 390, 395, 397, 432, 510, 527, 558, 600, 649, 657, 716, 724, 725, 727

on variation *per se* (including slow vs. rapid change): refs 83, 96, 140, 155, 175, 176, 212, 248, 253, 293, 303, 311, 322, 389, 394, 397, 432, 440, 503, 510, 524, 660, 666, 669, 716, 724, 727, 732

on varieties and species divergence: refs 39, 43, 58, 89, 96, 175, 212, 303, 322, 389, 390, 394, 410, 432, 527, 558, 648, 716, 724, 727, 729, 732

on mimicry: refs 72, 96, 98, 121, 123, 134, 175, 176, 177, 179, 238, 253, 272, 319, 353, 359, 535, 648, 715, 716, 719, 724, 725

on other kinds of protective and otherwise functional coloration: refs 96, 123, 129, 130, 134, 138, 139, 140, 176, 186, 205, 238, 257, 272, 304, 308, 318, 319, 338, 378, 389, 417, 424, 511, 535, 715, 716, 719, 724, 725, 727, 729

on other evolutionary implications of colour: refs 64, 96, 139, 257, 272, 293, 304, 308, 324, 338, 348, 378, 389, 424, 425, 432, 499, 716, 719, 724, 725, 732

on particular adaptations, especially as related to natural selection: refs 57, 64, 79, 83, 93, 136, 139, 140, 155, 164, 165, 197, 198, 220, 240, 241, 248, 257, 272, 279, 304, 322, 324, 352, 381, 389, 394, 397, 415, 477, 485, 499, 503, 510, 527, 558, 633, 660, 669, 715, 716, 719, 724, 725, 727, 732

on instinct, automatism, and related aspects of animal behavior: refs 57, 83, 94, 136, 139, 164, 180, 186, 216, 217, 220, 222, 227, 228, 245, 303, 304, 353, 424, 432, 470, 534, 716, 718, 724, 725, 727, 729 on sexual selection: refs 134, 139, 140, 173, 186, 253, 272, 304, 424, 459, 527, 712, 716, 719, 724, 725, 727

on heredity: refs 123, 134, 161, 175, 211, 248, 338, 432, 445, 468, 473, 499, 510, 596, 649, 660, 669, 716, 724, 725, 727, 732, 733, 737

on George Romanes's theory of physiological selection/the infertility of hybrids: refs 100, 389, 390, 395, 428, 429, 527, 724, 727, 729

on acquired characters, telegony, and prenatal influence: refs 43, 311, 394, 397, 415, 445, 468, 473, 476, 479, 502, 503, 510, 524, 534, 596, 649, 660, 716, 724, 725, 727, 733, 737

on evolutionary (and evolution as related to) systematics: refs 20, 28, 77, 79, 88, 89, 205, 239, 246, 279, 309, 394, 716, 718, 719, 725

on the relations between evolution and biogeography/palaeogeography/physical geography: refs 20, 38, 53, 79, 81, 82, 85, 89, 96, 97, 146, 181, 188, 244, 248, 257, 269, 272, 302, 316, 322, 340, 393, 469, 488, 503, 646, 715, 716, 718, 719, 720, 721, 724, 725, 727, 728, 729, 732

on the origin of life: refs 196, 211, 348, 602, 646, 700, 706, 728, 730, 732, 746, 751

on the evolution of the mind and body of mankind: see Section One, p. 64

on the 'function' of beauty in nature: refs 26, 139, 140, 213, 272, 424, 432, 470, 547, 716, 719, 725, 732

on evolutionary philosophy and cosmology: refs 93, 99, 136, 140, 146, 155, 165, 173, 210, 211, 221, 242, 253, 272, 303, 305, 322, 352, 490, 493, 504, 523, 574, 589, 600, 646, 649, 669, 695, 700, 706, 716, 719, 724, 725, 727, 728, 729, 732, 733, 742, 744, 746, 748, 750, 751, 753

on the history of evolutionary theory (including evaluations of Darwin's contributions): refs 134, 253, 311, 358, 493, 510, 516, 532, 589, 599, 649, 656, 657, 660, 666, 672, 673, 676, 716, 724, 725, 726, 727, 729, 742, 743

on miscellaneous topics: refs 107, 108, 146, 165, 196, 211, 221, 248, 324, 348, 419, 442, 445, 499, 589, 648, 669, 716, 725, 726, 727, 728, 729, 732, 733, 735, 738

additional related topics: see Section Eight, p. 471

book reviews and notices: refs 134, 140, 146, 155, 161, 180, 186, 196, 210, 211, 217, 220, 221, 237, 238, 239, 242, 253, 303, 304, 305, 311, 314, 324, 337, 338, 348, 352, 394, 397, 424, 432, 437, 461, 470, 490, 493, 504, 523, 524, 532, 534, 716, 727

secondary literature: Beddall (1968, 1969), Brooks (1972, 1984), Eiseley (1958), George (1964), Gould (1980), Henderson (1958), Kleiner (1985), Kottler (1974), Loewenberg (1959), McKinney (1966, 1972), Marchant (1916), C. H. Smith (1989), R. Smith (1972), Vorzimmer (1970)

§ 7 · PHYSICAL GEOGRAPHY AND GEOLOGY

. . . These various facts as to the distribution of alpine lakes—their almost total absence in all parts of the world outside of glaciated districts, and within glaciated districts their prevalence in the newer and more easily denuded rocks—are what have to be explained by the advocates of the theory of earth-movements, and this, so far as I am aware, they have never attempted to do. Equally important, and equally difficult to explain on the earth-movement theory, is the fact that alpine lakes are almost always situated just at those spots where, by means of converging valleys, the glaciers would become heaped up and attain their maximum thickness, or where there is good evidence that they have been very thick; and it is the grinding power of this enormous weight of ice, acting differentially as regards the softer and harder rocks, that has worn out hollows in pre-existing valleys now occupied by lakes. In almost every case, too, it will be seen that there is a constriction or narrowing of the valley towards or beyond the lower end of the lake, which, by preventing the free escape of the ice, has increased its thickness and grinding power . . .

('The glacier theory of alpine lakes,' 1893: ref 462, p. 438)

WALLACE'S contributions to the physical sciences were neither so varied nor so important as those he made to biogeography and evolutionary theory. None the less, he frequently wrote on themes involving physical geography, and occasionally on geological subjects as well. It is now little known that, all told, he produced several dozen works in these fields, some of which featured significant ideas, and all of which exhibit a clarity and forcefulness of argument paralleling that found in his biological writings.

In *My life*, Volume 1, Wallace writes that his interest in the physical aspects of natural history developed during his early surveying years. At that time, he began to take notice of geological phenomena (such as fossils) and, as already mentioned in Section Three, became acquainted with various aspects of agriculture and the economics of land use. About 1845, he read Charles Lyell's *Principles of geology*, the uniformitarian perspective of which profoundly influenced all his subsequent thoughts on cognate subjects. Lyell's anti-catastrophism arguments, for example, certainly pervade Wallace's ideas on the nature of biological change, especially in so far as Wallace ascribed species modification to continually and gradually acting processes.

Had things gone otherwise, Wallace might well first have made his scientific reputation as a physical geographer rather than as a biologist. He had planned to write a book on the physical geography of the Amazon upon returning to England, but this project fell through when he lost all pertinent notes in the fire at sea. Wallace was, in fact, a fully expert physical geographer (his only regular employment after 1862, interestingly, consisted of over twenty years of spending several weeks each summer correcting examination papers as an Assistant Examiner on this subject, primarily for the British Government's Science and Art Department). His main original contributions within this domain revolve generally around: (1) the relation of physical geography and geology to biogeographic and evolutionary processes, and (2) glacial theory.

Although many of Wallace's early writings contain observations on the physical geography and geology of the areas where he was collecting, his first organized treatment of these subjects came in the form of a regional survey of the Malay Archipelago prepared after his return to England:

On the physical geography of the Malay Archipelago (ref 78)

. . . On the present occasion I propose to give a sketch of what is most interesting in the physical geography of [the 'Malay or Indian Archipelago'], including in that term the general relations of the organic world to the present and past conditions of the earth's surface.

1. *Definition of the Archipelago, Position, Extent, and Magnitude of the chief Islands.*—It first becomes necessary to define accurately the limits of the Archipelago, pointing out exactly what islands we include within it; for, though "all the islands between south-eastern Asia and Australia" seems pretty definite, yet to the eastward this region blends insensibly into the vast extent of the Pacific islands. According to my views, the Malay—or, as I should prefer to name it, the Indo-Australian—Archipelago extends from the *Nicobar Islands* on the north-west to *St. Christoval*, one of the Solomon Islands, on the south-east, and from *Luzon* on the north to *Rotti*, near Timor, on the south. The eastern boundary is drawn at this particular point for reasons which will be explained further on. Though not geographically correct to include any part of a continent in an archipelago, it is necessary for our purpose to consider the Malay peninsula as not only almost but quite an island, since it cannot be physically separated from the region of which we are now treating.

Thus limited, the Archipelago is of a somewhat triangular form, with an extreme length of about 5000, and a breadth of rather more than 2000 English miles. The mere statement of these dimensions, however, will give but an imperfect idea of the extent and geographical importance of this region . . . Educated persons . . . can hardly bring themselves to imagine that this sea of islands is really in many respects comparable to the great continents of the earth. The traveller, however, soon acquires different ideas. He finds himself sailing for days or even for weeks along the shores of one of these great islands, often so great that the inhabitants believe it to be a boundless continent. He finds that voyages among these islands are commonly reckoned by weeks and months, and that the inhabitants of the eastern and western portions of the Archipelago are as mutually unknown to each other as are the native races of

North and South America. On visiting the coasts of one of the larger islands, he hears of the distinct kingdoms which lie along its shores, of the remote north or east or south of which he can obtain little definite information, and of the wild and inaccessible interior, inhabited by cannibals and demons, the haunt of the charmed deer which bears a precious jewel in its forehead, and of the primaeval men who have not yet lost their tails. The traveller, therefore, soon looks upon this region as one altogether apart. He finds it possesses its own races of men and its own aspects of Nature. It is an island-world, with insular ideas and feelings, customs, and modes of speech; altogether cut off from the great continents into which we are accustomed to divide the globe, and quite incapable of being classed with any of them. Its dimensions, too, are continental. You may travel as many thousand miles across it, in various directions, occupying as many weeks and months as would be necessary to explore any of the so-called quarters of the globe. It contains as much variety in its climate, in its physical phenomena, its animate and inanimate life, and its races of mankind as some of those regions exhibit . . .

I will now endeavour to give you a clearer idea of its extent and magnitude by comparing it with some regions nearer home. If, first, we bring the Malay Archipelago to Europe, keeping the meridians parallel, and place the western extremity of the island of *Sumatra* upon the *Land's End*, *New Guinea* will then spread over *Turkey* and a good deal of *Persia*, and the *Solomon Islands* will reach to the borders of the *Punjaub*; while the northern extremity of *Luzon* will be near the *White Sea*, and the islands of *Timor* and *Rotti* in *Syria*. The area of the whole Archipelago is, however, much less than its dimensions would seem to imply, being, on a fair calculation, about equal to that portion of Europe which lies south of St. Petersburg and the Shetland Islands . . .

. . . This great region of mingled land and water is, then, as a whole, comparable in its dimensions with the primary divisions of the earth, while its component parts are on an equally extended scale,—two of the islands, *Borneo* and *New Guinea*, being the largest on the globe. They are nearly equal in extent, and the only other island which approaches them is *Madagascar*. Borneo would contain within its vast area the whole of Great Britain and Ireland, with all their islets from Scilly to Shetland in their true relative positions, and still leave boundless forests stretching out like an ocean beyond them. Then comes *Sumatra*, about equal to Great Britain; after which follow *Java*, *Luzon*, and *Celebes*, either of which may compare in size with *Ireland* or one of the larger *New Zealand Islands*. After these succeed eighteen islands, which average as large as Jamaica; more than *one hundred* about the size of the isles of Wight and Man, with many thousands of isles and islets below these, and which are practically innumerable.

In their physical constitution and attendant phenomena the islands of the Archipelago offer us some remarkable and instructive contrasts. Active and extinct volcanoes are abundant in many of the islands, in others they are altogether absent. The former, as a general rule, are subject to frequent earthquakes, which in the others are quite unknown. In the greater part of the Archipelago one vast, ever-verdant forest covers hill and valley, plain and mountain, up to the very loftiest summits; whereas in another and much smaller portion such dense and gloomy forests are altogether unknown, the country consisting of arid hills and plains, with a comparatively scanty covering of shrubs and trees. Again, over some extensive districts the monsoons, or periodical winds, with their attendant rains or drought, divide the year into a well-defined and regularly-recurring wet and dry season. Over other scarcely less extensive districts no such regularity exists; the inhabitants themselves can hardly tell you when their rainy or dry season usually begins, and the traveller soon finds the climate to be almost as variable and the skies as inconstant as in our own much-abused island. Even in districts where the season is regular, there are no less striking contrasts; one portion of an island having its wet weather while the remainder is parched up, and islands within sight of each other having very different seasons.

There is yet another contrasting aspect in which the Archipelago may be viewed, less obvious but leading to far more important results than any I have yet mentioned, namely, that one large portion of it is connected by a very shallow sea to the continent of Asia, another part is similarly joined to Australia, while the remaining islands are surrounded by a practially unfathomable ocean. We shall consider the chief islands of the Archipelago, therefore, under the heads of,—1st. *Volcanic* and *Non-Volcanic*; 2nd. *Forest Country* and *Open Country*; 3rd. *Well-marked Seasons* and *Undefined Seasons*; and 4th. *The Western or Indo-Malayan Region*, and the *Eastern or Austro-Malayan Region*.

Looking at a map on which the volcanic regions of the Archipelago are marked out—those which are subject to earthquakes, which are of volcanic origin, and which abound more or less in extinct as well as active volcanoes—we see at a glance that the great islands of Borneo and Celebes form the central mass around which the volcanic islands are distributed so as rudely to follow their outline and embrace them on every side but one in a vast fiery girdle. Along this great volcanic band (about 5000 miles in length) at least fifty mountains are continually active, visibly emitting smoke or vapour; a much larger number are known to have been in eruption during the last 300 years; while the number which are so decidedly of volcanic origin that they may at any moment burst forth again, must be reckoned by hundreds.

In the whole region occupied by this volcanic belt, and for a considerable breadth on each side of it, earthquakes are of continual recurrence, slight shocks being felt at intervals of every few weeks or months, while more severe ones, shaking down whole villages, and doing more or less injury to life and property, are sure to happen in one part or another of this district almost every year. In many of the islands the years of the great earthquakes form the chronological epochs of the native inhabitants, by the aid of which the ages of their children are remembered, and the dates of many important events are determined.

It is not now my object to describe the many fearful eruptions that have taken place in this region. In the amount of injury to life and property, and in the magnitude of their effects, they have not been surpassed by any upon record. Forty villages were destroyed by the eruption of Papaudayang in Java, where the whole mountain was blown up by repeated explosions, and a large lake left in its place. By the great eruption of Toruboro in Sumbawa 12,000 people were destroyed, and the ashes darkened the air, and fell thick upon the earth and sea for 300 miles round. Even quite recently, since I quitted the country, a mountain which had been quiescent for more than 200 years suddenly burst into activity. The island of Makiau, one of the Moluccas, was rent open in 1646 by a violent eruption which left a huge chasm on one side, extending into the heart of the mountain. It was, when I last visited it, clothed with vegetation to the summit, and contained twelve populous Malay villages. On the 29th of December, 1862, after 215 years of perfect inaction, it again suddenly burst forth, blowing up and completely altering the appearance of the mountain, destroying the greater part of the inhabitants, and sending forth such volumes of ashes as to darken the air at Ternate, 40 miles off, and almost entirely to destroy the growing crops on that and the surrounding islands.

The island of Java contains more volcanoes, active and extinct, than any other known district of equal extent. They are about forty-five in number, and many of them exhibit most beautiful examples of the volcanic cone on a large scale, single or double, with entire or truncated summits, and averaging about 10,000 feet high.

It is now well ascertained that almost all volcanoes have been slowly built up by the accumulation of the matter—mud, ashes, and lava—ejected by themselves. The openings or craters, however, frequently shift their position; so that a country may be covered with a more or less irregular series of hills in chains and masses only here and there rising into lofty cones, and yet the whole may be produced by true volcanic action. In this manner the greater part of Java has been formed. There has been some elevation, especially on the south coast, where are extensive cliffs of raised coral limestone; and there may be a substratum of older stratified rocks, but still essentially Java is volcanic; and that noble and fertile island—the very garden of the East, and perhaps upon the whole the richest, the best cultivated, and the best governed tropical island in the world—owes its very existence to the same intense volcanic activity which still occasionally devastates its surface.

The great island of Sumatra exhibits in proportion to its extent a much smaller number of volcanoes, and a considerable portion of it has had probably a non-volcanic origin.

To the eastward the long string of islands from Java passing by the north of Timor and away to Banda are probably all due to volcanic action. Timor itself consists of ancient stratified rocks, but is said to have one volcano near its centre.

Going northward, Amboyna, a part of Bouru, and the west end of Ceram, the north part of Gilolo and all the small islands of Siau and Sauguir are wholly volcanic. The Philippine Archipelago contains many active and extinct volcanoes, and has probably been reduced to its present fragmentary condition by subsidences attending on volcanic action.

All along this great line of volcanoes are to be found more or less palpable signs of great upheaval and depression of land. The range of islands south of Sumatra, a part of the south coast of Java, and of the islands east of it, the west and east end of Timor, portions of all the Moluccas, the Ké and Aru Islands, Waijiou, and the whole south and east of Gilolo consist in a great measure of upraised coral-rock, exactly corresponding to that now forming in the adjacent seas. In many places I have observed the very surfaces of the upraised reefs, with the great masses of coral standing up in their natural position and hundreds of shells, so fresh-looking that it was hard to believe they had been more than a few years out of the water; and, in fact, it is very probable that such changes have occurred within a few centuries.

In striking contrast with this region of subterranean fires, the island of Celebes in all its southern peninsulas, the great mass of Borneo, and the Malay peninsula, are not known to contain a single volcano, active or extinct. To the east of the volcanic band is another quiescent area of 1000 miles wide, the great island of New Guinea being free from volcanoes and earthquakes. Towards its eastern extremity, however, these reappear in some small islands off its coast, and in New Britain, New Ireland, and the Solomon Islands, which contain active volcanoes.

The difference between the aspect of the volcanic and the non-volcanic regions is by no means so striking as might be imagined. Where active volcanoes or true volcanic peaks exist, a peculiar character is at once given to the islands, which are also in almost every case characterised by excessive fertility. In many of the adjoining districts, however, though volcanic products may be everywhere visible, the general aspect of the country, the outline of the hills, and the character of the vegetation, does not differ materially from those of many parts of Borneo and New Guinea. The island of Amboyna, for instance, consists principally of raised coral-rock almost everywhere covered with deep-red volcanic clays and gravels, and in places capped with basalt and lavas; yet the soil is by no means fertile, and where the native forest vegetation is cleared off, the ground bears only a scanty covering of dwarf shrubs and rigid herbage. The chief characteristic of the non-volcanic regions appears to me to consist in the great flat valleys that line the coasts, and penetrate far inland between the mountain ranges—the result of the long and uninterrupted action of rivers and tropical rains (combined probably with a slow elevation of the land) in filling up the gulfs that once intervened between the mountain ranges. A subsidence of a few hundred feet would reduce Borneo into a shape very similar to that of Celebes, which island may be considered to be now in the state that Borneo has just passed out of, and to be still engaged in filling up and converting into swampy plains the deep gulfs that at present occupy the spaces between her radiating lines of mountains.

The very extraordinary forms of Celebes and Gilolo have been imputed by some authors to sudden and capricious elevation. Mr. Windsor Earl speaks of the volcanic action where it was strongest "throwing the islands into fantastic forms." Celebes, however, is free from volcanoes except at its northern extremity, and its southern peninsula consists of mountains of basalt and limestone. From peculiarities in its natural productions, the shallowness of some of the gulfs between its peninsulas, and the number of coral islets that surround the southern portion of it, Celebes was once probably much more extensive, perhaps equal to Borneo, at a time when Borneo was just rising above the ocean . . . As in every part of the world of which we have accurate geological knowledge, risings and sinkings of the land to the amount of several hundreds of feet have repeatedly occurred, these two islands may each have successively assumed the form of the other without any violent convulsion. From the vast, swampy, level plains which stretch into the very heart of Borneo, allowing vessels to ascend its southern rivers about 300 miles in a straight line, it has probably been for a long time stationary, and thus been enabled to fill up the gulfs that formerly penetrated it. At a still earlier period it must have been much more deeply submerged, when the extensive coal-beds found in almost every part of it were being formed. This, however, was at no very remote period, geologically speaking, for the coal of Borneo is all tertiary. Instead of the ferns and lepidodendra, and other plants of extinct genera which abound in our coal-shales, those of Borneo contain only impressions of leaves of exogenous trees which can hardly be distinguished from those growing in the surrounding forests.

The contrasts of vegetation and climate in the Archipelago may best be considered together, the one being to some extent dependent on the other.

Placed immediately upon the Equator, and surrounded by extensive oceans, it is not surprising that the various islands of the Archipelago should be almost always clothed with a forest vegetation from the level of the sea to the summits of the loftiest mountains. This is the general rule. Sumatra, New Guinea, Borneo, the Philippines and the Moluccas, and the uncultivated parts of Java and Celebes, are all forest countries, except a few small and unimportant tracts, due perhaps, in some cases, to ancient cultivation or accidental fires. To this, however, there is one important exception in the island of Timor and all the smaller islands opposite, in which there is absolutely no forest such as exists in the other islands, and this character extends in a lesser degree to Flores, Sumbawa, Lombock, and Bali.

In Timor the most common trees are *Eucalypti* of several species, so characteristic of Australia, with sandalwood, acacia, and other sorts in less abundance. These are scattered over the country more or less thickly, but never so as to deserve the name of a forest. Coarse and scanty grasses grow beneath them on the more barren hills, and a luxuriant herbage in the moister localities. In the islands between Timor and Java there is often a more thickly wooded country, but thorny and prickly trees abound. They seldom reach any great height, and during the force of the dry season they almost completely lose their leaves, allowing the ground to be parched beneath them, and contrasting strongly with the damp, gloomy, ever-verdant forests of the other islands. This peculiar character, which extends in a less degree to the southern peninsula of Celebes and the east end of Java, is most probably owing to the proximity of Australia. The south-east monsoon, which lasts for about two-thirds of the year (from March to November), blowing over the northern parts of that country, produces a degree of heat and dryness which assimilates the vegetation and physical aspect of the adjacent islands to its own. A little further eastward in Timorlaut and the Ké Islands, a moister climate prevails, the south-east winds blowing from the Pacific through Torres Straits, and as a consequence every rocky islet is clothed with verdure to its very summit. Further west again, as the same winds blow over a wider and wider extent of ocean, they have time to absorb fresh moisture, and we accordingly find the island of Java possessing a less and less arid climate in the dry season, till in the extreme west near Batavia rain occurs more or less all the year round, and the mountains are everywhere clothed with forests of unexampled luxuriance.

The changes of the monsoons and of the wet and dry seasons in some parts of the Archipelago, are very puzzling; and an accurate series of observations in numerous localities is required to elucidate them.

Speaking generally, the whole south-western part of the Archipelago, including the whole range of islands from Sumatra to Timor, with the larger half of Borneo and the southern peninsula of Celebes, have a dry season from April to November, with the south-east monsoon. This same wind, however, bends round Borneo, becoming the south-west monsoon in the China sea, and bringing the rainy season to Northern Borneo and the Philippines.

In the Moluccas and New Guinea the seasons are most uncertain. In the south-east monsoon from April to November, it is often stormy at sea, while on the islands it is very fine weather. There is generally not more than two or three months of dry hot weather about August and September. This is the case in the northern extremity of Celebes and in Bouru, whereas in Amboyna July and August are the worst months in the year. In Ternate, where I resided at intervals for three years, I never could find out which was the wet and which the dry season. The same is the case at Banda, and a similar uncertainty prevails in Menado, showing probably that the proximity of active volcanoes has a great disturbing meteorological influence. In New Guinea a great amount of rain falls more or less all the year round. On the whole the only general statement we can make seems to be that the countries within about three degrees on each side of the Equator have much rain and not very strongly contrasted seasons; while those with more south or north latitude, have daily rains during about four months in the year, while for five or six months there is almost always a cloudless sky and a continual drought.

We have next to consider the Malayan Archipelago in its geological and zoological relations to Asia and to Australia, it being now a well established fact that one portion of it is almost as much Asiatic in its organic productions as the British Isles are European, while the remainder bears the same relation to Australia that the West India Islands do to America.

It was first pointed out by Mr. George Windsor Earl, in a paper read before this Society eighteen years ago, that a shallow sea connected the great islands of Sumatra, Borneo, and Java, to the Asiatic continent, with which they generally agreed in their natural productions; while a similar shallow sea connected New Guinea and some of the adjacent islands to Australia. Owing, however, to that gentleman's imperfect knowledge of the natural history of the various islands, he did not fully appreciate the important results of this observation, and in fact in the same paper argued in favour of the former connection of Asia and Australia—a connection to which the whole bearing of the facts in physical geography and natural history is plainly opposed.

. . . in all cases where we have independent geological evidence, we find that those islands, the productions of which are identical with those of the adjacent countries, have been joined to them within a comparatively recent period, such recent unity being in most cases indicated by the very shallow sea still dividing them; while in cases where the natural productions of two adjacent countries is very different, they have been separated at a more remote epoch—a fact generally indicated by a deeper sea now dividing them. The reason of this is obvious. For example: let a subsidence take place, cutting off any portion of a continent, and forming an island. The organic productions of the two portions are at first identical, but they are not permanent. The changes that have always gone on still go on. Some species slowly die out, new ones take their place, and thus in time the animals and plants of the island come to differ from those of the country from which it was severed; and if the subsidence which first separated them goes on widening and deepening the sea between them, there will come in time to be such a marked difference in their productions as we see between Madagascar and Africa.

This general principle is of almost universal application, so that when we find an island whose animals and plants exactly agree with those of an adjacent land, we look for evidence of its recent separation from that land; while, on the other hand, any remarkable diversity of natural productions forces on us the conclusion that the watery barrier which now exists has existed for a very long geological period; and when the diversity is almost total, not only in species but in larger groups such as general families and orders, we conclude that these countries could never have been connected since our continents and oceans had assumed their present general outlines.

Returning now to the Malay Archipelago, we see that the whole of the seas connecting Java, Sumatra, and Borneo with Malacca and Siam are under 50 fathoms deep, so that an elevation of only 300 feet would add this immense district to the Asiatic continent. The 100 fathom line will also include the Philippine Islands and the island of Bali, east of Java. From this we should naturally conclude that the subsidence breaking up this portion of Asia had recently taken place, and we have a very sufficient cause for such subsidence in the vast range of volcanoes in Sumatra and Java, whose elevatory action must have been counterbalanced by some adjacent depression. On examining the zoology of these countries this opinion is confirmed, for we find the most overwhelming evidence that these great islands must have once formed a part of the continent, and could only have been separated at a very recent geological epoch . . .

The general result . . . at which we arrive is, that the great islands of Java, Sumatra, and Borneo resemble in their natural productions the adjacent parts of the continent, almost as much as such widely-separated districts could be expected to do even if they still formed a part of Asia; and this close resemblance, joined with the fact of the wide extent of sea which separates them being so uniformly and remarkably shallow, and lastly, the existence of the extensive range of volcanoes in Sumatra and Java, which have poured out vast quantities of subterranean matter and have built up extensive plateaux and lofty mountain ranges, thus furnishing a "*vera causa*" for a parallel line of subsidence—all lead us irresistibly to the conclusion that at a very recent geological epoch the continent of Asia extended far beyond its present limits in a south-easterly direction, including the islands of Java, Sumatra, and Borneo, and probably reaching as far as the present 100 fathom line of soundings.

The Philippine Islands agree in many respects with Asia and the other islands, but present some anomalies, which seem to indicate that they were separated at an earlier period, and have since been subject to many revolutions in their physical geography.

Turning our attention now to the remaining portion of the Archipelago, we shall find that all the islands from Celebes and Lombock eastward, exhibit almost as close a resemblance to Australia and New Guinea as the Western Islands do to Asia. It is well known that the natural productions of Australia differ from those of Asia more than those of any of the four ancient quarters of the world do from each other. Australia, in fact, stands alone: it possesses no apes or monkeys, no cats or tigers, wolves, bears, or hyenas, no deer, or sheep, or oxen; no elephant, horse, squirrel, or rabbit; none, in short, of those familiar types of quadruped which are met with in every other part of the world. Instead of these, it has marsupials only, kangaroos and opossums, wombats and the duck-billed *platypus*. In birds it is almost as peculiar. It has no woodpeckers and no pheasants, families which exist in every other part of the world; but instead of them it has the mound-making brush-turkeys, the honey-suckers, the cockatoos, and the brush-tongued lories, which are found nowhere else upon the globe . . .

Now all these peculiarities exist also in the Australian portion of the Malay Archipelago, as may be seen by the . . . characteristic mammals and birds of the two regions. The contrast is nowhere so abruptly exhibited as on passing from the island of Bali to that of Lombock, where the two regions are in closest proximity. In Bali we have barbets, fruit-thrushes, and woodpeckers; on passing over to Lombock these are seen no more, but we have abundance of cockatoos, honeysuckers, and brush-turkeys, which are equally unknown in Bali and every island further west. The strait here is 15 miles wide, so that we may pass in two hours from one great division of the earth to another, differing as essentially in their animal life as Europe does from America. If we travel from Java or Borneo, to Celebes or the Moluccas, the difference is still more striking. In the first, the forests abound in monkeys of many kinds, wild cats, deer, civets, and otters, and numerous varieties of squirrels are constantly met with. In the latter none of these occur; but the prehensile-tailed opossum is almost the only terrestrial animal seen, except wild pigs, which are found in all the islands, and deer (which have probably been recently introduced) in Celebes and the Moluccas. The birds which are most abundant in the Western Islands are woodpeckers, barbets, trogons, fruit-thrushes, and leaf-thrushes: they are seen daily, and form the great ornithological features of the country. In the Eastern Islands these are absolutely unknown, honeysuckers and small lories being the most common birds; so that the naturalist feels himself in a new world, and can hardly realise that he has passed from the one region to the other in a few days, without ever being out of sight of land.

The inference that we must draw from these facts is undoubtedly that the whole of the islands eastwards from Java and Borneo do essentially form a part of a former Australian or Pacific Continent, from which they were separated, not only before the Western Islands were separated from Asia, but probably before the extreme south-eastern portion of Asia was raised above the waters of the ocean; for a great part of the land of Borneo and Java is known to be geologically of quite recent formation, while the very great difference of species and in many cases of genera also between the productions of the Eastern Malay Islands and Australia, as well as the great depth of the sea now separating them, point to a comparatively long isolation and an early epoch of separation. It is interesting to observe among the islands themselves how a shallow sea always intimates a recent land-connection. The Aru Islands, Mysol, and Waigiou, as well as Jobie, agree with New Guinea in the mammalia and birds much more closely than they do with the Moluccas, and we find that they are all united to New Guinea by a shallow sea. In fact, the 100 fathom line round New Guinea marks out accurately the range of the true Paradise birds.

The existence of a Pacific continent was long ago indicated by Mr. Darwin's researches on the structure and origin of coral-reefs, the numerous atolls and barrier-reefs in the whole of this district being shown to depend upon the subsidence of land for long periods. This so exactly agrees with the singular unity now existing among the organic productions of a vast number of islands, which at the same time are very different from those of any other part of the world, that we must accept it as a fair deduction from the only evidence we can ever hope to obtain of this class of changes.

I would particularly call attention to the fact that the division of the Archipelago here pointed out, into two regions characterised by a striking diversity in their natural productions, does *not* correspond to any of the physical or climatal divisions of the surface. The great volcanic chain runs through both parts: Borneo closely resembles New Guinea, not only in its vast size but in its climate and the general aspect of its vegetation; the Moluccas are the counterpart of the Philippines in their volcanic origin, their extreme fertility, their luxuriant forests, and their frequent earthquakes; and the east end of Java has a climate almost as dry as that of Timor. Yet between these corresponding groups of islands, constructed as it were after the same pattern, there is the greatest possible contrast in the animal productions. Nowhere does the ancient doctrine—that the peculiar animal and vegetable productions of the various countries of the globe are directly dependent on the physical conditions of those countries (such as climate, soil, elevation, etc.)—meet with a more direct and palpable contradiction. Borneo and New Guinea, as physically alike as two distinct countries can be, are zoologically wide as the poles asunder; while Australia, with its dry winds, its open plains, stony deserts, and temperate climate, yet produces the quadrupeds and birds which are most nearly allied to those inhabiting the hot, damp forests which everywhere clothe the plains and mountains of New Guinea.

We can now give the reason for limiting the Malay Archipelago on the east by the Solomon Islands in the Pacific Ocean. Certain groups of birds, which have their metropolis in New Guinea and extend over the Moluccas to the eastward, are found also as far as the Solomon Islands to the eastward, but do not extend to New Caledonia (which is much more Australian in its productions), or to the Fejee Islands, where the peculiar Pacific fauna commences. These groups are the *scarlet lories* and the *white cockatoos*, and the occurrence of a new species of cassowary in New Britain is a further indication of these islands being as closely allied to New Guinea on the one hand as are the Moluccas on the other.

The nature of the contrast between these two great divisions of the Malay Archipelago will be best understood by considering what would take place if any two of the primary divisions of the earth were brought into equally close contact. Africa and South America, for example, differ very greatly in all their animal forms. On the one side we have baboons, lions, elephants, buffaloes, and giraffes; on the other spider-monkeys, pumas, tapirs, ant-eaters, and sloths; while among birds, the hornbills, turacos, orioles, and honey-suckers of Africa contrast strongly with the toucans, macaws, chatterers, and humming-birds of America.

But let us endeavour to imagine (what it is very probable may occur in future ages) that a slow upheaval of the bed of the Atlantic should take place, while at the same time earthquake-shocks and volcanic action on the land should cause increased volumes of sediment to be poured down by the rivers, so that the two continents should gradually spread out by the addition of newly-formed lands, so as to reduce the Atlantic which now separates them to an arm of the sea a few hundred miles wide. At the same time we may suppose islands to be upheaved in mid-channel, and, as the subterranean forces varied in intensity and shifted their points of greatest action, these islands would sometimes become connected with the land on one side or other of the strait, and at other times again be separated from it. Several islands would at one time be joined together, at another would be broken up again, till at last, after many long ages of such intermittent action, we might have an irregular archipelago of islands filling up the ocean channel of the Atlantic, in whose appearance and arrangement we could discover nothing to tell us which had been connected with Africa and which with America. The animals and plants inhabiting these islands would, however, certainly reveal this portion of their former history. On those islands which had ever formed a part of the South American continent we should be sure to find such common birds as chatterers and toucans and humming-birds, and some of the peculiar American quadrupeds; while on those which had been separated from Africa, hornbills, orioles, and honey-suckers would as certainly be found. Some portion of the upraised land might at different times have had a temporary connection with both continents, and would then contain a certain amount of mixture in its living inhabitants. Such seems to have been the case with the islands of Celebes and the Philippines. Other islands, again, though in such close proximity as Bali and Lombock, might each exhibit an almost unmixed sample of the productions of the continents of which they had directly or indirectly once formed a part.

In the Malayan Archipelago we have, I believe, a case exactly parallel to that which I have here supposed. We have indications of a vast continent, with a peculiar fauna and flora, having been gradually and irregularly broken up; the island of Celebes probably marking its furthest westward extension, beyond which was a wide ocean. At the same time Asia appears to have been extending its limits in a south-east direction, first in an unbroken mass, then separated into islands as we now see it, and almost coming into actual contact with the scattered fragments of the great southern land.

In dwelling upon this subject—which I trust I have succeeded in making intelligible—my object has been to show the important bearing of researches into the natural history of every part of the world upon the study of its past history. An accurate knowledge of any group of birds or of insects, and of their geographical distribution, may assist us to map out the islands and continents of a former epoch; the amount of difference that exists between the animals of adjacent districts being closely dependent upon preceding geological changes. By the collection of such minute facts alone can we hope to fill up a great gap in the past history of the earth as revealed by geology, and obtain some indications of the existence of those ancient lands which now lie buried beneath the ocean, and have left us nothing but these living records of their former existence.

It is for such inquiries the modern naturalist collects his materials; it is for this that he still wants to add to the apparently boundless treasures of our national museums, and will never rest satisfied as long as the native country, the geographical distribution, and the amount of variation of any living thing remains imperfectly known. He looks upon every species of animal and plant now living as the individual letters which go to make up one of the volumes of our earth's history; and, as a few lost letters may make a sentence unintelligible, so the extinction of the numerous forms of life which the progress of cultivation invariably entails will necessarily obscure this invaluable record of the past. It is, therefore, an important object, which governments and scientific institutions should immediately take steps to secure, that in all tropical countries colonised by Europeans the most perfect collections possible in every branch of natural history should be made and deposited in national museums, where they may be available for study and interpretation.

If this is not done, future ages will certainly look back upon us as a people so immersed in the pursuit of wealth as to be blind to higher considerations. They will charge us with having culpably allowed the destruction of some of those records of Creation which we had it in our power to preserve; and while professing to regard every living thing as the direct handiwork and best evidence of a Creator, yet, with a strange inconsistency, seeing many of them perish irrecoverably from the face of the earth, uncared for and unknown.

* * * * *

The preceding selection, first presented as a paper before the Royal Geographical Society in mid-1863, was, to say the least, a notable success, and perhaps more than any other work sealed Wallace's reputation as both an all-round naturalist and the leading authority on the East Indian region. His future physical science publications, however, would include relatively little in the way of travels-based observations. Turning more and more to the writings of others for his facts, he increasingly became an agent of synthesis between the physical and biological sciences.

As is apparent from materials already presented, Wallace made extensive use of the facts of animal and plant distribution to argue particulars of past connection of land and water. His studies eventually convinced him that the relative spatial distribution of ocean basins and continents had always been more or less what it is today. His first separate treatment of this matter was published in 1877 as 'The comparative antiquity of continents' (ref 269). The complementary argument appeared in Natural Science *in 1892:*

The permanence of the great oceanic basins (ref 453)

That the great Oceanic Basins, as well as the land areas of the globe, have persisted throughout a large portion, if not the whole, of known geological time, is a proposition which has been accepted by writers of such eminence, and is supported by so many distinct lines of evidence, that it seemed likely to become one of the established teachings of geology. Professor Dana was led to it by a study of the development of the North American Continent; Darwin upheld it from his study of Oceanic Islands, and the facts he adduced have since been strengthened by the discovery that the two supposed exceptions to the generalisation that no ancient sedimentary rocks occur on such islands—Rodriguez and St. Paul's Rocks—are no exceptions at all. Two successive heads of the Geological Survey of Great Britain, Sir Andrew Ramsay and Sir Archibald Geikie, have advocated similar views; while Mr. John Murray, of the "Challenger," holds that the vast mass of evidence now accumulated as to the nature of the deposits on the floors of the great oceans, indicates that they are distinct in character and origin from any of the widespread formations which make up the series of the sedimentary rocks. Coming to the subject from a totally different point of view, that of the physicist and mathematician, the Rev. Osmond Fisher arrives at similar results. In the latest issue of his important work, *Physics of the Earth's Crust*, he gives as one of his conclusions—"and lastly, that the great oceanic and continental areas have never changed places;" and, in summary of the whole work, he says:

> The occupation of an entire hemisphere by one great ocean is a remarkable circumstance, and we have seen reason for believing that this is a very ancient division of the surface, and that it is probably a mistake to suppose that every part of it has been sometimes raised above the sea, and sometimes depressed beneath it. The truth seems to be, that the region subject to these alterations of conditions does not extend very far away from the present coast lines.

When studying the causes which have brought about the geographical distribution of animals, I was compelled to deal with this question, because I found that it had been the custom of many writers to solve all anomalies of distribution by the creation of hypothetical lands, bridging across the great oceans in various directions and at many different epochs; and, having arrived at the conclusion that the distribution of organisms could be more harmoniously and consistently explained without such changes of sea and land, which usually created greater difficulties than those they were intended to explain, I gave, in my *Island Life*, a brief statement of the evidence which appeared to me to render such changes exceedingly improbable. This evidence was mainly a summary of facts and arguments adduced by the eminent men referred to above, and to this I added in my *Darwinism* a difficulty founded on mechanical considerations which seemed to me to furnish a preliminary reason why we should not accept the doctrine of the interchange of continental and oceanic areas without very clear and cogent reasons. Since then some other arguments of this nature have occurred to me, and as the theory of permanence has been recently attacked, by Mr. W. T. Blanford in his presidential address to the Geological Society in 1890, and by Mr. Jukes-Browne in his *Building of the British Isles*, it may be as well to consider these difficulties, which furnish, in my opinion, a very powerful argument against the interchange of oceanic and continental areas, and which has the advantage of not requiring any knowledge of the higher mathematics in order to estimate its validity.

And, first, it is necessary to clear away some misconceptions as to the proposition I really uphold, since arguments have been adduced which in no way affect that proposition. Thus, Mr. Jukes-Browne quotes Professor Prestwich as saying, "It is only the deeper portions of the great ocean troughs that can claim the high antiquity now advocated for them by many eminent American and English geologists." But this is all that is claimed. For practical purposes I at first took the 1,000 fathom line as, generally and roughly, indicating the separation between the oceanic and the continental areas, because at that time it did accurately divide the continental from the oceanic islands, as defined by a combination of geological and biological characters. It has, however, been since shown that two ancient continental islands—Madagascar and New Zealand—are separated from their respective continents by depths of more than 1,000 fathoms. We must, therefore, go as far as 1,500-, or, perhaps, in a few cases, to the 2,000-fathom line, and this will surely mark out "the deeper portions of the great ocean basins," since only isolated areas exceed 3,000 fathoms.

Now, if we look at the deep ocean basins marked out by the 2,000-fathom line, not on Mercator's projection which greatly exaggerates the shallower portion situated in the temperate and polar regions, but on an equal-area projection . . . we shall see that by far the larger part of all the great oceans are included by this line, and that, for the purpose of indicating the isolation of the continents from each other throughout the equatorial and most of the temperate zones, there is very little to choose between the 1,000-fathom or the 2,000-fathom boundary. The latter, however, allows more scope for possible land extensions between the three southern continents and the Antarctic lands, which, during mild epochs, and by means of intervening islands, may, perhaps, have served as a means of communication between these continents.

All that is necessary to maintain, therefore, is that existing continents with their included seas, and their surrounding oceanic waters as far as the 1,500-fathom, or in some extreme cases the 2,000-fathom line, constituting, according to Mr. Murray's data ('On the Height of the Land and the Depth of the Ocean,' *Scottish Geographical Magazine*, 1888), 71 per cent of the whole ocean, have almost certainly been ocean throughout all known geological time. (Mr. Murray gives his results for oceans and inland seas together. The above refers to the oceans only as usually understood; but the difference is not great.)

It will now be seen that this is a problem which deals with the very broadest contrasts of the earth's surface, and that its fundamental data are on so vast a scale as not to be materially affected by the smaller details of physical geography, or by differences of opinion as to the exact meaning of certain terms. Whether a particular island is more correctly classed as oceanic or continental, whether a certain portion of the ocean should be placed within the oceanic or the continental area, and whether certain rocks were formed in very deep or in comparatively shallow water, are of slight importance, except in so far as they may throw light on the real question, which is, whether the vast expanses of ocean beyond the 1,000-fathom line, forming about 92 per cent of the whole oceanic area, have ever been occupied, or extensively bridged over, by continental land. It is towards the solution of this great problem that I now propose to submit certain general considerations which appear to me to lie at the root of the whole matter.

Comparison of Oceanic and Continental Masses.—In the paper already referred to, Mr. John Murray has carefully estimated both the area of the land and of the water on the earth's surface, and their bulk as deduced from the best available data. Taking the whole area of the globe as 100, he finds the land surface to be 28, the water surface 72. But the mean height of the land above sea-level is 2,250 feet, while the mean depth of all the oceans and seas is 12,456 feet. In this estimate, however, all the inland seas and shallow coast waters are included, and as these, at least as far as the 100-fathom line, are universally admitted to be within the "continental area," we omit them in our estimate of the mean depth of the oceans proper, which are thus brought to something over 15,000 feet, or nearly seven times as much as the mean height of the land . . .

The mean depths of the several oceans and the mean heights of the several continents do not differ enough from each other to render this [understanding] . . . a very inaccurate representation of the proportion between any of the continents and their adjacent oceans; and it will therefore serve, roughly, to keep before the mind what must have taken place if oceanic and continental areas had ever changed places. It will, I presume, be admitted that, on any large scale, elevation and subsidence must nearly balance each other, and, thus, in order that any area of continental magnitude should rise from the ocean floor till it formed fairly elevated dry land, some corresponding area must sink to a like extent. But if such subsiding area formed a part or the whole of a continent, the land would entirely disappear beneath the waters of the ocean (except a few mountain peaks) long before the corresponding part of the ocean floor had approached the surface. In order, therefore, to make any such interchange possible, without the total disappearance of the greater portion of the subsiding continent before the new one had appeared to take its place, we must make some arbitary assumptions. We must suppose either that when one portion of the ocean floor rose, some other part of that floor sank to greater depths till the new continent approached the surface, or, that the sinking of a whole continent was balanced by the rising of a comparatively small area of the ocean floor. Of course, either of these assumed changes are conceivable and, perhaps, possible; but it seems to me that they are exceedingly improbable, and that to assume that they have occurred again and again, as part of the regular course of the earth's history, leads us into enormous difficulties. Consider, for a moment, what would be implied by the building up of a continent the size of Africa from the mean depth of the ocean. By comparing the area of Africa with that of the whole of the land, and the depth of the ocean with the mean height of the land, we shall find that if all the land of the globe above sea-level could be transferred to mid-ocean, it would not be sufficient to form the new continent, but would still leave it nearly 2,000 feet beneath the surface.

It thus appears that, if the elevation of the ocean floor, and the corresponding sinking of whole continents, constitute a portion of the regular change and development of the earth's surface, there would be not only a chance but a great probability of entire continents disappearing beneath the waters before new continents had risen to take their place. Even the total disappearance of all the large land masses might easily happen; for we see from the diagram that they might one after the other disappear with a corresponding rise of the adjacent portion of the ocean bed and still leave the ocean over the whole earth almost as deep as it is now. But, as will be shown further on, the geological record, imperfect as it is, teaches us that no such submergences have ever taken place.

Contour of the Ocean Floor as indicating Permanence.—Before extensive soundings revealed the depth of the ocean and the form of its floor, it was supposed that it would exhibit irregularities corresponding to those of the land, such as mountain-ranges, great valleys, escarpments, ravines, etc. But we now know that the main characteristic of the ocean floor is, that it is a vast undulating plain, the slopes rarely exceeding a hundred feet in a mile except near the margins of the continental areas, while usually the gradients are so slight that they would be hardly perceptible. Contrast this with the forms of all mountain ranges whose general rise for long distances is often five hundred feet in a mile, while slopes of from 20 degrees to 60 degrees are by no means uncommon.

Now if we suppose that considerable portions of the ocean depths have been formed by the subsidence of continents, we should certainly expect to find some indication of those surface features which characterise all continents, but which appear to be absent from all deep oceans. In order to account for the actual contours of the ocean on this theory, we must suppose that, during subsidence, all the mountain ranges, peaks, valleys, and precipices were reduced to an almost uniform level surface by marine denudation, which, unless the process of subsidence were incredibly slow, seems most improbable. Mr. Jukes-Browne, however, does not hold the view that they have been thus denuded, for he approvingly quotes Mr. Crosby as saying that—"the oceanic islands are, of course, merely the tops of submerged mountains, and it is only with the highest points of continents that they can properly be compared." If this is correct, then we ought to find in the vicinity of such islands all the chief features of submerged mountain ranges—precipices, deep valleys and ravines, arranged in diverging groups as they always occur in nature. But in no single case, that I am aware of, have any such features been discovered.

But a still greater difficulty remains to be considered. If oceanic and continental areas are interchangeable, it can only be because the very same causes (whatever they may be) that produce elevations and subsidences in the one produce them also in the other, and at first sight it appears probable that this would be the case. But if these causes have been at work upon the ocean floor throughout geological epochs, they would have produced irregularities of surface not less but far greater than on subaerial land. This must be so, because subaerial denudation continually neutralises much of the effect of upheaval in the continental areas, while in the ocean depths no such cause or anything analogous to it is in operation.

The forces which have been at work in every mountainous region have sometimes tilted up great masses of rock at high angles, upheaved them into huge anticlinal curves, or crushed them by lateral pressure into repeated folds, which in some cases appear to have fallen over so as to reverse the succession of the strata. But, notwithstanding these various forms of upheaval, involving vertical displacements which are sometimes several miles in extent, the surface of the land often shows no corresponding irregularities, owing to the fact that subaerial denudation has either kept pace with upheaval or has even exceeded it, so that the position of an anticlinal ridge may be, and often is, represented by a valley. Now, if we suppose that similar forces have been at work on that portion of the earth's surface forming the bed of the ocean, where there are no such counteracting agencies, we should expect to find irregularities in the ocean floor far greater than those which occur upon the land surface.

Still more difficult to explain would be the absence of precipitous escarpments due to faults, which, though frequently showing an upthrow or downthrow of the strata to the amount of many hundreds and sometimes thousands of feet, rarely exhibit any difference of level on the land surface, owing to the fact that subaerial denudation has kept pace with slow and intermittent elevation. But in the ocean depths no such denudation is going on; and we can therefore only account for its very uniform surface on the supposition that it is not subject to the varied and complex subterranean movements which have certainly acted within the continental areas throughout known geological time. (The Rev. O. Fisher has arrived at this conclusion from his own researches. He says: "The compression which has caused the thickening accompanied by corrugation, such as characterises most elevated tracts, is properly a continental phenomenon, and has no analogue beneath the oceans." –*Physics of the Earth's Crust*, p. 253)

Equal Range of the Geological Record in all the Continents.—There is one other general consideration which indicates the permanence and continuity of the Continental Areas, and which renders it very difficult, if not impossible, to suppose that they have ever changed places with the great oceans. It is, that on all the present continents we find either the same or a closely parallel series of geological formations, from the most ancient to the most recent. Not only do we find Palaeozoic, Mesozoic, and Cainozoic rocks everywhere present, but, in proportion as the continents are explored geologically, we find a tolerably complete series of the chief formations. From Laurentian to Carboniferous and Permian, from Trias to Cretaceous, and from Eocene to Quaternary, the geological series appears to be fairly represented not only in continents, but also to a considerable extent in the large continental islands, such as Great Britain and New Zealand.

Now this is certainly not what we should expect if the present continental areas had, at different epochs, risen out of the deep oceans. In that case some would have commenced their geological history at a later period than others, having either a late Palaeozoic or some Mesozoic formation, or even an early Tertiary for their lowest stratified rock. Others, which had become oceanic for the first time at a later epoch, would exhibit an enormous gap in the series, either several of the Mesozoic formations, for example, being absent, or some considerable portion of both Palaeozoic and Mesozoic, or Mesozoic and Tertiary. This would necessarily be the case, because we cannot believe that so vast a change as the subsidence of an entire continent till its site became a deep ocean, and its subsequent elevation till it again became dry land, could possibly be effected in any less extended periods, if at all.

Whenever such gaps, or smaller ones, now occur locally, they are generally held to imply the existence of terrestrial conditions, as in the case of China, which, according to Richthoven, has been continental since the Carboniferous epoch. In many cases there is distinct evidence of such conditions in lacustrine or freshwater deposits, dirt-beds, etc. But if a gap of such vast extent both in space and time as that here referred to were caused by the interchange of a continent and a deep ocean, the fact that it was so produced would be clearly evidenced by an almost uniform deposit either of organic or clayey ooze, similar to those now everywhere forming over the oceanic area. Even if we make the fullest allowance for denudation during elevation, sufficient indications of so widespread a formation should be detected. Such a deposit would, in fact, have every chance of being largely preserved, because, long before it rose to the level where it would be subject to denudation by waves or currents, it would, almost everywhere, be overlain by a series of shore deposits, and wherever these latter were preserved on the land surface the oceanic formation would necessarily be found under them. That no such enormous deficiency in the geological series characterises any of the continents, and that no widespread deposit of organic or clayey ooze at some definite horizon has been anywhere detected, though such a deposit must have been formed and largely preserved if the whole or any considerable part of a continent had risen from ocean depths at any period of its history, constitutes of itself a very strong argument against any such interchange of oceanic and continental areas having occurred.

I have now shown that there are three distinct groups of phenomena which are either altogether inconsistent with any general interchange of oceanic and continental areas, or which render it exceedingly difficult to understand how such interchange could have been brought about. These phenomena are:—(1) The enormous disproportion between the mean height of the land and the mean depth of the ocean, which would render it very difficult for new land to reach the surface till long after the total submergence of the sinking continent. (2) The wonderful uniformity of level over by far the greater part of the ocean floor, which indicates that it is not subject to the same disturbing agencies which throughout all geological time have been creating irregularities in the land-surface, irregularities which would be far greater than they are were they not continually counteracted by the lowering and equalising effects of subaerial denudation. (3) The remarkable parallelism and completeness of the series of geological formations in all the best known continents and larger continental islands, indicating that none of them have risen from the ocean floor during any portion of known geological history, a conclusion enforced by the absence from any of them of that general deposit of oceanic ooze at some definite horizon, which would be at once the result and proof of any such tremendous episode in their past history.

I submit that these facts, and the conclusions to be logically deduced from them, form a very powerful *à priori* argument as against those who maintain the interchange of continents and oceans as a means of explaining certain isolated geological or biological phenomena; such, for instance, as the much-disputed origin of the chalk, or the supposed necessity for land-communication to explain the distribution of certain groups of reptiles or fishes in remote geological times. Before postulating such vast revolutions of the terrestrial surface in order to cut the gordian knot of difficulties which may be mainly due to imperfect knowledge, it will be necessary to show that the considerations here adduced, as well as the great body of facts which have caused many eminent geologists, naturalists, and physicists to hold the doctrine of oceanic permanence, are either illogical or founded on incorrect data. For, surely, no one will suggest that so vast a problem of terrestrial physics can be held to be solved till we have exhausted all the evidence at our command, and have shown that it largely preponderates on one side or the other.

The present article is intended to supply a hitherto unnoticed class of arguments for oceanic permanence, and these must of course be taken in connection with the other evidence which has been summarised in the 6th chapter of the writer's *Island Life*, and also with the admirable "Summary" of the purely physical argument in the 2nd edition of Mr. O. Fisher's *Physics of the Earth's Crust*. It is certainly a remarkable fact that writers approaching the subject from so many distinct points of view—as have Professor Dana, Mr.

Darwin, Sir Archibald Geikie, Dr. John Murray, Rev. O. Fisher, and myself—should yet arrive at what is substantially an identical conclusion; and this must certainly be held to afford a strong presumption that that conclusion is a correct one.

* * * * *

The 'permanence' arguments presented by Wallace and his supporters were persuasive enough to remain the most generally accepted model until new facts began to surface in the late 1950s, and continental drift became a fully respected hypothesis. Some workers now accuse Wallace of retarding progress in this direction, but it must be remembered—as the preceding selection indicates—that the 'permanence' argument had objectives specific to the period. The complication of moving *continents was a matter wholly beyond the level of the discussion at that time.*

Yet it is still true that in this instance Wallace was guilty of 'over-synthesis.' As Fichman (1977) has shown, Wallace's work in biogeography, physical geography, and evolutionary biology took shape in a mutualistic fashion leaving the 'permanence' outlook itself a somewhat expedient resolution. Aggravating the problem was Wallace's adoption of mammals and birds as his 'standards' upon which to base regional biogeographic systemization: the relatively recent divergence of the major mammalian and avian lines makes it less necessary to resort to continental drift to explain their present distribution.

Another biology-related physical science discussion into which Wallace entered concerned the age of the earth, and the associated calibration of geological time. Not long after the publication of On the origin of species *the famous physicist Lord Kelvin produced an argument (based on a flawed pre-nuclear physics interpretation of solar thermodynamics) suggesting that the age of the earth might not be great enough to support the process of organic differentiation depicted by the Darwinian model. Wallace's first attempt to cope with this issue appeared in a book review of two works by Charles Lyell in April 1869 (ref 146); less than a year later he presented a separate treatment in two parts in* Nature*:*

The measurement of geological time (ref 159)

Modern geological research has rendered it almost certain, that the same causes which produced the various formations with their imbedded fossils, have continued to act down to the present day. It has therefore become possible that, by means of changes which are known to have occurred in a given number of years, some measurement of the time represented by the whole series of geological formations might be obtained. It is true, that changes in the earth's surface, the records of which constitute the materials for geological research, occur very slowly, yet not so slowly as to be quite imperceptible in historical time. Land has risen or sunk beneath the sea, rivers have deepened their channels and have brought down sediment which has converted water into land, cliffs have been eaten away and the surface of the earth has been, in many ways, perceptibly and measurably altered during an ascertained number of centuries. But it is found that these changes are too minute, too limited and too uncertain, to afford the basis of even an approximate measurement of the time required for those grand mutations of sea and land, those contortions of rocky strata many thousands of feet thick, those upheavals of mountain-chains and that elaborate modelling of the surface into countless hills and valleys, with long inland escarpments and deep rock-bound gorges, which form the most prominent and most universal characteristics of the earth's superficial structure. Another deficiency in this mode of measurement arises from the fact, now universally admitted, that the record of past changes is excessively imperfect, so that even if we could estimate with tolerable accuracy the time required to deposit and upheave the series of strata of which we have any knowledge; still that estimate would only represent an unknown proportion, perhaps a minute fraction of the whole time which has elapsed since the strata began to be formed.

But there is another class of geological phenomena which enable us to measure those very gaps in the record of which we have just spoken, and it is now generally admitted that the continual change of the forms of animal and vegetable life which each succeeding formation presents to us, affords the best means of estimating the proportionate length of geological epochs. Though we have no reason to think that this change was at all times effected by a uniform and regular process; yet believing, as we now do, that it was due to the action of a vast number and variety of natural causes acting and reacting on each other, according to fixed general laws, it seems probable that, with much local and temporary irregularity, there has been on the whole a considerable degree of uniformity in the rate at which organic forms have become modified. It may indeed be the case that this rate of variation has continually increased or diminished from the first appearance of life upon the earth until the present day, or has been subject to temporary changes; but so long as we have no proof that such was the case, we shall be safer in considering that the change has been tolerably uniform.

To measure geological time, therefore, all we require is a trustworthy unit of measurement for the change of species: but this is exactly what we have not yet been able to get; for the whole length of the historical period has not produced the slightest perceptible change in any living thing in a state of nature. Moreover, though, the much longer time that has elapsed since the Neolithic or Newer Stone age, has been sufficient for some changes of physical geography and has, to some extent, altered the distribution of animals and plants, it has not effected any alteration in their form. It is only when we get back to the Palaeolithic or Older Stone age, when men used chipped flints for weapons and Europe was, probably, either just emerging from the severity of the glacial epoch, or in some of the intercalated milder periods, that we meet with a decided change in the forms of life. Elephants and rhinoceroses, bears, lions and hyenas then inhabited Europe; but they were nearly all of species slightly different from any now existing, while the reindeer, the musk-sheep, the lemming and some other animals, were the same as those that still live in the Arctic regions: all the mollusca, however, were identical with living species.

In the newer Pliocene Crag, on the other hand, which seems to have been deposited just as the glacial epoch was coming on, there are 11 per cent of extinct species of shells and about 55 per cent of extinct mammalia. What we want, therefore, is to be able to estimate, by means of the physical changes before alluded to, the time since the beginning or the end of the glacial epoch. Then we should have the unit we require for measuring geological time by the repeated changes in the forms of life as we go further and further back into the past; but before showing how this may perhaps be done, something must be said about physical and astronomical determinations of the age of our globe.

A few years ago, Sir W. Thomson startled geologists by placing a limit to the time at their disposal, which they had been in the habit of regarding as practically infinite. He showed, from the known laws of heat and the conservation of energy, that there are determinable limits to the age of the sun. Then, applying the same principles to the earth, he showed that, from the known increase of heat towards its interior and from experiments on the rate of cooling of various rocks, it cannot have existed in habitable state for more than about one hundred million years. It is within that time, therefore, that the whole series of geological changes, the origin and development of all forms of life, must be comprised. But, geologists had been accustomed to demand a much vaster period than this for the production of the series of fossiliferous deposits in the crust of the earth; while the researches of Mr. Darwin render it almost certain that, however vast the time since the Silurian and Cambrian epochs, yet anterior to these, at least an equal, and probably a much longer, series of ages must have elapsed since life first appeared upon the earth, in order to allow for the slow development of the varied and highly organised forms which we find in existence at those early epochs. Sir Charles Lyell is not disposed to admit the accuracy of these calculations, and Professor Huxley has criticised them in detail, with a view of showing that they are, in many respects, unsound; while Mr. [James] Croll as strenuously maintains that they are sound in principle and accurate within certain limits.

We have now to consider the bearing of Astronomy upon the problem. In a series of admirable papers in the *Philosophical Magazine*, Mr. Croll has fully discussed the question, how far variations in the excentricity of the earth's orbit, together with the precession of the equinoxes, have produced variations of climate in past ages. He has endeavoured to show that the date of the last glacial epoch and those preceding, may be determined by such considerations. With this view he has laboriously calculated tables showing the amount of excentricity for a period of three million years, at intervals of 10,000 years for a large portion of that time, and 50,000 for the remainder. These tables show that the amount of excentricity is alternately great and small at intervals of 50,000 or 100,000 years, as represented with sufficient accuracy in the diagram, which I have constructed by means of his figures. Owing to the precession of the equinoxes, combined with the revolution of the apsides, either pole will be presented towards the sun (constituting summer in that hemisphere and winter in the opposite one) at a different point in the earth's orbit on each succeeding year, the motion being such as to cause a complete

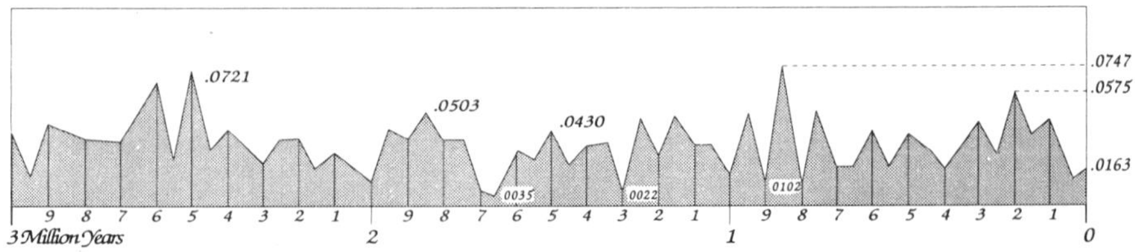

revolution in 21,000 years. If, therefore, at any one period, winter in the northern hemisphere occurs when the earth is nearest the sun or in *perihelion* (as is the case now), in 10,500 years it will occur in *aphelion*; at the one period the winters will be shorter and warmer, at the other longer and colder. When the excentricity is great (say two, three, or four times what it is now), Mr. Croll shows that, from the known laws of heat in reference to air and water, winter in *aphelion* will lead to an accumulation of snow, in the polar regions, which the summer will not be able to melt. This will go on increasing for many thousand years, till winter occurs near the *perihelion*, when the snow will be melted and transferred to the opposite pole. When the excentricity was very great a glacial epoch would occur in each hemisphere for more or less than 10,500 years, the other portion of the period of 21,000 years being occupied by an almost perpetual spring, with two transition periods from that to the glacial epoch. By examining the diagram of excentricity, we see that during the last three million years there have been more than twelve periods of great excentricity, each long enough to admit two or three, and several of them eight or ten, complete revolutions of the equinoctial points, thus sufficing for the production of not less than fifty or sixty glacial epochs in each hemisphere, with intervening phases of perpetual spring or summer.

The diagram also shows us (and this is of very great importance) that the present amount of excentricity is exceptionally small. During the last three million years there have only been five occasions, always of very short duration, when it has been less than it is now, while periods of high excentricity have often lasted for two hundred thousand years at a time. This period of three million years probably includes a large portion of the tertiary period, which therefore should have mainly consisted of alternations of warm and cold climates in each hemisphere, the latter generally forming true glacial epochs. This seems the legitimate deduction from Mr. Croll's reasoning and from the tables of excentricity with which he has furnished us; but, as he very justly argues, we cannot expect to find geological evidence of all these changes of climate. The warm and temperate periods will naturally leave the best records, while the cold epochs will generally be characterised only by an absence of organic remains. Besides, we must consider 10,500 years as a very small fragment of time in geology and we have good reason for thinking that several such periods might pass away without the occurrence of those exceptional conditions which Mr. Darwin and Sir C. Lyell have shown to be necessary for the preservation of any geological record. As to physical proofs of ice-action, very few could survive the repeated denudations, upheavals and subsidences, which the surface must have undergone since any of the earlier glacial epochs; so that it may be fairly argued that these repeated changes of climate may have occurred and yet have left no distinct record by which the geologist could interpret their history.

Throughout the whole of his argument, Mr. Croll considers astronomical causes to be the most important and effective agents in modifying climate, while Sir Charles Lyell maintains that the distribution of land and water, with their action on each other by influencing marine and aerial currents, are of prepondering importance. He has certainly shown that these causes have an immense influence at the present time. The effects which, on Mr. Croll's theory, ought to be produced by the existing phase of precession combined with even the small amount of excentricity that now exists, is not only neutralised, but actually reversed by terrestrial causes. Dove has shown that the whole earth is really warmer when it is furthest from the sun in June, than when it is nearest in December, a fact which is to be explained by the northern hemisphere (turned toward the sun in June) having so much more land than the southern. So, the northern hemisphere being three millions of miles nearer the sun in winter than in summer, while the southern hemisphere is the reverse, the northern winter ought to be warmer and the northern summer cooler than the southern; but this, too, is the opposite of the fact, for the southern summer is more than 11 degrees Fahr. cooler than ours, while its winter is nearly 5 degrees Fahr. warmer. The immense differences of temperature of places in the same latitude, sometimes amounting to nearly 30 degrees Fahr., can also be traced, in almost every instance, to the distribution of land and water and of winds and currents. Sir Charles Lyell further argues that the existing distribution of land is so extremely irregular—such an undue proportion being near the poles, while there is such a deficiency at the equator and in the south temperate zone—that whatever differences may have occurred in past time, they can hardly fail to have often been such as to cause a more uniform climate. Therefore he believes that if the poles were tolerably free from land, so as to admit of the uninterrupted circulation of the warmer equatorial waters and to afford no lodgment for great accumulations of snow and ice, a glacial epoch would be impossible even during the most extreme phases of excentricity.

We have now much evidence to show that three distinct modifications in physical geography occurred just before or during the Glacial epoch, which would each tend to lower the temperature. The first is the submergence of the Sahara, which would have caused the southerly winds to be charged with aqueous vapour, condensing on the Alps into snow instead of being, as now, dry and heated and acting powerfully to melt the glaciers. The second is the submergence of Lapland, which would have admitted the cold iceberg-laden waters of the Arctic Sea into the very heart of Europe. The third is the probable submergence of part of Central America, causing the Gulf Stream to be diverted into the Pacific. The only proof of this is the fact that one-third of the known species of marine fishes are absolutely identical on the two sides of the isthmus of Panama; but it is impossible to conceive any means by which such amount of identity could have been brought about except by a recent, if only a temporary, communication. A subsidence and elevation no greater than what occurred in Wales about the same time—as proved by Arctic shells of existing species in drift 1,300 feet above the sea—would have effected the communication by a broad and deep channel. Now if any two of these changes of physical geography occurred together, we may be sure that a very small increase of excentricity would have led to a more severe glacial epoch than would be possible, under existing conditions, with a much larger excentricity. We must keep this in mind when attempting to fix the most probable date for the last glacial epoch.

We have now to consider an entirely distinct set of facts which have an important bearing on the probable time elapsed since the last glacial epoch. Messrs. A. Tylor, Croll, and Geikie have shown that the amount of denudation now taking place is much greater than has generally been supposed. The quantity of water discharged by several rivers and the quantity of sediment carried down by those rivers have been measured with tolerable accuracy, and allowing for the difference of specific gravity between sediment and rock, it can be easily calculated, from the known area of each river basin, what average thickness has been removed from its whole surface in a year, since all the matter brought down by the river must evidently have come from some part of its basin. In this way it is found that the Mississippi has its basin lowered 1/6000 of a foot per annum; the Ganges, 1/2358; the Rhone, 1/1526; the Hoang-Ho, 1/1464; the Po, 1/729.

But it is evident that this amount will be distributed very unequally over different parts of the basin, according as the surface is flat or sloping, whether the slopes are of loose soil or of rock, whether the rock is solid or friable. The perfectly flat alluvial plains that form a considerable part of many river basins, will not only suffer no denudation, but will generally receive deposits of sediment during floods, and all such flat lands should therefore be deducted from the area of the river. Slightly undulating lands, especially if well covered with forest, will also suffer scarcely any denudation, as is well seen in the case of the Rio Negro branch of the Amazon and other black water rivers of South America, which hardly carry down any perceptible sediment even when in full flood. Again, wherever lakes occur, they receive all the sediment from the basin above them, which portion should therefore be treated by itself, since it contributes no sediment to the main river. If we look at a physical map of North America we see that a large extent of the Mississippi basin consists of alluvial flats and slightly undulating prairie, sufficiently explaining its small proportionate denudation. Even the Rhone, which has a high rate of denudation, flows through a great extent of low lands and perfectly flat meadows, while the upper portion of its valley which produces most sediment is cut off by the Lake of Geneva. In order, therefore, to arrive at any fair estimate of the amount of denudation in the upland and mountainous portions of the Rhone valley (which is what we require for our purpose) we have considerably to reduce the area of its basin by taking away the flat lands in all its valleys, and considerably to increase the amount of sediment by adding all that is now poured into the Lake of Geneva. We shall probably not be far wrong in adding one third to its denuding powers on these grounds, which will lead us to the startling fact that the Rhone basin is being lowered at the rate of a foot in a thousand years; but even this is considerably less than in the case of the Po. Mr. Croll takes the Mississippi denudation of a foot in six thousand years as a measure for that of Europe; but for reasons above stated I conceive this to be quite out of the question, and I maintain, that if we are to use his measure of denudation for any practical purpose, we must apply that of European rivers to European phenomena, that of Alpine rivers to Alpine phenomena, and must further make the necessary corrections for alluvial flats and intercepting lakes.

Mr. Croll and Sir Charles Lyell were at first both inclined to adopt the period of high excentricity which occurred from 750,000 to 950,000 years ago as that of the glacial epoch, but Mr. Croll, in consideration of the proofs of rapid denudation above given, now believes that the period beginning about 240,000 and ending about 80,000 years ago, is the more likely one. Even this, however, offers difficulties. Denudation in Wales and Scotland must probably have gone on as rapidly as the average rate of the Rhone valley, especially during the period when the old glaciers were disappearing; and eighty thousand years will therefore imply eighty feet of average denudation over the whole surface of the country, if less in one part then correspondingly more in another; but how is this consistent with the preservation of ice-ground rock-surfaces and glacial furrows in so many situations, as well as numberless heaps of loose matter, the moraines of ancient glaciers, apparently just as they were left by the ice? There are, it is true, a few considerations that go to diminish though not to remove the difficulty. The amount of denudation is now abnormally large, because the large quantities of glacial drift left over the surface of the country, supply much of the sediment carried down by the streams of Alpine countries. Many glacial markings were at first covered up and preserved by drift or alluvium, and have been since exposed by denudation: those earliest exposed are obliterated, but new surfaces are being continually uncovered. The amount of denudation of a solid rock surface may not be a tenth part of that which now obtains in glaciated districts; a fact which can only be ascertained by determining the amount of sediment brought down by streams the basins of which are *free from drift or gravel*, and consist almost wholly of *compact rock surfaces*. We still have to deal with the difficulty of the moraines, whose form and aspect are often so fresh that we can hardly believe them to have been much changed since the ice left them, although it is impossible to understand how they have escaped the denudation which has lowered the whole surface of the country eighty, or even a much smaller number of feet.

It is true Mr. Geikie, in his paper on Modern Denudation (*Transactions of the Geological Society of Glasgow*, vol. III, p. 153), suggests that all the effects of ice-action, now visible, are merely the few examples which have been preserved owing to a concurrence of favourable conditions, while a much larger number have been destroyed; and I learn from him that there are in Scotland moraines in all stages of decay. If this be the true explanation of the difficulty, it follows that if one valley or hill-side has remained unaltered during 80,000 years, another must have been denuded to double the average amount.

Having thus shown the difficulty there is in accepting even the shorter period of 80,000 years for the date of the end of the glacial epoch in Europe, let us see what other modes of measurement are available. In Sir Charles Lyell's *Antiquity of Man*, 2nd ed., p. 28, we have three different calculations of the age of the bronze and stone periods in Switzerland, which would place the latter at about from 5,000 to 7,000 years ago. At page 321 of the same work, we have an estimate of the age of the upper delta of the Tinière by M. Morlot. The lower delta (by the presence of Roman remains in one of the upper strata) is calculated with tolerable certainty to be 10,000 years old, while the upper delta, 150 feet above the lake, is ten times as large as the lower one, and is therefore supposed to be 100,000 years old. From its fossil remains it is believed to be post-glacial; but it is evident that, during the melting of the ice, the torrent might have been more powerful, and have accumulated a delta much more rapidly than now. The peat mosses of Denmark, indicating that the present beech-tree vegetation of that country, which was also characteristic of it in the Roman period, was in the Bronze age replaced by oaks, and in the still earlier Stone age by fir-trees, imply a very long lapse of time; yet this only takes us back to the Neolithic age, when all the shells and all the mammalia were of existing species. The 8,000 or 10,000 years of the Swiss Stone age may, however, have sufficed for this change. There seems to be no doubt, that the time which elapsed from the close of the glacial epoch (when man used the rudest flint weapons, and was coeval with many extinct animals, when, moreover, the climate and physical features of the country were considerably different from what they are now) up to the Neolithic age, was much greater than from the latter date to the present day, but how much greater it is impossible to determine. The position of many of the tool-bearing gravels shows that rivers then flowed at much higher levels, but from the known rate of denudation, a valley might be deepened even 50 or 100 feet in as little as 50,000 years, since it is in valleys that the effects of denudation would be greatest; and the extinction of the various animals might certainly take place in an equal time under such conditions as are not unlikely to have occurred at a period of great climatic change.

It does not appear, therefore, that any of the estimates of time founded on an actual basis of observed change in a known period, require us to assume more than 80,000 years since the close of the glacial epoch, while the measurement of the existing rate of denudation renders it almost certain that it was less rather than more. We may fairly assume, that even if a large excentricity has been an essential condition of a glacial epoch, the ice would maintain itself into a period of less excentricity than would be required to bring one on. Now 74,000 years ago the excentricity was about double what it is now, and the winter of the northern hemisphere then occurred in *aphelion*, so that the glacial epoch would at that time probably have been in full force, and we may assume that it might continue 3,000 or 4,000 years longer. But when we come to

65,000 years back, we find the excentricity scarcely more than it is at present, and winter nearly in *perihelion*; so that we must conclude, if excentricity has anything to do with it, that the last glacial period came to an end not less than 70,000 years ago.

Now it is most important to observe that, for the last 60,000 years, the excentricity has been very small—for three-fourths of the time less than it is now. During this time the opposite phases of precession, each lasting 10,500 years, will have produced scarcely any effect on climate, which in every part of the earth will have been *nearly uniform for that long period.* But this is quite an exceptional state of things; for the curve of excentricity shows us that, during almost the whole of the last three million years, the excentricity has been high—almost always twice, and sometimes three and four times as much as it is now. If, therefore, Mr. Croll's theory be correct, there will have been a change each 10,500 years during this vast period (in all the extra-tropical regions at least) from a very cold to a very mild climate. This will necessarily have caused much migration both of plants and animals, which would inevitably result in much extinction and comparatively rapid modification. Allied races would be continually brought into competition, altered physical conditions would induce variation, and thus we should have all the elements for natural selection and the struggle for life, to work upon and develop new races. High excentricity would therefore lead to a rapid change of species, low excentricity to a persistence of the same forms; and, as we are now, and have been for 60,000 years, in a period of low excentricity, *the rate of change of species during that time may be no measure of the rate that has generally obtained in past geological epochs.* Thus we should have explained the extraordinary persistence of organic forms during the preceding Neolithic age, although slight changes of climate and of physical geography have undoubtedly taken place; and it would prove to be not so much the *usually* slow rate of organic change, as the fact of our living in the midst of an *exceptionally uniform climatic epoch*, that has hitherto prevented us from obtaining a measure of the average duration of species.

These considerations have an important bearing on our estimate of the duration of the glacial epoch itself, and on our calculation of geological time from the change of species since its commencement. If it terminated 70,000 years ago, and if each 10,500 years before that date, there was alternately a warm period and a glacial epoch, there would necessarily occur a series of northern and southern migrations of animals and plants, and thus deposits formed at times not geologically remote, might contain very distinct groups of animals. These might even meet and be confounded in the same strata, and thus lead to that extraordinary mixture of northern and southern forms which occurs in some of the more recent formations, like the hippopotamus and mammoth in the Norfolk crag and the lower brick-earths.

Geologists seem hardly to have attached sufficient importance to the great gap that intervenes between the Palaeolithic and Pre-historic ages. Mr. Boyd Dawkins has shown, from a careful study of their mammalian remains, that the whole of the post-glacial river deposits and cave-beds of this country (148 in number) are of the same age, being characterised by about twenty species of extinct or arctic mammalia, and this was the age of Palaeolithic man. In the Pre-historic or Neolithic age all these have disappeared, while the sheep, goat, dog, and *Bos longifrons* are first met with. Now, on the theory that this Palaeolithic age was entirely post-glacial, and that the climate and physical geography of Britain have been since slowly approaching their present condition, how is this great gap to be accounted for? The large number of places where remains have been found, shows that the conditions requisite for preserving them very frequently occurred, and there must therefore have been some special cause which has prevented any record being left of the long period during which they were becoming extinct. Mr. J. Scott Moore (*Pre-glacial Man and Geological Chronology*. Dublin, 1868.) maintains that they were all pre-glacial, and that the gap was the glacial epoch itself. He adduces in corroboration the striking fact that none of the supposed post-glacial gravels ever rest on the boulder clay, but always on an older rock, which could hardly have been the case in every instance were they all pre-glacial. Again, Mr. Boyd Dawkins tells us that the identity of such a large proportion of the species of pre-glacial and post-glacial mammals "forbids the idea of the existence of any gap or lacuna which would warrant the classification of the one as tertiary and the other as quaternary." But if we admit the occurrence, during the last glacial epoch, not only of one or two, but of a series of alternate cold and warm periods, we may make the Palaeolithic age *inter-glacial*, and suppose it to have occupied several of these alternations of climate. If we further place the last submergence which separated Britain from the Continent, during one of the later phases of extreme cold, when most of the extinct mammalia, as well as man, had migrated southwards, we shall sufficiently account for the great gap that intervenes between the Palaeolithic and Pre-historic ages.

In the *Principles of Geology*, 10th ed., vol. i, p. 300, Sir C. Lyell has given an estimate of the duration of geological epochs, from the proportionate change in the species of marine mollusca, taking as a basis a million years since the beginning of the glacial epoch. Of the marine shells then living, six per cent have become extict, while at the close of the glacial epoch they were all of the existing species, but this does not necessarily imply that the former are many times older than the latter. The glacial period itself may have been the cause of their extinction independently of mere time; so that the Bridlington beds, where the above-mentioned proportion of extinct species occurs, need not on this account be more than twice as old as those glacial or post-glacial drifts which contain only living species, or, according to our previous estimate, about 140,000 years. The Norfolk crag, which contains eleven per cent of extinct shells, may be from 40,000 to 60,000 years older; this will allow for two or three alternations of warm and cold periods, which, at a time of such high excentricity, must have been strongly contrasted, and have led to a correspondingly rapid change of species. From these considerations it becomes evident that the time, measured by the occurrence of five per cent of extinct species of marine shells, is not necessarily the whole number of years which has elapsed since they existed, but only that number *minus* the last

60,000 years of uniform climate and specific immobility; and we may be even too lavish of time if we allow so much as 100,000 years for this amount of change under the influence of those repeated alternations of climate which have characterised the last three million years and which have probably more or less characterised all past geological time. If now we take this number as our datum instead of a million years, all Sir Charles Lyell's figures will be reduced to a tenth, and will stand thus: the time elapsed since the beginning of the—

Lower Miocene	2,000,000 years
Eocene	6,000,000 years
Cretaceous	10,000,000 years
Triassic	14,000,000 years
Permian	16,000,000 years
Carboniferous	18,000,000 years
Devonian	20,000,000 years
Silurian	22,000,000 years
Cambrian	24,000,000 years

These figures will seem very small to some geologists who have been accustomed to speak of "millions" as small matters; but I hope I have shown that, so far as we have any means at present of measuring geological time, they may be amply sufficient. Taking Sir William Thomson's allowance of a hundred million years for the time during which the earth can have been fit for life, it yet allows Mr. Darwin, for the process of development from the primordial germ, three times as many years anterior to the Cambrian epoch as have elapsed since that date, an amount of time which, I believe, will fully satisfy him, by whatever scale we may measure it.

The evidence of the rapidity of denudation would indeed tend to the still further shortening of the estimate here given; and it is not impossible that a concurrence of geographical conditions might have brought down the glacial epoch into a period when the excentricity was no greater, or even less, than it is now. This, however, is hardly probable, and I am inclined to think that the considerations already alluded to will, to a considerable extent, explain how it is that so many signs of glacial action still remain, in spite of such denudation. The only argument I consider new in this paper, is that derived from the uniformity of climate during the last 60,000 years, and the alternations of heat and cold for a long time previously, leading to a slower change of species since the glacial epoch than at any former period, thus allowing us to suppose change of form in the organic world to go on more rapidly than we had before thought possible. If this be a sound deduction, it will, I believe, more than anything else, enable us to bring the period required for the development of the whole organic world within that which modern natural philosophy assigns as the age of the habitable earth.

Much of the force of my argument appears to depend upon the accuracy of Mr. Croll's view, that, during a time of great excentricity, there will be in each hemisphere alternately a glacial epoch of about 10,500 years, and a perpetual spring or summer for about an equal period. But Sir Charles Lyell argues, with great force, for the opposite view, that the cold of one period would be continued through the other, and that during the whole continuance of a phase of high excentricity both hemispheres would be in a state of glaciation. Supposing this view to be the true one, it will not very materially affect my argument, for the diagram shows many comparatively rapid alternations from a very high to a very low excentricity, which would also be from a glacial to a temperate climate and would certainly tend to comparative rapidity of specific change; while in each 10,500 years there would, no doubt, be some retreat and advance of the snow line, followed by a less amount of migration, competition, and variation. During the last 60,000 years, on the other hand, the change of excentricity has been hardly perceptible, and the change of organic forms may be supposed to have been far below the average.

* * * * *

Wallace continued the 'denudation rates' argument in subsequent works, most notably Island life *in 1880 (ref 721). In 1883 he responded to one related criticism as follows:*

Mr. Wallace's reply to Mr. T. Mellard Reade on the age of the earth (ref 367)

I have just received from Mr. T. Mellard Reade, F.G.S., a copy of his paper on the 'Age of the Earth' (which appeared in your *Magazine* of July last), in which I am asked to put that gentleman right as regards what he calls his "analysis" of some figures and estimates given in my *Island Life*; and I gladly seize the first opportunity of doing so . . .

The first statement of Mr. Reade's which I have to "put right" is the following:—"It is evident, if the figures mean anything at all, that three millions of square miles 177,200 feet thick represent the whole of the rock removed by denudation in all forms since the geological history of the earth began. Spread this over 57 million square miles of land and we get a deposit 9326 feet thick deposited in all geological time." This is not quite an accurate representation of my statements. The figures quoted represent, not the whole matter denuded, but only that portion of which a record still exists in the rocks; and this matter has been deposited, not "in all geological time," but only in that portion of geological time indicated by the known series of stratified rocks; unconformities and other breaks representing unknown intervals of which we have no record. With these corrections the figures used by me do imply what Mr. Reade says they do; which is, in other words,—that the average thickness of that portion of the earth's crust formed by the known stratified rocks does not probably exceed nine or ten thousand feet.

Mr. Reade, however, without directly impugning these figures, attempts to show that they lead to absurd or incredible results, and he does this by manipulating them in a way which is altogether beyond my comprehension. He first says, that these rocks have been made and destroyed over and over again; and then argues that, *because* the exposed igneous rocks cover about 1/12 of the land surface, *therefore* "each particle of rock, on the average, has been denuded and laid down at least twelve times." I have in vain tried to see any connection between these two statements, but what follows is still more unintelligible. Mr. Reade adds:—"From this it follows that the actual thickness of the sedimentary crust of the earth, if there were no sedimentary rocks except on the site of the present land areas, *would be* 12/9326 = 777 feet." Correcting the clerical error of 12/9326 instead of 9326/12, this means that, because the stratified rocks have been successively formed from the denudation of older rocks (stratified and igneous), therefore their *actual* thickness would be many times less than by estimates founded on direct measurement it is known they *actually* are! It is, I think, evident that, from Mr. Reade's point of view, he should have here multipled instead of divided by 12. For if the older rocks have been reduced in thickness by denudation, and their debris has gone to form newer rocks in each successive epoch, it is clear that when first deposited all the rocks *would have been* thicker than now, though there is no definite relation between the number of successive formations and their greater thickness, as Mr. Reade seems to suppose. For example, if half the original Palaeozoic rocks have been denuded to form the Mesozoic and parts of the later rocks, and half of the original Mesozoic to form the Tertiary, and half these again to form glacial and recent deposits, each would have been at first about twice as thick as it is now,—not one-fourth the thickness, as Mr. Reade's mode of calculation would make them; and as the whole problem is one of the *time* taken to produce these various deposits, the greater original thickness would have to be used in the calculation.

But, even if Mr. Reade's figures are thus corrected, his whole criticism is radically unsound; for, as I have explained in my original discussion of the subject, denudation is so unequal in its action and occurs so generally on the edges of uplifted strata not over their surfaces of deposit, that it would be quite possible for 9/10 or even 99/100 of a formation to be destroyed by denudation, and yet for the remaining 1/10 or 1/100 to give a fairly accurate measure of the *average* thickness or even sometimes of a *maximum* thickness of the original deposit. Our measures of the thickness of the sedimentary rocks will, therefore, not be seriously affected by the fact that by far the larger portion of all of them have been destroyed by denudation, and again and again laid down to form newer rocks; and as I have used measures of the *maximum* thicknesses, I have considered that these would in all probability not differ much from the original *average* thicknesses of the same rocks before they had suffered denudation. No doubt some rocks may have been wholly destroyed by denudation, or are so covered up by later deposits as to be beyond our reach, and to allow for these I am willing to admit that my estimate of the whole thickness of the rocks, and therefore of the time taken to produce them, may have to be considerably increased; but this would bring my figures nearer to those usually arrived at, not enormously further from them as Mr. Reade endeavours to prove.

Yet again, Mr. Reade points out that continents have fluctuated, and have sometimes been larger than now. To allow for this he doubles the land surface and reduces the corresponding thickness of the strata to one-half! But, surely, if the continents have been sometimes larger, they have also been sometimes smaller, and I see no reason to think we can take any fairer *average* than that of the present area; and even if the average *had* been double, then the denudation and the deposit would presumably have been double also, not half as Mr. Reade suggests.

With regard to my fundamental position—that the areas of deposition are (and always have been) very much smaller than the areas of denudation, and that, in making any estimate of geological time founded on the thickness of the sedimentary rocks and the known rate of denudation, this fact must be taken account of, Mr. Reade makes no objection; and, whatever "confusion of ideas" may have pervaded my estimate, the subject has certainly not been rendered clearer by his criticism . . .

* * * * *

Wallace had been studying the then still-new ideas on continental glaciation advanced by Louis Agassiz and others even before his consideration of the 'geological time' problem, having published a review paper on the subject some three years earlier (ref 124). In this study his experience as a surveyor aided him considerably. He well understood the basic principles of geodesy, as is demonstrated by the following letter, printed in Reader *in 1866 (also see Tuan 1963 and Schadewald 1978):*

Is the earth an oblate or prolate spheroid? (ref 115)

In Dr. Pratt's letter in your journal of April 28, he seems to argue that modern astronomers and geometers are in error as to the true form of the earth. His words are: "In conformity with the assumed oblate figure of the earth, arcs of the meridian should progressively *diminish* from the equator to the poles. *In fact*, these arcs become *longer* with advance in this direction." And he goes on to advance a theory of some polar attraction in space which has drawn out the earth at the poles instead of flattening it, as is commonly, but he thinks erroneously, assumed. Von Gumpach has been long asserting the very same thing, and has importunately called the attention of our Government to the fact, that numbers of vessels are annually lost owing to the impossibility of calculating their true position, so long as navigators mistake the very figure of the globe they are travelling over. But his warnings have been all in vain. The Admiralty persist in refusing to alter the Nautical Almanack, and the philosopher thinks he has just cause of complaint because the Astronomer Royal will neither accept his conclusions nor point out the flaw in his argument.

Now that a mathematician and astronomer like Dr. Pratt takes up the very same ground as Von Gumpach, it seems time that the matter should be clearly explained; and, with your permission, though neither an astronomer nor mathematician, I will endeavour to do so; and I have the more hope of succeeding because I once felt a difficulty as to the very same point myself.

The fact (universally stated in works on astronomy and geodesy) that degrees of the meridian *increase* in length towards the poles, on account of the earth's *compression* at the poles, is, indeed, one well calculated to mystify a mere mathematician, though it is clear enough to anyone who reflects on the various conditions involved in the problem. If we look at the diagram of a sphere, and the space from the equator to the pole be divided into equal parts subtending angles of one degree each at the centre, and we then flatten the poles by cutting off a portion with a curve of greater radius, it is evident that the distance from the pole to the centre of the sphere will be shorter than before, and therefore, that degrees of latitude, *measured angularly from that centre*, would really diminish in length from the equator towards the poles.

But in our actual rotating globe, the unequally curved surface is one of equilibrium, owing to the varying centrifugal force at different latitudes; and, as degrees of a meridian can only be measured upon the surface by *tangents* or *perpendiculars to it* (obtained by the *spirit-level* or the *plumb-line*), it follows that a degree at the pole, measured by an angular instrument from the earth's centre, would not represent a degree of latitude, because the curvature of the polar regions has its centre much further off than the earth's centre of gravity, and a degree measured on the surface would therefore be longer. The *centre of curvature* of the earth's surface rarely coincides with the centre of gravity, and a plumb-line will therefore not always point directly to that centre. It will do so only at the equator and the pole. Everywhere else adjacent plumb-lines will meet at points within or beyond the centre, according as the curvature of the surface is less or greater than the mean curvature of the globe. The flattened polar regions are, for the geometer, portions of a larger sphere; the protuberant equator (as far as latitude is concerned) is part of a smaller one; and degrees of the meridian measured on these parts must be respectively longer and shorter than what would be due to the mean curvature of the globe.

These considerations seem so very obvious, that I am almost afraid I have mistaken Dr. Pratt's meaning. I hope, however, that the explanation here given may be useful to some young astronomers, as I do not recollect seeing it in any popular work.

* * * * *

Dr. Pratt produced a response to this criticism, but Wallace ended the discussion a week later with the following decisive remarks (ref 116):

Dr. Pratt maintains two propositions, which are incompatible with each other:—

1. That a plumb-line everywhere points to the centre of the earth.
2. That the earth, at sea level, is not a sphere.

If both are true, it follows that there are places where the plumb-line is not at right angles to the ocean surface, so that the water must there stand permanently out of level. In other words, the forces that determine the direction of the plumb-line, and those that determine the level of fluids, are not the same at the same points on the earth's surface. Will he explain this little difficulty in the way of his peculiar view?

* * * * *

Glacial theory would have interested Wallace on the basis of its relevance to biogeography alone. There was, for example, the problem of how to explain the presence of otherwise arctic forms in equatorial mountains such as the Andes. Darwin believed that the equatorial lowlands had been cooler during the glacial epoch than they are today, and that these forms had evolved from ancestors then inhabiting the nearby low-lying areas. Wallace, however, favored the explanation of dispersal along (now discontinuous) ecological corridors in north-to-south-extending elevated zones. This and other questions might be resolved, he thought, were an understanding reached regarding the root causes of widespread glaciation episodes.

Wallace's first significant attempts to isolate the causes of the glacial periods took the forms of the Lyell review (ref 146) referred to earlier, a paper read at the 1870 meeting of the British Association for the Advancement of Science (ref 171), and 'The measurement of geological time' (ref 159), presented earlier. The problem became one of his long-term interests. The most extensive exposition of Wallace's views on ice-age causation can be found in Island life *(ref 721), but most of the important points were also raised in an anonymous 'notice' printed in* The Quarterly Review *in 1879. From this long work (which also reviews several important glacial geomorphology features) I have reproduced the better part of this discussion:*

Glacial epochs and warm polar climates (ref 313)

. . . it is a very remarkable fact, that the only theory which affords a distinct explanation of the changes which brought about the glacial period, leads to the conclusion that . . . alternations of warm and cold climates must have occurred during its continuance; and as the same theory also explains, more or less completely, the other changes of climate of which we have geological evidence, it will be most convenient to our readers at once to give an outline of that theory, and to show as clearly as possible by its aid how the glacial epoch was brought about. In doing so we shall have the opportunity of clearing up some of the misconceptions with which the subject has been encumbered; and we shall then proceed to give an account of the proofs of warm arctic climates, and to show how far the same theory is able to explain them.

Astronomical causes of periodical changes of Climate.—The elliptical orbit in which the earth moves round the sun has an eccentricity of about one and a half millions of miles, which causes us to be sometimes three millions of miles nearer the sun than at others. Strange as it may seem, we in the northern hemisphere are nearest to the sun in winter, and furthest off in summer, while in the southern hemisphere it is the reverse; and this peculiarity must have some effect in making our northern winters less severe than those of the

Antarctic lands. But the earth moves more rapidly in that part of its orbit which is nearest the sun, so that our winter is not only milder, but several days shorter, than that of the southern hemisphere. There can be no doubt that if this state of things were reversed, our winters occurring when we were furthest from the sun, or in *aphelion*, and our summers when we were nearest the sun, or in *perihelion*, we should experience a decided difference in climate, our winters being more severe and longer, and our summers hotter but shorter. Now there is a combination of astronomical causes (the precession of the equinoxes and the revolution of the apsides) which actually brings this change about every 10,500 years, so that after this interval the condition of the two hemispheres is reversed as regards nearness to the sun in summer and comparative duration of summer and winter, and this change has been going on throughout all geological periods. As we shall continually have to refer to these changes, we shall speak of them as the 'opposite phases of precession,' when winter in the northern hemisphere occurs at *aphelion* or *perihelion*, that is when the earth is farthest from or nearest to the sun.

But the amount of eccentricity itself varies very largely, though very slowly, and it is now nearly at a minimum. It also varies very irregularly, but its amount has been calculated by Mr. Croll and others for 3,000,000 years back, during which time it has much more frequently been of greater than of less amount than it is now. Going backward from the present time, we find that for about 25,000 years it was slightly greater, and for 30,000 years more rather less, than it is now, so that for a period of nearly 60,000 years back we can impute no important change of climate to this cause. But it then increases rapidly till about 72,000 years ago, when it was double, while 100,000 years ago it was more than two-and-a-half times, its present amount. Going back still further, it diminishes for 50,000 years to double, and then again increases for 60,000 years till it becomes nearly three-and-a-half times, its present amount. Going back still further, it diminishes for 50,000 years to double, and then again increases for 60,000 years till it becomes nearly three-and-a-half times what it is now. It then diminishes rapidly, and after one more considerable rise sinks to almost exactly its present amount 400,000 years ago. We thus find that for a period of about 140,000 years, beginning 70,000 years back, the eccentricity was always more than double its present amount, with two maxima, of more than two-and-a-half and nearly three-and-a-half times that quantity. It is to this long period of high eccentricity that geologists are pretty generally agreed to refer the glacial epoch, and to this we shall accordingly for the present confine our remarks.

Let us then endeavor to ascertain what would have been the climatic conditions of greatest eccentricity 210,000 years ago. The mean distance of the sun may be taken at 92,500,000 of miles, and the eccentricity at the period we are referring to was, in round numbers, five millions of miles. The sun's distance would therefore be 97,500,000 miles when in *aphelion*, and 87,500,000 when in *perihelion*—a difference of ten millions of miles. The question is, what difference would this make in our climate, when our winters occurred in *aphelion* or at a time when we were furthest from the sun? And in estimating this we must remember that the quantity of heat received from the sun is in inverse proportion to the square of the distance, so that instead of being about one-tenth less, as the simple proportion of the above numbers would give, the difference will be very nearly one-fifth—that is, the earth would be receiving only four-fifths of the heat in *aphelion* that it received in *perihelion*.

In order to understand what effect this would have on climate, we must first obtain a clear idea of the power of the sun in raising the earth's surface-temperature; and on this subject there is a very general misconception, owing to the unscientific manner in which all thermometers are graduated and figured, so that we reckon degrees of temperature either from the freezing-point of water or from a point thirty-two degrees below it. The true zero from which we ought to measure terrestrial temperatures is, however, about 271° F. below the freezing point, or − 239° F., this being the temperature to which the surface of the earth would sink if the heat of the sun were entirely withdrawn from it. When therefore we speak of the mean temperature of the equator being 80° F., while that of England is about 50° F., these numbers convey no real information, either as to the total amount of sun-heat received or as to its proportions in the two latitudes. But if we add to both of these numbers 239° F., we obtain 319° F. for the absolute temperature of the equator, and 289° F. for that of England, and these numbers are truly proportional to the amount of heat in each place due, directly or indirectly, to the solar influence. We may here mention that it is the universal opinion of modern physicists, that the internal heat of the earth escapes outward so slowly as to contribute no sensible portion of that which we perceive as climatic temperature, and that it may therefore be neglected in all calculations as to comparatively recent changes of climate. Two facts may be adduced, which, if they do not prove, are at all events quite in accordance with this view, while they are opposed to the view that the internal heat makes itself felt in any important degree. The one is, that in the very deepest tropical oceans the bottom temperature is close to the freezing-point; the other, that in many parts of Siberia, after passing below the surface-soil, there are found seventy and in some places several hundred feet of permanently frozen ground.

The mean January temperature of England may be taken at 39° F., which is equivalent to 278° F. of absolute temperature; and if we calculate what would be the mean temperature of the same month, when the sun was distant 97,500,000, instead of 91,000,000 of miles as it is now, we find it comes out 242° F., which is equivalent to 3° F. of our thermometers, or 29 degrees of frost. Calculated in the same way, the summer would be excessively hot, the July mean temperature being 125° F. The winter, however, would be 26 days longer than the summer, and the total quantity of sun-heat received by us during the year would be exactly the same as it is now. How then, it may be asked, could this produce a glacial period? Would not the snow that fell in the winter be all melted by the intensely hot summer? The answer to these questions involves considerations of great interest, which must be discussed in some detail; but we may first point out, that the figures above given are not to be taken as implying that these temperatures would actually occur, but rather as quantities of heating power, which might be distributed over wider areas or longer times, so as to produce results differing considerably from the maxima and minima indicated. Our actual climate does not depend so much on the amount of sun-heat we ourselves receive, as on the way in which the heat of adjacent regions is distributed. Owing to our insular position and the influence of the Gulf-stream, we have a much warmer and more uniform climate than is due to our latitude, yet the cold arctic winds often lower our temperature to the freezing-point, at times when we are receiving an amount of sun-heat which would give us genial warmth if its effect were not thus counteracted.

At the time we are speaking of, the winter of the whole northern hemisphere would be as much colder on the average as our own would be, that is, about 36 degrees F. colder than now. The tropics and parts of the southern hemisphere would also be colder, so that it is almost certain that our climate in winter would have been quite as bad if not worse than these figures represent, and would probably have been something like that of Yakoutsk in Northern Siberia, the coldest place in the world outside the Arctic circle. It is hardly possible, therefore, to exaggerate the degree of winter cold of Britain and of all the north temperate zone at that period; and the winter would not only be thus intensely cold, but all its effects would be exaggerated by its being very long. Then would follow a summer of great heat, with a sun probably as fierce as that of India or Australia, and which, though shorter than our present summers by nearly a month, would still be long as compared with that of the Arctic regions. And during this summer, we must remember, all the world would be warmer; evaporation would be greater than now, and the surface waters of the ocean, being kept at a high temperature, would give us warmer and moister breezes than we now possess. Again, therefore, we have to ask, could the winter do anything that the summer could not undo? In order to be in a position to answer this question, we must consider how the heat received from the sun is disposed of, and the influence of water and air, snow and ice, in distributing or storing up heat or cold.

The great aerial ocean which surrounds the earth has the wonderful property of allowing the heat-rays from the sun to pass through it without being warmed by them. When the earth is heated, however, the air gets warm by contact with it, and the vapour and carbonic acid in the air are also impervious to the radiant heat given out by the earth, and are therefore warmed by radiation. But the air thus warmed is in continual motion, owing to change of density. It is lifted up and pushed aside by cooler and heavier air; and thus heat can never *accumulate* in the atmosphere, or warm it beyond a very moderate degree, so that the long-continued sun-heat of the tropics is in great part carried away to give warmth to colder regions. Water also is mobile; and though it receives and stores up a great deal of heat, it is for ever dispersing it over the earth. The rain, which brings down a certain portion of heat from the atmosphere or absorbs it from the earth on which it falls, flows away in streams to the ocean; while the ocean itself, constantly impelled by the winds, forms great currents, which carry off the surplus heated water of the tropics to the temperate and even to the Polar regions. An immense quantity of sun-heat is also used up in evaporating water, and the vapour is conveyed by the aerial currents to distant countries, where, on being condensed into rain, it gives up much of this heat to the earth and atmosphere. The power of water in carrying away heat is well seen in the abnormally high temperature of arid deserts, while the still more powerful influence of air can be best understood by considering how rapidly a few hours of our northern sunshine will heat a tightly-closed glasshouse far above the temperature produced even by the vertical sun of the equator where the air is free to circulate. We can quite understand, then, that, were not a very large proportion of the sun's heat carried away from the tropics by air and water, those parts of the earth would be uninhabitable furnaces, as would indeed *any* part of the earth where the sun shone brightly throughout a summer's day.

But though sun-heat cannot be stored up beyond a very moderate amount, yet cold can be stored up to an almost unlimited extent, owing to the peculiar property of water in becoming solid at a very moderately low temperature; and as this is a point of the highest importance in the enquiry we are now pursuing, it must be carefully considered.

Let us then examine the very different effects produced by water falling as rain, or as snow, which may yet not differ from each other more than two or three degrees in temperature. The rain, however much of it may fall, runs off rapidly into rivers, and soon reaches the ocean. If cold, it cools the air and the earth temporarily, but can produce no permanent effect, because a few hours or days of sunshine restore to the air and the surface soil all the heat they had lost. But if snow falls for a long time, it remains where it fell, becomes compacted into a mass, and keeps the earth below it and the air above it at or very near the freezing-point till all is melted. If the quantity is great, it may take days, or weeks, or months, to melt it all; and if more falls during winter than can be melted during the succeeding summer, then the snow will be perpetual, as we find it above a certain height on all great mountains, and in some parts of the Arctic and Antarctic regions, and then no amount of sun-heat warms the air much above the freezing-point. In N. latitude 80°, Captain Scoresby had the pitch on one side of his ship melted by the heat of the sun, while salt water was freezing on the other side, owing to the coldness of the air.

It is very important to remember that this storing up of cold depends entirely on the amount of snowfall and the want of sufficient heat to melt it, not at all on the actual cold of the winter or the average cold of the year. A place may be intensely cold in winter and may even have a very short summer, yet if so little snow falls that it is quickly melted, there is nothing to prevent that summer being warm and the earth producing a rich and luxuriant vegetation. As examples of this, we have great forests in the north of Asia and America, where the winters are colder and the summers shorter than in southern Greenland, or in Heard Island, both of which are almost wholly covered with perpetual snow and ice. At the Jardin, on the Mont Blanc range, surrounded by perpetual snow, the lowest winter temperature is − 6° F., while at Upernivik in Greenland it is − 49 F., yet in summer the thermometer often rises to 50° F.

or even to 58° F., and the country is covered with a luxuriant herbaceous and shrubby vegetation. Even in the very highest latitudes reached by our last Arctic expedition, there is very little perpetual snow or ice; for Captain Nares tells us, that north of Hayes Sound, in latitude 79° N., the mountains were remarkably free from snow during the summer (*Proc. Royal Geogr. Soc.* vol. xxi, p. 276). And the reason of this is evidently the scanty snowfall, which rendered it sometimes difficult to obtain enough to form protecting banks round the ships; and this was north of 80 degrees latitude, where the sun was absent for 142 days.

It is a very remarkable and most suggestive fact, that nowhere in the world at the present time are there any extensive lowlands covered with perpetual snow. The Tundras of Siberia and the barren grounds of North America are all clad with some kind of summer vegetation; and it is only where there are lofty mountains or high plateaux, as in Greenland, Spitsbergen, and Grinnell Land [in the north of Ellesmere Island], that glaciers, accompanied by perpetual snow, descend to the sea-level. In the Antarctic regions there are extensive highlands, and these are everywhere exposed to the influence of the moist sea air; and it is here accordingly that we find what appears to be a veritable belt of permanent ice covering the whole circumference of the Antarctic continent, and forming a girdle of ice-cliffs which almost everywhere descend into the sea. Such Antarctic islands as South Georgia, South Shetland, and Heard Island, are often said to have perpetual snow at the sea-level; but they are all mountainous and send glaciers down into the sea, and as they receive moist air on every side, the precipitation, which almost all takes the form of snow even in summer, is of course unusually large.

Lieutenant Payer of the Austrian Arctic expedition has furnished us with a clue to these phenomena, by his observation of the rate of evaporation of ice during the Arctic summer. He found that no less than four feet thick of ice was thus evaporated. Now to replace this, a quantity of snow equivalent to about forty-five inches of rain must fall; and as the precipitation rarely reaches this amount outside the tropics, except in islands and in mountainous districts, and it is generally less than half as much over the greater part of the northern continents, we can quite understand why perpetual snow should be so rare a phenomenon on lowlands. Mountains, however, acting as condensers, are exposed to great precipitation, and when covered with snow their condensing power becomes greater, and this tends to increase the accumulation of snow; while the rarefied air has less capacity for heat, and is thus unable to melt or evaporate the snow as it does on the lowlands. We can thus understand how it is that only where there are extensive highlands, to act both as condensers of moisture and accumulators of snow, do we find anything approaching the condition of things which appears to have prevailed in our country during the glacial epoch.

These considerations seem to render Mr. Croll's theory of polar ice-caps, constantly increasing in thickness to the pole itself, quite untenable. He believes that the Antarctic continent is now covered by such an ice-cap, the thickness of which at the South Pole he calculates at twelve miles, giving a slope at an angle of only half a degree,—and less than this, he says, will not produce motion. To the objection, that the outer band of ice will condense all the vapour, he replies that however small the annual amount of snowfall may be, if more falls than is melted, the ice must continue to accumulate year by year, till its thickness in the centre of the continent be sufficiently great to produce motion ('On the Glacial Epoch,' *Geological Magazine*, July & August, 1864). But all the evidence we have of the rapid diminution of snowfall and rainfall inland, and especially when the coast consists of lofty mountains, as in the Antarctic regions, combined with the enormous evaporating power of the four months' polar sun, render such continuous increase as he supposes quite impossible. He has himself told us that during summer, from May 10th to August 3rd, a period of eighty-five days, the quantity of heat received from the sun in consequence of his remaining above the horizon is actually greater at the North Pole than at the equator (*Philosophical Magazine*, February, 1870). The greatest amount of snowfall that occurs on lowlands is perhaps on the shores of Lake Superior, where, according to Alexander Agassiz (as quoted by Sir C. Lyell), the average annual snowfall for fifteen years was seventy-two feet. Yet the snow never lay more than six feet thick on the ground, and disappeared completely in summer; although for four winter months the temperature was 5° F. below zero (*Principles of Geology*, Eleventh edition, i, p. 290). We may be quite sure, then, that beyond the mountain barrier of the Antarctic continent there are either lowlands and plateaux free from snow in summer, or an icy sea like that around the North Pole; but certainly no continuous ice-cap like that imagined by Mr. Croll.

It is in accordance with all these facts that the more remarkable indications of glaciation are always found in the vicinity of mountain ranges of sufficient altitude and extent to form condensers of the aqueous vapour and accumulators of snow, and that such indications point to a maximum of glaciation wherever the geographical position is such as to favour excessive rainfall. Thus in America the White Mountains of New England and the plateau north of the St. Lawrence have been the centres of the greatest glaciation on that continent; while in Europe, the Scandinavian mountains, and those of Ireland, Wales, and Scotland, together with the great central mass of the Alps, furnish the most remarkable evidences of extensive glaciation.

From these various considerations it is clear, that the increased cold of winter, when the eccentricity was great and the sun at its greatest distance during that season, would not of itself produce a glacial epoch, unless the amount of vapour to be condensed was also exceptionally great. Now the greatest quantity of snow falls in the Arctic regions in summer and autumn, and with us the greatest quantity of rain falls in the autumnal months. It is probable, therefore, that in all northern lands glaciation would commence at the time when autumn occurred in *aphelion*. All the rain which falls on our mountains at that season would then fall as snow, and, being further increased by the snow of winter, would form accumulations which the summer would not melt. As time went on and the *aphelion* occurred in winter, the perennial snow on the mountains would chill the summer vapours, so that they too would fall as snow, and thus increase the amount of deposition; and this would be still further increased from causes which have been admirably discussed by Mr. Croll.

The trade winds owe their existence to the great difference between the temperature of the equator and the poles, which causes a constant flow of air towards the equator. The strength of this flow depends on the difference of temperature, and this difference is now greatest between the South Pole and the equator, owing to the much greater accumulation of ice in the Antarctic regions. The consequence is, that the south-east trades are stronger than the north-east, the division between them being considerably on the north side of the equator. But just in proportion to the strength of the trade winds is the strength of the anti-trades—the upper current which carries the warm moisture-laden air of the tropics towards the poles. These are now strongest in the southern hemisphere, and thus they supply the moisture necessary to produce the enormous quantity of ice in the Antarctic seas. But during the period we are discussing—of high eccentricity and the northern winter in *aphelion*—this state of things would be reversed. The South Pole, having its winter in *perihelion*, would be nearly free of ice, while the north-temperate zone would be largely snow-clad, and the north-east trades would also be much stronger, and would bring with them the increased quantity of moisture, which is the chief thing required to produce a condition of glaciation.

The increased force of the north-east trades will have, however, another and totally distinct effect, which will still further increase the tendency to an accumulation of perpetual snow and ice. It is now generally admitted that we owe our mild climate and our comparative freedom from snow to the influence of the Gulf-stream, which equally ameliorates the climate of Scandinavia and Spitzbergen, as shown by the northward curvature of the isothermal lines, so that Trondhjem in N. lat. 63 degrees 25 minutes has the same mean temperature as Halifax (Nova Scotia) in N. lat. 45 degrees. The quantity of heat brought into the North Atlantic by the Gulf-stream depends mainly on the superior strength of the south-east trades. When the north-east trades were the more powerful, its strength would certainly be much less, while it is possible, as Mr. Croll thinks, that almost the whole of it might be diverted southward, owing to the peculiar form of the east coast of South America, and would go to swell the Brazilian current and ameliorate the climate of the southern hemisphere.

That some effects of this nature would follow from any increase of the arctic and decrease of the antarctic ice, may be considered certain; and Mr. Croll has clearly shown that in this case cause and effect act and react on each other in a remarkable way. The increase of snow and ice in the northern hemisphere is the cause of an increased supply of moisture being brought by the more powerful anti-trades, and this greater supply of moisture leads to an extension of the ice, which reacts in still further increasing the supply of moisture. The same increase of snow and ice diminishes the power of the Gulf-stream, and this diminution makes both summer and winter colder, and thus helps on still further the formation and preservation of the icy mantle. At the very same time these agencies are acting in the reverse way in the opposite hemisphere, diminishing the supply of moisture carried by the anti-trades, but increasing the temperature by means of southward ocean-currents.

We have now sufficiently answered the question, why the short hot summer would not melt the snow which accumulated during the long cold winter produced by high eccentricity and autumn or winter in *aphelion*, even though the amount of heat received from the sun in the whole year was exactly the same as now. The reasons are mainly two: first, that heat cannot be permanently accumulated, being continually carried away by winds and flowing water, while cold can be so accumulated, owing to the comparative immobility of snow and ice; and, secondly, because there are two great agencies, the winds and the ocean-currents, which are so affected by an increase of the snow and ice towards one pole and its diminution at the other, as to help on the process when it has once begun, and to produce by their action and reaction a maximum of effect, which without their aid would be altogether unattainable. To these we may add a third agency which we have not yet mentioned—that snow and ice reflect heat (as they do light) to a much greater degree than do land or water. The heat, therefore, of the short summer would be partly lost by reflection, and partly used up in melting or evaporating a certain portion of snow. Much of it too would be reflected from the upper surface of the clouds which, owing to intense evaporation at this season, would be very prevalent in the north temperate zone. Combining these various causes and effects, we have little difficulty in comprehending how the wonderful phenomena of glaciation in our own country, in the Alps of Central Europe, and over extensive regions in North America, were brought about.

Before quitting this part of our subject, we must briefly refer to the changes of level of the sea and land which occurred during the glacial period, and which Mr. Croll believes were mainly due to a shifting of the earth's centre of gravity owing to a transfer of the ice-cap from one pole to the other, thereby causing an accumulation of water around the heavier pole. There is evidence of many distinct periods of submergence. Mr. James Geikie believes that during the greatest extension of the ice, the land was somewhat lower than it is now; for he says that as the ice melted away in Scotland the sea occupied its place, following its retreating footsteps and laying down the boulder-clay; but this submergence did not exceed 200 or 260 feet. At a later period, the south of Scotland and Wales were submerged to a depth of about 1300 feet or more, up to which height marine shells are found in the drift. The land then rose again, with several pauses, till Britain became joined to the Continent, as it had been before the glacial epoch. Another submergence afterwards took place to a less amount, and then a slight elevation, bringing the land into the state in which it is now. A somewhat similar series of changes appear to have taken place in North America, but owing to the absence of marine shells the amount of subsidence is doubtful. Mr. James D. Dana believes that the land stood higher during the glacial period, and we have seen how important a factor this is in initiating glaciation and increasing its intensity.

On the assumption that an ice-cap, continually increasing in thickness, extended from lat. 50 degrees in Europe and 40 degrees in North America up to the pole, it has been calculated that a rise of the sea-level would be produced in the latitude of Edinburgh of about 800 or 1000 feet. But if, as we have endeavoured to prove, no such general ice-cap ever existed, but merely local ice-sheets collected over all high lands which were areas of great precipitation, then probably not more than one-fifth of this amount of ocean-rise would occur, sufficient to account for the submergence at the period of the formation of the boulder-clay, and for the various raised beaches at lower levels. The greater subsidences were probably local, and were perhaps due to the enormous weight of the accumulations of ice over given areas. Owing to the earth's crust giving way slowly to such strains, subsidence would only begin when the ice-sheet had nearly attained its maximum extent, and would probably continue for some time while it was diminishing; and this seems to accord very well with the ascertained facts.

It will now be seen that the theory, by which the glacial epoch is accounted for, explains also those curious indications of intercalated warm periods to which we have already referred. For after a lapse of about 10,500 years the *aphelion*, or time of minimum sun-heat in the northern hemisphere, would have gradually changed from summer to winter. The result would be, that our winters would then be much warmer and our summers somewhat cooler than they are now, and we should thus have a kind of perpetual spring. At the same time the south-westerly winds would be less strong, and would bring less vapour, while the Gulf-stream would be much stronger, and this would still further ameliorate our climate and help to get rid of all the snow and ice, till at the height of this mild epoch both would be probably unknown. We find then that our theory necessitates the existence of these warm interglacial periods, the evidences for which were once such a puzzle to geologists: and we cannot but admit that this circumstance is an additional support to the theory. We have now, however, to consider a series of facts of the most interesting nature, which bring out this aspect of the value of our theory in a still more convincing manner.

Evidence of the former occurrence of Warm Climates in Arctic Regions.—That a milder climate prevailed in the Arctic regions, at a not very remote epoch, is proved by two distinct kinds of evidence. We have, first, the remains of large herbivorous mammalia, such as the mammoth, woolly rhinoceros, bison, and horse, found in the icy alluvial deposits of Northern Siberia, and sometimes preserved with the skin and carcass entire. These remains occur over the whole area of Northern Siberia, and are abundant in the Liakhov Islands, nearly ten degrees north of the Arctic circle; and though the animals may have been overwhelmed by floods and buried in icy deposits, it is generally admitted that, in order to supply such large beasts with suitable food, the climate of Northern Siberia must have been less severe than it is now. Less doubtful proofs of a mild climate are however afforded by the numerous discoveries of unfossilized trees in the arctic lands of North America. Remains of large pine-trees were found abundantly by Sir Robert M'Clure in a ravine on the west coast of Baring Island (about 73° N. lat.). A piece of one of these trees which was brought home was twelve inches in diameter, and was estimated to have seventy rings of annual growth. Similar remains have been found in Prince Patrick Island, and in Wellington Channel, much further north . . . Here we have undoubted proofs of a considerably milder climate in the extreme north . . .

Very much more remarkable, however, are the remains of a luxuriant flora of the Miocene period, which have been found in great abundance in many arctic localities, especially in Iceland, Greenland, and Spitzbergen, with smaller deposits at the Mackenzie River, in Banks's Land, and in Grinnell Land, lat. 81 degrees 45 minutes N. This flora consisted of conifers in great abundance, more than fifty species being known, about two hundred and thirty flowering plants (Monocotyledons and Dicotyledons), and seventeen species of ferns and horsetails. Taking the plants found in Greenland, in 70° N. latitude . . . we find . . . in all one hundred and thirty-seven species of trees and shrubs . . . we are led to conclude that the flora of Greenland was probably richer then than that of any part of Europe is now, and perhaps as rich as that of the Northern United States or Japan . . .

In the other localities referred to, as well as in Alaska and the Liakhov Islands, less extensive remains of a similar flora have been found. Another occurs near Dantzic in 55° N. latitude, containing the swamp-cypress, sequoias, oaks, and poplars; along with a few more southern forms, as laurels, figs, and cinnamons. In the Isle of Mull, in Scotland, a fragmentary deposit of the same age has been preserved, with a hazel, a plane, and a sequoia, and similar plants occur at Bovey Tracey in Devonshire. It is however in Central Europe . . . near Breslau, north of the Carpathians, that the richest deposits of Miocene plants have been found, and some of these are identical with Greenland species, while many others are closely allied. There is in fact the clearest evidence that one characteristic flora then covered the whole of the north temperate and Arctic zones, with just about as much difference as now exists between the vegetation of France and Norway or the Southern United States and Canada. This flora bears more resemblance to that now inhabiting Eastern North America and Eastern Asia, than to the flora of Europe; and though it clearly shows differences of climate according to latitude, these differences are far less than such as now exist, indicating a much greater uniformity of temperature . . .

We have already explained that, if glacial epochs were mainly due to great eccentricity of the earth's orbit, they must always have alternated in each hemisphere with correspondingly mild periods, each lasting about 10,500 years; and we have shown that evidence of such intercalated mild periods is actually found, both amid the records of the last glacial epoch in Scotland, Switzerland, and North America, and in the still existing remains of large unfossilized fir-trees in the highest Arctic regions. This glacial epoch has been assumed to have been caused by the first period of high eccentricity which we meet with in our backward survey—that which occurred 200,000 years ago. Still further back, we meet with no equally high eccentricity till 850,000 years ago, when it considerably exceeded that of the last glacial epoch, amounting to near 6¾ instead of 5 millions of miles, and causing a difference of about thirty-six days between the length of winter and summer. At this time there must have been a correspondingly severe glacial epoch, with alternating periods of very mild spring-like climates, and it is to one or the other of these alternating mild periods that we impute the various Miocene floras that have been discovered in so many places in the Arctic regions and in the North temperate zone.

As we have already sufficiently explained how the mild climate, approaching to a perpetual spring, was caused in the northern hemisphere at the time when the southern hemisphere was highly glaciated, we need only say here that the greater degree of eccentricity would produce at least a correspondingly greater effect. But we know that considerable changes of land and sea have occurred since the Miocene period, and we may well suppose that these were sometimes favourable to the climate of the Arctic regions. Such would be the case if the extent of land in the Antarctic regions were greater than now, admitting of a much greater extension of the snow and ice. This would increase the flow of the Gulf-stream, and of other currents bringing warm water to the north to an enormous extent,—and if we also suppose that there was less high land near the north pole, and some difference in the American coast line, admitting the warm currents more freely into the Arctic regions; and if we further remember that during winter the whole north temperate zone would be kept warm by the nearness of the sun at that season,—it is not difficult to understand that the cold of the Arctic winter might be so ameliorated that, notwithstanding the four months' absence of the sun, the winter's cold of Spitzbergen might never exceed that of the north of Scotland at the present time.

It is now generally admitted by botanists, that the three or four months' polar night would be no serious obstacle to the growth of a luxuriant forest vegetation, if the temperature and general climate were otherwise suitable. The hothouses in the Botanical Gardens of St. Petersburg are thickly matted up for six months, yet the prolonged darkness does not prevent even tropical palms from being successfully cultivated; while in Arctic lands many shrubby plants are covered up with snow for more than that period. A far more important element in the growth of broad-leaved dicotyledonous trees is a tolerably calm atmosphere. Long-continued cold winds and gales are highly prejudicial, and it is to these that the absence or scarcity of trees in many of the stormy islands of the southern hemisphere is probably due. It is this too that dwarfs the Arctic trees; and it is probably their power of resisting long-continued winds, that enables tall coniferae to extend further north than equally lofty deciduous trees. Now it is an interesting fact, which appears to have escaped the attention even of Mr. Croll, that during the mild periods caused by high eccentricity there would be a great diminution of wind, owing to the much greater uniformity of temperature over wide areas and to the weakness of the trades and anti-trades of the northern hemisphere. That there must have been such an absence of violent winds during the summer, is evident to any one who looks at the foliage of the Arctic trees figured in Heer's work [*Flora Fossilis Arctica*, 1869–1878], specimens of which may be seen in the geological gallery of the British Museum. The leaves of the oaks, planes, chestnuts, vines, and poplars, are larger than those of most European species, and plainly tell of a comparatively tranquil atmosphere; and it is certainly a most suggestive fact, that the same theory, which accounts for so many of the past changes of our earth, explains also this undoubted feature of the Arctic climate in the Miocene warm period.

The difficulty of associating such luxuriant vegetation with the long polar night and its accompanying severe cold, has led many writers to maintain the necessity of a considerable decrease in the obliquity of the ecliptic, while others have supposed that the position of the pole itself may have changed. But astronomers and physicists both deny that such changes have occurred to anything like the requisite amount, while the latter change would not produce the desired effect, since there are proofs of a nearly simultaneous mild climate all round the polar area (Rev. Samuel Haughton's 'Notes on Physical Geology.' Read before the Royal Society, and published in *Nature*, vol xviii, p. 266.). The changes of eccentricity, and of the phases of perihelion and aphelion, are, on the other hand, universally admitted and calculable phenomena; while their effects on the accumulation of snow and ice, on the winds and on the ocean current, have been worked out in so masterly a manner by Mr. Croll, that they are rapidly taking their place among the established deductions of physical science. If we add to these such moderate changes in the distribution of sea and land, the outline of coasts, and the elevation of mountains and plateaux, as we know to have occurred again and again during the Tertiary period, we obtain a combination of causes which seem fully adequate to have brought about those wonderful changes of climate, manifested on the one hand by the recent glaciation of our own islands, and on the other by the luxuriant Arctic vegetation of the Miocene age.

Indications of Glacial and Mild Climates throughout Geological Time.—It will of course occur to the reader that, if the extreme eccentricity 850,000 years ago was the cause of the mild climate in the Arctic regions, and its accompanying luxuriant flora, there must have been also, each alternate 10,500 years, a glacial epoch of extreme intensity; and it may be asked, where are the proofs of such periods of glaciation? The answer is that there are some indications of such a glacial epoch, and, though very scanty, they are such as we can alone expect to find. The evidence of the last glacial epoch, which is more especially convincing, is that of the superficial effects produced by the ice—the striations, the *roches moutonnées*, the moraines, the travelled blocks, and the "till." But these have only been preserved to us here and there, because their formation is so recent, and because they once covered the whole country. The greater part of them have been destroyed; and that any traces of them still remain is probably due to the fact that, since the last glacial epoch passed away, there has been a period of very low eccentricity, and consequently great stability of climates, and comparatively little denudation. No fragment of any such evidence from the remote Miocene glacial epoch could possibly reach us, because there has been, in almost all the time intervening between it and our last glacial epoch, an amount of eccentricity always much greater than now, and on the average nearly double. The whole lapse of time has therefore been a continued series of alternate periods of mild and severe climates, often culminating in lesser glacial epochs and necessarily leading to a great denudation and an almost complete remodelling of the earth's surface. There are only two kinds of evidence of these remote glacial epochs that can possibly remain to us,—beds containing glacial fossils, and travelled angular blocks imbedded in marine or lacustrine deposits. The former have not yet been recognized, though they probably exist, and it is quite possible that some deposits of Arctic shells, classed as glacial, may really belong to this period. The shells of warm seas would undoubtedly suffer modification and extinction by change of climate, but those of cold regions might perhaps undergo little change; for though the polar seas of mild epochs might never be frozen, neither would they ever rise in temperature much above the freezing point, except in comparatively shallow water. Many of the Arctic molluscs live at considerable depths and have a wide range, and it is not improbable that the Arctic shell-fauna has continued almost unchanged from the Miocene period. It is therefore quite possible that some of the numerous deposits with shells, which have been found in the Arctic regions at elevations of from 100 to 500 feet above the sea, may belong to the Miocene period. The same difficulty occurs in the interpretation of the crag and associated deposits of our own country. The presence of a large proportion of living and Arctic shells has been held to prove a more recent origin, whereas it may indicate an intercalated cold period at an earlier date. Geologists almost always form their conclusions as to the age of these fragmentary deposits on the assumption that there was a continuous deterioration of climate from the Miocene through the Pliocene to the Glacial epoch. If, however, as we believe, there have *always* been alternations of warm and cold periods, a very different interpretation may be placed on the facts, and some difficulties may be overcome.

The other kind of evidence, that of travelled blocks, is however found in the Miocene deposits of Central and Southern Europe. In Northern Italy, near Turin, there is a sandstone formation full of Miocene shells, intercalated among whose strata are beds of conglomerate, containing huge angular blocks of serpentine and greenstone, sometimes more than twenty feet long. Some of them are partially striated and polished, and similar rocks occur in the Alps about twenty miles off.

When we go back to the Eocene period, we find indications of a decidedly more tropical climate than the Miocene, in fact quite as warm as that of any earlier period, so far as we can judge by organic remains, for palms, turtles, large snakes, and crocodiles, then inhabited England. Yet in this period we also have indications of ice-action, in an extensive deposit of finely stratified sandstone several thousand feet in thickness, extending from Switzerland to Vienna, and quite destitute of organic remains, but which contains in several places enormous blocks, either angular or partly rounded, and composed of oolitic limestone or granite. Near the Lake of Thun one of the granite blocks of this formation is of huge dimensions, being 105 feet long, 90 feet wide, and 45 feet thick. The granite is red, and of a peculiar kind, which cannot now be matched anywhere in the Alps. Similar erratics have been found in beds of the same age in the Carpathians and in the Apennines, indicating probably an extensive inland European sea, into which glaciers descended from the surrounding mountains.

Going back from the Eocene all through the Secondary formations, the organic remains are such as to indicate warm European seas, and there are no satisfactory proofs of ice-action. But when we reach the remote Palaeozoic formations, we lose all clear evidence of very high temperatures, since even the wonderful coal-flora is now generally admitted to be indicative of a mild or warm uniform climate, but by no means necessarily of a tropical one. Here again we meet with unmistakable signs of ice-action in the Lower Permian conglomerates of the west of England. These contain partially-rounded or angular fragments of various rocks, with striated or polished surfaces just like the stones of the 'till.' These blocks lie confusedly bedded in a red unstratified marl, and can often be traced to Welsh rocks, from twenty to fifty miles distant. This remarkable deposit was first adduced by Professor Ramsay as indicating a remote glacial epoch, and, after a personal examination of it on the ground, Sir Charles Lyell agrees that this is the only possible explanation that, with our present knowledge, we can give of it.

Numerous examples of erratic blocks, which seem to indicate ice-action, occur in the Carboniferous formation in Scotland, in France, in Nova Scotia, in Ohio, in India, and in Australia. The Old Red Sandstone and Silurian formations contain similar blocks and boulders; while Principal J. W. Dawson, of Montreal, who is not an extreme glacialist, and whose opinion is therefore unbiassed, believes that we have evidence of ice-action in temperate latitudes as far back as the Huronian age—that is beyond our Cambrian formation.

The only evidence wanting to complete the proof of glacial epochs having occurred repeatedly throughout all geological time, is the discovery of deposits of arctic marine shells similar to those of the drift, which have sometimes been raised more than a thousand feet above the sea. But here again it is a question whether such deposits can be recognized if they exist. The alternate periods of about ten thousand years of mild, and of glacial conditions, are so short geologically, that the marine deposits formed during a series of such changes may be represented by alternate bands or strata in one deposit, and the fossils of both periods may be more or less mingled together. Even if the deposit formed during a phase of glacial conditions is sufficiently distinct in composition to be separated from adjacent beds, its comparative poverty in organic remains, or their small size, will not be imputed to cold, because geologists have not yet recognized the constant alternation of short periods of mild and glacial climates as an established fact, but are accustomed to consider a glacial epoch to be one long-continued period of generally glacial conditions. Again, it must be remembered, that during the cold periods denudation will be greatly checked, so far as the carrying of the denuded matter to the sea by running water is concerned; while during the succeeding warm period it will be at a maximum, the melting of snow and ice being added to the normal rainfall, while there will be abundance of loose materials ground off by the ice, to load the rivers with great volumes of sediment. It follows, that marine deposits representing periods of glaciation will always be very scanty, as compared with those of the succeeding warm periods, and this is another reason why any deposits of this kind which do exist may easily have been overlooked.

But if the evidence of remote glacial epochs is rare and fragmentary, that of the occurrence of warm periods in the Arctic regions is frequent and ample. Besides the wonderful Miocene flora already described, there is a somewhat older one of the Upper Cretaceous age in Greenland, containing besides abundance of dicotyledons—such as figs, magnolias, myrtles, and the sub-tropical genus Myrsine-cycads, conifers, and numerous ferns, one of which is a tree-fern, with a thick stem which has been found in the Greensand of England.

In the same locality in Greenland (70 degrees 33 minutes N. latitude, and 52 degrees W. longitude), and also in Spitzbergen, a more ancient flora of the Lower Cretaceous age has been found, differing widely from the preceding in the great abundance of ferns, cycads, and conifers, and the scarcity of dicotyledons, which are represented by a single species of poplar. Among the ferns many belong to the genus *Gleichenia*, now entirely tropical.

Proofs of a mild Arctic climate in Jurassic times are found in the rich flora of this age in East Siberia and Amur-land, with less productive deposits in Spitzbergen, and at Ando in Norway within the Arctic circle. But more remarkable are the ammonites and the vertebrae of *Ichthyosaurus* and *Teleosaurus* found in the Jurassic beds of the Parry Islands in 77 degrees N. latitude.

In the still earlier Triassic deposits of Spitzbergen, species of nautili and ammonites also occur.

True coal of the Carboniferous period has been found at Spitzbergen and at Bear Island, on the north coast of East Siberia, and it contains Calamites and Lepidodendrons, with large spreading ferns. Marine deposits of the same age contain large stony corals; while the more ancient Silurian limestones, which are widely spread in the high Arctic regions, contain abundance of corals and shells of Cephalopodous mollusca, like those of the same formation in the temperate zones.

It must undoubtedly be admitted, that this connected series of records of the animal and vegetable productions of the Arctic regions, extending over the whole vast period from the Silurian to the Miocene, inclusive, and widely scattered over the Arctic and sub-arctic zones, indicates rather a constant mild climate than an alternation of warm and cold periods; and it is therefore not surprising that this view should be generally adopted. We must remember, however, that there are yet vast gaps in the record, representing long ages of which we know nothing; while the condition of the Arctic regions during cold periods would certainly be such as not to favour the formation of stratified deposits, or the preservation of animal or vegetable remains. Denudation by water would be almost wholly checked, except perhaps where great northward-flowing rivers brought down the products of warmer lands; while the masses of debris carried by glaciers and icebergs, even if preserved to our times, would rarely contain fossils.

In view of the important part played by extensive tracts of high land in producing glaciation, it is worthy of consideration whether, in the absence of such conditions, anything like a severe Arctic climate could exist, even during periods of high eccentricity with winter in aphelion. Mr. Croll has well remarked, that the influence of the lofty ice-clad mass of Greenland on the climate of the northern hemisphere is overwhelmingly great, and that were it wholly removed the resulting amelioration of climate, even now, would be almost magical. Some portion of the milder climate in the Arctic regions may then be due to the absence of any high land in districts of great precipitation; and we may easily imagine such an arrangement of the land as to concentrate all the heating power of the oceanic currents in a limited area, and thus to produce a mild or even warm climate on one side of the pole, while the other side experienced an Arctic winter. And if at this time high land prevailed, as now, round the south pole, the powerful influences of winds and currents, already described, would be kept constantly at work, so as to intensify the glaciation of the Antarctic, while ameliorating the climate of the Arctic regions. Such favourable geographical conditions might possibly keep up a warm arctic climate during an entire phase of high eccentricity, and if these conditions prevailed in several geological periods, the abundant indications of luxuriant floras and sub-tropical faunas in high northern latitudes would be to a great extent explained.

Let us now briefly summarize the facts of this strange history. The geological record is best known to us in temperate latitudes, and the series of extinct animals and plants it has brought to light indicates almost always warmer climates than at present. The Pliocene was a very little warmer, the Miocene considerably warmer, the Eocene almost or quite tropical. Further back we have no proofs of any increasing warmth; and it is generally admitted that even the carboniferous flora does not imply a climate at all warmer than the Eocene. But at a comparatively very recent period, just at the close of the Pliocene, we have irresistible proofs of intense cold in the northern hemisphere, which reduced the northern half of our own islands, and much of Europe and North America, to the condition in which Greenland is now. But there are also indications that this arctic climate alternated with milder intervals; and further, that in Miocene, Eocene, and many older periods, distinct glacial epochs occurred, which may have been as severe as that we have recently gone through. Then we have another series of still more startling climatic changes, in the warm climates of the Arctic regions. These have been proved to occur, first, probably, at or near the time of the glacial period; then in Miocene, Upper Cretaceous, Lower Cretaceous, Jurassic, Triassic, Carboniferous, and Silurian times. There is also evidence that some similar changes occurred in the southern hemisphere, of which, however, our limited space has not permitted us to give any account.

Now the whole series of these wonderful changes can be explained by a full consideration of the influence of certain astronomical facts—the eccentricity of the earth's orbit, and the precession of the equinoxes—combined with the physical properties of snow and ice in storing up cold, and the action and reaction of these on the winds and the ocean currents; the whole being further modified and intensified by changes in the distribution, and especially in the elevation, of the land in temperate and polar regions.

The objection to the theory is, that it accounts for too much. It not only explains all the changes of climate of which we have evidence, but it necessitates a whole series of changes of which we have no direct evidence. Some attempt has been here made to explain why the record of such changes is unlikely to have been preserved, and why, in cases where it has been preserved, it may nevertheless have been overlooked. The imperfection of all our records of the past is too well known to geologists, for this difficulty to have much weight with them; but we may further point out that none of the alternative hypotheses yet suggested at all remove this difficulty. If the pole had shifted its place any number of times to bring Greenland or Spitzbergen into warm latitudes, the Arctic regions must still have been *somewhere*, and the difficulty is, that no Arctic remains are *anywhere* found beyond recent times. And if we postulate any amount of change in the obliquity of the ecliptic (as advocated by the late Mr. Belt), we still have to trust to differences of eccentricity, and of winter or summer in perihelion, to produce glacial epochs and warm arctic climates alternately, and this leaves the problem exactly where it is now. As to the theory of a cooling earth, even if it were not totally inadmissible on physical grounds, it would leave the glacial epoch itself—the great starting-point in the complex problem of terrestrial climates—totally unaccounted for. We claim, therefore, that the known facts of "eccentricity," when properly applied, do serve to explain the known changes of our climate in past time; and that this is really the only working hypothesis now available, since all others have to make assumptions which either astronomers, physicists, or geologists will not grant. It were much to be wished that palaeontologists would keep this theory in view when studying in detail the subdivisions of any formation, with the object of ascertaining whether such evidence of changes of climate as it requires may not sometimes have been overlooked.

* * * * *

Wallace's model of the causes of the ice age drew its share of critical reaction. Over the 1880–1 period he responded (refs 331, 333, 335, and 336) to various objections that were printed as letters to the Editor in Nature. *Two further responses to criticism, also printed in* Nature *(refs 520 and 521), show that he was still standing his ground as late as 1896.*

Considering the complexity of the topic (and allowing for some obviously incorrect deductions based in part on inaccurate or incomplete data), it is remarkable how closely Wallace's model of the causes of the ice age approaches current views on the subject. In bringing together several previously unconnected and vital lines of thought he served an important function, and it seems an oversight that his name is now seldom connected with the history of this problem.

Wallace also attempted to come to grips with the properties of ice movement and transport, most effectively applying what he worked out to a defence of the theory of the glacial origin of alpine lake basins. The arguments are most forcefully stated in Pt. II of 'The ice age and its work' (ref 481). This paper, published in Fortnightly Review *in 1893, became one of Wallace's most widely read and cited works.*

II. Erosion of lake basins (ref 481)

Lakes are distributed very unequally over the various parts of the world, and they also differ much in their position in relation to other physical peculiarities of the surface. Most of the great continents have a considerable number of lakes, many of great size, situated on plateaux or in central basins; while the northern parts of Europe are thickly strewn with lakes of various dimensions, some on the plains, others in subalpine valleys, others again high up among the mountains, these latter being of small size and usually called tarns. The three classes of lakes last mentioned occur in the greatest profusion in glaciated districts, while they are almost absent elsewhere; and it was this peculiarity of general distribution, together with the observation that all the valley lakes of Switzerland and of our own country occurred in the track of the old glaciers, and in situations where the erosive power of the ice would tend to form rock-closed basins, that appears to have led the late Sir Andrew Ramsay to formulate his theory of ice-erosion to explain them. He was further greatly influenced by the extreme difficulty or complete inadequacy of any possible alternative theory—a difficulty which we shall see remains as great now as at the time he wrote.

This question of the origin of the lake-basins of the glaciated regions is especially interesting on account of the extreme divergence of opinion that still prevails on the subject. While the general facts of glaciation, the extent and thickness of the old glaciers and ice-sheets, and the work they did in distributing huge erratics many hundred miles from their sources and in covering thousands of square miles of country with thick layers of boulder-clay and drift, are all admitted as beyond dispute, geologists are still divided into two hostile camps when the origin of lake-basins is concerned; and the opposing forces seem to be approximately equal. Having for many years given much attention to this problem, which has had for me a kind of fascination, I am convinced that the evidence in favour of glaciation has not been set forth in all its cumulative force, while many of the arguments against it seem to me to be either illogical or beside the point at issue. I have also to adduce certain considerations which have hitherto been overlooked, but which appear to me to afford very strong, if not conclusive evidence for erosion as against any alternative theory yet proposed. I shall, therefore, first set forth, as fully as the space at my command will allow, the general evidence in favour of the ice-origin of certain classes of lakes, and the special conditions requisite for the production of lakes by this agency. The objections of the best authorities will then be considered and replied to, and the extreme difficulties of the alternative theories will be pointed out. I shall then describe certain peculiarities, hitherto unnoticed, which clearly point to erosion, as opposed to any form of subsidence and upheaval, in the formation of the lakes in question. Lastly, the special case of the Lake of Geneva will be discussed, as affording a battle-ground that will be admitted to be highly favourable to the anti-glacialists, since most of them have adduced it as being entirely beyond the powers of the ancient glaciers to have produced.

1. *The different kinds of Lakes and their distribution.*—To clear the ground at the outset, it may be well to state that the great plateau lakes of various parts of the world have no doubt been formed by some kind of earth movements occurring subsequent to the upheaval and partial denudation of the country . . . Scanty rainfall and great evaporation seem . . . to be almost essential to the formation of the larger plateau lakes. Rarely, such lakes may have been formed in comparatively well-watered districts, but the earth-movements must in these cases have been exceptionally rapid and extensive, and they are accordingly found most often in countries subject to volcanic disturbances. Such are the lakes of Southern Italy, of Macedonia, of Asia Minor, and perhaps those of Central Africa.

Quite distinct from these are the sub-alpine lakes of those mountain groups which have been subject to extreme glaciation. These are characteristically valley-lakes, occurring in the lower portions of the valleys which have been the beds of enormous glaciers, their frequency, their size, and their depth bearing some relation to the form and slope of the valleys and the intensity of the glaciation to which they have been subject. In our own country we have in Wales a small number of valley-lakes; in the Lake District, where the ice-sheet can be proved to have been much thicker and to have lasted longer, we have more numerous, larger, and deeper lakes; and in Scotland, still more severely glaciated, the lakes are yet more numerous, many of those in the west opening out to the sea and forming the lochs and sounds of the Western Highlands. Coming to Switzerland, which . . . bears indications of glaciation on a most gigantic scale, we find a grand series of valley-lakes both on the north and south, situated for the most part in the tracks of those enormous glaciers whose former existence and great development is clearly proved by the vast moraines of Northern Italy and the travelled blocks of Switzerland and France. In Scandinavia, where the ice-age reigned longest and with greatest power, lakes abound in almost all the valleys of the eastern slope, while on the west the fiords or submerged lakes are equally characteristic.

In North America, to the south of the St. Lawrence River and of Lakes Ontario and Erie, there are numbers of true valley-lakes, as there are also in Canada, besides innumerable others scattered over the open country, especially in the north, where the ice-sheet must have been thickest and have lingered longest. And in the southern hemisphere we have, in New Zealand, a reproduction of these phenomena—a grand mountain-range with existing glaciers, indications that these glaciers were recently much more extensive, a series of fine valley-lakes forming a true lake district, rivalling that of Switzerland in extent and beauty, with fiords on the south-west coast comparable with those of Norway.

Besides these valley-lakes there are two other kinds of lakes always found in strongly glaciated regions. These are alpine tarns—small lakes occurring at high elevations and very often at the heads of valleys under lofty precipices; and small or large plateau or low-level lakes which occur literally by thousands in Northern Canada, in Sweden, Finland, Lapland, and North-western Russia. The valley-lakes and the alpine tarns are admitted by all geologists to be mostly true rock-basins, while the plateau and low-country lakes are many of them hollows in the drift with which much of the country is covered, though rock-basins are also not unfrequent.

Here, then, we see a remarkable association of lakes of various kinds with highly glaciated regions. The question is whether there is any relation of cause and effect in the association; and to determine this we must take a rapid survey of other mountain regions where indications of ice-action are comparatively slight or altogether wanting, and see whether similar lakes occur there also . . .

Spain and Portugal are pre-eminently mountainous countries . . . yet there is not a single valley-lake in the whole peninsula, and but very few mountain tarns. Sardinia and Corsica are wholly mountainous, but they do not appear to possess a single valley-lake. Nor does the whole range of the Appenines, though there are many large plateau lakes in Southern Italy. Farther south we have the lofty Atlas Mountains, but giving rise to no subalpine valley-lakes. The innumerable mountains and valleys of Asia Minor have no lakes but those of the plateaux; neither has the grand range of the Lebanon, a hundred miles long, and giving rise to an abundance of rivers. Turning to the peninsula of India we have the ranges of the Ghauts [Ghats], eight hundred miles long, the mountain mass of the Neilgherries and that of Ceylon, all without such lakes as we are seeking, though Ceylon has a few plateau lakes in the north. The same phenomenon meets us in South Africa and Madagascar—abundance of mountains and rivers, but no valley-lakes. In Australia, again . . . not a single true valley-lake. Turning now to the New World, we find no valley-lakes in the southern Alleghanies, while the grand mountains of Mexico and Central America have a few plateau lakes, but none of the class we are seeking. The extremely mountainous islands of the West Indies—Cuba, Hayti, and Jamaica—are equally deficient. In South America we have on the east the two great mountain systems of Guiana and Brazil, furrowed with valleys and rich in mountain streams, but none of these are adorned with lakes. And, lastly, the grand ranges of the equatorial Andes, for ten degrees on each side of the Equator, produce only a few small lakes on the high plateaux, and a few in the great lowland river plains—probably the sites of old river channels—but no valley-lakes in any way comparable with those of Switzerland or even of our own insignificant mountains.

Having thus roughly surveyed the chief mountain regions of the whole world, we find that true subalpine valley-lakes, that is, lakes in the lower parts of the valleys descending from mountain ranges or groups, filling up those valleys for a considerable distance, usually very deep, and situated in true rock-basins—that such lakes as these are absolutely unknown anywhere but in those mountain regions which independent evidence shows to have been subject to enormous and long-continued glaciation. No writer that I am acquainted with has laid sufficient stress on this really marvellous fact of lake-distribution. Professor Bonney passes it by with the remark, that there is a perfect gradation of lakes, from the smallest tarns to those of North America and Central Africa; and Mr. Douglas Freshfield says, that, wherever on the surface of our globe there are heights there must be hollows; and other writers think that lakes are general results of the process of mountain-making. But none of these writers have apparently even noticed the fact, that glacier valley-lakes have a distinctive character which separates them broadly from the lakes of all non-glaciated countries, and that they are totally absent from such countries.

But besides the mountains which possess true valley-lakes, there are a number of ranges which have been glaciated yet do not possess them, and this absence of lakes has been used as an argument against the connection of valley-lakes with glaciation. A little examination, however, shows us that these cases greatly strengthen our argument. Comparatively large and deep valley-lakes are the result of excessive glaciation, which has occurred only when conditions of latitude, altitude, and moisture combined to produce it. In regions where glaciation was of diminished intensity, from whatever causes, valley-lakes diminish in size and number, till at last only tarns are found in moderately glaciated districts. Thus, the Pyrenees were far less severely glaciated than the Alps; they consequently possess no large valley-lakes, but numerous small high lakes and tarns. As we go eastward in the Alps, the diminished rain and snow-fall led to less severe glaciation, and we find the valley lakes diminish in size and numbers till far east we have only tarns. The Carpathians have no valley-lakes, but many tarns. The Caucasus has no lakes, and very few tarns, and this may be partly due to the steepness of the valleys, a feature which is, as we shall see, unfavourable to lake formation. In the South Island of New Zealand the lakes are small in the north, but increase in size and number as we go south where the glaciation was more intense. These numerous facts, derived from a survey of the chief mountains of the world, are amply sufficient to show that there must be *some* causal connection between glaciation and these special types of lakes. What the connection is we shall inquire later on.

2. *The conditions that favour the production of Lakes by Ice-erosion.*—Those who oppose the production of lake basins by ice-erosion often argue as if the size of the glacier was the only factor, and urge that, because there are no lake-basins in one valley where large glaciers have been at work, those which exist in another valley where the glaciers were no larger, could not have been produced by them. But this by no means follows, because the production of a lake-basin depends on a combination of favourable conditions. In the first place it is evident that ice-erosion to some extent must have taken place along the whole length of the glacier's course, and that in many cases the result might be simply to deepen the valley all along, not quite equally, perhaps, but with no such extreme differences as to produce a lake-basin. This would especially be the case if a valley had a considerable downward slope, and was not very unequal in width or in the nature of the rocks forming its floor. The

first essential to lake-erosion is, therefore, a differential action, caused locally either by increased thickness of the ice, a more open and level valley-floor, or a more easily eroded rock, or by any combination of these.

If we look at the valley-lakes of our own country and of Switzerland, the first thing that strikes us is their great length, and their situation, usually at the lower end of the valley where it emerges from the higher mountains into comparatively low country. Windermere is over ten miles long, Ullswater nearly eight miles, and the larger lakes of Switzerland and North Italy are very much longer. The first essential condition, therefore, was a valley, the lower part of which was already nearly level for several miles, and with a considerable width to the base of the mountain slopes. In the non-glaciated districts of our own country, the Dart and the Tamar are examples of rivers which have cut their valleys down nearly to sea-level while still among the hills; and in South Wales the Wye, the Usk, and the Severn have a similar character.

It must always be remembered that glacial erosion is produced by the tremendous vertical pressure of the ice, by its lower strata being thickly loaded with hard rocks frozen into its mass, and by its slow but continuous motion. In the lower part of its course a glacier would be most charged with rocky debris in its under strata, since not only would it have been continually breaking off and absorbing, as it were, fresh material during every mile of its onward course, but more and more of its superficial moraines would be engulphed by crevasses or *moulins*, and be added to the grinding material below. That this was so is proved by the great quantity of stones and grit in the "till," which is thought by Professor James Geikie to consist, on the average, of as much stony matter as clay, sometimes one material preponderating, sometimes the other. The same thing is indicated by the enormous amount of debris often found on the lower parts of large glaciers. The end of the great Tasman glacier in New Zealand is thus completely hidden for five miles, and most of the other glaciers descending from Mount Cook have their extremities similarly buried in debris. Dr. Diener found the Milam glacier in the Central Himalayas completely covered with moraine rubbish; and Mr. W. M. Conway states that the lowest twenty miles of the Hispar glacier (forty miles long) are "entirely covered with a mantle of moraine." If these glaciers extended to over a hundred miles long, as did the Rhone glacier when it reached the Lake of Geneva, much of this debris would probably have found its way to the bottom, and thus supply the necessary grinding material and the abundant stones of the "till" found everywhere in the tracks of the old glaciers.

Again, although ice is viscous and can slowly change its shape to almost any extent, yet it takes a considerable time to adapt itself to continually changing outlines of the valley bottom. Hence, where great inequalities occur portions of the rocky floor might be bridged over for a considerable space, and where a valley had a narrow V-shaped bottom the sub-glacial stream might eat away so much of the ice that the glacier might rest wholly on the lateral slopes, and hardly touch the bottom at all. On a tolerably wide and level valley bottom, however, the ice would press with its fullest intensity, and its armature of densely packed stones and rock-fragments would groove and grind the rocky floor over every foot of its surface, and with a rate of motion perhaps greater than that of the existing Greenland and Alaskan glaciers, owing to the more southern latitude and therefore higher mean temperature of the soil and the ice. At the same time sub-glacial streams, forced onward under great hydrostatic pressure, would insinuate themselves into every vacant groove and furrow as each graving tool successively passed on and the one behind it took a slightly different position; and thus the glacial mud, the product of the erosion, would be continually washed away, finally escaping at the lower extremity of the glacier, or in some cases getting embayed in rocky hollows where it might remain permanently as masses of clayey "till," packed with stones and compressed by the weight of the ice to the hardness of rock itself. The continual lubrication of the whole valley floor by water forced onward under pressure, together with the ever-changing form of the under surface of the glacier as it slowly moulded itself to the varying contours of the rocks beneath, would greatly facilitate the onward motion. Owing to these changes of form and the great upward pressure of the water in all the hollows to which it gained access, it seems probable that at any one time not more than half the entire bottom surface of the glacier would be in actual contact with the rock, thus greatly reducing the friction; while, as the process of erosion went on, the rock surfaces would become continually smoother and the inequalities less pronounced, so that even when a rock-basin had been ground out to a considerable depth the onward motion might be almost as great as at the beginning of the process.

If, now, we consider that the erosion I have attempted to describe was going on during a large part of the glacial period, under a weight of ice varying from one to three or four thousand feet in thickness; that the huge grinding tool was at work day and night, winter and summer, century after century, for whatever number of thousands of years we give to the glacial period; that—as innumerable other facts prove—the ice moved irresistibly over hill and dale, and up slopes far steeper than any formed by the upward slopes of the bottom of our deepest lakes, what is there of impossible, or even of improbable in the belief, that lake-basins were produced by such differential erosion? To the ordinary observer it seems impossible that a mountain valley, half a mile wide and bounded by rocky slopes and precipices two or three thousand feet high, can have been formed without any "convulsion of nature," but merely by the natural agencies he sees still in action—rain and frost, sun and wind—and that the small rock-encumbered stream now flowing along its bottom can have carried away the whole of the cubic miles of solid rock that once filled up the valley. But the geologist knows that these apparently insignificant forces *have* done the work, through their continuous action always in one direction for thousands, or even for millions, of years; and, therefore, having before him so many proofs of the eroding power of ice, in planed and rounded rocks, and in the grooves and furrows which are the latest marks left by the ice-tool, and bearing in mind the long duration and possibly recurrent phases of the ice age—to be measured certainly by tens, perhaps by hundreds, of thousands of years—he can have little difficulty in accepting the erosion of lake-basins as the most satisfactory explanation of their origin.

3. *Objections of modern writers considered.*—Professor Bonney and many other writers ask, why lakes are so few though all the chief valleys of the Alps were filled with ice; and why, for instance, there is no great lake in the Dora Baltea valley whose glacier produced the great moraines of Ivrea opposite its outlet into the plains of Italy, and which form a chain of hills fifteen miles long and fifteen hundred feet high? The answer, in the case of the Dora Baltea, is not difficult, since it almost certainly has had a series of lake-basins at Aosta, Verrex, and other places where the broad level valley is now filled with alluvial gravel. But the more important point is the extreme narrowness of the lower part of the valley above Donnas and again near its entrance into the valley of the Po. The effect of this would be that the great glacier, probably two thousand feet thick or more, would move rapidly in its upper layers, carrying out its load of stones and debris to form the terminal moraine, while the lower strata, choked in the defiles, would move very slowly. And once out in the open valley of the Po, then a great inlet of the warm Mediterranean Sea, the ice would rapidly melt away in the water and in the warm moist atmosphere, and therefore have no tendency to erode a lake basin.

The Lake of Lugano, with its curious radiating arms, is said to be another difficulty. But each of these arms is the outlet of a valley or series of valleys, which were no doubt reduced to nearly level plains by sub-aerial denudation before the ice began its work. The basin of these valleys comprises about two hundred square miles and the watershed to the north is moderately high; but there can be no doubt that a large overflow from the Como glacier poured into it; and the difficulty seems to me to be purely imaginary if we simply recognise the fact that an essential preliminary to lake-erosion is a pre-existing nearly level valley bottom.

Another difficulty is said to be the frequent presence of islands in the lakes: but here again the answer is easy. The islands, always ground down to *roches moutonnées*, were craggy hills in the pre-existing valleys, and such hills existed because they had for ages resisted the sub-aerial denudation which had hollowed out the valleys. The same characters of density or toughness that enabled them to resist ordinary denudation, enabled them also, to some extent, to resist destruction by ice-erosion; just as the character of the rocks which enabled ordinary denudation to bring them down to a nearly level surface in the valley-bottom, also facilitated the ice-erosion which converted the level valley floor into a rock-basin, and after the ice left it into a lake.

Every writer brings forward the well-known fact that the ends of glaciers pass over beds of gravel or moraine matter, without destroying or even disturbing it. But there is no reason why they should do more than compress such beds of loose material and roughly level their surfaces. It is the old delusion of a glacier acting like a scoop or a plough that leads to the idea that if it can erode rock slowly it must altogether demolish gravel or boulder clay. But if we turn to the description I have given of how a glacier erodes a rock-basin and apply this to its passage over a bed of gravel or boulder clay, we shall see that in the latter case the erosion would be much more difficult, because each ice-embedded stone or rock would press into the yielding material, which would close up instantly behind it under pressure of the ice and thus leave no result. Where the sub-glacial water accumulated, channels would be cut in a gravel or clay, but elsewhere there would probably be no erosion at all. Some writers maintain that the lakes were all filled up with alluvium previous to the glacial epoch, and that the ice cleared out this incoherent matter; but it is almost certain that no such clearance would have taken place, because the glacier would pass over such a surface, the stones temporarily furrowing it, while the sub-glacial water would cut for itself one or more deep channels, and there would thus be no water under pressure acting over the whole surface of the basin, which must be so great an aid to erosion in solid rock.

These considerations apply to the equally common objection, that the great masses of boulder clay left behind by the ice-sheet and over which it must have passed, prove that it could have had little eroding power. The product of the erosion of irregular rock-surfaces in an undulating tract of country, where not carried away by water, would necessarily, by the pressure of the ice, be forced into the more or less sheltered or land-locked hollows, thus tending to equalise the surface-contours and facilitate the onward motion of the ice. In such hollows it would be pressed and compacted by the weight of the ice, but would be neither eroded nor forced away until, by the continued process of rock-erosion, it became exposed to unequal lateral pressure, when it would be gradually removed to some other sheltered hollow, perhaps to again undergo the same process of removal at a later period, and finally rest in the positions in which we find it. During the later stages of the ice age when, notwithstanding the onward motion of the middle portions of the glacier, the lower portion was melting away both above and below, and the terminal ice-cliff was permanently retreating, almost the whole of the eroded matter except what was carried away by the sub-glacial torrents, would remain behind; and it is this final product of glacial erosion that forms the huge deposits of boulder clay which encumber the surface of the lowlands in most highly glaciated countries. When, however, the moving ice changed its direction, as it often did during the varying phases of the ice age, it sometimes acted most energetically in crushing, dragging, and contorting both the boulder clay and other superficial beds, often causing the wildest confusion in the deposits and sometimes embedding huge sheets of tertiary strata or chalk in the midst of the boulder-clay. But this is a very different mode of action from that by which hard rocks are ground down or lake-basins eroded.

In reply to the continual assertions of Professor Bonney and of most of the Alpine explorers, that the action of glaciers is entirely superficial, and that they actually preserve the surfaces they cover from denudation, a few facts may be here given. From a large number of gaugings by Dollfus-Ausset, Dr. Penck has calculated that the solid matter in the torrent which issues from the Aar glacier annually amounts to six hundred and thirty-eight cubic metres for each square kilometre of the surface of the glacier, a quantity sufficient to lower the bed of the glacier one metre in sixteen hundred and sixty-six years, or one foot in five hundred and five years; and the same writers calculate that the same amount of erosion in a valley by water alone would require two and a half times as long (Falsan, *La Période Glaciaire*, p. 90). Other writers have made estimates less favourable to ice as an agent of erosion; but even if the amount annually be but small, the cumulative effect was undoubtedly very great in the case of the enormous glaciers of the ice age. The very wide areas covered with boulder clay and drift in North America, and its great average depth, have already been referred to in my previous article (*Fortnightly Review*, November, 1893, p. 625); but a still more striking estimate has been made of the amount of rock debris in Northern Europe which can be traced to Scandinavia. Dr. Amund Helland states that about eight hundred thousand square miles are covered with such drift to an average depth of one hundred and fifty feet, of which about one hundred feet are of Scandinavian origin, the remainder being local. The area of Scandinavia and Finland, from which this debris has been derived, is very much less than the area over which it is distributed, so that to produce it an amount equal to an average thickness of two hundred and fifty-five feet must have been removed from those countries. To this must be added the amount which has gone into the Baltic and North Seas, and also that which has been carried away by rain and rivers since the ice age passed away, and yet further, the enormous amount that still remains on the lowlands of Scandinavia, and we shall then arrive at an amount probably twice as great as the above estimate, that is, something like five hundred feet as the average amount of ice-erosion of Scandinavia during the glacial period (*Fragments of Earth Lore*, by James Geikie, F. R. S. (1893), p. 167). Now, unless this estimate is wildly and extravagantly erroneous—and Professor Geikie adopts it as, *primâ facie*, not extravagant—we have an amount of ice-erosion so enormous as to put completely out of court all the allegations of those who attempt to minimise it as a mere smoothing off of sharp angles and rugged surfaces. I am not aware that Professor Bonney denies the Scandinavian origin of the greater part of the northern drift, and unless he can show that its quantity is something like a fiftieth part only of the estimate of Dr. Helland, I cannot understand how he can still maintain that the glaciers and ice-sheets of the ice age were agents of abrasion, not of erosion, and that they were therefore impotent to grind away the comparatively small amount of rock removed, under the most favourable conditions, from the basins of the valley-lakes whose origin we are discussing.

4. *The Alternative Theory and its Difficulties.*—There is really only one alternative theory to that of ice-erosion for the origin of the class of lakes we have been discussing, viz., that they were formed before the glacial epoch, by earth-movements of the same nature as those which are concerned in mountain formation, that is, by lateral pressure causing folds or flexures of the surface; and where such flexures occurred across a valley a lake would be the result. This is Professor Bonney's theory given in his paper in the *Geographical Journal*, and it is also that of Desor, Forel, Favre, and other eminent geologists. It is explained fully in the work of M. Falsan (already quoted), who also adopts it; and it may be considered, therefore, that if this theory can be shown to be

untenable that of glacial erosion will hold the field, since there is no other that can seriously compete with it. Professor Bonney considers this theory completely satisfactory, and he complains that the advocates of glacial erosion have never discussed it, intimating that they "deemed silence on this topic more prudent than speech."

As this theory is put forward with so much confidence, and by geologists of such high reputation, I feel bound to devote some space to its consideration, and shall, I think, be able to show that it breaks down on close examination.

In the first place, it does not attempt to explain that wonderful absence of valley-lakes from all the mountain regions of the world except those which have been highly glaciated. It is, no doubt, true that during the time the lakes were filled with ice instead of water they would be preserved from filling up by the influx of sediment; and this may be fairly claimed as a reason why lakes of this class should be somewhat more numerous in glaciated regions, but it does not in any way explain their total absence elsewhere. We are asked to believe that in the period immediately preceding the glacial epoch—say, in the Newer Pliocene period—earth-movements of a nature to produce deep lakes occurred in every mountain range without exception that was about to be subject to severe glaciation, and not only so, but occurred on both sides of each range, as in the Alps, or all round a mountain range, as in our Lake district, or in every part of a complex mountain region, as in Scotland from the Firth of Clyde to the extreme north coast—all in this very limited period of geological time. We are further asked to believe that during the whole period from the commencement of the ice age to our day such earth-movements have never produced a single group of valley-lakes in any one of the countless mountain ranges and hilly regions throughout the whole of the very much more extensive non-glaciated regions of the globe! This appears to me to be simply incredible. The only way to get over the difficulty is to suppose that earth-movements of this nature occurred only at that one period, just before the ice age came on, and that the lakes produced by them in all other regions have since been filled up. But is there any evidence of this? And is it probable that *all* lakes so produced in non-glaciated regions, however large and deep they might be, and however little sediment was carried down by their inflowing streams, should yet all have disappeared. The theory of the pre-glacial origin of these lakes thus rests upon a series of highly improbable suppositions entirely unsupported by any appeal to facts. There is, however, another difficulty which is perhaps even greater than those just considered. Whatever may be the causes of the compression, elevation, folding, and other earth-movements which have led to the formation of mountain masses, there can be no doubt that they have operated with extreme slowness; and all the evidence we have of surface movements now going on show that they are so slow as to be detected only by careful and long-continued observations. On the other hand, the action of rivers in cutting down rocky barriers is comparatively rapid, especially when, as in all mountainous countries, they carry in their waters large quantities of sediment, and during floods bring down also abundance of sand, gravel, and large stones . . .

It is, in fact, only on account of this powerful agency that we do not find valley lakes abounding in every mountainous country, since it is quite certain that earth-movements of various kinds must have been continually taking place. But if rivers have always been able to keep their channels clear, during such movements, among the mountains of the tropics and of all warm countries, some reason must be found for their inability to do so in the Alps and in Scotland, in Cumberland, Wales, and southern New Zealand; and as no reason is alleged, or any proof offered, that sufficiently rapid and extensive earth-movements actually did occur in the subalpine valleys of these countries, we must decline to accept such a hypothetical and unsatisfactory explanation.

Nothing is more easy, and nothing seems at first sight more plausible, than to allege these "earth-movements" to account for any one lake whose origin may be under discussion. But it ceases to be either easy or plausible when we consider the great number of the lakes to be accounted for, their remarkable positions and groupings, and their great depths. We must postulate these movements, all about the same time, in every part of the Highlands of Scotland, everywhere in the Lake district, and on both sides of the Alps. Then, again, the movements must have been of greater extent just where we can prove the glaciation to have been most severe. It produced lakes from 100 feet to 270 feet deep in Cumberland and Westmoreland; in Scotland, where the ice was much thicker, the lakes are from over 300 to over 1,000 feet deep; while in the Alps of Switzerland and North Italy, with its vast glaciers and ice-sheets, many are over 1,000 feet, and one reaches the enormous depth of over 2,500 feet. It may be said that the depth is in proportion to the height of the mountains; but in equally high mountains that have not been glaciated there are no lakes, so this cannot be the true explanation. One more remarkable coincidence must, however, be pointed out. The two largest Swiss lakes—those of Geneva and Constance—are situated just where the two greatest West European rivers, the Rhone and the Rhine, get beyond the mountain ranges; while on the south, one of the largest and by far the deepest of the lakes—Lago Maggiore—collected into its basin the glacier streams from a hundred miles of the high Alps, extending from Monte Rosa on the west to the peaks above San Bernardino on the east. Throughout this great curve of snowy peaks the streams converge, with an average length of only 30 miles, to unite in a valley only 646 feet above the sea level. No such remarkable concentration of valleys is to be found anywhere else in the Alps, and no other lake reaches to nearly so great a depth. On the theory of glacial erosion we have here cause and effect; on that of earth-movements we have another mere coincidence added to the long series already noticed . . . We see, then, that on the theory of erosion, the size, depth, and position of the chief lakes are all intelligible, while on that of earth movements they have no meaning whatever, since the deep-seated agencies producing subsidence, upheaval, or curvature of the surface would be as likely to act in the small as in the large valleys, and to produce deep lakes in other places than those where, at a later epoch, the thickest glaciers accumulated.

5. *The Contours and Outlines of the Lakes indicate Erosion rather than Submergence.*—While collecting facts for the present articles, it occurred to me that the rival theories of lake-formation—erosion and submergence—were so different in their mode of action that they ought to produce some marked difference in the result. There must be some criteria by which to distinguish the two modes of origin. Under any system of earth movements a valley-bottom will simply become submerged, and be hardly more altered than if it had been converted into a lake by building an artificial dam in a convenient situation. We should find, therefore, merely a submerged valley with all its usual peculiarities. If, however, the lake-basin has been formed by glacial erosion, then some of the special valley features will have been destroyed, and we shall have a distinct set of characters which will be tolerably constant in all lakes so formed. Now I find that there are three such criteria by which we ought to be able to distinguish the two classes of lakes, and the application of these tests serves to show that most of the valley-lakes of glaciated countries were not formed by submergence.

The first point is that valleys in mountainous countries often have the river-channel forming a ravine for a few miles, afterwards opening out into a flat valley, and then again closing, while at an elevation of a hundred or a few hundred feet, at the level of the top of the ravine, the valley walls slope back on each side, perhaps to be again flanked by precipices. Now, if such a valley were converted into a deep lake by any form of subsidence, these ravines would remain under water and form submerged river channels. But neither in the lakes which have been surveyed by the Swiss Government, nor in the Atlas des Lacs Françaises of M. Delebecque, nor in those of the German Alps by Dr. Alois Geistbeck, nor in the lakes of our own country, can I find any indications of such submerged river-channels or ravines, or any other of the varied rock features that so often occur in valleys. Almost all these lakes present rather steeply sloping sides with broad, rounded, or nearly level bottoms of saucer shape, such as are certainly not characteristic of sub-aerial valley bottoms, but which are exactly what we might expect as the ultimate result of thousands of years of incessant ice-grinding . . .

The next point is, that Alpine lake-bottoms, whether large or small, frequently consist of two or more distinct basins, a feature which could not occur in lakes due to submergence unless there were two or more points of flexure for each depression, a thing highly improbable even in the larger lakes and almost impossible in the smaller. Flexures of almost any degree of curvature are no doubt found in the rocks forming mountain chains; but these flexures have been produced deep down under enormous pressure of overlying strata, whereas the surface beds which are supposed to have been moved to cause lakes are free to take any upward or downward curves, and as the source of motion is certainly deep-seated those curves will usually be of very gradual curvature. Yet in the small lake of Annecy there are two separate basins; in Lake Bourget also two; in the small lake of Aiguebellette, in Savoy, there are three distinct basins of very different depths; and in the Lac de St. Point, about four miles long, there are also three separate flat basins. In Switzerland the same phenomenon is often found. In the Lake of Neufchatel there are three basins separated by ridges from twenty to thirty feet above the deeper parts. The small Lac de Joux, at the head of a high valley in the Jura, has also three shallow basins. Lake Zurich consists of three well-marked basins. The exceedingly irregular Lake of Lucerne, formed by the confluence of many valleys meeting at various angles hemmed in by precipitous mountains, has eight distinct basins, mostly separated by shallows at the narrow openings between opposing mountain ridges. This is exactly what would result from glacier action, the grinding power of which must always be at a maximum in the wider parts of valleys, where the weight of the ice could exert its full force and the motion be least impeded. On the subsidence or curvature theory, however, there is no reason why the greatest depth should occur in one part rather than in another, while separate basins in the variously diverging arms of one lake seem most improbable . . . [additional examples omitted]

The third point of difference between lakes of erosion and those of submersion is the most important and most distinctive, and furnishes, I think, what may be termed a diagnostic character of lakes of erosion. In most river valleys through a hilly or mountainous country outside of the glaciated districts, the tributary streams entering more or less at right angles to the main valley are seen to occupy small valleys of their own, which usually open out for a short distance at the same level before joining the main valley. Of course there are also torrents which rush down steep mountain slopes directly to the main river, but even these have usually cut ravines more or less deeply into the rock. Now if in such a valley we could mark out a contour line two hundred, three hundred, or five hundred feet above the level of the main stream, we should see that line continually turning up each side valley or ravine till it reached the given level at which to cross the tributary stream, and then turning back to the main valley. The contour line would thus form a series of notches or loops of greater or less depth at every tributary stream with its entering valley or deeply cut ravine, and if the main valley were filled with water this line would mark out the margin of the lake. As an illustration of this feature we may take the south-west coast of England, which has never been glaciated, but which has undergone a slight recent subsidence as indicated by the submerged forests which occur at several places. The result of this submergence is that the lower parts of its larger river valleys have been converted into inland tidal lakes, such as Poole Harbour, Dartmouth Harbour, Kingsbridge River, Plymouth and Devonport Harbours, and Carrick Road above Falmouth. The Dart River is an excellent example of such a submerged valley . . . [wherein] . . . the characteristic outline of such a valley is well indicated, the water running up every tributary stream as described above . . .

If we now turn to the valley-lakes in glaciated districts we shall find that they have a very different contour . . . In these . . . it will be seen that the water never forms inlets up the inflowing streams, but all of these without exception form an even junction with the lake margin, just as they would do if flowing into a river . . . [This] is equally a characteristic of every lake in the Lake district, and of all the Swiss and Italian lakes. On looking at the maps of any of these lakes one cannot but see that the lake *surface*, not the lake *bottom*, represents approximately the level of the pre-glacial valley, and that the lateral streams and torrents enter the lake in the way they do because they could only erode their channels down to the level of the old valley before the ice overwhelmed it. Of course this rule does not apply to large tributary valleys carrying separate glaciers, since these would be eroded by the ice almost as deeply as the main valley.

The three features of the valley-lakes of glaciated regions now pointed out—the absence of submerged ravines or river-channels either of the main river or of tributary streams; the basin-forms of the lake bottoms and the frequent occurrence of two or more separate basins even in small lakes; and the simple form of surface contour of all this class of lakes, so strongly contrasting with that of valleys known to have been recently submerged, as well as with the contour lines of valleys in non-glaciated districts and in those which are known to be post-glacial—seem to afford, as nearly as the case admits, a demonstration that the lakes presenting these features have been formed by erosion and not by submergence.

In connection with this subject may be noticed the many cases in which Alpine valleys present indications of having been greatly deepened by glacial erosion, although, owing either to the slope of the ground or the uniformity of the ice-action, no lake has been produced. In some valleys, as in that of Lauterbrunnen, the trough between the vertical rock-walls was probably partly formed before the ice age, but was greatly deepened by glacial erosion, the result being that the tributary streams have not since had time to excavate ravines of equal depth with the main valley, and therefore form a series of cascades over the lateral precipices, of which the Staubbach is the finest example. In many other cases, however, the side streams have cut wonderfully narrow gorges by which they enter the main valley. This work was probably begun by a sub-glacial stream, and the action of the atmosphere being shut out by the superincumbent ice and all variation of temperature avoided, the torrent cut for itself a very narrow groove, sometimes with overhanging sides, as it found layers of somewhat softer rock to eat away; and the upper surface of the rock being ground smooth by the ice, the atmosphere has had little effect since, and the gorge, while deepened below, has remained as restricted above as when first eroded. Such are the gorges of the Trient, Leuk, Pfäffers, and many others well known to Alpine tourists. I am not aware whether such extremely narrow winding gorges, often only two or three feet between the rock walls, are to be found in countries which have never been glaciated. I do not myself remember reading of any, though of course tremendously deep ravines are common, but these are of quite a different character. Should it be found that these extremely narrow rock-walled gorges are peculiar to glaciated districts they will afford us a means of estimating the amount of glacial erosion in valleys where no lake basins have been formed.

The Lake of Geneva as a test of the Rival Theories.—When I recently began to study this question anew, I was inclined to think that the largest and deepest of the Alpine lakes, such as Geneva, Constance, Lago Maggiore, and Lago di Garda, might perhaps have originated from a combination of earth-movements with ice-erosion. But on further consideration it appears that all the characteristic features of erosion are present in these as fully as in the smaller lakes. They are situated in the largest river-valleys or in positions of greatest concentration of the glacier-streams; their contours and outlines are those of eroded basins; while all the difficulties in the way of an origin by earth-movements are as prominent in their case as in that of any other of the lakes. I will therefore discuss, first, some of the chief objections to the erosion theory as applied to the above-named lake, and then consider the only alternative theory that has obtained the acceptance of modern writers.

One of the first objections made was, that the lake did not lie in the direction of the greatest action of the glacier, which was straight across to the Jura where the highest erratic blocks are found. This was urged by Sir Charles Lyell immediately after Ramsay's paper was read, and as it has quite recently been put forth by Professor Bonney, it would appear to be thought to be a real difficulty. Yet a little consideration will show that it has not the slightest weight. No lake was eroded in the line of motion of the central and highest part of the old glacier, because that line was over an elevated and hilly plateau, which is even now from five hundred to a thousand feet above the lake, and was then even higher, since the ice-sheet certainly effected some erosion. The greatest amount of erosion was of course in the broad and nearly level valley of the pre-glacial Rhone, which followed the great curve of the existing lake, and had produced so open a valley *because the rocks in that direction were easily denuded.* Objectors invariably forget or overlook the indisputable fact that the existence of a broad, open, flat-bottomed valley in any part of a river's course, proves that the rocks were there either softer or more friable, or more soluble, or by some combination of characters more easily denuded. A number of favourable conditions were combined to render ice-erosion easy in such a valley. The rock was, as we have shown, more easy to erode; owing to the low level the ice was thicker and had greater weight there than elsewhere; owing to the flatness and openness of the valley the ice moved more freely there; owing to the long previous course of the glacier its under surface would be heavily loaded with rock and grit, which, during its whole course would, by mere gravitation, have been slowly working its way downward to the lowest level; and, lastly, all the sub-glacial torrents would accumulate in this lowest valley, and, as erosion went on, would, under great hydrostatic pressure, wash away all the ground-out material, and so facilitate erosion. To ask why the lake was formed in the valley, where everything favoured erosion, rather than on the plateau where everything was against it, is to make mere verbal objections which have no relation to the conditions that actually existed.

Another objection almost equally beside the real question is to ask why the deepest part of the lake is near the south or convex side, whereas a stream of water always exerts most erosive force against the concave side (Falsan, *La Période Glaciaire*, p. 153; Fabre, *Origine des Lacs Alpins*, p. 4). The answer is, that ice is not water, and that it moves so slowly as to act, in many respects, in quite a different manner. Its greatest action is where it is deepest—in the middle of the ice stream—while water acts least where it is deepest, and more forcibly at the side than in the middle. The lake is, no doubt, deepest in the line of the old river, where the valley was lowest; and that may well have been nearer the southern than the northern side of the lake.

Another frequently urged objection is, that as the glacier has not widened the narrow valley from Martigny to Bex it could not have eroded a lake nearly a thousand feet deep. This seems to me a complete *non sequitur.* As a glacier erodes mainly by its vertical pressure and by the completeness of its grinding armature of rock, it is clear that its grinding power laterally must have been very much less than vertically, both on account of the smaller pressure because it would mould itself less closely to the ever-varying rocky protuberances, and mainly, perhaps, because at the almost vertical sides of the valley it would have a very small stony armature, the blocks continually working their way downward to the bottom. Thus, much of the ice in contact with the sides of narrow ravines might be free of stones, and would therefore exert hardly any grinding power. It is also quite certain that the ice in this narrow valley rose to an enormous height, and that the chief motion and also the chief erosion would be on the lateral slopes, while the lower strata, wedged in the gorge, would be almost stationary.

The most recent researches, according to M. Falsan, show that the thickness of the ice has been usually underestimated. A terminal moraine on the Jura at Chasseron is 4,000 feet above the sea, or 2,770 feet above Geneva. In order that the upper surface of the ice should have had sufficient incline to flow onward as it did, it was probably 5,000 or 6,000 feet thick below Martigny and 4,000 or 5,000 feet over the middle of the lake. It is certain, at all events, that whatever thickness was necessary to cause onward motion that thickness could not fail to be produced, since it is only by the onward motion to some outlet or lowland where the ice can be melted away as fast as it is renewed that indefinite enlargement of a glacier is avoided. The essential condition for the formation of a glacier at all, is, that more ice should be produced annually than is melted away. So long as the quantity produced is on the average more than melted, the glaciers will increase; and as the more extended surface of ice, up to a certain point, by forming a refrigerator helps its own extension, a very small permanent annual surplus may lead to an enormous extension of the ice. Hence, if at any stage in its development the end of a glacier remains stationary, either owing to some obstacle in its path or to its having reached a level plain where it is unable to move onward, the annual surplus of ice produced will go to increase the thickness of the glacier and its upper slope till motion *is* produced. The ice then flows onward till it reaches a district warm enough to bring about an equilibrium between growth and dissolution. If, therefore, at any stage in the growth of a glacier a thickness of six, seven, or even eight thousand feet is needed to bring about this result, that thickness will inevitably be produced. We know that the glacier of the Rhone *did* move onward to the Jura and beyond it; that the northward branch flowed on beyond the Soleure till it joined the glacier of the Rhine; and that its southern branch carried Alpine erratics to the country between Bourg and Lyons, 250 miles from its source. We know, too, that throughout this distance it moved at the bottom as well as at the top, by the rounded and polished rocks and beds of stiff boulder clay which are found in almost every part of its course.

In view, therefore, of the admitted facts, all the objections alleged by the best authorities are entirely wanting in real force or validity; while the enormous size and weight of the glacier and its long duration, as indicated by the great distance to which it extended beyond the site of the lake, render the excavation by it of such a basin as easy to conceive as the grinding out of a small alpine tarn by ice not one-fourth as thick, and in a situation where the grinding material in its lower strata would probably be comparatively scanty.

We have now to consider the theory of Desor, adopted by M. Favre, and set forth in the recent work of M. Falsan as being "more precise and more acceptable" than that of Ramsay. We are first made acquainted with a fact which I have not yet alluded to, and which most writers on the subject either fail to notice or attempt to explain by theories, as compared with which that of Ramsay is simple, probable, and easy of comprehension. This fact is, that around Geneva at the outlet of the lake, as well as at the outlets of the other great lakes, there is spread out an *old alluvium* which is always found *underneath the boulder-clay and other glacial deposits*. This alluvium is, moreover, admitted to be formed in every case of materials largely derived from the great Alpine range. Now here is a fact which of itself amounts to a demonstration that the lakes *did not exist* before the ice age; because, in that case all the Alpine debris would be intercepted by the lake (as it is now intercepted) and the alluvium below the glacial deposits would be, in the case of Geneva, that formed by the wash from the adjacent slopes of the Jura; while in every case it would be local not Alpine alluvium.

Professor James Geikie informs me that he considers the so-called "old alluvium" to be probably only the fluvio-glacial gravels and sands swept out from underneath the advancing glacier, and therefore to be no older, geologically, than the moraine matter which overlies it. The Swiss geologists, however, do not appear to hold this view, since they have recourse to a very remarkable hypothesis in order to overcome what they evidently believe to be a real difficulty in the way of the pre-glacial origin of the lake. The suggested explanation is as follows: At the beginning of the ice age the glacier of the Rhone crept on down its valley past Martigny and St. Maurice till it reached the lake; it is then supposed not to have marched on with an ice-wall, say five hundred or more feet high, but to have at once spread out like so much soft pitch, and to have filled the lake to its present water-level or thereabouts. Then, over this great plain of ice, the sub-glacial torrent of the Rhone is supposed to have flowed, carrying with it and depositing at the end of the lake that ancient alluvium which, somehow, has got to be accounted for! (A. Falsan, *La Période Glaciaire*, p. 135, 137)

Having thus filled the lake with ice instead of water, the main body of the glacier is supposed to start afresh and to travel over the ice, and thus obviate the imaginary difficulty of a glacier moving up hill, though every student of glaciers now admits that they did so, and though it is universally admitted that this very glacier of the Rhone moved over higher, steeper, and more irregular hills on its way to the Jura and to Soleure.

Now this extraordinary theory involves two difficulties which are passed by in silence, but which seem to entirely contravene all that we know of the nature of glaciers, and to be entirely unsupported by facts. The first is, the glacier ceasing to move onward as a glacier, but spreading out to fill up a lake basin, as if the lake were simply frozen to the bottom. Is this conceivable or possible? I think not. When glaciers come down to a fiord or to the sea they do not spread out laterally but move on till the water is deep enough to buoy them up and break off icebergs, and no reason is given why anything different should have happened in the case of the great Swiss and Italian lakes, supposing they existed before the ice age came on. That the glacier should afterwards slide over this level plain of ice is equally inconceivable, in view of the property of relegation of ice under pressure. Owing to this property the glacier and the lake ice would become one mass, and would move on together under the law of decreasing velocity with depth. This, however, is of little importance if, as I conceive, the supposition of the formation of an ice-sheet at the water level for fifty miles in advance of the glacier is an impossible one. The only other theory is, that the lake was filled up by alluvium before the ice age, and that the glacier re-excavated it. I have, however, already given reasons why the glacier would not have done so, and the very existence of this ancient alluvium in the course of the ancient glacier is a proof that it did not do so. This theory seems now to have no supporters . . .

On the whole, I venture to claim that the facts and considerations set forth in this paper show such a number of distinct lines of evidence, all converging to establish the theory of the ice-erosion of the valley lakes of highly glaciated regions—a theory first advocated by the late Sir Andrew Ramsay—that that theory must be held to be established, at all events provisionally, as the only one by which the whole body of the facts can be explained and harmonized.

* * * * *

Wallace played another role in the development of glacial theory by alerting his colleagues to studies that were being published on the evidence for glaciation in the southern hemisphere. Notable examples of this work include refs 172, 456, 480, and 482.

To a certain degree, Wallace's involvement in landscape and glacial climatology studies was a logical extension of his interest in the causes of biological evolution. But it was also something of a hobby for him, in some cases a chance to 'stretch out' in a purely intellectual fashion, in others an opportunity to try to infuse the ordinary with new meaning. At the same time, however, Wallace firmly believed that one should never lose sight of the connections between land and labour; this, of course, was the whole point of the land nationalization movement: only a strategy of land use consistent with higher objectives would allow us the material stability permitting a truly 'progressive' future.

Related writings

regional physical geography descriptions: refs 3, 4, 6, 11, 14, 18, 19, 22, 29, 31, 41, 44, 51, 53, 64, 65, 78, 85, 91, 103, 104, 105, 128, 288, 301, 320, 406, 441, 464, 529, 615, 710, 714, 715, 719, 720, 721, 725, 727, 729

on mathematical geography and geodesy: refs 11, 115, 116, 159, 162, 163, 171, 229, 253, 455, 526, 727, 728, 729

on the 'calibration' of geological time and age of the earth: refs 146, 159, 171, 233, 367, 458, 460, 513, 721

on the permanence of continents and ocean basins: refs 81, 146, 188, 269, 275, 340, 367, 393, 453, 457, 718, 719, 721, 724, 727, 728

on the formation of mountains and the interior of earth: refs 78, 295, 298, 300, 455, 514, 727

on the physical geography, geology, and classification of islands: refs 38, 78, 131, 188, 316, 317, 340, 381, 393, 453, 465, 715, 720, 721, 727

on the causes of glacial epochs: refs 146, 159, 171, 214, 313, 325, 331, 335, 336, 520, 521, 721

on the theories of glacial motion and the origin of alpine lakes: refs 124, 184, 233, 462, 472, 481, 484, 486, 487, 489, 727

on southern regions' glacial epochs: refs 124, 172, 238, 456, 472, 480, 482, 720, 724

on climatology: refs 41, 78, 146, 159, 171, 288, 313, 325, 331, 333, 335, 336, 520, 521, 547, 646, 714, 719, 720, 721, 725, 726, 728, 730

on the relations between physical geography and geology and biogeography: see Section Five, p. 282

on the relations between physical geography and geology and evolutionary theory: see Section Six, p. 364

on miscellaneous topics in physical geography and geology: refs 20, 53, 143, 146, 224, 267, 268, 288, 291, 464, 481, 526, 547, 602, 646, 714, 715, 716, 718, 719, 720, 724, 725, 726, 727, 728, 729, 730, 732

on miscellaneous topics in glaciology: refs 124, 146, 267, 292, 313, 406, 481, 529, 689, 721, 726, 727

book reviews and notices: refs 131, 146, 172, 184, 229, 233, 238, 313, 381, 455, 456, 727

secondary literature: Brooks (1984), Fichman (1977), George (1964), Schadewald (1978), Tuan (1963)

§ 8 · SOME MISCELLANEOUS SUBJECTS: SYSTEMATICS, GEOGRAPHIC EDUCATION, AND ASTRONOMY

⁂

> . . . Can we believe that we are fulfilling the purpose of our existence while so many of the wonders and beauties of the creation remain unnoticed around us? While so much of the mystery which man has been able to penetrate, however imperfectly, is still all dark to us? While so many of the laws which govern the universe and which influence our lives are, by us, unknown and uncared for? And this not because we want the power, but the will, to acquaint ourselves with them. Can we think it right that, with the key to so much that we ought to know, and that we should be the better for knowing, in our possession, we seek not to open the door, but allow this great store of mental wealth to lie unused, producing no return to us, while our highest powers and capacities rust for want of use? . . .
>
> ('The advantages of varied knowledge' 1843: ref 1, p. 203)

IN THIS last Section, we shall notice Wallace's contributions to three areas that do not fall neatly within the subjects covered by the other Sections. These range from the most technical aspect of his work—systematic revision—to the most synthetic and speculative—astronomy/cosmology—to the socially conscious nature of his promotion of geographic education.

⁂

Considering his status as one of history's most productive natural history collectors, it would be rather strange had Wallace had no general position on the goals of systematic description and revision: for some fifteen years, his very livelihood depended on his ability to recognize rare—and thus more valuable—forms. Relatively little has been written on Wallace's ideas on systematics per se, *yet in this area too he was well ahead of his time. Largely as a function of his interest in evolution, he was an early promoter of 'natural' forms of classification. Consider the following selection from 'Attempts at a natural arrangement of birds' (ref 28, pp. 195–6), published in 1856:*

. . . The method [of classification] illustrated at the commencement of this paper, of marking off certain groups from the general mass, has been satisfactory, because the portions so severed have been not only capable of definition, but have contained only species which have agreed in all essential points of their structure and oeconomy. They have therefore met with general acceptance, and in all the different systems of ornithologists these groups have scarcely suffered any variation. But in attempting to carry out this system in a further division of the Passeres, no such satisfactory and generally accepted results have been produced. No systematist has been satisfied with the arrangements of his predecessors, and, after an endless variety of divisions and subdivisions, we are as far off from any generally accepted system of arrangement as ever.

The reason of this we conceive to be, that we have to deal with a mass of species in which the series is so nearly complete, that there are no more of those great chasms separating considerable portions from each other, and that the affinities are so intricate and minutely varied, and so cut up as it were by minor gaps between genera and families, that any attempt to form great and well-marked subdivisions must fail, for the simple reason that such are not marked out by nature. In such a case an *arrangement* may be possible, but a *classification* may not be so. We must therefore give up altogether the principle of *division*, and employ that of *agglutination* or juxtaposition. We shall best explain our meaning by pointing out the errors we conceive to have been produced by the former method in most modern classifications.

The system of Cuvier, as modified by Vigors, Swainson, and G. R. Gray, may be fairly taken as that most generally in use, at least in this country. The tribes of the Conirostres, Dentirostres, Fissirostres, Tenuirostres, and Scansores, are said to be the natural divisions of the Passeres, the main difference of opinion being as to whether the Scansores should or should not form a separate order, a question we believe to be of no importance whatever. These divisions being accepted, every bird is forced into one of them, and the result has been the most incongruous and unnatural combinations. For instance, in the Tenuirostres are combined the Humming Birds and the Sun Birds (*Nectarinia*), families which in a natural arrangement would have, in our opinion, the mass of the other Passeres intervening between them. In the case of these two families, a mere outward resemblance appears to have been universally mistaken for an affinity. A similarity in size, in the prevalence of metallic colours, and in the slenderness of a very variable bill, has been taken to over-balance the most important structural differences. The universal characteristics of the Hummers are, excessively long wings and as excessively small feet, with more or less united toes. They take their food exclusively on the wing. Every motion is made upon the wing. The feet are solely used as means of support, never for locomotion. The Sun Birds and their allies, the Caerebidae of America, have on the other hand long legs and toes, the hinder toe especially being very long and powerful; they are therefore as capable of hopping and perching as any of the most highly developed Passeres. Their wings, too, are short and round, quite incapable of any powerful flight, and their tail almost invariably short and even. Much stress has been laid upon the similar form and structure of the tongue. But the extensile tubular tongue only exists in some genera (*Anthreptes*, *Arachnothera*), while in others it is short, flat, and not extensible (*Dicaeum*, *Cinnyris*, Caerebidae). There is therefore no general agreement of structure to unite these groups, except the solitary and trivial one of an elongate and slender bill.

On similar principles, we believe the Conirostres and Dentirostres to be equally untenable and unnatural. They are professedly founded on one character only, and not on general structure; and it is therefore not to be wondered at, that in their attempts to pay some little regard to natural affinities, while forcing the genera and families into these divisions, no two naturalists should be able to arrive at the same results. The association of the Hornbills with the Crows, the separation of the Larks from the Wagtails, and the necessity of putting Dentirostral birds (Tanagers and Jays) in the Conirostral tribe, are some of the inconsistent results of the system.

The remaining groups, the Scansores and the Fissirostres, we believe to be much more natural, and in fact to be the only ones which can be distinctly separated from the Passeres, of which they form an abnormal development . . .

* * * * *

In January 1864, Wallace continued the preceding line of discussion in 'Remarks on the value of osteological characters in the classification of birds' (ref 88), a criticism of studies published by M. E. Blanchard:

. . . No one can be more convinced than myself of the utility of osteology, and especially of the *sternum*, in the classification of birds . . . I cannot, however, allow that *osteological* characters are an all-sufficing guide. Like every other character taken singly, osteology is a very uncertain and irregular test of affinity, and is, moreover, in almost every case accompanied by parallel external characters. Sometimes one, sometimes another part of the bird's organization has varied more rapidly, so that one group exhibits the most striking constancy of a part which in another group is subject to extreme modifications. The sternum is no exception to this rule, and by following it alone we should make the greatest errors in classification. For example, the sterna of the Finches and the Flycatchers are scarcely distinguishable, notwithstanding the great dissimilarity in almost every part of the structure of these birds—their bills, their feet, their plumage, their habits, food, and digestive organs. On the other hand, the sterna of the several genera of the Caprimulgidae differ from each other more than do those of the most distinct families of the restricted Passeres. The Bee-eaters, the Barbets, and the Woodpeckers, again, are three very distinct families, which, in a classification founded upon all parts of a bird's organization, cannot be brought in close contact; and yet their sterna, according to M. Blanchard, much resemble each other. It is evident, therefore, that the whole structure of a bird and its corresponding habits may be profoundly modified, and yet the sternum may retain a very close resemblance to a common form; and, on the other hand, the sternum may undergo important changes, while the general organization and habits are but little altered . . .

With regard to M. Blanchard's determination of affinities from the body of the *sternum* only, without its appendages, I must remark that it often leads to erroneous results. For example, he says that the *sterna* of *Merops* and *Tamatia* do not differ enough to deserve a separate description; and he includes *Megalaema* and *Tamatia* in one section, as having the same form of sternum. He notices some differences in the Picidae, but remarks on their resemblance to *Megalaema* and to the Toucans. Now in all these points an examination of the entire sternum, with the *furcula*, coracoids, and clavicles attached, leads me to very different results. The sterna of *Merops* and *Nyctiornis*, compared with those of two species of *Megalaema*, seem to me to show no resemblance whatever: in almost every part they present important differences of form, surface-texture, and proportions, while the *furcula* and coracoids are so different in the two, that I should unhesitatingly place them far apart, in at least different tribes or primary divisions of the Passeres. On the other hand, the sternum of the Toucans (*Pteroglossus*) resembles that of *Megalaema* most closely in every particular, and especially in the extreme weakness and complete separation of the two arms of the *furcula*—a character which I am not aware exists in any other families of birds. The sternum of the Picidae presents many important differences from those of all these families, and fully bears out the isolation which their external characters exhibit. It differs much from *Megalaema* and *Pteroglossus* in its general form, as well as in details of structure, and still more from *Merops*. It seems to approach the typical Passeres more than either of the other groups to which M. Blanchard compares it; but its peculiar pyramidal shape, so remarkably narrowed at the anterior extremity, and its very short clavicles distinctly separate it as a characteristic and isolated form. It will, therefore, I think, be admitted that the affinities indicated by the complete sternum and appendages are much more in accordance with those derived from external form and structure, and from habits and economy, than those which M. Blanchard deduces from the body of the sternum alone.

These remarks are made in no spirit of depreciation of this very interesting and valuable work, but for the purpose of showing that isolated characters may lead to erroneous conclusions from whatever part of the organism they are chosen, and that in this respect osteological have no positive superiority over external characters. M. Blanchard tells us, in the introduction to this first instalment of his work, that he proposes to examine successively each separate part of the bird's skeleton. His future researches may therefore seriously modify the conclusions he has hitherto arrived at. I cannot but think, however, that he would have produced a more satisfactory work, if he had based it upon the comparison of the entire sternum, with its appendages attached, and also on the cranium, these two parts being of the greatest importance in classification.

It has been well observed by Professor [Richard] Owen that those parts of an animal which have the least immediate connexion with its habits and economy are exactly those which best exhibit deep-seated and obscure affinities. The wings, the feet, and the beak in birds may undergo the most extraordinary modifications in the same group in accordance with differences of habits and of external conditions, while at the same time such apparently insignificant characters as the general colouring, the texture of the plumage, the scaling of the tarsi, or the colour and texture of the eggs remain constant, and reveal the true relations of the species. Thus it is that the form of the sternum is of such importance, since it has no immediate dependence on external form and habits. The Sparrow, the Flycatcher, the Wren, and the Sunbird, all have one characteristic form of sternum; while between those of the Swallow and the Swift there is the greatest diversity.

It is evident also that the modifications of form immediately dependent on habits and external conditions are generally to be seen in the skin even better than in the skeleton of a bird. These are principally changes of form, size, and proportion in the bill, the feet, the wings, which are excellent characters for distinguishing genera and even families; while for determining the true affinities of isolated groups we must have recourse to those characters which, having no direct dependence on habits, etc., are often persistent in a remarkable degree. Of these, no doubt, the sternum is of the highest value; but there are many others of almost equal importance. Such are the texture of the plumage; the form of the feathers and their arrangement over the surface of the body; the form of the nostrils; the scutellation of the tarsi; the mode of nidification, with the form, texture, and colour of the eggs; the covering of the young bird, and its changes of plumage; peculiarities of food, characteristic habits, and peculiar attitudes and actions.

As an instance of the value of such apparently trifling characters as the last, I may mention that the first time I saw a Roller (*Coracias temmincki*) alive, I was at once satisfied it was a Fissirostral bird, from a peculiar jerking motion of the head and tail when it alighted, which is common to Kingfishers, Trogons, Bee-eaters, and Motmots, but never seen in the typical Passeres. In like manner the motions of the *Eurylaemi* convinced me that they were not Fissirostres, but typical Passeres . . .

Now that true principles of classification are becoming so much better understood, we may, I think, hope that the chaos which has so long existed in ornithology will soon give way to a truly natural system which must obtain general acceptance.

* * * * *

In 1860 Wallace made the suggestion, later to become a reality, that an international review board be organized to settle questions regarding competing claims of priority in zoological nomenclature. This occurred as an aside in a letter sent to his collections agent Samuel Stevens (ref 63):

. . . Allow me here to make a remark on the constant changes of specific names by yourself and Mr. Gray. It strikes me that, by forcing the law of priority to its extreme limits, you create a complicated synonomy, instead of settling it. Was not that law made to decide among several names already in use—not to introduce diversity where uniformity of nomenclature has hitherto existed? What is gained by changing *Eclectus linnaei* into *E. cardinalis*, and *Paradisea superba* into *P. atra*, when it is almost certain that such changes will not be generally adopted? I believe the synonymy of Natural History will never be settled till a tribunal shall be appointed by general assent, from whose decrees there shall be no appeal. It matters absolutely nothing whether a bird has one name or another; but it is of the utmost importance that it should not have two or three at once. A synonymical catalogue, which should be authoritative and final by the general consent of naturalists in congress assembled, would be a work worthy of the century. Let ornithologists be the first in the field, and the other -ologists will soon follow . . .

* * * * *

Wallace's feelings on the subject of nomenclatural priority found another outlet when he reviewed entomologist David Sharp's The object and method of zoological nomenclature *in* Nature *in 1874 (ref 239):*

Zoologists and botanists universally adopt what is termed the binomial system of nomenclature invented by Linnaeus. The essential principle of this system is, that every species of animal or plant is to have a name made up of two words, the second word—which is called the specific or trivial name, having exclusive reference to the species itself, the first word—which is called the generic name, indicating the genus, or small natural group, which comprises the species in question along with others. Thus the cat, the tiger, and the lion, belonging to one genus . . . are called respectively, *Felis cattus*, *Felis tigris*, and *Felis leo*. The name of each species, therefore, shows us what group it belongs to, and thus gives us a clue to its affinities; and the system of nomenclature is to this extent classificatory. But, as the true natural grouping of species has not yet been agreed upon by naturalists, and genera have been in a state of incessant change from the time of Linnaeus to the present hour (or for about a century), the names of an immense number of species have been repeatedly altered; and one of the first requisites of a good system of nomenclature—that the same object shall always be known by the same name—has been lost, in the attempt to make the name a guide to classification, while the classification itself has ever been fluctuating and still remains unsettled . . .

In order to determine in every case which of the names which are or have been in use is the right name, and so arrive at uniformity of nomenclature, certain rules have been pretty generally agreed upon, the most important of which is that of "priority." This means that the first name given to a species is to be the name used, even when it has never come into general use, but is now discovered in some scarce volume dated 80 or 100 years ago. But this absolute law of priority only applies to the specific or trivial name; in the case of the generic name no such absolute priority has been thought possible, because the genera of the old authors were very extensive groups, which have now been divided, in some cases into hundreds of genera. This process of division has, however, gone on step by step, one author dividing an old genus into three or four new ones, with new names; another dividing some of these still further, with more new names; another dividing some of these still further, with more new names; another perhaps discovering that these genera were not natural, and grouping the species into genera on altogether different principles, and again giving new names. Genera have been thus subdivided to such an extent that the owls, for example, which Linnaeus classed as one genus, now number more than fifty; and the ten British owls have to be placed in nine distinct genera.

In the very ingenious and careful essay which has led to these remarks, Mr. David Sharp, a well-known entomologist, advocates a mode of attaining the great desideratum of naturalists—a fixed and uniform nomenclature of species—which has not, so far as we are aware, been suggested before, although it is at once simple and logical. He proposes that, not merely one-half, but the entire name of every species once given, should be inviolable, until by general consent some permanent classificatory system of naming species, analogous to that used in chemistry, is arrived at. The insect named by Linnaeus *Papilio dido* should, for example, retain that name, although it must find its classificatory place in the genus *Coloenis* and the family Nymphalidae; while the glossy starling of the East should retain the name *Turdus cantor*, given to it by Gmelin, although it is no thrush, and belongs to the genus *Calornis*. The name would thus remain fixed, however the place of the species in our classifications might be changed; and the very errors of the original describers might help us to remember the object referred to by directing our attention to the cause of their error in classifying it . . .

A great evil of the present system is, that while professing to keep the specific or trivial name inviolable, it often compels an entire change of name. This happens whenever, by a new arrangement, a species has to be placed in a genus which already contains the same trivial name. Two species thus come to have the same name, and one of these must be wholly changed. The evil of this system of perpetually changing names is not so much the trouble it gives us to find out what object a name really refers to (though that is serious) as the enormous waste of labour involved in the elaborate working out of synonomy, rendered many fold more difficult by the complication of changes in both the generic and specific names, from a variety of causes. These difficulties are much greater in the case of genera than in that of species; and this portion of synonomy would be almost got rid of if it were decided that the first binomial name given to a species should never be changed. We should then avoid the absurdity of having hundreds of familiar names abolished, because a mere compiler of an early catalogue, who had perhaps never seen the objects themselves, divided them up almost at random into a number of named groups, or because some modern student thinks it advisable to split up every large genus into dozens of smaller ones.

These appear to be weighty arguments in favour of Mr. Sharp's proposal, yet we are far from thinking that it will be adopted. For, after all, the changed names are but few in comparison with those which remain unchanged for considerable periods; and the charm of a nomenclature which is to a considerable extent classificatory is so great, that most naturalists will strongly object to giving it up. So long as the old name keeps within the bounds of the modern family (which is in most cases a stable and well-defined group) there might be little objection to retaining it; but when it leads to the use of a name indicating a distinct and often quite unrelated family—as *Silpha scabra* for one of the Lamellicornes, (*Trox scabra*) in the example given by Mr. Sharp—the system will, we apprehend, be almost unanimously rejected . . .

Although it appears to us pretty certain that the plan of returning to the first generic name given to a species will not be adopted, the proposal to do so may lead to a reconsideration of the practice of applying the law of priority to generic names, as all are agreed it must be applied to specific or trivial names. If the generic part of the name may be altered any number of times in accordance with altered views as to classification, the principle of priority in the mere name is so totally given up, that it seems absurd to use it for the purpose of resuscitating the obsolete appellations of early writers. When an author is admitted to have defined a natural genus, he should have full power to give a name to that genus, because it is really a new thing; and it is both illogical and inconvenient to reject his name because some former writer has given another name to a group, not the same, but which merely happened to contain some one or more of the same species. Again, we think Mr. Sharp's arguments suggest the advisability of opposing the splitting up of large genera into many smaller ones otherwise than provisionally; the old generic name continuing to be used till there is a concurrence of opinion as to the necessity of adopting the new ones. The older authors were often modest enough to do this; indicating natural divisions of large genera, but not naming them; whereas modern naturalists, as a rule, feel bound to give a new name to every fragment they can split off an established genus.

It appears, then, to the present writer, that the plan best adapted to lead speedily to a fixed nomenclature, and at the same time one that will least offend the prejudices of zoologists, is as follows:—

1. To adopt, absolutely and without exception, the principle of priority as regards specific or trivial names.

2. To adopt the same principle for genera only so long as the generic character or definition of the genus remains unaltered; but whenever an original investigator defines a genus more completely than has been done before, he is to be left free to name it as he pleases. Every consideration of utility and common sense will of course lead him to retain a name already in use when the new genus does not materially differ from an older one; but of that he is alone the judge, and it should be absolutely forbidden to any third party to say that a name so given must be changed.

3. Whenever genera which are widely recognised are split up into a number of proposed smaller ones, the old generic name should continue in use till further investigation determines whether the new groups are sufficiently well defined and natural to supplant the old one. . .

* * * * *

Wallace wrote on the theory of systematics on several other occasions as well. In addition he was, as noted earlier (in Section Five), a persistent critic of those who failed to record geographical data pertinent to their collections. But the bulk of his writings on systematics consisted of taxonomic revisions—over five hundred published pages' worth, primarily of birds and insects. And, it must be remembered, he was the source of materials that led to a vast number of similar efforts by specialists whose names are now known only in connection with the histories of particular subfields of zoology.

Though Wallace fully supported the work of scientific specialists, he none the less believed that their efforts had to be placed within the context of societal priorities. He applauded 'science for science's sake', but not if this meant the diversion of government funds to activities that did not directly serve the needs of the average taxpayer. In a highly controversial, aggressively libertarian letter printed in Nature *in early 1870, he spoke out on the relation between science and education:*

Government aid to science (ref 157)

. . . The public mind seems now to be going mad on the subject of education; the Government is obliged to give way to the clamour, and men of science seem inclined to seize the opportunity to get, if possible, some share in the public money. Art education is already to a considerable extent supplied by the State,—technical education (which I presume means education in "the arts") is vigorously pressed upon the Government,—and Science is now urging her claims to a modicum of State patronage and support.

Now, sir, I protest most earnestly against the application of public money to any of the above specified purposes, as radically vicious in principle, and as being in the present state of society a positive wrong. In order to clear the ground let me state that, for the purpose of the present argument, I admit the right and duty of the State to educate its citizens. I uphold national education, but I object absolutely to all sectional or class education; and all the above-named schemes are simply forms of class education. The broad principle I go upon is this,—that the State has no moral right to apply funds raised by the taxation of all its members to any purpose which is not directly available for the benefit of all. As it has no right to give class preferences in legislation, so it has no right to give class preferences in the expenditure of public money. If we follow this principle, national education is not forbidden, whether given in schools supported by the State, or in museums, or galleries, or gardens, fairly distributed over the whole kingdom, and so regulated as to be equally available for the instruction and amusement of all classes of the community. But here a line must be drawn. The schools, the museums, the galleries, the gardens, must all alike be *popular* (that is, adapted for and capable of being fully used and enjoyed by the people at large), and must be developed by means of public money to such an extent only as is needful for the highest attainable *popular* instruction and benefit. All beyond this should be left to private munificence, to societies, or to the classes benefited, to supply.

In art, all that is needed only for the special instruction of artists, or for the delight of amateurs, should be provided by artists and amateurs. To expend public money on third-rate prints or pictures, or on an intrinsically worthless book, both of immense value on account of their rarity, and as such of great interest to a small class of literary and art amateurs and to them only, I conceive to be absolutely wrong. So, in science, to provide museums such as will at once elevate, instruct, and entertain all who visit them is a worthy and a just expenditure of public money; but to spend many times as much as is necessary for this purpose in forming enormous collections of all the rarities that can be obtained, however obscure and generally uninteresting they may be, and however limited the class who can value or appreciate them, is, as plainly, an unjust expenditure. It will, perhaps, surprise some of your readers to find a naturalist advocating such doctrines as these; but though I love nature much I love justice more, and would not wish that any man should be compelled to contribute towards the support of an institution of no interest to the great mass of my countrymen, however interesting to myself.

For the same reason I maintain that all schools of art or of science, or for technical education, should be supported by the parties who are directly interested in them or benefited by them. . . . if our men of science want more complete laboratories, or finer telescopes, or more expensive apparatus of any kind, who but our scientific associations and the large and wealthy class now interested in science should supply the want? They have hitherto done so nobly, and I should myself feel that it was better that the march of scientific discovery should be a little less rapid (and of late years the pace has not been bad), than that Science should descend one step from her lofty independence and sue *in formâ pauperis* to the already overburthened taxpayer . . .

I maintain further, not only that the money spent for the purposes here indicated is wrongly spent, but also that it is in a great measure money wasted. The best collectors are usually private amateurs, the best workers are usually home students or the employés of scientific associations, not of governments . . . Experience shows that public competition ensures a greater supply of the materials and a greater demand for the products of science and art, and is thus a greater stimulus to true and healthy progress than any Government patronage. Let it but become an established rule that all institutions solely for the advancement of science and art must be supported by private munificence, and we may be sure that such institutions would be quite as well supported as they are now, and I believe much better. If they were not, it would only prove more clearly how unjust it is to take money from the public purse to pay for that which science-and-art-amateurs would very much like to have, but are not willing themselves to pay for.

The very common line of argument which attempts to prove the wide-spread uses and high educating influences of art and of science, are utterly beside the question. Every product of human intellect is more or less valuable; but it does not therefore follow that it is just to provide any particular product for those who want it, at the expense of those who either do not want, or are not in a condition to make use of it. Good architecture, for instance, is a very good thing, and one we are much in want of; but it will hardly be maintained that architects should be taught their profession at the public expense. . . . if we once admit the right of the Government to support institutions for the benefit of any class of students or amateurs however large and respectable, we adopt a principle which will enable us to offer but a feeble resistance to the claims of less and less extensive interests whenever they happen to become the fashion.

If it be asked (as it will be) what we are to do with existing institutions supported by Government, I am at once ready with an answer. Taking the typical examples of the National Gallery and the British Museum, I maintain that these institutions should be reorganised, so as to make them in the highest degree entertaining and instructive to the mass of the people;—that no public money should be spent on the purchase of specimens, but what they already contain should be so thoroughly cared for and utilised as to make these establishments the safest, the best, and the most worthy receptacles for the treasures accumulated by wealthy amateurs and students, who would then be ready to bestow them on the nation to a much greater extent than they do at present. From the duplicates which would thus accumulate in these institutions, the other great centres of population in the kingdom should be proportionately supplied, and from the Metropolitan centres trained officers should be sent to organise and superintend local institutions, such a proportion of their salaries being paid by Government as fairly to equalise the expenditure of public money over the whole kingdom, and thus not infringe that great principle of equality and justice which I maintain should be our guide in all such cases . . .

* * * * *

Wallace's attention to the problems of public education is perhaps best illustrated by his long-term interest in the manner of exhibition of museum collections. The first of his several writings on this subject appeared in 1869 under the title:

Museums for the people (ref 143)

. . . Public museums of natural objects being such entirely modern institutions, we can hardly wonder that no generally accepted principles have yet been laid down for their construction or arrangement. They most frequently originated with private collectors, whose plan was naturally followed in their enlargement; and when they outgrew their original domicile, an architect was called in, who, according to his special tastes, designed a temple or a palace for their reception. However inconvenient or unsuitable the original mode of exhibition might turn out, or however ill adapted to its purpose the new building might prove, it would, of course, be exceedingly difficult and expensive to alter either of them, more especially as the modified plan might be found, after trial, to have defects as great as that which it replaced . . .

Accepting as a basis the few essential principles that seem now to be agreed upon, I propose to follow them out into some rather important details.

I shall consider, in the first place, what should be the scope of a Typical Popular Museum, and then sketch out the arrangements best adapted to make it both entertaining and instructive to the young and ignorant, and a means of high intellectual culture and enjoyment to such as may be disposed to avail themselves fully of its teachings.

Museums are well adapted to illustrate all those branches of knowledge whose subject-matter consists mainly of definite movable and portable objects. The great group of the natural history sciences can scarcely be taught without them; while mathematics, astronomy, physics, and chemistry make use of observatories and laboratories rather than museums. The fine and mechanical arts, as well as history, can also be illustrated by extensive collections of objects; and we are thus led to a broad division of museums, according as they deal mainly with natural objects or with works of art.

A museum of natural objects appears, for a variety of reasons, best fitted to interest, instruct, and elevate the middle and lower classes, and the young. It is more in accordance with their tastes and sympathies, as shown by the universal fondness for flowers and birds, and the great interest excited by new or strange animals. It enables them to acquire a wide and accurate knowledge of the earth and of its varied productions; and if they wish to follow up any branch of natural history as an amusement or a study, it leads them into the pure air and pleasant scenes of the country, and is likely to be the best antidote to habits of dissipation or immorality. Such museums, too, offer the only means by which the mass of the working classes can obtain any actual knowledge of the wonderful productions of nature in present or past ages; and such knowledge gives a new interest to works on geography, travel, or natural history. Owing to the wide disconnection of these subjects from the daily pursuits of life, they are so much the better adapted for the relaxation of those who earn their bread by manual labour. The inexhaustible variety, the strange beauty, and the wondrous complexity of natural objects, are pre-eminently adapted to excite both the observing and reflective powers of the mind, and their study is well calculated to have an elevating and refining effect upon the character.

Works of art, on the other hand, though in the highest degree instructive and elevating to some minds, are not so universally attractive; and, what is more important, do not exercise so many faculties, and do not offer such wide and easily-reached fields of study for the working classes. Some previous training or special aptitude is required in order to appreciate them; and it may even be asserted with truth that the study of nature is a necessary preliminary to the appreciation of art. It does not seem improbable that, even if our object were to make artists and lovers of art, good museums of natural objects might be the most useful first step. We have further to consider that objects of art are already widely spread, and more or less accessible. Our great public buildings contain their art-decorations. The houses of the wealthy and the shops of our streets are full of art, and the artisan has frequent opportunities of seeing them, while local exhibitions of art are not uncommon, and will no doubt become more and more frequent. The very young and the very ignorant would learn nothing in an art museum, while they would certainly gain both knowledge and pleasure in such an one as I am about to describe.

A Typical Museum of Natural History should contain a series of objects to illustrate all the sciences which treat of earth, nature, and man. These are—1, Geography and Geology; 2, Mineralogy; 3, Botany; 4, Zoology; 5, Ethnology. I will briefly sketch what seems to be the best mode of illustrating these sciences in a museum for the people.

Geography and Geology.—Some knowledge of the earth and its structure is so essential a preliminary to any acquaintance with natural history, and the working classes have so few opportunities of seeing large maps, globes, or models, that a good series of these should form a part of the museum. In particular, relief-maps, models, and maps to illustrate physical geography and geology, large sections and diagrams, and large globes, should be so exhibited that they could be conveniently examined and studied in detail.

The country around the museum should be shown on a large scale by a model or relief-map, in which the undulations of the ground and the hills, valleys, mountains and streams should be shown on a natural scale of heights, so as not to exaggerate the slopes to three or four times their actual steepness, as is usually done. The more important mountainous regions of our islands, as well as some portions of the Alps and Himalayas, should be shown in the same manner, and all on the same scale, so as to exhibit their true relations to each other. Only in this way can the erroneous ideas derived from maps on different scales and models on an exaggerated vertical scale, be counteracted.

Geological maps and sections should also be exhibited; and some geological models on a large scale, built up in separate layers to show the actual position of the stratified rocks in some district not far removed from the museum, would give a more accurate notion of the geological structure of the earth than could be obtained by any number of maps and descriptions. In the same way the extent of the ice during the glacial period in the north of England might be shown, by movable layers of white cement or papier-maché, while very large-scale models of portions of the same area might show the results of the ice-action in the rounded and smoothed rocks, the moraines and the deposits of boulder-clays, as well as the erratic blocks coloured to show the parent-rock from which they had been brought by the ice and with lines to show, the course they had travelled. Ice-ground boulders from the moraines and slabs showing glacial striae and polish should also be exhibited, with special models of remarkable perched blocks and erratics, moraines and extensive striated rocks. Clearly printed labels should explain how these various objects enable us to determine with considerable accuracy the height to which different areas were buried in ice and the directions in which it flowed. By these various illustrations the reality and the marvellous character of the Great Ice Age would be brought home to all intelligent visitors, and would so interest them that they would long to see the proofs of it in some glaciated country which they could reach by a holiday excursion; and the evidences of ice-action are so widely spread over all the northern and western portions of our islands that for most persons the opportunity could without difficulty be found.

Mineralogy.—A series of the most important and best marked minerals should be exhibited, with tables and diagrams explaining the principles of their classification. Their number should not be too large, and every specimen should be accompanied by a label containing a brief account of all that is most interesting connected with it—its chemical constitution, its affinities, its distribution, and its uses. Combined with this collection there should be a series of specimens illustrating the mode of manufacture of the more important minerals, and their application to the arts and sciences. To give a local interest, all British specimens should be placed on tablets of one distinct colour, so as at once to catch the eye, and enable the student to form some idea of the comparative productiveness of his own country.

Botany.—The series of specimens to illustrate the science of botany in a popular museum may be of two kinds: such as show the main facts of plant-structure and classification; and others to teach something of the variety, the distribution, and the uses of plants.

By means of specimens, dissections, drawings, and models, the important radical differences of the great primary divisions of plants—cellular and vascular—acrogenous, endogenous, and exogenous—might be made clearly manifest. Alongside of the drawings and dissections there should be cheap fixed microscopes, showing the main structural differences, thus giving a reality and intensity to the characters which drawings or descriptions alone can never do.

Each of the most important natural orders of plants should next be illustrated by specimens of various kinds. Their structure and essential characters should first be shown, in the same way as the higher groups. Their geographical distribution should be marked out on small maps. Good dried specimens, and, if necessary, drawings or models of flowers or fruit, of the more characteristic and remarkable species, should then be exhibited; and along with these, samples of whatever useful products are derived from them. Where remarkable forest-trees occur in an order, good coloured drawings of them should be shown, as well as longitudinal and cross sections of their wood. In the same, or an adjoining case, specimens or casts of the most important fossil-plants of the same order may be exhibited, illustrating their range backward into past time.

By such a scheme as this, in a comparatively small space and with a small number of specimens, all that is of most importance in the vegetable kingdom would be shown. . . . Botany would lose much of its supposed difficulty and repulsiveness, and . . . many might be thereby induced to devote their leisure to this most useful and attractive study.

In order to assist those who are really students, a separate room should be provided, containing a Herbarium of British plants, as well as one illustrative of the more important exotic families and genera; and to this should be attached a collection of the more useful botanical works.

Zoology.—Owing to the superior numbers and greater variety of animals, their more complicated structure and more divergent habits, the higher interest that attaches to them, and their greater adaptability for exhibition, this department must always be the most extensive and most important in a Natural History Museum.

The general principles guiding the selection and exhibition of animals are the same as have been applied to plants, subject to many modifications in detail. The great primary divisions, or sub-kingdoms (Vertebrata, Mollusca, etc.), as well as the classes in each sub-kingdom (Mammalia, Birds, etc. and Cephalopoda, Gasteropoda, etc.), should be defined, by means of skeletons and anatomical preparations or models, so as to render their fundamental differences of structure clear and intelligible. At the head of each order (or subdivision of the class) a similar exposition should be made of essential differences of structure; and in every case the function or purpose of these differences should be pointed out by means of clearly-expressed tables and diagrams.

We now come to the specimens of animals to be exhibited, in order to give an adequate idea of their variety and beauty; their strange modifications of form and structure, their singular habits and mode of life, their distribution over the surface of the earth, and their first appearance in past time. To do this effectively requires a mode of exhibition very different from that which has been usually adopted in museums.

Throughout the animal kingdom, at least one or more species of every important family group should be exhibited; and in the larger and more interesting families, one or more species of each genus. The number of specimens is not, however, so important as their quality and the mode of exhibiting them. A few of the more important species in each order, well illustrated by fine and characteristic specimens, would be far better than ten times the number if imperfect, badly prepared, and badly arranged. Let any one look at an artistically mounted group of fine and perfect quadruped or bird skins, which represent the living animals in perfect health and vigour, and by their characteristic attitudes and accessories tell the history of the creature's life and habits; and compare this with the immature, ragged, mangy-looking specimens one often sees in museums, stuck up in stiff and unnatural attitudes, and resembling only mummies or scarecrows. The one is both instructive and pleasing, and we return again and again to gaze upon it with delight. The other is positively repellent, and we feel that we never want to look upon it again.

I consider it therefore an important principle, that in a museum for the people nothing should be exhibited that is not good of its kind, and mounted in the very best manner. Fortunately, specimens of a large number of the most beautiful and extraordinary animals are now exceedingly common, and every well-marked group in nature may be illustrated without having recourse to the rarer and more costly species. Carrying out these views, we should exhibit our animal in such a way as to convey the largest amount of information possible. The male, female, and young should be shown together, the mode of feeding or of capturing its prey, and the most characteristic attitudes and motions, should be indicated; and the accessories should point out the country the species inhabits, or the kind of locality it most frequents. A descriptive tablet should of course give further information; and in the immediate vicinity, specimens showing any remarkable points of its anatomy, and any useful products that are derived from it, should be exhibited.

Each group of this kind would be a study of itself, and should therefore be kept quite distinct and apart from every other group. It should be so placed that it could be seen from several points of view, and every part of each individual composing it closely examined. To encourage such examination and study, seats should be placed conveniently near it—a point strangely overlooked in most museums, where it seems to be taken for granted that visitors will pass on without any desire to linger, or any wish for a more close examination. It would add still further to the interest of these typical groups, if it were clearly shown how much they represented, by giving a list of all the well-known species of the genus or family, with their native country and proportionate size, and indicating, by means of a coloured line, which of them were exhibited in the museum. This would be an excellent and most intelligible guide to the collection itself, and would enable the visitor to judge how far it gave any adequate notion of the variety and exuberance of nature.

It would also, I think, be advisable, that as far as possible each well-marked and important group of any considerable extent should occupy one room or compartment only, where it would be separated from all others, where the attention could be concentrated upon it, and where the extent to which it was illustrated could be seen at a glance. This has not, I believe, been yet attempted in any museaum; and when I come to speak of the building arrangements, I will explain how it can be easily managed. In this room, a department would also be devoted to the comparative anatomy of all the more important species and groups exhibited; and a large map should be suspended, showing in some detail their geographical distribution. Here, too, we should place specimens or casts of the fossil remains of the family, with restorations of some of the more important species; and along with these, diagrams, showing the progress of development of the group throughout past time, as far as yet known.

This mode of attractive and instructive exhibition might be well carried out in the Mammalia, Birds, and Insects; less perfectly in the Reptiles and Fishes, whose colours can hardly be well preserved except in spirits. Even here, however, by using oblong earthenware vessels with glass fronts, instead of the usual bottles, many fishes and marine animals could be exhibited in life-like attitudes and with their colours well preserved. Mollusca may be well illustrated by means of models of the animals, as also may the marine and fresh-water Zoophytes. The more minute and delicate animals should be shown by means of a series of cheap microscopes or large lenses, fixed in suitable positions; and with a careful outline of the animal's history on a tablet or card, close by.

Connected with this, as with the botanical division of the museum, there should be a students' department, to which all should have free access who wished to obtain more detailed knowledge. Here would be preserved, in the most compact and accessible form, in cabinets or boxes, all specimens acquired by the museum and which were not required or were not adapted for exhibition in the popular department. Here, too, should be . . . the best zoological library of reference that could be obtained. In this department, donations of almost any kind would be acceptable; for, when not required for popular exhibition, an immense number of specimens can be conveniently and systematically arranged in a very limited space, and for purposes of study or for identification of species are almost sure to be of value . . .

Ethnology.—We now come to the last department of our ideal museum, and it is one to which a large or a small proportion of space may be devoted, according to the importance that may be attached to it. In accordance with the plan already sketched out for the other departments, the following would be a fair representation of Ethnological science.

The chief well-marked races of man should be illustrated either by life-size models, casts, coloured figures, or by photographs. A corresponding series of their crania should also be shown; and such portions of the skeleton as should exhibit the differences that exist between certain races, as well as those between the lower races and those animals which most nearly approach them. Casts of the best authenticated remains of prehistoric man should also be obtained, and compared with the corresponding parts of existing races.

The arts of mankind should be illustrated by a series, commencing with the rudest flint implements, and passing through those of polished stone, bronze, and iron—showing in every case, along with the works of prehistoric man, those corresponding to them formed by existing savage races. Implements of bone and of horn should follow the same order.

Pottery would furnish a most interesting series. Beginning with the rude forms of prehistoric races, and following with those of modern savages, we should have the strangely-modelled vessels of Peru and of North America, those of Egypt, Assyria, Etruria, Greece, and Rome, as well as the works of China and of mediaeval and modern Europe.

The art of sculpture and mode of ornamentation should be traced in like manner, among savage tribes, the Oriental nations, Greece, and Rome, to modern civilization. Works in metal and textile fabrics would admit of similar illustration. Characteristic weapons should also be exhibited; and painting might be traced in broad steps, from the contemporary delineation of a Mammoth up to the animal portraiture of Landseer.

This comprises a series of Ethnological illustrations that need not occupy much space, and would, I think, be eminently instructive. The clothing, the houses, the household utensils, and the weapons of mankind, can hardly be shown with any approach to completeness in a Popular Museum; and many of these objects occupy space quite disproportionate to their intrinsic interest or scientific value. They could in most cases be sufficiently indicated by drawings or models.

Situation and Plan of museum

. . . This brings me to say a few words on the kind of building best adapted to such an institution as is here sketched out.

In his President's address to the British Association at Norwich, Sir Joseph Hooker made some admirable remarks on the situation of museums. He observed:

> Much of the utility of museums depends on two conditions often strangely overlooked, viz. their situation, and their lighting and interior arrangements. The provincial museum is too often huddled away almost out of sight, in a dark, crowded, dirty thoroughfare, where it pays dear for ground rent, rates, and taxes, and cannot be extended. Such localities are frequented by the townspeople only when on business, and when they consequently have no time for sight seeing. In the evening, or on holidays, when they would visit the museum, they naturally

> prefer the outskirts of the town to its centre . . . The museum should be in an open grassed square or park, planted with trees, in the town or its outskirts; a main object being to secure cleanliness, a cheerful aspect, and space for extension. Now, vegetation is the best interceptor of dust, which is injurious to the specimens as well as unsightly, whilst a cheerful aspect, and grass and trees, will attract visitors, and especially families and schools.

Evidently, then, the proper place for the museum is the centre of the park or public garden . . . With how much greater pleasure the workman and his family could spend a day at the museum, if at intervals they could stroll out on to the grass, among flowers and under shady trees, to enjoy the refreshments they had brought with them. They would then return to the building with renewed zest, and would probably escape the fatigue and headache that a day in a museum almost invariably brings on . . .

In designing museums, architects seem to pay little regard to the special purposes they are intended to fulfil. They often adopt the general arrangement of a church, or the immense galleries and lofty halls of a palace. Now, the main object of a museum-building is to furnish the greatest amount of well lighted space, for the convenient arrangement and exhibition of objects which almost all require to be closely examined. At the same time they should be visible by several persons at once without crowding, and admit of others freely passing by them. None except the very largest specimens should be placed so as to rise higher than seven feet above the floor, so that palatial rooms and extensive galleries, requiring proportionate altitude, are exceedingly wasteful of space, and otherwise ill adapted and unnecessary for the real purposes of a museum . . .

Following out this view, a simple and economical plan for a museum would seem to be, a series of long rooms or galleries, about thirty-five or forty feet wide, and twelve or fourteen feet high on each floor, the four or five feet below the ceiling on both sides being an almost continuous series of window openings, while at rather wide intervals large windows might descend to within three feet of the floor. At such distances apart as were found most convenient for the arrangement of the collections, movable upright cases might be placed transversely, leaving a central space of about five feet for a continuous passage; and the compartments thus formed might be completed by partitions and doors connecting opposite cases, wherever it was thought advisable to isolate any well-marked group of animals, or other division of the museum. By this means the proportion between wall-cases and floor space might be regulated exactly according to the requirements of each portion of the collection; and abundant light would be obtained for the perfect examination of every specimen.

Two of the great evils of museums are, crowding and distraction. By the crowding of specimens, the effect of each is weakened or destroyed; the eye takes in so many at once that it is continually wandering towards something more strange and beautiful, and there is nothing to concentrate the attention on a special object. Distraction is produced also by the great size of the galleries, and the multiplicity of objects that strike the eye. It is almost impossible for a casual visitor to avoid the desire of continually going on to see what comes next, or wondering what is that bright mass of colour or strange form that catches the eye at the other end of the long gallery. These evils can best be avoided, by keeping, as far as possible, each natural group of objects in a separate room, or a separate compartment of that room—by limiting as much as possible the number of illustrative groups of species, and at the same time making each group as attractive and instructive as possible. The object aimed at should be, to compel attention to each group of specimens. This may be done by making it so interesting or beautiful at first sight as to secure a close examination; by carefully isolating it, so that no other object close by should divide attention with it; and by giving so much information and interesting the mind in so many collateral matters connected with it, as to excite the observant and reflective as well as the emotional faculties.

The general system of arrangement and exhibition here pointed out does not at all depend on the building. It can be applied in any museum, and is, I believe, already to some extent adopted in our best local institutions. It has, however, never yet been carried out systematically; and till this is done, we can form no true estimate of how popular a Natural History Museum may become, or how much it may aid in the great work of national education . . .

* * * * *

Wallace found a variety of ways to draw attention to the need for forms of education that stressed self-involvement over inculcation. He took especial interest in the problems of geographic education. His essays on the organization of museum exhibits, already discussed, feature much commentary on the portrayal of geographical relationships and their meaning to regional character. In 1878 he composed an essay describing how a public forest tract in Epping might be turned into a self-maintaining botanical garden exhibiting the products of the world's temperate forests (ref 292). In 1896 he 'revised' French geographer Elisée Reclus's suggestion that a large model of the globe be built and kept on permanent public display (ref 526). In My life, *Volume 2, he spent fourteen pages discussing geographical illiteracy under the guise of describing his experiences as an examiner in geography. In* The geographical distribution of animals *(ref 718) his arrangement of the subject matter and use of plates conveying the appearance of 'typical regional faunas' set the example for the 'faunal diorama' museum-exhibit device, now routinely employed, in which the biotic elements of a given location are arranged into a single, 'lifelike' semblance of their actual ecological associations. (In recent years, this innovation has found a new incarnation in the increasing number of 'biome' exhibits appearing in zoological parks.) And there were, of course, his popular regional geography writings—especially* The Malay Archipelago, *an important source of inspiration for a whole generation of tropical naturalists (and novelists too—this was, for example, Joseph Conrad's favourite book).*

Wallace's concern over the public's ignorance of geographical relationships can again be linked to his belief that human evolution was now proceeding as a function of the expansion and integration of intellectual and moral horizons. Like Immanuel Kant, another thinker who took a strong interest in geography, he felt that an appreciation of places and people remote from personal contact was critical to an insightful picture of one's own existence. This position was taken to its extreme in his forays into astronomy.

*Wallace produced several writings focusing on what were considered at his time the extremes of time and space. Malinchak (1987) and R. Smith (1972) have argued that most of these efforts were inspired by his belief in spiritualism, and there can be little doubt that a connection existed. But even as a child Wallace had been fascinated by astronomy (*My life, *Vol. 1), and his interest in the subject was renewed when he began to investigate the causes of glaciation in the 1860s. He became further involved while researching the material for the astronomy review sections of the 1898 and 1903 editions of* The wonderful century *(ref 726). The relationship of astronomy to the origin of life is the central focus of the books* Man's place in the universe *(ref 728) and* Is Mars habitable? *(ref 730), and figures in several of his other writings as well. Wallace's peculiarly teleological—and rather naïve, considering the state of the information available at that time—approach to the subject is apparent in his controversial 1903 essay:*

Man's place in the universe (ref 602)

To the early astronomers the earth was the center of the visible universe, sun, moon, planets and stars all alike revolving around it in more or less eccentric and complex orbits, and all were naturally thought to exist as appendages to our globe and for the sole use and enjoyment of man—"the sun to rule by day, the moon and the stars to rule by night." But when the Copernican system became established, and it was found that our earth was not specially distinguished from the other planets by any superiority of size or position, it was seen that our pride of place must be given up. And, later, when the discoveries of Newton and of the many brilliant astronomers who succeeded him, together with the ever-widening knowledge derived from the growing power and perfection of the telescope and of improved astronomical instruments, showed us the utter insignificance even of our sun and solar system among the countless hosts of stars and the myriads of clusters and nebulae, we seemed to be driven to the other extreme and to be forced to recognize the fact that this vast, stupendous universe could have no special relation to ourselves any more than to any other of the million suns and systems, many of which were probably far grander and more important than ours, and perhaps fitted to be the abode of more highly organized beings.

During the last half century, and perhaps much longer, popular writers have often dealt with the problem of the habitability of the planets by intelligent beings, and the probability of other suns being attended by other trains of planets similarly inhabited, and the most diverse and even opposing views have been held as to the inferences to be drawn from these supposed facts. Sir David Brewster held them to be almost essential to an adequate conception of the power and wisdom of the Deity, and in some way bound up with the doctrines of Christianity, and this has been the view of many of the teachers of religion. On the other hand, the tendency has been to give us wider views of the vastness, the variety and the marvelous complexity of the stellar universe, and proportionally to reduce the importance of our little speck of earth almost to the vanishing point, and this has been made use of by the more aggressive among modern skeptics to hold up religious creeds and dogmas to scorn and contempt. They point out the irrationality and absurdity of supposing that the Creator of all this unimaginable vastness of suns and systems, filling for all we know endless space, should have any *special* interest in so pitiful a creature as man, the degraded or imperfectly developed inhabitant of one of the smaller planets attached to a second or third rate sun; while that He should have selected this little world for the scene of the tremendous and necessarily unique sacrifice of his Son in order to save a portion of these "miserable sinners" from the natural consequences of their sins, was, in their view, a crowning absurdity too incredible to be believed by any rational being. And it must be confessed that the theologians had no adequate reply to this rude attack; while many of them have felt their position to be untenable, and have renounced the idea of a special revelation and a supreme Saviour for the exclusive benefit of so minute and insignificant a speck in the universe.

But during the last quarter of the past century the rapidly increasing body of facts and observations leading to a more detailed and accurate knowledge of stars and stellar systems have thrown a new and somewhat unexpected light on this very interesting problem of our relation to the universe of which we form a part; and altho these discoveries have, of course, no bearing upon the special theological dogmas of the Christian, or of any other religion, they do tend to show that our position in the material universe is special and probably unique, and that it is such as to lend support to the view, held by many great thinkers and writers to-day, that the supreme end and purpose of this vast universe was the production and development of the living soul in the perishable body of man.

The Agnostics and Materialists will no doubt object that the want of all proportion between the means and the end condemns this theory from its very foundation. But is there any such want of proportion? Given infinite space and infinite time, and there can be no such thing as want of proportion if the end to be reached were a great and worthy one, and if the particular mode of attaining that end were the best, or perhaps even the only possible one; and we may fairly presume that it was so by the fact that it *has* been used and *has* succeeded.

The development of man as a spiritual being with all his intellectual powers and moral possibilities is certainly a great end in itself, so great and so noble that if a universe of matter and ether as large as that of which we have now obtained some definite knowledge were required for the work, why should it not be used? Of course I am taking the view of those who believe in some Intelligent Cause at the back of this universe, some Creator or creators, Designer or designers. For those who take the other view, that matter and ether, with all the laws and forces without which they could not exist for a moment, are in their essential nature eternal and self-existent, no such objection is tenable. For the production of life and of man then becomes merely a question of chance—of the right and exact combination of matter and its complex forces occurring after an almost infinite number of combinations that led to nothing. On this view the argument as to our unique position, derived from the discoveries of the New Astronomy, is even more forcible, tho hardly so satisfactory, because it also teaches us that if man is a product of blind forces and unconscious laws acting upon non-living matter, then, as he has been produced by physical law, so he will die out by the continued operation of the same laws, against which there is no appeal . . .

It is the object of the present paper to set forth the nature of the evidence bearing upon man's position in the universe and to summarize the various lines of research that converge to render it at least a thinkable and rational hypothesis. Altho most of the facts and conclusions are well known separately, and have been set forth by both scientific and popular writers, I am not aware that they have been combined as I now attempt to combine them, or the conclusions drawn from them which seem to me to be the obvious ones.

Are the Stars Infinite in Number?—It has often been suggested that the stars are infinite in number and that the stellar universe is therefore infinite in extent; and if the preponderance of evidence pointed in this direction our inquiry would be useless, because as regards infinity there can be no difference of position. In whatever part of it we may be situated that part can be no nearer the center than any other part. Infinite space has been well defined as a circle, or rather a sphere, whose center is everywhere and circumference nowhere.

As the telescope increased in efficiency through the labors of Dollond and Herschel it was found that every increase of power and of light due to increased diameter of object-glass or mirror greatly increased the number of visible stars, and this increase went on with approximate equality of rate till the largest modern telescopes were nearly reached. But latterly increased size and power has revealed new stars in a smaller and smaller proportion, indicating that we are approaching the outer limits of the starry system. This conclusion is further enforced by the fact that the numerous dark patches in the heavens where hardly any stars are visible, and those seen are projected on intensely dark background, as in the "coal-sacks" of the southern hemisphere and rifts and channels in the Milky Way itself, continue to present the same features in telescopes of the very highest powers as they do in those of very moderate size. This could not possibly happen if stars were infinite in number, or even if they extended in similar profusion into spaces very much greater than those to which our telescopes can reach, because in that case these dark backgrounds could be illuminated by the light of millions of stars so distant as to be separately invisible, as in the case of the Milky Way itself. The only other explanation would be that the star-system is penetrated in several directions by perfectly straight tunnels of enormous length compared with their diameter in which no stars exist, and this is considered to be so improbable as to be unworthy of consideration.

The same conclusion is reached by means of that powerful engine of research, the photographic plate. When this is exposed in the focus of a telescope for three hours a much greater number of stars are revealed than any telescope can detect, but longer exposures add less and less to the number, again indicating that the limit of stars in that direction is nearly reached.

Yet again, the method of counting the stars of the various astronomical magnitudes gives a similar result. At each lesser magnitude the number of stars is about three times greater than that of the next higher magnitude, and this rule applies with tolerable accuracy down to those of the ninth magnitude. The total number of visible stars from the first to the ninth magnitude is about 200,000. Now if this rate of increase continued, the number down to the seventeenth magnitude, the faintest visible in the best modern telescopes, would be about 1,400 millions. But both telescopic observation and photographic charts show that there is nothing approaching this number, it being estimated that the total number thus visible does not exceed 100 millions—again proving that as our instruments reach further and further into space they find a continuous diminution in the number of stars, thus indicating an approach to the outer limits of the stellar universe.

But perhaps the most striking proof of the limited extent of the universe of luminous stars is that dependent on the laws of light. This has been long known to astronomers and it has been very clearly and briefly stated by Professor Simon Newcomb, one of the profoundest of mathematical astronomers. He tells us to imagine a series of concentric spheres, each the same distance apart from the first, which includes only the stars visible to the naked eye. The space between each pair of these spheres will be in extent proportional to the squares of the diameters of the sphere that limit it; and as the light we receive from each star is inversely proportional to its distance from us, it follows that if each region were equally strewn with stars of the same average brightness then we should receive the same amount of light from each region, the diminution of light from each star being exactly compensated by the vastly greater numbers in each successively larger sphere. Hence it follows that if these concentric spheres were infinite we should receive an infinite amount of light from them, and even if we make an ample allowance for stoppage of light by intervening dark bodies, or by cosmic dust, or by imperfect transparency of the ether, we should at least receive quite as much light from them as the sun gives us at noonday. But the amount we actually receive is so immensely less than this as to prove that the concentric spheres of stars beyond those visible to the naked eye cannot be very numerous. For the total light of all the stars is estimated to be not more than about one-fortieth of moon-light, which is itself only about one five hundred thousandth of sunlight. This proof of the limited extent of the stellar universe is therefore a very forcible one, and taken in connection with that afforded by telescopic research, as already described, is altogether conclusive.

We have next to consider the facts known as to the distribution and arrangement of the stars, and the conclusions to be drawn therefrom.

The Distribution of the Stars in Spacc.—The first great fact bearing upon this subject is that a large number of stars are not "fixed," as was universally believed down to the eighteenth century, but that many of them, and probably all, have proper motions of their own. These motions are very small and can only be detected by observations continued for many years. The most rapid motion yet observed is that of a small star of 6½ magnitude in the constellation Ursa Major, which moves seven seconds of arc per annum, while others move only this amount in a century, and all but a few less than a second per annum. The proper motions of several thousand stars have now been determined. These motions are in every possible direction, but it has been recently discovered that considerable groups of stars often move in the same direction and at the same rate. The Pleiades exhibit this phenomenon, but much larger groups have the same kind of motion, and this has led to the theory that in certain parts of the heavens there is a star-drift in fixed directions. Our sun is now known to have its own "proper motion," the direction and rate of which has been determined approximately. This will, of course, produce an apparent movement in all the stars except those situated exactly in the line of our motion, and the displacement thus caused has to be allowed for in determining the true motion of the stars in space. Should any of the stars be moving obliquely toward us we shall only perceive that portion of the motion which is at right angles to the direction of the star from us, but the beautiful method of determining motion in the line of sight by means of the spectroscope has overcome this difficulty, and by its means we now know the real motion of many stars, both in direction and velocity, when we have been able to measure their distance from us.

This measurement of distances of the stars is the most difficult of all the instrumental determinations of modern astronomy, both on account of the extreme remoteness of most of them, and because owing to the motions of the stars themselves we have no fixed point from which to determine changes of position.

Most people know that by means of a measured base-line the distances of very remote and inaccessible objects can be determined with considerable accuracy, depending upon the length of the base and its careful measurement, and equally upon the extremely accurate measurement of the angles taken at each extremity of the base. It is in this way that the position of mountain peaks is determined, as well as the distances across narrow seas, while all civilized countries have been trigonometrically surveyed in this manner.

In the case of the stars the base line used is the diameter of the earth's orbit, more than one hundred and eighty millions of miles. Every six months we are at opposite ends of this base, and if we had any absolutely fixed point in the heavens in the right position from which to take our angles we could in this way determine the distance of some of the stars. But as almost all the stars are moving at various rates and in various directions, as our sun itself is moving, and as the proper motions of the stars can only be determined in relation to other stars, there is everywhere a complication of opposing motions and nowhere the assured fixity we require for such delicate measurements. But, notwithstanding all these difficulties, astronomers have by various ingenious methods now measured the distances of a number of stars with considerable precision, notwithstanding the failures of their predecessors for nearly two centuries. The nearest of all the stars are so remote that the distance between the earth and the sun as seen from the star would subtend an angle of considerably less than one second of arc, while most of those measured are so excessively distant that this angle is often one-tenth of a second, or even considerably less. To understand how small a quantity this is and what a distance it implies, it may be stated that, viewed at a mile distant, the small letter o in this page would subtend an angle of about one-tenth of a second. From a star of an average distance from us, therefore, the earth and sun, if they could be seen, would appear only as far apart as the opposite sides of the letter o when a mile away from us. But stars twice as far as these have been measured, it is believed, with some degree of certainty and the distances of about sixty stars have now been satisfactorily ascertained.

It was long supposed that the brightest stars were the nearest to us, but it is now known that there is little or no relation between brightness or magnitude and distance. The nearest star yet measured is indeed a very bright one in the southern hemisphere, Alpha Centauri; but one almost as near, 61 Cygni, is of the fifth magnitude only, and another still nearer in the constellation Piscis Australis is of the seventh magnitude. Those stars of the first magnitude which have had their distances measured have a parallax of considerably less than one-tenth of a second, and are therefore among the remoter stars.

The true relation, as was long suspected theoretically, is between proper motion and distance, those which move fastest being nearest to us . . . All may have, and probably have, real motions which do not differ *very* greatly in rapidity, but only in those which are comparatively near us can we detect any motion at all. This theoretical conclusion being confirmed by all the stars whose distances have been measured, we have a most valuable and trustworthy means of ascertaining their comparative distances from us, since those whose proper motions are either exceedingly small or cannot be detected at all are certainly very much further from us than those which have well marked and large proper motions. It is by such indications that we are enabled to arrive at some definite conclusions as to the real form and structure of the stellar universe, as we will proceed to show.

The Galaxy, or "Milky Way." By far the most prominent feature in the starry heavens is that vast irregular nebulous ring which in all ages has attracted the attention and excited the admiration of observers. This great ring divides the whole heavens into two hemispheres, making an angle of about 63 degrees with the equinoctial, so that portions of it pass not far from the north and south poles. Its nebulosity is now believed to be almost wholly due to the massing together of myriads of minute stars, since each increase in the power of the telescope shows more and more of these stars, while the best photographic plates show them everywhere closely packed, but still with a luminous haze between them indicating yet more stars beyond.

But besides these minute stars which give us the cloudy or milky appearance, it is found that stars of all degrees of brilliancy are more numerous in the milky way and in its vicinity than elsewhere. The two poles of the Galaxy are the regions where stars are scantiest. Each 15 degrees nearer to it they increase in numbers, at first slowly, then more rapidly, till we reach its borders.

The following series of numbers give the average number of stars in a square of 15 minutes at each 15 degrees from the pole of the galaxy, as determined by Sir John Herschel: 4–5–8–13–24–53.

Later observations have fully confirmed this, while it has been shown by the late Mr. Proctor that all stars down to the tenth magnitude, more than 324,000, when carefully mapped, mark out the Milky Way in all its details by their greater density. Later still the Italian astronomer Schiaparelli, by using all the materials now available, arrives at the same result, and Professor Simon Newcomb, of Washington, after a close examination of his maps assures us that the Milky Way can be fairly traced out by the region of maximum agglomeration of stars.

These facts lead to the conclusion that the Galaxy is a vast annular agglomeration of stars forming a great circle round the heavens, altho in places very irregular, being split in two for about one-third of its circumference and being besides full of irregular dark streaks and patches, where the most powerful telescopes show very few stars. So that, as Sir John Herschel says, we are irresistibly led to the conclusion that in those regions "we see fairly through the starry stratum;" and this is further shown by the fact that in these parts "the ground of the heavens seen between the stars is for the most part perfectly dark, which would not be the case if multitudes of stars, too minute to be individually discernible, existed beyond." This great ring is, therefore, evidently not very much extended in the direction of its own plane—that is, the ring is not flat or greatly compressed (as is Saturn's ring, for example) or we should nowhere see through it.

But what is more important is that we must be situated not *in* any part of it, as was once supposed, but at or near the very central point in the plane of the ring—that is, nearly equally distant from every part of it. This must be the case because from any other position the ring would not appear to us so uniform as it does. If we were much nearer to one side of it than to the other, the nearer side would appear broader, the more remote side narrower, and these two directions would show a decided difference in the numbers of the visible stars. Sir John Herschel, indeed, thought, the southern portion was nearer to us than the northern because of its greater *brightness*, which, he says, is very striking and conveys strongly the *idea* of greater proximity. But this may be deceptive, because the whole Milky Way shows great irregularities and variations in brightness, and it is a remarkable fact that the portions near the north and south poles are *both* equally narrow, while the parts 90 degrees from them are *both* very broad, rather suggesting equality of distance in all directions. Nearness would be indicated by a widening out of stars of all magnitudes, not necessarily by any general increase of brilliancy. The facts, therefore, seem to show that we are about equally distant from all parts of the Milky Way.

Very important, however, is Sir John Herschel's testimony to the close correspondence of the Galaxy as a whole to a great circle. He tells us that, following the line of its greatest brightness, it conforms as nearly as may be to that of a great circle inclined about 63° to the equinoctial and cutting that circle in R. A. 6 h. 47 m., and 18 h. 47 m., while its poles are in R. A. 12 h. 47

m., N. Decl. 27°, and R. A. 0 h. 47 m., S. Decl. 27°. He therefore determines it, *by the figures he gives*, to lie in an *exact* great circle as seen from the earth, as nearly as so irregular an object can be defined. But neither he nor any other astronomer so far as I am aware makes any remark on the extraordinary nature of this fact, which proves that we are placed *exactly* in the plane of the greatest density of the ring. The fact of the Galaxy forming a great circle as seen from the earth being so familiar no one seems to have thought it worth while to ask why it is so. If we could look at such a fact from the outside, as it were, we should certainly impute to it some causal connection between our system and the Galaxy. But before speculating what this relation may mean we must consider another point of equal importance in our relation to the system of stars.

Our Star Cluster.—It has long been observed that the brighter stars seem scattered over the whole heavens with no special abundance in or near the Milky Way, and this was thought to be due to their being much nearer to us. It is now known, however, that brightness is no indication of nearness, so that this fact has little significance. But, as we have seen, we do possess a real test of nearness in the amount of the proper motion of stars, and this leads us to a very definite and most suggestive conclusion. For the stars which are nearest to us, judged by this test, not only have no apparent relation to the Milky Way, but are spread over *every* part of the heavens with tolerable uniformity. The most recent examination of this class of stars is by Professor Simon Newcomb, who states the result in the following words: "If we should blot out from the sky all the stars having no proper motion large enough to be detected, we should find remaining stars of all magnitudes, but they would be scattered almost uniformly over the sky, and show no tendency toward the Milky Way."

Professor Kapteyn, of Groningen, appears to have been the first to draw the obvious conclusion from these facts that these nearer stars, spread around us in *every* direction, constitute a globular mass which he termed the "solar cluster," nearly concentric with the Milky Way, and that our sun is "deeply immersed" in this cluster.

Other astronomers have adopted this view, which seems to be almost indisputable if the facts are as stated. For if the cluster were *not* globular its component stars would not appear to be so uniformly spread over the *whole* heavens; and if our sun were *not* situated at or near its center but much nearer to one side of it than to the other, then we should inevitably find the stars of this type (those with measurable proper motions) much more numerous in one direction than in a direction exactly opposite. But altho there may be some irregularities in their distribution, it has not been pointed out that there is any such regular inequality as this, and if there is not, then we must be situated very near to the center of this "solar cluster."

The results so far reached by astronomers as the direct logical conclusion from the whole mass of facts accumulated by means of those powerful instruments of research which have given us the New Astronomy is that our sun is one of the central orbs of a globular star-cluster, and that this star-cluster occupies a nearly central position in the exact plane of the Milky Way.

But I am not aware that any writer has taken the next step, and, combining these two conclusions, has stated definitely that our sun is thus shown to occupy a position very near to, if not actually at, the center of the whole visible universe, and therefore in all probability in the *center* of the *whole material universe.*

This conclusion is no doubt a startling one, and all kinds of objections will be made against its being accepted as a proved fact. And yet I am not acquainted with any great inductive result of modern science that has been arrived at so gradually, so legitimately, by means of so vast a mass of precise measurement and observation and by such wholly unprejudiced workers. It may not be proved with minute accuracy as regards the actual mathematical center. That is of not the least importance. But that it is substantially correct in the terms I have stated there seems no good reason to doubt, and I therefore hold it to be right and proper to have it so stated, and provisionally accepted, until further accumulations of evidence may show to what extent it requires modification.

This completes the first part of our inquiry, but an equally important part remains to be considered—our position in the Solar System itself as regards adaptability for organic life. Here, too, I am not aware that the whole facts have been sufficiently considered. Yet they are facts that indicate our position in this respect to be in all probability as central and unique as is that of our sun in the stellar universe.

The Earth as Adapted for Life.—Among the many workers who have more or less seriously discussed the question of the adaptability of other planets for the development of organic life and of the higher forms of intellectual beings, I have not met with any who have considered the problem in all its bearings. They have usually been content to show that certain planets *may possibly* be *now* in a condition to support life in forms not very dissimilar from those upon our earth; but they have never adequately considered the question, Could such life have originated and have been developed upon these planets? This is the real *crux* of the problem, and I believe that a full consideration of the required conditions will satisfy us that, so far as we can judge, no other planet can fulfil them. Let us, therefore, consider what these conditions are.

The earlier writers on this subject could give free play to their imaginations and overcome difficulties of temperature, moisture, etc., by supposing that in other worlds there might be other elements which had different properties from any we possess, and which might render life possible under conditions very unlike those which are essential here. But the revelations of spectrum-analysis have shown us the unity of the constitution of matter throughout the whole material universe, so that not only are the planets of the solar system all composed of the same elements, but that the furthest stars and remotest nebulae alike consist of the very same elements with which we are so familiar, while the same physical and chemical laws undoubtedly prevail. We may be confident, therefore, that wherever organized life may have been developed it must be built up out of the same fundamental elements as here on earth.

The essential features of the structure of organized beings are continuous growth and repair of tissues, nutrition by the absorption of dead or living matter from without and its transformation into the various unstable compounds of which their bodies are built up. For these purposes a double system of circulation, gaseous and liquid, has to be constantly in operation, and this is carried on by means of minute tubular or cellular vessels which permeate every part of the body. These wonderfully complex and exquisitely adjusted circulating systems are entirely dependent on the continuous maintenance of a very narrow range of temperatures somewhere between the extremes of the boiling and the freezing points of water, but really within much narrower limits, since if the whole of the water at any time became solidified all the higher forms of life would be destroyed, while a temperature very much below the boiling point if permanently maintained would be almost equally detrimental.

When we consider that the temperature of space is about − 273° C., while that of the outer surface of the sun is about 9,000 C., we realize what a combination of favorable conditions must exist to preserve on the surface of a planet a degree of heat which shall never for any considerable time fall below 0° C., or rise above, say, 75° C., and that these narrow limits must be *continuously maintained* not for hundreds or thousands only, but for millions, perhaps for *hundreds of millions of years*, if life is to be developed there. It is the maintenance of this comparatively uniform surface temperature for such enormous periods—during, in fact, the whole time covered by the geological record—that most writers have overlooked as among the necessary conditions for the development of the higher forms of life on a planet; and this omission vitiates all their reasoning, since they have to show not only that the requisite conditons of temperature *may* exist now, but that there is even a probability that they have existed or will exist for a sufficiently extended period to allow of the development of a complex system of organic life comparable with our own. Let us, then, enumerate the chief favorable conditions, which in their combination appear to have rendered this development possible on our earth. These are:

(1) A distance from the sun such as to keep up the temperature of the soil to the required amount, by sun-heat alone, and to evaporate sufficient water to produce clouds, rain and a system of river circulation.

(2) An atmosphere of sufficient extent and density to allow of the production and circulation of aqueous vapor in the form of clouds, mists and dews, and to serve also as an equalizer of sun heat during day and night, winter and summer, and also between the tropical and temperate zones. This amount of atmosphere is held to be largely dependent upon the mass of a planet, and this one feature alone probably renders Mars quite unsuitable, since its mass is less than one-eighth that of the earth.

(3) The very large proportion of the surface covered by deep oceans, so that they surround and interpenetrate the land, and by their tides and currents keep up a continuous circulation, and are thus the chief agents in the essential

equalization of temperatures. This, again, is largely dependent on our possessing so large a satellite capable of producing a regular but not excessive tidal action. The want of such a satellite may alone render Venus quite unsuitable for the development of high forms of life, even if other conditions were favorable, which seems in the highest degree improbable.

(4) The enormous average depth of these oceans, so that the bulk of water they contain is about thirteen times that of the land which rises above their level. This indicates that they are *permanent features of the earth's surface*, thus insuring the maintenance of continuous land areas and of uniform temperatures during the whole period of the development of life upon the earth. (The evidence which demonstrates this permanence is set forth in my *Island Life*, Chap. VI, and enforced by additional arguments in my *Studies Scientific and Social*, Vol. I, Chap. 2.) It is extremely improbable that this remarkable condition obtains in any other planet.

(5) Lastly, one of the most peculiar and least generally considered features of our earth, but one which is also essential to the development and maintenance of the rich organic life it possesses, is the uninterrupted supply of atmospheric dust, which is now known to be necessary for the production of rain-clouds and beneficial rains and mists, and without which the whole course of meteorological phenomena would be so changed as to endanger the very existence of a large portion of the life upon the earth. How and why this is so is fully explained in my *Wonderful Century*. Now, the chief portion of this fine dust, distributed through the upper atmosphere from the equator to the poles with wonderful uniformity, is derived from those great terrestrial features which are often looked upon as the least essential, and even as blots and blemishes on the fair face of nature—deserts and volcanoes. Most persons no doubt think they could both be very well spared, and that the earth would be greatly improved, from a human point of view, if they were altogether abolished. Yet it is almost a certainty that the consequences of doing so would be to render the earth infinitely less enjoyable, and perhaps altogether uninhabitable by man. We must therefore reckon a due proportion of deserts and active volcanoes, with sufficiently constant winds to distribute the dust from them, as among the permanent essentials of a globe fitted for the development of intelligent life. This utility of deserts and volcanoes is, I think, now stated for the first time.

Now if we consider that these five distinct conditions or sets of conditions, many of them dependent on a delicate balance of forces acting at the origin of our planet, appear to be absolutely essential for the existence of high types of organic life, we shall at once see how peculiar and unique is our place and condition within the solar system, since we know with almost complete certainty that they do *not* all co-exist in any of the other planets. And when we consider further that, even if they do happen to exist now, that would be nothing to the purpose unless we had reason to believe that they had also existed, as with us, *in unbroken continuity* for scores or perhaps hundreds of millions of years. All the evidence at our command goes to assure us that our earth alone in the Solar System has been from its very origin adapted to be the theater for the development of organized and intelligent life. Our position within that system is therefore as central and unique as that of our sun in the whole stellar universe.

But, it may be asked, even if it be conceded that both by position, by size and by its combination of physical features we really do stand alone in the solar system in our adaptation for the development of intelligent life, in what way can the position of our sun at or near the center of the stellar universe, as it certainly appears to be, affect that adaptation? Why should not one of the suns on the confines of the Milky Way or in any other part of it possess planets as well adapted as we are to develop high forms of organic life?

These are questions which involve the most difficult problems in mathematical physics, and only our greatest thinkers possessing the highest mathematical and physical knowledge could be expected to give any adequate answer to them. In the meantime I will briefly indicate what seems to me to be the probable nature of the reply. Accepting the proof astronomers have given us, that, so far from the material universe of which our sun forms a part extending infinitely into space, we can actually *see* beyond its outer boundaries and can even approximately give a maximum limit to its magnitude, we are confronted with the problem of how a limited universe of matter and ether with the motions and forces which everywhere pervade it can conserve those forces at and near its furthest limits. Is it, in fact, necessarily becoming dissipated into outer space? Do any of its constituent suns, like those comets which have hyperbolic or parabolic orbits, continually fly out beyond its range and become lost to it forever? Comparing the stars of the Milky Way to the molecules of a gas, must not a certain proportion of these stars continually escape from the attractive powers of their neighbors as a result of collisions or in other ways, and wandering into outer space soon become dead and cold and lost forever to the universe? Will not the whole of the outer margins of the stellar universe be therefore unstable, always being liable to pass into regions where they would be dissipated as we see comets dissipated before our eyes? If such results are certain it will follow that the outer portions of the universe, at all events, and for an unknown extent inward, will be entirely unfitted to insure that *continuity of uniform conditions* which is the first essential for the development of life.

But this is only a small portion of the problem. A still more difficult question is, how will the ether behave near the outer borders of the universe? Can gravitation maintain its influence on the confines of a finite universe in the same degree as near its center? If, as now generally believed, gravitation is really produced by pressure of some kind, which must be equal in all directions, then it is almost certain that at any considerable distance beyond the central portion of the universe gravitation would vary in intensity in different directions. Whether this variation could possibly be detected by means of the motions of remote binary stars, or in any other way, it must be left for mathematicians and astronomers to determine.

But, leaving this question of variation of the force of gravity as beyond our powers at present, we may give a little consideration to those wonderful radiant forces, other than light and heat, the very existence of some of which we have only recently discovered. Such are electricity, magnetism, the Röntgen rays, the Hertzian, the Goldstein, the Becquerel rays and some others. That electrical forces bear an important part in the development of living organisms there can be little doubt, while the other forms of radiation here referred to, some of which produce curious physiological effects, can hardly be supposed to have been wholly without influence in the formation of the marvelous living machine, the substance of which in its complexity both of structure and constituent elements is a true microcosm—an epitome of matter and its forces. But if all these radiant forces, or several of them, have combined in the development of life, we may feel sure that they can only have done so under conditions which limit their energy to that gentle and imperceptible action which has caused them to remain so long hidden even from the most inquisitive seekers of the past century. And it is at least a possible, and I think not improbable, supposition that this imperceptibility and continuity may exist only in the more central portions of the universe, while in its outer regions less regularity may prevail; and while some of these necessary radiant forces may be wanting, others may be too abundant or be manifested in so irregular or excessive a manner as to be antagonistic to the delicate and nicely balanced forces which are essential to the orderly development of life.

Returning now for a moment to the consideration of our position in the stellar universe, it will assume a somewhat different aspect in view of the possibilities or probabilities just set forth. We can hardly suppose any longer that *three* such remarkable coincidences of position and consequent physical conditions should occur in the case of the one planet on which organic life *has* been developed without any causal connection with that development. The three startling facts—that we *are* in the center of a cluster of suns, and that that cluster *is* situated not only precisely in the *plane* of the Galaxy, but also *centrally* in that plane, can hardly now be looked upon as chance coincidences without any significance in relation to the culminating fact that the planet so situated *has* developed humanity.

Of course the relation here pointed out *may* be a true relation of cause and effect and yet have arisen as the result of one in a thousand million chances occurring during almost infinite time. But, on the other hand, those thinkers may be right who, holding that the universe is a manifestation of Mind, and that the orderly development of Living Souls supplies an adequate reason why such a universe should have been called into existence, believe that we ourselves are its sole and sufficient result, and that nowhere else than near the central position in the universe which we occupy could that result have been attained.

* * * * *

The book of the same title (ref 728) that appeared only shortly after the publication of the preceding essay was itself in effect extended in 'case study' fashion when Wallace wrote Is Mars habitable? *(ref 730) in 1907. The reasons for his undertaking this work, and the general nature of the analysis, are succinctly set out in its Preface:*

This small volume was commenced as a review article on Professor Percival Lowell's book, *Mars and its Canals*, with the object of showing that the large amount of new and interesting facts contained in this work did not invalidate the conclusion I had reached in 1902, and stated in my book on *Man's Place in the Universe*, that Mars was not habitable.

But the more complete presentation of the opposite view in the volume now under discussion required a more detailed examination of the various physical problems involved, and as the subject is one of great, popular, as well as scientific interest, I determined to undertake the task.

This was rendered the more necessary by the fact that in July last Professor Lowell published in the *Philosophical Magazine* an elaborate mathematical article claiming to demonstrate that, notwithstanding its much greater distance from the sun and its excessively thin atmosphere, Mars possessed a climate on the average equal to that of the south of England, and in its polar and sub-polar regions even less severe than that of the earth. Such a contention of course required to be dealt with, and led me to collect information bearing upon temperature in all its aspects, and so enlarging my criticism that I saw it would be necessary to issue it in book form.

Two of my mathematical friends have pointed out the chief omission which vitiates Professor Lowell's mathematical conclusions—that of a failure to recognise the very large conservative and *cumulative* effect of a dense atmosphere. This very point however I had already myself discussed in Chapter VI., and by means of some remarkable researches on the heat of the moon and an investigation of the causes of its very low temperature, I have, I think, demonstrated the incorrectness of Mr. Lowell's results. In my last chapter, in which I briefly summarise the whole argument, I have further strengthened the case for very severe cold in Mars, by adducing the rapid lowering of temperature universally caused by diminution of atmospheric pressure, as manifested in the well-known phenomenon of temperate climates at moderate heights even close to the equator, cold climates at greater heights even on extensive plateaux, culminating in arctic climates and perpetual snow at heights where the air is still far denser than it is on the surface of Mars. This argument itself is, in my opinion, conclusive; but it is enforced by two others equally complete, neither of which is adequately met by Mr. Lowell.

The careful examination which I have been led to give to the whole of the phenomena which Mars presents, and especially to the discoveries of Mr. Lowell, has led me to what I hope will be considered a satisfactory physical explanation of them. This explanation, which occupies the whole of my seventh chapter, is founded upon a special mode of origin for Mars, derived from the Meteoritic Hypothesis, now very widely adopted by astronomers and physicists. Then, by a comparison with certain well-known and widely spread geological phenomena, I show how the great features of Mars—the "canals" and "oases"—may have been caused. This chapter will perhaps be the most interesting to the general reader, as furnishing a quite natural explanation of features of the planet which have been termed "non-natural" by Mr. Lowell . . .

* * * * *

In passing we may briefly note some of Wallace's other interests, all of which he wrote on at one point or another during his long lifetime: gardening, photography, literature and poetry (he was an avid reader of novels, a fair poet, and not a bad sketch artist), phrenology, and the history of science. Writing came easy to Wallace, and so did charging all he set down with that same unique Wallaceian combination of honesty and tenacity. Our concluding selection here consists of excerpts from an interview (ref 751) published in 1912 in which he criticizes the mechanistic assertions of the chemist E. A. Schaefer regarding the origin of life:

. . . If the chemists do produce life, it is not they who produce it. The chemist never goes into the ultimate cause. He does not deal with the directing power. What does force come from? Matter itself, when you get down to its fundamentals, is becoming almost as complex as living matter. In the mere atom there have been shown to exist countless minute things, every part imbued with force. Whence comes the force?

I maintain you cannot explain the smallest portion of dead matter without a series of forces which imply mind, which imply direction. One of my great points, going back and back and back through life, is the matter of the universe in its bulk. You have got to consider the origin of what is called the dead universe, the cosmos, as full of complex directions and laws not quite so complex, but nearly so, as those of living matter. Whence do these forces come?

Professor Schaefer never attempts to give any idea of how feeling, sense, and the power of perception can possibly arise out of dead matter. He says distinctly it arises from the nervous system. He would produce a nervous system, but is it likely that the nervous system, which is a machine for the manifestation of consciousness, should produce consciousness? Huxley said that life is a cause and not a consequence of organization. It is not organization that produces the life, as is assumed all through Professor Schaefer's lectures. If you assume that the directing power is essentially a spiritual power, then you can understand all this, but without it you cannot understand it . . .

If living matter had been immortal from the beginning, development would have stopped. All the forces of life are directed in a way that is utterly distinct from chemistry. Anything chemistry can do is quite beside the question. In chemistry only certain things will produce certain results. In life the most diverse things will produce the same results. One man may feed entirely on animal food; another entirely on vegetable. The machinery is the same, yet this same machinery so differently fed produces identical results in bone, muscle, nerve, skin, hair—everything. The organism is like an enormous engine, but an engine which can reproduce itself. *There* is the directing power.

Related writings

on museums and geographical education: refs 143, 157, 158, 168, 170, 226, 349, 401, 402, 404, 405, 526, 727, 729, 747

narratives concerning bird-collecting/collections: refs 6, 15, 32, 33, 35, 37, 45, 48, 57, 58, 59, 60, 61, 62, 64, 66, 67, 714, 715, 729

narratives concerning insect-collecting/collections: refs 2, 3, 4, 6, 14, 17, 19, 21, 25, 31, 32, 33, 35, 40, 44, 45, 47, 50, 54, 61, 154, 714, 715

on collecting as an occupation, specimen preparation and art, etc.: refs 7, 8, 21, 33, 40, 45, 49, 58, 254, 357, 363, 535, 715, 733

avian revision/systematics: refs 5, 28, 37, 46, 55, 63, 68, 69, 70, 71, 72, 73, 75, 77, 79, 80, 84, 88, 102, 112, 114, 141, 142, 246, 279, 718, 719, 725

non-avian vertebrate revision/systematics: refs 8, 12, 24, 718

invertebrate revision/systematics: refs 9, 16, 36, 42, 50, 56, 74, 96, 109, 117, 127, 135, 145, 147, 148, 154, 156, 203, 205, 309, 716, 718

on the theory, rules and method of systematics: refs 28, 63, 75, 77, 88, 181, 205, 239, 246, 279, 718, 719, 725

on astronomical subjects: refs 155, 288, 547, 602, 604, 606, 608, 609, 642, 646, 706, 719, 721, 725, 726, 728, 730, 741, 745

on the origin of life: see Section Six, p. 364

on literature and poetry: refs 475, 574, 577, 594, 612, 614, 615, 631, 644, 654, 704, 708, 729, 738, 743, 744, 745

book reviews and notices of miscellaneous travel and natural history works: refs 160, 166, 178, 181, 182, 183, 185, 187, 189, 190, 192, 193, 194, 195, 200, 201, 203, 205, 215, 223, 230, 232, 238, 239, 247, 249, 250, 252, 254, 274, 299, 307, 314, 315, 321, 326, 327, 328, 330, 332, 342, 351, 355, 357, 362, 363, 372, 381, 388, 421, 444, 447, 463, 517, 727

book reviews of elementary mathematics, physics, and mechanics texts: refs 354, 373, 377, 392

secondary literature: Brooks (1984), Durant (1979), George (1964), McKinney (1972), Malinchak (1987), Marchant (1916), Schwartz (1984*b*)

BIBLIOGRAPHY OF THE WRITINGS OF ALFRED RUSSEL WALLACE

Scope

THIS Bibliography of Wallace's writings is divided into three Parts. The first and longest Part lists those essays, empirical studies, reviews, letters to the Editor, notes, etc., that Wallace published during his lifetime (and a few relevant items appearing after his death). The second Part treats his monographs, including all editions published in English during his lifetime (and those few which continued to be reprinted unchanged after his death). Also included in Part II are separately reprinted excerpts from these books which I have encountered. In the third Part, I have listed some significant interviews he gave.

Each unique entry (i.e., disregarding reprintings) has been assigned a number; these are of course the numbers referred to in the Selections and Index portions of this work. The entries in each of the three Parts are listed chronologically according to their first date of publication or formal oral presentation (there are some unavoidable exceptions to this format, however). I have opted for ease of reference over chronological continuity *across* the three Parts; thus, each Part is individually organized chronologically, but the numbering scheme continues from Part to Part without resettings. In all three Parts short explanatory notes have been inserted *in brackets* where I have deemed such information useful. I have physically examined almost every entry (including reprintings).

As the line must be drawn somewhere, this list does not include all Wallace-originated material I am aware was published. With but few exceptions I have not, for example, included second-person synopses of public lectures he gave. I have also made no effort to trace personal letters that may have appeared in collections of other individuals' correspondence. About two thousand of Wallace's personal letters have survived; some of these were printed in *My life* (ref 729) and Marchant (1916). (McKinney (1972) provides the most complete list of unpublished sources.) I should in addition note that Brooks (1984) reprints most of the Wallace letters that were published in natural history journals before 1858.

Abbreviations and conventions

To save space, the following abbreviations of book titles and organizations are employed throughout Parts I and II:

ASL: Anthropological Society of London
A-VL: Anti-Vaccination League

BAAS: British Association for the Advancement of Science
CTNS: *Contributions to the theory of natural selection* (ref 716)
ESL: Entomological Society of London
LNS: Land Nationalisation Society
LSL: Linnean Society of London
NSTN: *Natural selection and tropical nature* (ref 725)
RGS: Royal Geographical Society
SPR: (British) Society for Psychical Research
SSS: *Studies scientific and social* (ref 727)
TNOE: *Tropical nature and other essays* (ref 719)
ZSL: Zoological Society of London

In addition, journal titles are often abbreviated, 'n.s.' is used in place of 'new series' (or 'o.s.' for 'old series', '2nd s.' for 'second series', etc.), 'n.d.' in place of 'no date', and 'n.p.' in place of 'no pagination'. Further, the acronym 'lttE' replaces the frequently appearing 'letter to the Editor.' The meanings of the few other abbreviations employed should be self-evident.

Printed abstracts and the like are labelled as appropriate, and followed by an oblique (/) and one of four letters to indicate whether the available evidence suggests to me that the summary was: not prepared by Wallace himself (/N), not likely to have been prepared by Wallace (/U), quite possibly prepared by Wallace(/P), or prepared by Wallace (/Y). Titles of monographs and pamphlets are printed in italics; titles of all shorter selections are given within single quotation marks (or, when the main entry, with no special designation).

Miscellaneous conventions: In Parts I and III, each main entry (consisting of the first and/or most complete place of publication or presentation) is followed (if pertinent) by a list of reprintings, partial reprintings, summaries, etc., separated by semicolons. In Part II, each entry is subdivided according to *revised* editions, with reprintings of each edition following within brackets (and separately published excerpts listed at the end of the overall entry). In all Parts, except where noted otherwise, it can be assumed that reprintings carry the same title as the original (for monographs, this treatment extends to publication and pagination data). Obliques (/) are used in all Parts to signify parallel conditions of publication; for instance where an article was published in successive numbers of a journal, or an edition (or reprint of an edition) of a book was published by more than one firm. When in capitals titles are exactly as listed; lower-case 'titles' convey generic information ('letter', 'review', etc.) when actual titles are lacking. When the entry is part of a larger article as noted, pagination of entry is *directly* followed by pagination of the overall article *in parentheses*.

Starred (*) entries are not specifically listed in the bibliography of Wallace's publications found in Marchant (1916).

The bibliographic form is wholly of my own making, and I hereby absolve Oxford University Press of any responsibility for its many idiosyncracies.

Part I: Papers, essays, reviews, letters, etc.

1.* The Advantages of Varied Knowledge [excerpts from this essay, composed in late 1843, and possibly presented as a public lecture at that time]. In *My Life*, Vol. 1: 201–204 (Oct. 1905).

2.* Capture of *Trichius fasciatus* near Neath [excerpt from a letter sent from Neath, Wales]. *Zoologist* 5: 1676 (April? 1847).

3.* Journey to Explore the Province of Pará [extract from a letter dated 23 Oct., Pará, from Wallace and Henry W. Bates to Samuel Stevens, their collections sales agent]. *Ann. & Mag. Nat. Hist.* 3 (2nd s.): 74–75 (Jan. 1849: no. 13, 2nd s.).

4.* Journey to Explore the Natural History of South America [extracts from a letter dated 12 Sept. 1849, Santarem, to Samuel Stevens]. *Ann. Mag. Nat. Hist.* 5 (2nd s.): 156–157 (Feb. 1850: no. 26, 2nd s.).

5. On the Umbrella Bird (*Cephalopterus ornatus*), "Ueramimbé," L. G. [from a letter dated 10 March 1850, Barra do Rio Negro (Manaus); communicated to the ZSL meeting of 23 July 1850]. *Proc. ZSL* 18: 206–207 (1850).
—reprint: *Ann. & Mag. Nat. Hist.* 8 (2nd s.): 428–430 (Nov. 1851: no. 47, 2nd s.).

6.* Journey to Explore the Natural History of the Amazon River [extracts from letters dated 15 Nov. 1849, Santarem, and 20 March 1850, Barra do Rio Negro, to Samuel Stevens]. *Ann. & Mag. Nat. Hist.* 6 (2nd s.): 494–496 (Dec. 1850: no. 36, 2nd s.).

7.* letter [concerning the ship fire dated 19 Oct. 1852, 43 Upper Albany St., Regent's Park]. *Zoologist* 10: 3641–3643 (Nov.? 1852).

8. On the Monkeys of the Amazon [a paper read at the ZSL meeting of 14 Dec. 1852]. *Proc. ZSL* 20: 107–110 (1852).
—reprints: *Ann. & Mag. Nat. Hist.* 14 (2nd s.): 451–454 (Dec. 1854: no. 84, 2nd s.); in Portuguese as 'Os Símios (Macacos) da Amazônia' in *Boletim do Museu Paraense de Historia Natural e Ethnografia, Belém* 1(1/4): 375–381 (1896).

9. Some Remarks on the Habits of the Hesperidae [dated March 1853]. *Zoologist* 11: 3884–3885 (May? 1853).

10. On the Insects Used for Food by the Indians of the Amazon [a paper read at the ESL meeting of 6 June 1853]. *Trans. ESL* 2 (n.s.), part VIII: 241–244 (April 1854).

11. On the Rio Negro [a paper read at the RGS meeting of 13 June 1853]. *J. RGS* 23: 212–217 (includes fold-out map of the Rio Negro) (1853).
—map reprinted in *My Life*, Vol. 1, opposite p. 320 (Oct. 1905).

12. On some Fishes Allied to *Gymnotus* [a paper read at the ZSL meeting of 12 July 1853]. *Proc. ZSL* 21: 75–76 (1853).
—reprint: *Ann. & Mag. Nat. Hist.* 14 (2nd s.): 398–399 (Nov. 1854: no. 83, 2nd s.).

13. On the Habits of the Butterflies of the Amazon Valley [a paper read at the ESL meetings of 7 Nov. and 5 Dec. 1853]. *Trans. ESL* 2 (n.s.), part VIII: 253–264 (April 1854).

14.* letter [concerning collecting dated 9 May 1854, Singapore]. *Zoologist* 12: 4395–4397 (Aug.? 1854).

15. On the Ornithology of Malacca. *Ann. & Mag. Nat. Hist.* 15 (2nd s.): 95–99 (Feb. 1855: no. 86, 2nd s.).

16. Description of a New Species of *Ornithoptera. Ornithoptera brookiana*. Wallace. [communicated to the ESL meeting of 2 April 1855]. *Proc. ESL, 1854–1855*: 104–105.
—reprint: *Zoologist* 13: 4819–4820 (Sept.? 1855).

17. The Entomology of Malacca [dated 25 Nov. 1854, Sarawak, Borneo]. *Zoologist* 13: 4636–4639 (April? 1855).

18.* letter [concerning Wallace's collecting environs in Sarawak dated 1854, Sarawak, Borneo; anonymous, but easily associated with Wallace's activities at this time and place]. *Literary Gazette* (London) no. 2003: 366b–c (9 June 1855).

19. Extracts of a Letter from Mr. Wallace [from a letter concerning collecting dated 10 Oct. 1854, Singapore]. *(Hooker's) J. Botany* 7(7): 200–209 (July? 1855).

20. On the Law which has Regulated the Introduction of New Species [dated Feb. 1855, Sarawak, Borneo]. *Ann. & Mag. Nat. Hist.* 16 (2nd s.): 184–196 (Sept. 1855: no. 93, 2nd s.).
—reprints: in *CTNS*: 1–25 (April 1870); in *NSTN*: 3–19 (May 1891).

21.* letter [concerning collecting dated 8 April 1855, Si Munjon Coal Works, Borneo]. *Zoologist* 13: 4803–4807 (Sept.? 1855).

22.* Borneo [letter concerning Wallace's collecting environs in Sarawak dated 25 May 1855, Si Munjon Coal Works, Borneo]. *Literary Gazette* (London) no. 2023: 683b–684a (27 Oct. 1855).
—edited version reprinted in *My Life*, Vol. 1: 342–343 (Oct. 1905).

23. Some Account of an Infant "Orang-utan." *Ann. & Mag. Nat. Hist.* 17 (2nd s.): 386–390 (May 1856: no. 101, 2nd s.).
—reprint: incorporated, in large part, into Chapter 4 of *The Malay Archipelago* (Feb. 1869).

24. On the Orang-utan or Mias of Borneo [dated Dec. 1855, Sarawak]. *Ann. & Mag. Nat. Hist.* 17 (2nd s.): 471–476 (June 1856: no. 102, 2nd s.).

25. Observations on the Zoology of Borneo [dated 10 March 1856, Singapore]. *Zoologist* 14: 5113–5117 (June? 1856).

26. On the Habits of the Orang-utan of Borneo. *Ann. & Mag. Nat. Hist.* 18 (2nd s.): 26–32 (July 1856: no. 103, 2nd s.).
—reprint: portions incorporated into Chapter 4 of *The Malay Archipelago* (Feb. 1869).

27. On the Bamboo and Durian of Borneo [part of a letter to W. J. Hooker]. *(Hooker's) J. Botany* 8(8): 225–230 (Aug.? 1856).
—reprints: in German as 'Ueber Bambus und Durian auf Borneo' in *Allgemeine Gartenzeitung* 24(37): 292–293 (13 Sept. 1856) / 24(39): 308–309 (27 Sept. 1856); incorporated, in large part, into Chapter 5 of *The Malay Archipelago* (Feb. 1869).

28. Attempts at a Natural Arrangement of Birds. *Ann. & Mag. Nat. Hist.* 18 (2nd s.): 193–216 (Sept. 1856: no. 105, 2nd s.).

29. Notes of a Journey up the Sadong River, in North-west Borneo [communicated to the RGS meeting of 10 Nov. 1856]. *Proc. RGS Lond.* 1 (1855–6 & 1856–7), no. 6: 193–205 (1857).
—reprint: incorporated, in large part, into Chapters 5 & 6 of *The Malay Archipelago* (Feb. 1869).

30. A New Kind of Baby [concerning a baby orangutan; anonymous, but in Wallace's style and listed in the bibliography of Marchant (1916)]. *Chambers's J.* 6 (3rd s.): 325–327 (22 Nov. 1856: no. 151, 3rd s.).

31.* letter [concerning collecting dated 21 Aug. 1856, Ampanam, Lombock]. *Zoologist* 15: 5414–5416 (Feb.? 1857).

32.* letter [concerning collecting dated 27 Sept. 1856, Macassar]. *Zoologist* 15: 5559–5560 (June? 1857).

33.* letter [concerning collecting dated 1 Dec. 1856, Macassar]. *Zoologist* 15: 5652–5657 (Aug.? 1857).

34.* The Dyaks [anonymous, but containing various details strongly suggesting it is by Wallace]. *Chambers's J.* 8 (3rd s.): 201–204 (26 Sept. 1857: no. 195, 3rd s.).
—reprint: *The National Magazine* (New York) 11: 511–516 (Dec. 1857).

35.* letter and postscript [concerning collecting to Samuel Stevens; dated 10 March 1857, Dobbo, Aru Islands, and 15 May, Dobbo; communicated to the ESL meeting of 5 Oct. 1857]. *Proc. ESL, 1856–1857*: 91–93.

36. On the Habits and Transformations of a Species of *Ornithoptera*, Allied to *O. priamus*, Inhabiting the Aru Islands, near New Guinea [communicated to the ESL meeting of 7 Dec. 1857]. *Trans. ESL* 4 (n.s.), part VII: 272–273 (April 1858).

37. On the Great Bird of Paradise, *Paradisea apoda*, Linn.; '*Burong mati*' (*Dead Bird*) of the Malays; '*Fanèhan*' of the Natives of Aru. *Ann. & Mag. Nat. Hist.* 20 (2nd s.): 411–416 (Dec. 1857: no. 120, 2nd s.).
—reprint: as 'On the Great Bird of Paradise *Paradisea apoda*, Linn.; Burong Mati (Dead Bird) of the Malays' in *Natuurkundig Tijdschrift voor Nederlandsch Indië* 17, 1858–59 (vierde serie, deel 3): 377–387.

38. On the Natural History of the Aru Islands. *Ann. & Mag. Nat. Hist., Supplement* to vol. 20 (2nd s.): 473–485 (Dec. 1857: no. 121, 2nd s.).
—reprint: *Natuurkundig Tijdschrift voor Nederlandsch Indië* 17, 1858–59 (vierde serie, deel 3): 356–376.

39. Note on the Theory of Permanent and Geographical Varieties. *Zoologist* 16: 5887–5888 (Jan.? 1858).

40. On the Entomology of the Aru Islands. *Zoologist* 16: 5889–5894 (Jan.? 1858).

41. On the Arru Islands [communicated to the RGS meeting of 22 Feb. 1858]. *Proc. RGS Lond.* 2 (1857–58), no. 3: 163–170 (selection followed by an account of related discussion on pp. 170–171) (1858).
—reprint: portions incorporated into Chapter 33 of *The Malay Archipelago* (Feb. 1869).

42.* A Disputed Case of Priority in Nomenclature [dated 1 Jan. 1858, Amboyna; communicated to the ESL meeting of 3 May 1858]. *Proc. ESL, 1858–59*: 23–24.

—reprint: *Zoologist* 16: 6117–6118 (July? 1858).

43. On the Tendency of Varieties to Depart Indefinitely from the Original Type [dated Feb. 1858, Ternate; third part of 'On the Tendency of Species to Form Varieties; and On the Perpetuation of Varieties and Species by Natural Means of Selection' by Charles Darwin and Alfred Wallace; communicated by Sir Charles Lyell and Joseph D. Hooker to the LSL meeting of 1 July 1858]. *J. Proc. Linn. Soc. Zool.* 3(9): 53–62 (45–62) (20 Aug. 1858).

—reprints: almost verbatim in *CTNS*: 26–44 (April 1870), and in turn in *NSTN*: 20–33 (May 1891); in *Popular Science Monthly* 60: 13–21 (5–21) (Nov. 1901); in *The Darwin-Wallace Celebration held on Thursday, 1st July, 1908, by the Linnean Society of London* (printed for the Linnean Society by Burlington House, Longmans, Green & Co., London, Feb. 1909): 98–107 (89–107); reprinted in German in *Charles Darwin und Alfred Russel Wallace. Ihre Ersten Publicationen über die "Entstehung der Arten" nebst einer Skizze ihres Lebens und einem Verzeichniss ihrer Schriften.* by Adolf Bernhard Meyer (Eduard Besold, Erlangen, 1870).

44.* letter [concerning collecting dated 20 Dec. 1857, Amboyna]. *Zoologist* 16: 6120–6124 (July? 1858).

45.* letters [extracts from letters concerning collecting dated 2 Sept. 1858, Ternate]. *Ibis* 1: 111–113 (Jan. 1859: no. 1).

—more extensive extracts printed in 'Extracts from a Letter of Mr. A. R. Wallace to Mr. S. Stevens' in *Zoologist* 17: 6409–6413 (March? 1859).

46. Correction of an Important Error Affecting the Classification of the Psittacidae. *Ann. & Mag. Nat. Hist.* 3 (3rd s.): 147–148 (Feb. 1859: no. 14, 3rd s.).

47.* letter [extract from letter concerning collecting dated 29 Oct. 1858, Batchian, Moluccas; communicated to the ESL meeting of 7 March 1859]. *Proc. ESL, 1858–1859*: 61.

—reprint: *Zoologist* 17: 6546–6547 (March 1859).

48.* letter [extract from letter concerning collecting dated 29 Oct. 1858, Batchian, Moluccas; communicated to the ZSL meeting of 22 March 1859]. *Proc. ZSL* 27: 129 (1859).

49.* Remarks on Enlarged Coloured Figures of Insects [dated Nov. 1858, Batchian, Moluccas; communicated to the ESL meeting of 2 May 1859]. *Proc. ESL, 1858–1859*: 66–67 (followed by an account of related discussion on p. 67).

—reprint: *Zoologist* 17: 6617–6618 (Aug.? 1859).

50.* letter [extracts from letter concerning collecting dated 28 Jan. 1859, Batchian; communicated to the ESL meeting of 6 June 1859]. *Proc. ESL, 1858–1859*: 70 (followed by an account of related discussion on pp. 70–71).

51. Notes of a Voyage to New Guinea [communicated to the RGS meeting of 27 June 1859]. *J. RGS* 30: 172–177 (1860).

—summary/N of paper and related discussion entitled 'Notes on a Voyage to

New Guinea' printed in *Proc. RGS Lond.* 3 (1858–59), no. 6: 358–359 (358–361) (1859).

52. Letter from Mr. Wallace concerning the Geographical Distribution of Birds [dated March 1859, Batchian]. *Ibis* 1: 449–454 (Oct. 1859: no. 4).
—reprinted in German as 'Ein Brief über die Geographische Verbreitung der Vögel' in *Journal für Ornithologie* 8: 47–53 (Januar 1860: heft 43).

53. On the Zoological Geography of the Malay Archipelago [communicated by Charles Darwin to the LSL meeting of 3 Nov. 1859]. *J. Proc. Linn. Soc. Zool.* 4: 172–184 (1860).

54. Note on the Habits of Scolytidae and Bostrichidae [communicated to the ESL meeting of 5 Dec. 1859]. *Trans. ESL* 5 (n.s.), part VI: 218–220 (July 1860).

55. Notes on *Semioptera wallacii*, Gray [extract from a letter dated 30 Sept. 1859, Amboyna; communicated to the ZSL meeting of 24 Jan. 1860]. *Proc. ZSL* 28: 61 (1860).
—reprints: *Ann. & Mag. Nat. Hist.* 6 (3rd s.): 75–76 (July 1860: no. 31, 3rd s.); as 'Habits of the Standard-wing' in *Zoologist* 19: 7710 (Oct.? 1861).

56. Note on the Sexual Differences in the Genus *Lomaptera* [communicated to the ESL meeting of 6 Feb. 1860]. *Proc. ESL, 1858–59*: 107.
—reprint: *Zoologist* 18: 6938–6939 (April? 1860).

57. The Ornithology of Northern Celebes [part of a lttE dated Oct. 1859, Amboyna]. *Ibis* 2: 140–147 (April 1860: no. 6).
—extract from letter printed as 'On the Nidification of Birds of the Family Megapodiidae' in *Zoologist* 19: 7435 (April? 1861).

58.* letters [extracts from letters concerning collecting dated 22 Oct. 1859, Amboyna]. *Ibis* 2: 197–199 (April 1860: no. 6).

59.* letters [extracts from letters concerning collecting dated 26 Nov. 1859, Awaiya, Ceram, 31 Dec. 1859, Passo, Amboyna, and 14 Feb. 1860, Passo]. *Ibis* 2: 305–306 (July 1860: no. 7).

60.* letters [note/N describing letters concerning collecting dated June 1860, Ceram]. *Ibis* 3: 118 (Jan. 1861: no. 9).

61.* letter [extracts from letter concerning collecting dated 7 Dec. 1860, Ternate]. *Ibis* 3: 211–212 (April 1861: no. 10).

62. On the Ornithology of Ceram and Waigiou [dated 20 Dec. 1860, Ternate]. *Ibis* 3: 283–291 (& 1 colour plate) (July 1861: no. 11).

63.* letters [note/N containing extracts from letter dated 10 Dec. 1860, Ternate, and describing letter dated 6 Feb. 1861, Delli, Timor, both concerning collecting]. *Ibis* 3: 310–311 (July 1861: no. 11).

64. Notes on the Ornithology of Timor [dated 20 April 1861, Delli, Timor]. *Ibis* 3: 347–351 (Oct. 1861: no. 12).

65. On the Trade of the Eastern Archipelago with New Guinea and its Islands [communicated to the RGS meeting of 13 Jan. 1862]. *J. RGS* 32: 127–137 (1862).
—summary/Y of paper and related discussion printed as 'On the Trade between the Eastern Archipelago and New Guinea and its Islands' in *Proc.*

RGS Lond. 6 (1861–62), no. 2: 43–44 (43–46) (1862); paper of same title concerning same subject read in Cambridge at the 4 Oct. 1862 meeting of Section E, Geography and Ethnology, of the BAAS: abstract/Y printed in "Notices and Abstracts of Miscellaneous Communications to the Sections" portion of the *Rept. BAAS* 32 (1862) (John Murray, London, 1863): 148–149.

66.* letters [extracts from letters concerning collecting dated 20 Sept. 1861, Batavia]. *Ibis* 4: 95–96 (Jan. 1862: no. 13).

67. Narrative of Search after Birds of Paradise [a paper read at the ZSL meeting of 27 May 1862]. *Proc. ZSL, 1862*: 153–161 (1862).
—reprints: *Ann. & Mag. Nat. Hist.* 10 (3rd s.): 464–472 (Dec. 1862: no. 60, 3rd s.); partial reprint in *My Life*, Vol. 1: 387–394 (Oct. 1905).

68. On some New and Rare Birds from New Guinea [a paper read at the ZSL meeting of 10 June 1862]. *Proc. ZSL, 1862*: 164–166 (& 3 colour plates) (1862).
—reprint: *Ann. & Mag. Nat. Hist.* 11 (3rd s.): 56–58 (Jan. 1863: no. 61, 3rd s.).

69. Descriptions of Three New Species of *Pitta* from the Moluccas [a paper read at the ZSL meeting of 24 June 1862]. *Proc. ZSL, 1862*: 187–188 (1862).
—reprint: *Ann. & Mag. Nat. Hist.* 11 (3rd s.): 133–134 (Feb. 1863: no. 62, 3rd s.).

70. On some New Birds from the Northern Moluccas. *Ibis* 4: 348–351 (Oct. 1862: no. 16).

71. List of Birds from the Sula Islands (East of Celebes), with Descriptions of the New Species [a paper read at the ZSL meeting of 9 Dec. 1862]. *Proc. ZSL, 1862*: 333–346 (& 3 colour plates) (1862).

72. List of Birds Collected in the Island of Bouru (One of the Moluccas), with Descriptions of the New Species [a paper read at the ZSL meeting of 13 Jan. 1863]. *Proc. ZSL, 1863*: 18–36 (& 3 colour plates) (1863).

73. Note on *Corvus senex*, Garn. & Less., and *Corvus fuscicapillus*, G. R. Gray. *Ibis* 5: 100–102 (Jan. 1863: no. 17).

74. Notes on the Genus *Iphias*; with Descriptions of Two New Species from the Moluccas. *J. Entomology* 2: 1–5 (& 1 colour plate) (Jan. 1863: no. 7).

75. On the Proposed Change in Name of *Gracula pectoralis*. *Ann. & Mag. Nat. Hist.* 11 (3rd s.): 15–17 (Jan. 1863: no. 61, 3rd s.).

76.* discussion/N [of three papers on Australia read by J. M. Stuart, "Mr. Landsborough" and J. M'Kinlay at the RGS meeting of 9 March 1863]. *Proc. RGS* 7 (1862–63), no. 3: 88–89 (82–90) (1863).

77. Who are the Humming Bird's Relations? *Zoologist* 21: 8486–8491 (April? 1863).

78. On the Physical Geography of the Malay Archipelago [a paper read at the RGS meeting of 8 June 1863]. *J. RGS* 33: 217–234 (& map) (1863).
—reprints: incorporated, in large part, into Chapter 1 of *The Malay Archipelago* (Feb. 1869); summary/U of paper and related discussion printed in *Proc. RGS* 7 (1862–63), no 5.: 206–210 (205–212) (1863); translated into Dutch by P. J. Veth and reprinted as pamphlet entitled *Over de Physische Geographie*

van den Indischen Archipel (Joh. Noman en Zoon, Zalt-Bommel, The Netherlands, 1865; pp. (1)–24); same paper read in Newcastle-upon-Tyne at the 1 Sept. 1863 meeting of Section D, Zoology and Botany, of the BAAS: abstract/U printed in the "Notices and Abstracts of Miscellaneous Communications to the Sections" portion of the *Rept. BAAS* 33 (1863) (John Murray, London, 1864): 107–108. [excerpts, summaries, etc., under various titles: *Reader* 2: 480a–c (24 Oct. 1863: no. 43); *Athenaeum* no. 1874: 404–405 (26 Sept. 1863); in *Annual of Scientific Discovery* (for 1864) ed. by David A. Wells (Gould & Lincoln, Boston, 1864): 338–341]

79. The Bucerotidae, or Hornbills. *Intellectual Observer* 3(5): 309–317 (& 1 colour plate) (June 1863).

80. Notes on the Fruit-pigeons of the Genus *Treron*. *Ibis* 5: 318–320 (July 1863: no. 19).

81. On the Geographical Distribution of Animal Life [a paper read in Newcastle-upon-Tyne at the 31 Aug. 1863 meeting of Section D, Zoology and Botany, of the BAAS]. abstract/P printed in "Notes and Abstracts of Miscellaneous Communications to the Sections" portion of the *Rept. BAAS* 33 (1863) (John Murray, London, 1864): 108–109.
—reprints: same abstract printed in *Reader* 2: 352c–353a (26 Sept. 1863: no. 39); paper printed in full as 'On some Anomalies in Zoological and Botanical Geography' in *Edinburgh New Philosophical J.* 19: 1–15 (7 Jan. 1864); this then reprinted in *Natural History Rev.* 4: 111–123 (Jan. 1864: no. 13).

82. On the Varieties of Men in the Malay Archipelago [a paper read in Newcastle-upon-Tyne at the 1 Sept. 1863 meeting of Section E, Geography and Ethnology, of the BAAS]. abstract/P printed in "Notices and Abstracts of Miscellaneous Communications to the Sections" portion of the *Rept. BAAS* 33 (1863) (John Murray, London, 1864): 147–148.
—reprints: same abstract, followed by an account of related discussion, printed in *Reader* 2: 483a-b (483a–c) (24 Oct. 1863: no. 43) and *Anthro. Rev.* 1: 441–442 (441–444) (Nov. 1863: no. 3); paper later presented as 'On the Varieties of Man in the Malay Archipelago' at the Ethnological Society of London meeting of 26 Jan. 1864 and printed in full in *Trans. Ethnol. Soc. Lond.* 3 (n.s.): 196–215 (1865); this incorporated, in large part, into Chapter 40 of *The Malay Archipelago* (Feb. 1869).

83. Remarks on the Rev. S. Haughton's Paper on the Bee's Cell, and on the Origin of Species. *Ann. & Mag. Nat. Hist.* 12 (3rd s.): 303–309 (Oct. 1863: no. 70, 3rd s.).

84. On the Identification of the *Hirundo esculenta* of Linnaeus, with a Synopsis of the Described Species of *Collocalia* [a paper read at the ZSL meeting of 10 Nov. 1863]. *Proc. ZSL, 1863*: 382–385 (1863).

85. A List of the Birds Inhabiting the Islands of Timor, Flores, and Lombock, with Descriptions of the New Species [a paper read at the ZSL meeting of 24 Nov. 1863]. *Proc. ZSL, 1863*: 480–497 (& 1 colour plate) (1863).
—reprint: portions incorporated into Chapter 14 of *The Malay Archipelago* (Feb. 1869).

86.* discussion/N [of 'On the Vitality of the Black Race,' a paper by Count Oscar Reichenbach read at the ASL meeting of 15 Dec. 1863]. *J. ASL* 2: lxxiii (lxv–lxxiii) (1864).

87.* discussion/N [of papers on the extinction of races by Richard Lee and T. Bendyshe read at the ASL meeting of 19 Jan. 1864]. *J. ASL* 2: cx–cxi (xcv–cxiii) (1864).

88. Remarks on the Value of Osteological Characters in the Classification of Birds. *Ibis* 6: 36–41 (Jan. 1864: no. 21).

89. Remarks on the Habits, Distribution, and Affinities of the Genus *Pitta*. *Ibis* 6: 100–114 (Jan. 1864: no. 21).

90.* discussion/N [of 'On Anthropological Desiderata,' a paper by James Reddie read at the ASL meeting of 2 Feb. 1864]. *J. ASL* 2: cxxix–cxxx (cxv–cxxxv) (1864).

91.* discussion/N [of paper on North Peru by Don Antonio Raimondy read at the RGS meeting of 8 Feb. 1864]. *Proc. RGS* 8 (1863–64), no. 3: 61–62 (58–62) (28 April 1864).

92.* discussion/N [of paper on the Neanderthal Man skull by C. Carter Blake read at the ASL meeting of 16 Feb. 1864]. *J. ASL* 2: clvi (cxxxix–clvii) (1864).

93. The Origin of Human Races and the Antiquity of Man Deduced from the Theory of "Natural Selection" [a paper read at the ASL meeting of 1 March 1864]. *J. ASL* 2: clviii–clxx (followed by an account of related discussion on pp. clxx–clxxxvii) (1864).

—reprints: summary/Y of paper including quotations entitled '"Natural Selection" Applied to Man' printed in *Natural History Rev.* 4 (n.s.): 328–336 (July 1864: no. 15); revised version entitled 'The Development of Human Races under the Law of Natural Selection' printed in *CTNS*: 303–331 (April 1870); this then reprinted in *NSTN*: 167–185 (May 1891); revised version from *CTNS* reprinted nearly verbatim as first half of pamphlet entitled *The Action of Natural Selection on Man* (University Series, No. 6: Charles C. Chatfield & Co., New Haven, Aug. 1871): 1–28 (No. 6 included with Nos. 7–10 as *Half Hours with Modern Scientists. Lectures and Essays. Second Series.* first published by Charles C. Chatfield & Co., New Haven, April 1873; No. 6 later reprinted, as part of pamphlet (1882 & 1886) and as part of collection (1877), by D. Van Nostrand, New York).

94.* discussion/N [of 'On the Place of the Sciences of Mind and Language in the Science of Man,' a paper by Luke Owen Pike read at the ASL meeting of 15 March 1864]. *J. ASL* 2: ccviii (cxcii–ccviii) (1864).

95.* discussion/N [of 'Notes on the Capabilities of the Negro for Civilisation,' a paper by Henry F. J. Guppy read at the ASL meeting of 15 March 1864]. *J. ASL* 2: ccxiii–ccxiv (ccix–ccxvi) (1864).

96. On the Phenomena of Variation and Geographical Distribution as Illustrated by the Papilionidae of the Malayan Region [a paper read at the LSL meeting of 17 March 1864]. *Trans. LSL* 25, part I: 1–71 (& 8 colour plates) (1865).

—reprints: edited version printed as 'The Malayan Papilionidae or Swallow-tailed Butterflies, as Illustrative of the Theory of Natural Selection' in

CTNS: 130–200 (April 1870); abstract/Y entitled 'Mr. Wallace on the Phenomena of Variation and Geographical Distribution as Illustrated by the Malayan Papilionidae' printed in *Reader* 3: 491b–493b (16 April 1864: no. 68); discussions/N of the same subject at the ESL meetings of 4 April 1864 and 2 May 1864 printed in *J. Proc. ESL, 1864–1865*: 15–16 & 21–23; 4 April 1864 discussion reprinted in *Zoologist* 22: 9080–9081 (June? 1864).

97. Bone-caves in Borneo [lttE]. *Reader* 3: 367 (19 March 1864: no. 64); also printed in *Natural History Rev.* 4: 308–311 (April 1864: no. 14).

98.* discussion/N [regarding mimicry in insects, at the ESL meeting of 4 April 1864]. *J. Proc. ESL, 1864–1865*: 14–15 (14–15).
—same account printed in *Zoologist* 22: 9079–9080 (9079–9080) (June? 1864).

99.* discussion/N [of 'On the Universality of Belief in God, and in a Future State,' a paper by Rev. F. W. Farrar read at the ASL meeting of 5 April 1864]. *J. ASL* 2: ccxx (ccxvii–ccxxii) (1864).

100.* discussion/N [of 'On Hybridity,' a paper by Rev. F. W. Farrar read at the ASL meeting of 5 April 1864]. *J. ASL* 2: ccxxvi–ccxxvii (ccxxii–ccxxix) (1864).

101. Note on *Astur griseiceps*, Schlegel. *Ibis* 6: 184 (& 1 colour plate) (April 1864: no. 22).

102. On the Parrots of the Malayan Region, with Remarks on their Habits, Distribution, and Affinities, and the Descriptions of Two New Species [a paper read at the ZSL meeting of 28 June 1864]. *Proc. ZSL, 1864*: 272–295 (includes map of the Malay Archipelago) (1864).

103.* discussion/N [of paper on the delta of the Amazons by Henry W. Bates read in Bath at the 17 Sept. 1864 meeting of Section E, Geography and Ethnology, of the BAAS]. *Athenaeum* no. 1928: 470 (470) (8 Oct. 1864).

104. On the Progress of Civilization in Northern Celebes [a paper read in Bath at the 19 Sept. 1864 meeting of Section E, Geography and Ethnology, of the BAAS]. abstract/P printed in "Notices and Abstracts of Miscellaneous Communications to the Sections" portion of the *Rept. BAAS* 34 (1864) (John Murray, London, 1865): 149–150.
—reprints: same abstract printed in *Athenaeum* no. 1929: 503 (15 Oct. 1864) and *Anthro. Rev.* 2: 332–334 (followed by an account of related discussion on p. 334) (Nov. 1864: no. 7); different abstract of paper consisting largely of many excerpts from the written text printed in *Reader* 4: 485c–486b (15 Oct. 1864: no. 94); paper later read at the Ethnological Society of London meeting of 24 Jan. 1865 and printed in full in *Trans. Ethnol. Soc. Lond.* 4 (n.s.): 61–70 (1866); this incorporated, in large part, into Chapter 17 of *The Malay Archipelago* (Feb. 1869).

105.* discussion/N [of paper on islands north of Flores by John Cameron read at the RGS meeting of 12 Dec. 1864]. *Proc. RGS* 9 (1864–65), no. 2: 31–32 (30–32) (1865).

106.* discussion/N [of 'Linga Puja, or Phallic Worship in India,' a paper by E. Sellon read at the ASL meeting of 17 Jan. 1865]. *J. ASL* 3: cxviii–cxix (cxiv–cxxi) (1865).

107.* The *British Quarterly* and Darwin [lttE]. *Reader* 5: 77c–78a (21 Jan. 1865: no. 108).

108.* The *British Quarterly* Reviewer and Darwin [lttE]. *Reader* 5: 173a–b (11 Feb. 1865: no. 111).

109. List of the Land Shells Collected by Mr. Wallace in the Malay Archipelago, with Descriptions of the New Species by Mr. Henry Adams [a paper read at the ZSL meeting of 25 April 1865]. *Proc. ZSL, 1865*: 405–416 (& 1 colour plate) (1865).

110.* Public Responsibility and the Ballot [lttE responding to comments by John Stuart Mill]. *Reader* 5: 517a–b (6 May 1865: no. 123).

111.* discussion/N [of 'On the Efforts of Missionaries among Savages,' a paper by Rev. J. W. Colenso read at the ASL meeting of 16 May 1865]. *J. ASL* 3: cclxxxviii (ccxlviii–cclxxxix) (1865).

112. Descriptions of New Birds from the Malay Archipelago [a paper read at the ZSL meeting of 13 June 1865]. *Proc. ZSL, 1865*: 474–481 (& 2 colour plates) (1865).

113. How to Civilize Savages [signed 'W.' but referred to in *My Life* (Vol. 2, pp. 24 & 52)]. *Reader* 5: 671a–672a (17 June 1865: no. 129).
—expanded version printed in *SSS*, Vol. 2: 107–114 (Nov. 1900).

114. On the Pigeons of the Malay Archipelago. *Ibis* 1 (n.s.): 365–400 (& 1 colour plate) (Oct. 1865: no. 4, n.s.).
—reprint: in German as 'Ueber die Tauben des Malayischen Archipels' in *Journal für Ornithologie* 14: 269–285 (Juli 1866: heft 82).

115.* Is the Earth an Oblate or a Prolate Spheroid? [lttE]. *Reader* 7: 497b–c (19 May 1866: no. 177).

116.* lttE [one of two concerning the nature of the sphericity of the earth printed as 'The Figure of the Earth (not Archdeacon Pratt's)']. *Reader* 7: 546b (546a--b) (2 June 1866: no. 179).

117. List of Lepidopterous Insects Collected at Takow, Formosa, by Mr. Robert Swinhoe [a joint paper by Wallace & Frederic Moore read at the ZSL meeting of 12 June 1866]. *Proc. ZSL, 1866*: 355–365 (1866).

118. The Scientific Aspect of the Supernatural. *The English Leader* 2(4): 59–60 / 2(5): 75–76 / 2(6): 91–93 / 2(7): 107–108 / 2(8): 123–125 / 2(9): 139–140 / 2(10): 156–157 / 2(11): 171–173 (11 Aug. 1866 to 29 Sept. 1866: nos. 52–59).
—reprints: as pamphlet entitled *The Scientific Aspect of the Supernatural: Indicating the Desirableness of an Experimental Enquiry by Men of Science into the Alleged Powers of Clairvoyants and Mediums* (F. Farrah, London, 1866; pp. (i)–(vi), (1)–57); translated into German by G. C. Wittig and reprinted as pamphlet entitled *Die Wissenschaftliche Ansicht des Uebernatürlichen* (Leipzig, 1874; pp. xiv, 128); long excerpt printed as 'Mesmerism, Clairvoyance & Spiritualism' in *The Spiritualist* (London) 1(19): 145a–148c (15 March 1871); expanded version printed in *On Miracles and Modern Spiritualism*: 29–137 (March 1875).

119.* Address [given in Nottingham on 23 Aug. 1866 as President of the Dept. of Anthropology, Section D, Biology, of the BAAS]. in "Notices and Abstracts of Miscellaneous Communications to the Sections" portion of the *Rept. BAAS* 36 (1866) (John Murray, London, 1867): 93–94.
—reprints: as part of an article/N entitled 'Anthropology at the British

Association' in *Anthro. Rev.* 4: 389–91 (386–408) (Oct. 1866: no. 15); untitled, in *Athenaeum* no. 2027: 278 (1 Sept. 1866); summary/N of address printed in *Reader* 7: 768c (1 Sept. 1866: no. 192).

120.* discussion/N [of various remarks made by others while serving as President of the Aug. 1866 meeting of the Dept. of Anthropology, Section D, Biology, of the BAAS; part of an article/N entitled 'Anthropology at the British Association']. *Anthro. Rev.* 4: 391–408 (386–408) (Oct. 1866: no. 15).

121.* On Reversed Sexual Characters in a Butterfly, and their Interpretation on the Theory of Modifications and Adaptive Mimicry (Illustrated by Specimens) [a paper read in Nottingham at the 27 Aug. 1866 meeting of Section D, Biology, of the BAAS]. abstract/P printed in "Notices and Abstracts of Miscellaneous Communications to the Sections" portion of the *Rept. BAAS* 36 (1866) (John Murray, London, 1867): 79.

—reprints: different abstract/U, followed by an account of related discussion, printed in *Athenaeum* no. 2028: 309 (309) (8 Sept. 1866), in *Reader* 7: 847a (847a) (6 Oct. 1866: no. 197), and in *Annual of Scientific Discovery* (for 1866 and 1867) ed. by Samuel Kneeland (Gould & Lincoln, Boston, 1867): 310–311 (310–311); discussion/N of the same subject at the ESL meeting of 19 Nov. 1866 printed in *J. Proc. ESL, 1866–1867*: xxxviii–xxxix (xxxviii–xli); this then reprinted in *Zoologist* 2 (2nd s.): 573–574 (573–576) (Feb. 1867).

122.* discussion/N [of 'On Flint Implements Recently Discovered in North Devon,' a paper by H. S. Ellis read in Nottingham at the Aug. 1866 meeting of the Dept. of Anthropology, Section D, Biology, of the BAAS]. *Reader* 7: 863c (13 Oct. 1866: no. 198).

123.* Natural Selection [lttE concerning mimicry]. *Athenaeum* no. 2040: 716–717 (1 Dec. 1866).

124. Ice Marks in North Wales (With a Sketch of Glacial Theories and Controversies). *Q. J. Science* 4: 33–51 (Jan. 1867: no. 13).

125.* Mr. Wallace on Natural Selection Applied to Anthropology [lttE]. *Anthro. Rev.* 5: 103–105 (Jan. 1867: no. 16).

126.* Postscript by Alfred R. Wallace [to an account of a séance written by Francis Sims, Wallace's brother-in-law, entitled 'A New Medium']. *The Spiritual Mag.* 2 (n.s.): 51–52 (49–52) (1 Feb. 1867).

—reprint: as part of note/N entitled 'Science and Spiritualism' in *Anthro. Rev.* 5: 242–243 (242–243) (April 1867: no. 17).

127. On the Pieridae of the Indian and Australian Regions [a paper read at the ESL meeting of 18 Feb. 1867]. *Trans. ESL* 4 (3rd s.), part III: 301–416 (& 4 colour plates) (Nov. 1867).

128.* discussion/N [of paper concerning a tributary of the Rio Purus in Peru by W. Chandless read at the RGS meeting of 25 Feb. 1867]. *Proc. RGS* 11 (1866–67), no. 3: 108–109 (100–110) (14 June 1867).

129.* discussion/N [Wallace's explanation of brilliant colours in caterpillar larvae, and others' comments thereon, presented at the ESL meeting of 4 March 1867]. *J. Proc. ESL, 1866–1867*: lxxx–lxxxi (lxxx–lxxxi).

130.* Caterpillars and Birds [lttE]. *The Field, The Country Gentleman's Newspaper* 29: 206a–b (23 March 1867: no. 743).

—reprint: in *My Life*, Vol. 2: 4–6 (Oct. 1905).

131. The Polynesians and their Migrations [review of *Les Polynésiens et leurs Migrations* by Armand de Quatrefages de Bréau, 1864]. *Q. J. Science* 4: 161–166 (April 1867: no. 14).
—expanded version printed in *SSS*, Vol. 1: 399–415 (Nov. 1900).

132.* Notes of a Séance with Miss Nicholl at the House of Mr. A. S——, 15th May. *The Spiritual Mag.* 2 (n.s.): 254–255 (1 June 1867).

133.* discussion/N [of 'On Physio-anthropology, its Aim and Method,' a paper by James Hunt read at the 4 June 1867 meeting of the ASL; Wallace's discussion presented at the 18 June 1867 meeting]. *J. ASL* 5: cclvi–cclvii (ccix–cclvii) (1867).

134. Mimicry, and other Protective Resemblances among Animals [notice of works by Henry W. Bates, Alfred R. Wallace, Andrew Murray, & Charles Darwin; anonymous, but referred to in a letter from Darwin to Wallace dated 6 July 1867 (Marchant, 1916, p. 154); written in 1865–1866, according to *My Life* (Vol. 1, p. 407)]. *Westminster Rev.* 32 (n.s.), no. 1: 1–43 (London ed.) (1 July 1867: no. 173).
—reprints: revised version printed in *CTNS*: 45–129 (April 1870); this then reprinted in *NSTN*: 34–90 (May 1891); extracts from *CTNS* (1871 ed.) version reprinted in *The Naturalist as Interpreter and Seer* ed. by George Iles (Doubleday, Page & Co., New York, 1902)· 71–100.

135. A Catalogue of the Cetoniidae of the Malayan Archipelago, with Descriptions of the New Species [a paper read at the ESL meeting of 1 July 1867]. *Trans. ESL* 4 (3rd s.), part V: 519–601 (& 4 colour plates) (May 1868).
—synopsis of the new species described printed in *J. Proc. ESL, 1866–67*: xciii–xcvii.

136. The Philosophy of Birds' Nests. *Intellectual Observer* 11(6): 413–420 (July 1867).
—revised version printed in *CTNS*: 211–230 (April 1870); this then revised further and printed in *NSTN*: 98–117 (May 1891).

137.* letter [concerning séance experiences to Benjamin Coleman; described and quoted from in Coleman's article 'Passing Events—The Spread of Spiritualism']. *The Spiritual Mag.* 2 (n.s.): 349–350 (342–354) (1 Aug. 1867).

138. The Disguises of Insects. *Hardwicke's Science-Gossip* 3: 193–198 (1 Sept. (1867).
—revised version printed in *SSS*, Vol. 1: 185–198 (Nov. 1900).

139. On Birds' Nests and their Plumage; or the Relation between Sexual Differences of Colour and the Mode of Nidification in Birds [a paper read in Dundee at the 9 Sept. 1867 meeting of Section D, Biology, of the BAAS]. abstract/U printed in "Notices and Abstracts of Miscellaneous Communications to the Sections" portion of the *Rept. BAAS* 37 (1867) (John Murray, London, 1868): 97.
—reprints: entire paper printed as 'Birds' Nests and Plumage, or the Relation between Sexual Differences of Colour and the Mode of Nidification in Birds' in *The Gardeners' Chronicle and Agricultural Gazette* 12 Oct. 1867 (no. 41 for 1867): 1047–1048; greatly expanded version entitled 'A Theory of Birds' Nests: Shewing the Relation of certain Sexual Differences of Colour in Birds to their Mode of Nidification' printed in *J. Travel & Nat. Hist.* 1(2): 73–89 (March? 1868); the revised version reprinted in French as 'Théorie

sur les Relations qui Existent entre certaines Différences Sexuelles de Couleur chez les Oiseaux et leur Mode de Nidification' in *Archives des Sciences Physiques et Naturelles* 33 (nouvelle période): 5–30 (Septembre 1868: no. 129); further revised version printed in *CTNS*: 231–263 (April 1870); this then reprinted in *NSTN*: 118–140 (May 1891).

140. Creation by Law [in effect a review of *The Reign of Law* by the Duke of Argyll, 1867]. *Q. J. Science* 4: 471–488 (& 1 plate) (Oct. 1867: no. 16).
—revised version printed in *CTNS*: 264–302 (April 1870); this then reprinted in *NSTN*: 141–166 (May 1891).

141. On the Raptorial Birds of the Malay Archipelago. *Ibis* 4 (n.s.): 1–27 (& 1 colour plate) (Jan. 1868: no. 13, n.s.).

142.* Corrections of, and Additions to, the Catalogue of the Raptorial Birds of the Malay Archipelago. *Ibis* 4 (n.s.): 215–216 (April 1868: no. 14, n.s.).

143. Museums for the People. *Macmillan's Mag.* 19: 244–250 (Jan. 1869: no. 111).
—slightly revised version printed in *SSS*, Vol. 2: 1–15 (Nov. 1900).

144.* lttE [one of two printed as 'A Scientific Club']. *Scientific Opinion* 1: 378b (378b–c) (17 March 1869: no. 20).

145. Notes on Eastern Butterflies [a paper read at the ESL meeting of 5 April 1869]. *Trans. ESL 1869*, part I: 77–81 (April 1869).

146. Sir Charles Lyell on Geological Climates and the Origin of Species [running title for review of *Principles of Geology* (10th ed.), 1867–1868, and *Elements of Geology* (6th ed.), 1865, both by Sir Charles Lyell; anonymous, but referred to in *My Life* (Vol. 1, p. 406)]. *Quarterly Rev.* 126: 359–394 (April 1869: no. 252).

147. Notes on Eastern Butterflies (continued) [a paper read at the ESL meeting of 3 May 1869]. *Trans. ESL 1869*, part IV: 277–288 (Aug. 1869).

148. Notes on Eastern Butterflies (continued) [a paper read at the ESL meetings of 7 June and 5 July 1869]. *Trans. ESL 1869*, part IV: 321–349 (Aug. 1869).

149.* questions [posed as member of the Committee on Spiritualism of the London Dialectical Society; concerning evidence presented by Manual Eyre and "Miss Douglass" at the Committee meetings of 8 June 1869 & 6 July 1869, respectively]. in *Report on Spiritualism, of the Committee of the London Dialectical Society, together with the Evidence, Oral and Written, and a Selection from the Correspondence* (Longmans, Green, Reader & Dyer, London, Nov. 1871): 183 (178–184), 210 (209–211).

150.* discussion/N [of 'On the Primitive Condition of Man,' a paper by Sir John Lubbock read in Exeter at the 23 Aug. 1869 meeting of Section D, Biology, of the BAAS; part of an article/N entitled 'Anthropology at the British Association, 1869']. *Anthro. Rev.* 7: 420–421 (414–432) (Oct. 1869: no. 27).

151.* discussion/N [of 'The Occasional Definition of the Convolutions of the Brain on the Exterior of the Head,' a paper by T. S. Prideaux read in Exeter at the late August meeting of the BAAS; part of an article/N entitled 'Anthropology at the British Association, 1869']. *Anthro. Rev.* 7: 430–431 (414–432) (Oct. 1869: no. 27).

152.* The Origin of Civilization [lttE]. *Spectator* 42: 1072–1073 (11 Sept. 1869: no. 2150).

153.* The Origin of Moral Intuitions [lttE]. *Scientific Opinion* 2: 336b–337a (15 Sept. 1869: no. 46).
—reprint: most of letter incorporated into ref 165.

154. Notes on the Localities Given in *Longicornia Malayana*, with an Estimate of the Comparative Value of the Collections Made at Each of Them. *Trans. ESL* 3 (3rd s.), part VII: 691–696 (Oct. 1869).

155. The Origin of Species Controversy [review of *Habit and Intelligence in their Connexion with the Laws of Matter and Force* by Joseph John Murphy, 1869]. I. *Nature* 1: 105–107 (25 Nov. 1869) / II. 1: 132–133 (2 Dec. 1869).
—reprint: in French as 'L'Origine des Espèces.—L'Intelligence dans la Nature.—Classification des Sciences' in *Revue des Cours Scientifiques* 7: 605–608 (20 Août 1870: no. 38).

156. Introductory Remarks. to *A Catalogue of the Aculeate Hymenoptera and Ichneumonidae of India and the Eastern Archipelago* by Frederick Smith [a paper read at the LSL meeting of 16 Dec. 1869]. *J. Linn. Soc. Zool.* 11: 285–302 (1873).

157. Government Aid to Science [lttE]. *Nature* 1: 288–289 (13 Jan. 1870).
—reprint: in *My Life*, Vol. 2: 55–59 (Oct. 1905).

158. Government Aid to Science [lttE]. *Nature* 1: 315 (20 Jan. 1870).

159. The Measurement of Geological Time [according to *Island Life* (3rd ed., p. 223), first presented in 1869 as a paper to Section C, Geology, at the annual BAAS meetings]. I. *Nature* 1: 399–401 (17 Feb. 1870) / II. 1: 452–455 (3 March 1870).

160.* review [of *A Geographical Handbook of all the Known Ferns with Tables to Show their Distribution* by Katharine M. Lyell, 1870]. *Nature* 1: 428 (24 Feb. 1870).

161. review [of *Hereditary Genius: An Inquiry into its Laws and Consequences* by Francis Galton, 1869]. *Nature* 1: 501–503 (17 March 1870).

162.* lttE [one of several printed as 'Experiments on the Convexity of Water']. *The Field, The Country Gentleman's Newspaper* 35: 305a–b (305a–c) (2 April 1870: no. 901).

163.* lttE [one of two printed as 'The Convexity of Water']. *The Field, The Country Gentleman's Newspaper* 35: 317b–c (317b–c) (16 April 1870: no. 903).

164.* On Instinct in Man and Animals. in *CTNS*: 201–210 (April 1870).
—reprint: in *NSTN*: 91–97 (May 1891).

165. The Limits of Natural Selection as Applied to Man [a "further development" of the last portion of ref 146; incorporates part of ref 153]. in *CTNS*: 332–371 (April 1870).
—reprints: in *NSTN*: 186–214 (May 1891); pp. 333–360 of original *CTNS* version reprinted as second half of pamphlet entitled *The Action of Natural Selection on Man* (University Series, No. 6: Charles C. Chatfield & Co., New Haven, Aug. 1871): 29–54 (No. 6 included with Nos. 7–10 as *Half Hours with Modern Scientists. Lectures and Essays. Second Series*. first published by Charles C. Chatfield & Co., New Haven, April 1873; No. 6 later reprinted, as part of pamphlet (1882 & 1886) and as part of collection (1877), by D. Van Nostrand, New York); pp. 333–360 of original *CTNS* version reprinted as 'What Natural Selection can not Do' in *Littell's Living Age* 109 (4th s., vol. 21): 686–694 (10 June 1871: no. 1410).

166. review [of *The Handy Book of Bees; Being a Practical Treatise on their Profitable Management* by A. Pettigrew, 1870]. *Nature* 2: 82–83 (2 June 1870).

167.* discussion/N [of 'On the Geographical Distribution of the Chief Modifications of Mankind,' a paper by Thomas Huxley read at the Ethnological Society of London meeting of 7 June 1870]. *J. Ethnol. Soc. Lond.* 2 (n.s.): 411–412 (404–412) (1870).

168. Twelve-wired Bird of Paradise [lttE]. *Nature* 2: 234 (21 July 1870).

169. review [of *Researches into the Early History of Mankind and the Development of Civilization* (2nd ed.) by Edward B. Tylor, 1870]. *Nature* 2: 350–351 (1 Sept. 1870).

170. discussion/N [of 'Certain Principles to be Observed in the Establishment of a National Museum of Natural History,' a paper by P. L. Sclater read in Liverpool at the 16 Sept. 1870 meeting of Section D, Biology, of the BAAS]. *Nature* 2: 465 (465–466) (6 Oct. 1870).

171. On a Diagram of the Earth's Eccentricity and the Precession of the Equinoxes, Illustrating their Relation to Geological Climate and the Rate of Organic Change [a paper read in Liverpool at the 20 Sept. 1870 meeting of Section C, Geology, of the BAAS]. abstract/P printed in the "Notices and Abstracts of Miscellaneous Communications to the Sections" portion of the *Rept. BAAS* 40 (1870) (John Murray, London, 1871): 89.
—same abstract printed in *Athenaeum* no. 2242: 500–501 (15 Oct. 1870).

172. The Glaciation of Brazil [review of *Thayer Expedition: Scientific Results of a Journey in Brazil. By Louis Agassiz and his Travelling Companions. Geology and Physical Geography of Brazil.* by Charles Frederick Hartt, 1870]. *Nature* 2: 510–512 (27 Oct. 1870).

173. Man and Natural Selection. *Nature* 3: 8–9 (3 Nov. 1870).

174.* An Answer to the Arguments of Hume, Lecky, and Others, against Miracles [a paper read in London as the first of a series of weekly "winter soirées" on 14 Nov. 1870]. *The Spiritualist* (London) 1(15): 113c–116b (15 Nov. 1870). [account/N of discussion following the paper's presentation printed as 'The Meetings at the Beethoven Rooms' in *The Spiritual News* (London) 1(1): 2 (1 Dec. 1870)]
—reprints: as a privately printed pamphlet (London, 1871; pp. (1)–24); as 'No Antecedent Impossibility in Miracles. A Reply to Modern Objectors.' in *The Spiritual Mag.* 7 (n.s.): 112–129 (March 1872); revised version of paper printed in *On Miracles and Modern Spiritualism*: 1–28 (March 1875); abridged version of paper printed as 'Modern Objections to Miracles' in *Report on Spiritualism, of the Committee of the London Dialectical Society, together with the Evidence, Oral and Written, and a Selection from the Correspondence* (Longmans, Green, Reader & Dyer, London, Nov. 1871): 83–90 (preceded by introductory letter on p. 82).

175. Natural Selection—Mr Wallace's Reply to Mr. Bennett. *Nature* 3: 49–50 (17 Nov. 1870).
—excerpt reprinted in *My Life*, Vol. 2: 8–9 (Oct. 1905).

176. lttE [one of several printed as 'The Difficulties of Natural Selection']. *Nature* 3: 85–86 (85–86) (1 Dec. 1870).

177. The Difficulties of Natural Selection [lttE]. *Nature* 3: 107 (8 Dec. 1870).

178.* review [of *Observations on the Geology and Zoology of Abyssinia* by W. T. Blanford, 1870]. *Academy* 2: 66–67 (15 Dec. 1870: no. 15).

179. lttE [one of two printed as 'Mimicry versus Hybridity']. *Nature* 3: 165–166 (165–166) (29 Dec. 1870).

180. review [of *The Intelligence and Perfectibility of Animals from a Philosophic Point of View. With a Few Letters on Man.* by Charles Georges Leroy, 1870]. *Nature* 3: 182–183 (5 Jan. 1871).

181. The President's Address [read at the ESL annual meeting of 23 Jan. 1871]. *Proc. ESL, 1870*: xliv–lxix.
—reprints: summary/U of address printed untitled in *Nature* 3: 312 (16 Feb. 1871); a portion of address printed as 'Mr. Wallace's Anniversary Address' in *Nature* 3: 435–438 (30 March 1871); a portion of address reprinted as 'The Coleoptera of Madeira as Illustrating the Origin of Insular Faunas' in *SSS*, Vol. 1: 250–266 (Nov. 1900).

182.* review [of *A Voyage Round the World* by the Marquis de Beauvoir, 1870]. *Nature* 3: 244 (26 Jan. 1871).

183.* review [of *A Ride through the Disturbed Districts of New Zealand; Together with some Account of the South Sea Islands. Being Selections from the Journals and Letters of Lieut. the Hon. Herbert Meade, R.N. Edited by his Brother.*, 1870]. *Nature* 3: 264 (2 Feb. 1871).

184. The Theory of Glacial Motion [comments on two papers by James Croll: 'On Ocean Currents' and 'On the Cause of the Motion of Glaciers']. *Nature* 3: 309–310 (16 Feb. 1871).

185. The Metamorphoses of Insects [review of *The Transformations (or Metamorphoses) of Insects (Insecta, Myriapoda, Arachnida, and Crustacea)* by M. Émile Blanchard, adapted to English by P. Martin Duncan, 1870]. *Nature* 3: 329–331 (23 Feb. 1871).

186. review [of *The Descent of Man and Selection in Relation to Sex* by Charles Darwin, 1871]. *Academy* 2: 177–183 (15 March 1871: no. 20).
—reprint: as pamphlet entitled *Mr. Wallace and Mr. Darwin. A Review and Criticism of Mr. Darwin's "Descent of Man."* (August Brentano, New York, Aug. 1871; pp. 36).

187. review [of *The Honey Bee: Its Natural History, Physiology, and Management* (rev. ed.) by Edward Bevan, 1870]. *Nature* 3: 385 (16 March 1871).

188.* discussion/N [of 'On Additions to the Atlantic Coleoptera,' a paper by T. Vernon Wollaston read at the ESL meeting of 20 March 1871]. *Proc. ESL, 1871*: xiii (xi–xiii) (1871).

189. review [of *A Monograph of the Alcedinidae: or, Family of Kingfishers* by Richard B. Sharpe, 1868–1871]. *Nature* 3: 466–467 (13 April 1871).

190.* review [of *A History of the Birds of Europe, Including all the Species Inhabiting the Western Palaearctic Region, Part 1* by Richard B. Sharpe & Henry E. Dresser, 1871]. *Nature* 3: 505 (27 April 1871).

191.* On the Attitude of Men of Science towards the Investigators of Spiritualism. in *The Year-book of Spiritualism for 1871* ed. by Hudson Tuttle & J. M. Peebles (William White & Co., Boston, April 1871): 28–31.

192. review [of *British Insects. A Familiar Description of the Form, Structure, Habits, and*

Transformations of Insects. by E. F. Staveley, 1871]. *Nature* 4: 22–24 (11 May 1871).

193.* review [of *Natural History of the Azores, or Western Islands* by Frederick Du Cane Godman, 1870]. *Academy* 2: 266–267 (15 May 1871: no. 24).

194.* review [of *Notes of a Naturalist in the Nile Valley and Malta* by Andrew Leith Adams, 1870]. *Academy* 2: 336–337 (1 July 1871: no. 27).

195.* review [of *Journeys in North China, Manchuria, and Eastern Mongolia; with some Account of Corea* by Rev. Alexander Williamson, 1870]. *Academy* 2: 337 (1 July 1871: no. 27).

196. Bastian on the Origin of Life [review of *The Modes of Origin of Lowest Organisms: Including a Discussion of the Experiments of M. Pasteur, and a Reply to some Statements by Professors Huxley and Tyndall* by H. Charlton Bastian, 1871]. *Nature* 4: 178–179 (6 July 1871).

197. lttE [one of two printed as 'A New View of Darwinism']. *Nature* 4: 181 (180–181) (6 July 1871).

198. lttE [one of several printed as 'Mr. Howorth on Darwinism']. *Nature* 4: 221 (221–222) (20 July 1871).

199. lttE [one of several printed as 'Recent Neologisms']. *Nature* 4: 222 (222) (20 July 1871).

200. review [of *At Last: A Christmas in the West Indies* by Charles Kingsley, 1871]. *Nature* 4: 282–284 (10 Aug. 1871).

201.* review [of *Insects at Home: Being a Popular Account of British Insects, their Structures, Habits, and Transformations* by Rev. John G. Wood, 1871]. *Nature* 5: 65–68 (23 Nov. 1871).

202. *Reply to Mr. Hampden's Charges against Mr. Wallace* [a privately printed pamphlet responding to John Hampden's slanderous remarks: see refs 162 & 163, *My Life* (Vol. 2, pp. 364–376), & Schadewald (1978)]. J. J. Tiver, London, Nov. 1871; pp. (1)–8.
—includes two letters to Hampden earlier printed by the latter in his pamphlet entitled *Is Water Level or Convex after All?* (1870).

203.* review [of *A Synonymic Catalogue of Diurnal Lepidoptera* by William F. Kirby, 1871]. *Academy* 2: 538 (1 Dec. 1871: no. 37).

204.* letter [to James Edmunds; included as part of a letter by Edmunds entitled 'The Paper by Dr. Edmunds']. *The Spiritualist* (London) 1(28): 221b (221b–c) (15 Dec. 1871).

205. The President's Address [read at the ESL's annual meeting of 22 Jan. 1872]. *Proc. ESL, 1871*: li–lxxv.
—reprints: extract from address printed as 'Wallace on the Origin of Insects' in *Nature* 5: 350–351 (29 Feb. 1872); another extract printed as 'Death of Professor Ratzeburg' in *The Entomologist* 6: 55–56 (March 1872: no. 101).

206.* Dr. Carpenter and Psychic Force [lttE originally submitted to, but not printed by, *The Daily Telegraph*]. *The Spiritualist* (London) 2(2): 13a (15 Feb. 1872: no. 30).

207.* review [of *Primitive Culture: Researches into the Development of Mythology, Philosophy, Religion, Art, and Custom* by Edward B. Tylor, 1871]. *Academy* 3: 69–71 (15 Feb. 1872: no. 42).

—extensive excerpts included in 'Mr. Alfred R. Wallace on Expectant Attention and Popular Superstitions, and on the Difference between Mesmeric and Spiritual Phenomena,' a notice printed in *The Spiritual Mag.* 7 (n.s.): 145–151 (April 1872).

208. Ethnology and Spiritualism [lttE responding to Edward Tylor's comments (*Nature* 5: 343) on ref 207]. *Nature* 5: 363–364 (7 March 1872).
—reprints: as 'The Objective Nature of Spiritual Phenomena' in *The Spiritualist* (London) 2(3): 17a–b (15 March 1872: no. 31); extensive excerpts included in 'Mr. Alfred R. Wallace on Expectant Attention and Popular Superstitions, and on the Difference between Mesmeric and Spiritual Phenomena,' a notice/N printed in *The Spiritual Mag.* 7 (n.s.): 145–151 (April 1872).

209.* review [of *The Debatable Land between this World and the Next* by Robert Dale Owen, 1871]. *Q. J. Science* 2 (n.s.; 9, o.s.): 237–247 (April 1872: no. 34, o.s.).
—reprints: excerpts printed as 'Mr. Wallace on Spiritual Phenomena' in *The Spiritualist* (London) 2(4): 27b-c (15 April 1872: no. 32); excerpt printed as 'Mr. Alfred Wallace on the Relation of Spiritualism to Science' in *The Spiritual Mag.* 7 (n.s.): 231 (May 1872).

210. The Last Attack on Darwinism [review of *An Exposition of Fallacies in the Hypothesis of Mr. Darwin* by Charles R. Bree, 1872]. *Nature* 6: 237–239 (25 July 1872).

211. review [of *The Beginnings of Life: Being some Account of the Nature, Modes of Origin, and Transformations of Lower Organisms* by H. Charlton Bastian, 1872]. I. *Nature* 6: 284–287 (8 Aug. 1872) / II. 6: 299–303 (15 Aug. 1872).

212. discussion/N [of 'On the Diversity of Evolution under Uniform External Conditions,' a paper by Rev. John T. Gulick read in Brighton at the 19 Aug. 1872 meeting of the Dept. of Zoology and Botany, Section D, Biology, of the BAAS]. *Nature* 6: 407 (406–407) (12 Sept. 1872).
—a shorter summary/N printed in *Athenaeum* no. 2340: 274 (274) (31 Aug. 1872).

213.* discussion/N [of 'On the Scientific Value of Beauty,' a paper by F. T. Mott read in Brighton at the 20 Aug. 1872 meeting of the Dept. of Zoology and Botany, Section D, Biology, of the BAAS]. *Athenaeum* no. 2340: 275 (275) (31 Aug. 1872).

214. lttE [one of two printed as 'Ocean Circulation']. *Nature* 6: 328–329 (328–329) (22 Aug. 1872).

215.* review [of *The Birds of Europe, Parts 11 & 12* by Richard B. Sharpe & Henry E. Dresser]. *Nature* 6: 390–391 (12 Sept. 1872).

216.* Instinct [lttE]. *The Spiritualist* (London) 2(9): 70c (15 Sept. 1872: no. 37).

217. Houzeau on the Faculties of Man and Animals [review of *Études sur les Facultés Mentales des Animaux Comparées à celles de l'Homme, par un Voyageur Naturaliste* by Jean Charles Houzeau, 1872]. *Nature* 6: 469–471 (10 Oct. 1872).

218. Misleading Cyclopaedias [lttE]. *Nature* 7: 68 (28 Nov. 1872: no. 161).

219. Spiritualism and Science [letter to the Editor]. *The Times* (London) no. 27578: 10e (4 Jan. 1873).

—reprints: as 'Mr. A. R. Wallace on Spiritual Phenomena' in *The Spiritualist* (London) 3(5): 73 (15 Jan. 1873: no. 44); as part of article/N entitled 'The *Times* and Spiritualism' printed in *The Medium and Daybreak* (London) 5: 18b–19a (18a–19a) (10 Jan. 1873: no. 145); excerpts reprinted as 'Testimony of Mr. A. R. Wallace' in *The Spiritualist* (London) 3(20): 307–308 (1 Sept. 1873); as 'Experiences at a Séance' in *Psychic Facts. A Selection from the Writings of Various Authors on Psychical Phenomena.* ed. by W. H. Harrison (W. H. Harrison, London, 1880): 80–82.

220.* review [of *The Expression of the Emotions in Man and Animals* by Charles Darwin, 1872]. *Q. J. Science* 3 (n.s.; 10, o.s.): 113–118 (Jan. 1873: no. 37).

221. Modern Applications of the Doctrine of Natural Selection [review of *Physics and Politics; or, Thoughts on the Application of the Principles of "Natural Selection" and "Inheritance" to Political Society* by Walter Bagehot, 1872; and *Histoire des Sciences et des Savants depuis Deux Siècles suivie d'autres Études sur les Sujets Scientifiques en Particulier sur la Sélection dans l'Espèce Humaine by* Alphonse de Candolle, 1873]. *Nature* 7: 277–279 (13 Feb. 1873: no. 172).
—reprint: as 'The Doctrine of Natural Selection' in *Popular Science Monthly* 3: 22–28 (May 1873).

222. lttE [one of several printed as 'Inherited Feeling']. *Nature* 7: 303 (303) (20 Feb. 1873: no. 173).

223. review [of *Harvesting Ants and Trap-door Spiders. Notes and Observations on their Habits and Dwellings.* by J. Traherne Moggridge, 1873]. *Nature* 7: 337 (6 March 1873: no. 175).

224. Cave-deposits of Borneo [lttE introducing remarks by A. Everett]. *Nature* 7: 461–462 (17 April 1873: no. 181).

225. Disestablishment and Disendowment: with a Proposal for a really National Church of England. *Macmillan's Mag.* 27: 498–507 (April 1873: no. 162).
—reprint: in *SSS*, Vol. 2: 235–253 (Nov. 1900).

226. lttE [one of two printed as 'East India Museum']. *Nature* 8: 5 (5–6) (1 May 1873: no. 183).

227. Perception and Instinct in the Lower Animals [lttE]. *Nature* 8: 65–66 (22 May 1873: no. 186).
—excerpts reprinted in *My Life*, Vol. 2: 391–392 (Oct. 1905).

228. Perception and Instinct in the Lower Animals [lttE]. *Nature* 8: 302 (14 Aug. 1873: no. 198).

229. review [of *Advanced Text-book of Physical Geography* (2nd ed.) by David Page, 1873]. *Nature* 8: 358–361 (4 Sept. 1873: no. 201).

230.* review [of *A History of the Birds of Europe, Parts 18, 19 & 20* by Henry E. Dresser]. *Nature* 8: 380–381 (11 Sept. 1873: no. 202).

231. Free-trade Principles and the Coal Question [lttE]. *The Daily News* (London) no. 8546: 6d–e (16 Sept. 1873).
—reprint: as 'Coal A National Trust' in *SSS*, Vol. 2: 138–144 (Nov. 1900).

232. African Travel [review of *The Lands of Cazembe. Lacerda's Journey to Cazembe in 1798, etc.* translated and annotated by Capt. R. F. Burton and published by the Royal Geographical Society, 1873; and *The African Sketch Book* by Winwood Reade, 1873]. *Nature* 8: 429–431 (25 Sept. 1873: no. 204).

233. Lyell's *Antiquity of Man* [review of *The Geological Evidences of the Antiquity of Man* (4th ed.) by Charles Lyell, 1873]. *Nature* 8: 462–464 (2 Oct. 1873: no. 205).

234. A Primeval Race [review of *A Phrenologist amongst the Todas; or, The Study of a Primitive Tribe in South India: History, Character, Customs, Religion, Infanticide, Polyandry, Language* by William E. Marshall, 1873; anonymous, but in Wallace's writing style and listed in the bibliography of Marchant (1916)]. *Athenaeum* no. 2403: 624–625 (15 Nov. 1873).

235. Meyer's Exploration of New Guinea [lttE]. *Nature* 9: 102 (11 Dec. 1873: no. 215).

236. Limitation of State Functions in the Administration of Justice. *Contemporary Rev.* 23: 43–52 (Dec. 1873).
—revised version printed in *SSS*, Vol. 2: 150–166 (Nov. 1900).

237. The Origin of Man and of Civilisation [review of *Man and Apes: An Exposition of Structural Resemblances and Differences Bearing upon Questions of Affinity and Origin* by St. George Mivart, 1873; and 'On the Origin of Savage Life: Opening Address Read before the Literary and Philosophical Society of Liverpool, Oct. 6th, 1873' by Albert J. Mott, President]. *Academy* 5: 66–67 (17 Jan. 1874: no. 89).

238. review [of *The Naturalist in Nicaragua* by Thomas Belt, 1873]. *Nature* 9: 218–221 (22 Jan. 1874: no. 221).

239. review [of *The Object and Method of Zoological Nomenclature* by David Sharp, 1873]. *Nature* 9: 258–260 (5 Feb. 1874: no. 223).

240. lttE [one of two printed as 'Animal Locomotion']. *Nature* 9: 301 (301) (19 Feb. 1874: no. 225).

241. Animal Locomotion [lttE]. *Nature* 9: 403 (26 March 1874: no. 230).

242. review [of *Darwinism and Design; or Creation by Evolution* by George St. Clair, 1873; anonymous, but in Wallace's writing style and appears at time listed in the bibliography of Marchant (1916)]. *Spectator* 47: 535–536 (25 April 1874: no. 2391).

243.* A Defence of Modern Spiritualism. *Fortnightly Rev.* 15 (n.s.; 21, o.s.): 630–657 (1 May 1874: no. 89, n.s.) / A Defence of Modern Spiritualism. Part II. Spirit-photographs. 15 (n.s.; 21, o.s.): 785–807 (1 June 1874: no. 90, n.s.).
—reprints: *Eclectic Mag.* 20(2) (n.s.; 83(2), o.s.): 143–160 (Aug. 1874) / 20(3) (n.s.; 83(3), o.s.): 348–362 (Sept. 1874); *Banner of Light* (Boston) 35(10): 1a–d (6 June 1874) / 35(11): 1a–d (13 June 1874) / 35(12): 1a–d (20 June 1874) / 35(14): 1a–d, 2a (4 July 1874) / 35(15): 1a–d (11 July 1874); as first part of pamphlet entitled *Rise, Progress, and Defence of Modern Spiritualism* (Dunedin, 1874): 2–39; expanded version printed in *On Miracles and Modern Spiritualism*: 139–223 (March 1875); reprinted with virtually no changes as pamphlet entitled *A Defence of Modern Spiritualism. With a Preface by Epes Sargent.* (Colby & Rich, Boston, July 1874; pp. (i)–iv, (5)–63) [known reprints: 6th ed. (1882); 7th ed. (1894); 9th ed. (Banner of Light Publishing Co., Boston, 1900)]. [excerpts, summaries, etc., under various titles: *The Spiritualist* (London) 4(19): 220–221 (8 May 1874) / 5(5): 53–55 (31 July 1874); *Every Saturday* 2 (4th s.), no. 1: 16–19 (4 July 1874); *The Spiritual*

Mag. 9 (n.s.): 289–299 (July 1874); in *Rifts in the Veil. A Collection of Inspirational Poems and Essays given through Various Forms of Mediumship; also of Poems and Essays by Spiritualists.* (W. H. Harrison, London, 1878): 189–190]

244. Migration of Birds [lttE]. *Nature* 10: 459 (8 Oct. 1874: no. 258).
—excerpt reprinted in *My Life*, Vol. 2: 392–393 (Oct. 1905).

245. Automatism of Animals [lttE]. *Nature* 10: 502–503 (22 Oct. 1874: no. 260).

246. On the Arrangement of the Families Constituting the Order Passeres. *Ibis* 4 (3rd s.): 406–416 (Oct. 1874: no. 16, 3rd s.).

247.* review [of *Supplement to Harvesting Ants and Trap-door Spiders* by J. Traherne Moggridge, 1874]. *Nature* 11: 245–246 (28 Jan. 1875: no. 274).

248.* Acclimatisation. in *Encyclopaedia Britannica* (9th ed., 25 vols., Adam & Charles Black, Edinburgh, 1875–1889), Vol. 1: 84–90 (Jan. 1875).
—revised version printed as 'Acclimatization' in *Encyclopaedia Britannica* (11th ed., 29 vols., Cambridge Univ. Press, Cambridge, 1910–1911), Vol. 1: 114–119 (Appendix by Frank Finn completing the entry follows on pp. 119–121) (1910).

249.* review [of *A History of the Birds of Europe, Parts 35 & 36* by Henry E. Dresser]. *Nature* 11: 485–486 (22 April 1875: no. 286).

250. review [of *Wanderings in the Interior of New Guinea* by John A. Lawson, 1875]. *Nature* 12: 83–84 (3 June 1875: no. 292).

251.* Wallace and Spiritualism [note/N including extract from letter to Henry S. Olcott concerning Wallace's continuing interest in spiritualism]. *Banner of Light* (Boston) 37(10): 2e (5 June 1875).

252.* review [of *The Handy-book of Bees; Being a Practical Treatise on their Profitable Management* (2nd ed.) by A. Pettigrew, 1875; and *A Manual of Bee-keeping* by John Hunter, 1874]. *Nature* 12: 395 (9 Sept. 1875: no. 306).

253. review [of *Lessons from Nature, as Manifested in Mind and Matter* by St. George Mivart, 1876]. *Academy* 9: 562–563 (10 June 1876: no. 214, n.s.) / 9: 587–588 (17 June 1876: no. 215, n.s.).

254.* review [of *Notes on Collecting and Preserving Natural-history Objects* ed. by John E. Taylor, 1876]. *Nature* 14: 168 (22 June 1876: no. 347).

255.* A Sitting with Dr. Slade. *The Spiritualist* (London) 9(4): 42 (25 Aug. 1876).
—reprints: as part of 'The Testimony of Professor Wallace' by "Experience" in *North Cheshire Herald* 2 Sept. 1876: 6f (6f–g); excerpt included in note/N entitled 'Author of *Miracles and Modern Spiritualism* on Dr. Slade's Mediumship' in *The Medium and Daybreak* (London) 7: 583a–b (15 Sept. 1876: no. 337).

256.* *Geographical Distribution of Animals* [lttE correcting error made by a reviewer]. *Athenaeum* no. 2549: 311 (2 Sept. 1876).

257. Address [given in Glasgow on 6 Sept. 1876 as President of Section D, Biology, of the BAAS; divided into parts entitled 'On some Relations of Living Things to their Environment' and 'Rise and Progress of Modern Views as to the Antiquity and Origin of Man']. in "Notices and Abstracts of Miscellaneous Communications to the Sections" portion of the *Rept. BAAS* 46 (1876)

(John Murray, London, 1877): 100–119 (first part: 101–110; second part: 110–119).

—reprints: as 'Opening Address by the President, Alfred Russel Wallace' in *Nature* 14: 403–412 (7 Sept. 1876: no. 358); as 'Address at the Glasgow Meeting of the British Association, by Alfred Russel Wallace, President of the Section of Biology' in *Amer. J. Science* 12 (3rd s.): 354–385 (Nov. 1876: no. 71, 3rd s.); as Essay VII, 'By-paths in the Domain of Biology,' in *TNOE*: 249–303 (April 1878); extract from first part printed as 'The Peculiar Relations of Plants and Insects as Exhibited in Islands' in *The Entomologist* 9: 221–229 (Oct. 1876: no. 160); second part reprinted as 'Difficulties of Development as Applied to Man' in *Popular Science Monthly* 10: 60–72 (Nov. 1876); second part reprinted as 'The Antiquity and Origin of Man' in *NSTN*: 416–432 (May 1891); second part reprinted in French as 'La Civilisation Préhistorique' in *La Revue Scientifique* 11 (2e s.): 529–536 (2 Décembre 1876).

258.* The British Association at Glasgow [text of the discussion of Prof. William F. Barrett's paper on spiritualism read at the 11 Sept. 1876 meeting of the Dept. of Anthropology, Section D, Biology, of the BAAS, with the remarks of Wallace as Chair of the session]. *The Spiritualist* (London) 9(8): 88–94 (22 Sept. 1876).

—another account including a summary of Barrett's paper as well as text of the discussion following its presentation printed as 'Spiritualism at the British Association.—Meeting at Glasgow.' in *The Medium and Daybreak* (London) 7: 594b–598a (22 Sept. 1876: no. 338).

259.* lttE [dated 18 Sept., Glasgow; one of several printed as 'A Spirit Medium']. *The Times* (London) no. 28738: 4f (4f) (19 Sept. 1876).

—reprints: as part of article/N entitled 'Spiritualism and the Newspapers' in *The Spiritualist* (London) 9(8): 96 (96) (22 Sept. 1876); with other letters in 'The *Times* Controversy on the Mediumship of Dr. Slade' in *The Spiritualist* (London) 9(9): 105–106 (104–108) (29 Sept. 1876); with other letters in 'Letters and Comments on the Glasgow Discussion' in *The Medium and Daybreak* (London) 7: 598b (598a–599b) (22 Sept. 1876: no. 338); with other letters as 'Controversy concerning Dr. Slade' in *Banner of Light* (Boston) 40(2): 8a (8a–b) (7 Oct. 1876).

260. Erratum in Mr. Wallace's Address [lttE]. *Nature* 14: 473 (28 Sept. 1876: no. 361).

261.* Evidence of Mr. A. R. Wallace, President of the Biological Section of the British Association for the Advancement of Science [transcript of Wallace's testimony in the Henry Slade fraud trial; part of an article/N entitled 'Evidence in Defence of Dr. Slade']. *The Spiritualist* (London) 9(14): 161, 164 (160–161, 164–165) (3 Nov. 1876).

—reprints: above reprinted verbatim in *Banner of Light* (Boston) 40(13): 1b–c (1a–d) (23 Dec. 1876); note/N entitled 'The Slade Prosecution' including summary of Wallace's testimony printed in *The Times* (London) no. 28773: 11d–f (30 Oct. 1876).

262. Mr. Wallace and his Reviewers [lttE]. *Nature* 15: 24 (9 Nov. 1876: no. 367).

263.* Dr. Carpenter on Mesmerism, etc. [lttE]. *The Daily News* (London) no. 9559: 2c (11 Dec. 1876).
--reprints: as 'Mr. A. R. Wallace on Psychological Phenomena' in *The Spiritualist* (London) 9(20): 236 (15 Dec. 1876); as 'Dr. Carpenter on Spiritualism' in *The Medium and Daybreak* (London) 7: 796b–797a (15 Dec. 1876: no. 350).

264.* Dr. Carpenter on Spiritualism [lttE]. *The Daily News* (London) no. 9566: 3f (19 Dec. 1876).
—reprint: *The Medium and Daybreak* (London) 7: 804b–805a (22 Dec. 1876: no. 351).

265.* Mr. G. H. Lewes's Exposure of Mrs. Hayden [lttE]. *Spectator* 49: 1608 (23 Dec. 1876: no. 2530).
—reprint: as Appendix O, 'Letter from Mr. A. R. Wallace to the Editor of *The Spectator*, Dec. 23, 1876,' of *Mesmerism, Spiritualism, etc., Historically & Scientifically Considered* by William B. Carpenter (Longmans, Green, & Co., London, 1877): 153–154.

266. review [of *The Races of Man and their Geographical Distribution* (2nd ed.) from the German of Oscar Peschel, 1876]. *Nature* 15: 174–176 (28 Dec. 1876: no. 374).

267. Glacial Drift in California [lttE]. *Nature* 15: 274–275 (25 Jan. 1877: no. 378).

268. The "Hog-wallows" of California [lttE]. *Nature* 15: 431–432 (15 March 1877: no 385).

269. The Comparative Antiquity of Continents, as Indicated by the Distribution of Living and Extinct Animals [a paper read at the RGS meeting of 25 June 1877]. *Proc. RGS* 21 (1876–77), no. 6: 505–534 (followed by an account of related discussion on pp. 534–535) (19 Sept. 1877).
—reprints: with introductory portion re-written as Essay VIII, 'The Distribution of Animals as Indicating Geographical Changes' in *TNOE*: 304–347 (April 1878); summary/P printed in *The Geographical Mag.* 4: 220–222 (1 Aug. 1877).

270. review [of *Mesmerism, Spiritualism, etc., Historically & Scientifically Considered* by William B. Carpenter, 1877]. *Q. J. Science* 7 (n.s.; 14, o.s.): 391–416 (July 1877).
—reprints: as 'Dr. Carpenter on Spiritualism' in Supplement V of *Popular Science Monthly Supplement I–VI*: 435–450 (1877); as same in *Eclectic Mag.* 26(5) (n.s.; 89(5), o.s.): 538–555 (Nov. 1877); as same in *The Psycho-Physiological Sciences, and their Assailants* (Colby & Rich, Boston, 1878): 7–36.

271.* Spiritualism and Conjurors [lttE]. *The Spiritualist* (London) 11(7): 78 (17 August 1877: no. 260).

272. The Colours of Animals and Plants. I.—The Colours of Animals. *Macmillan's Mag.* 36: 384–408 (Sept. 1877: no. 215) / The Colours of Animals and Plants. II.—The Colours of Plants. 36: 464–471 (Oct. 1877: no. 216).
—reprints: in Supplement VI of *Popular Science Monthly Supplement I–VI* (1877): 522–541 / Supplement VII of *Popular Science Monthly Supplement VII–XII* (1878): 43–49; *Littell's Living Age* 135 (5th s., vol. 20): 67–86 (13

Oct. 1877: no. 1739) / 135 (5th s., vol. 20): 352–358 (10 Nov. 1877: no. 1743); reorganized version of the two parts reprinted in *Amer. Naturalist* 11(11): 641–662 (Nov. 1877) / 11(12): 713–728 (Dec. 1877); in French as 'Les Couleurs des Animaux et des Plantes' in *La Revue Britannique* 6 (nouvelle série): 67–118 (Novembre 1877); expanded version of Part I entitled 'The Colours of Animals and Sexual Selection' printed as Essay V of *TNOE*: 158–220 (April 1878); this then reprinted with further additions in *NSTN*: 338–394 (May 1891); expanded version of Part II entitled 'The Colours of Plants and the Origin of the Colour-sense' printed as Essay VI of *TNOE*: 221–248 (April 1878); this then reprinted with further additions in *NSTN*: 395–415 (May 1891).

273.* Slate-writing Extraordinary [lttE]. *Spectator* 50: 1239–1240 (6 Oct. 1877: no. 2571).
—reprints: *The Spiritualist* (London) 11(15): 174 (12 Oct. 1877); *The Medium and Daybreak* (London) 8: 641a (12 Oct. 1877: no. 393); in *Psychography: A Treatise on One of the Objective Forms of Psychic or Spiritual Phenomena* by M. A. (Oxon.) [pseudonym of William Stainton Moses] (W. H. Harrison, London, 1878): 104–105; excerpt reprinted in 'The Possibilities of Mal-observation in Relation to Evidence for the Phenomena of Spiritualism' [a paper by C. C. Massey read at the SPR meeting of 5 July 1886] in *Proc. SPR* 4 (1886–87): 84 (75–99) (Part X, 1887).

274.* Hartlaub's Birds of Madagascar [review of *Die Vögel Madagascars und der Benachbarten Inselgruppen. Ein Beitrag zur Zoologie der Äthiopischen Region.* by Gustav Hartlaub, 1877; anonymous, but obliquely referred to by Alfred Newton (*Nature* 17: 9–10), to whose comments Wallace later responded in ref 280]. *Nature* 16: 498–499 (11 Oct. 1877: no. 415).

275. The Zoological Relations of Madagascar and Africa [lttE]. *Nature* 16: 548 (25 Oct. 1877: no. 417).

276. lttE [concerning remarks by William B. Carpenter; one of several printed as 'Mr. Wallace and Reichenbach's Odyle']. *Nature* 17: 8 (8–9) (1 Nov. 1877: no. 418).

277. lttE [concerning remarks by William B. Carpenter; one of several printed as 'The Radiometer and its Lessons']. *Nature* 17: 44 (43–44) (15 Nov. 1877: no. 420).
—reprint: as 'Mr. Wallace on Dr. Carpenter' in *Banner of Light* (Boston) 42(11): 8b (8 Dec. 1877).

278. Bees Killed by Tritoma [lttE]. *Nature* 17: 45 (15 Nov. 1877: no. 420).

279. Humming-birds. *Fortnightly Rev.* 22 (n.s.; 28, o.s.): 773–791 (1 Dec. 1877: no. 132, n.s.).
—reprints: *Littell's Living Age* 136 (5th s., vol. 21): 3–14 (5 Jan. 1878: no. 1751); expanded version printed as Essay IV of *TNOE*: 124–157 (April 1878); this then reprinted in *NSTN*: 312–337 (May 1891).

280. The Comparative Richness of Faunas and Floras Tested Numerically [lttE]. *Nature* 17: 100–101 (6 Dec. 1877: no. 423).

281. Mr. Crookes and Eva Fay [lttE concerning William B. Carpenter's comments on the medium Eva Fay]. *Nature* 17: 101 (6 Dec. 1877: no. 423).

282.* Test Materialisation Séance with Mr. W. Eglinton [signed "Alfred R. Wallace, William Tebb, William Williams Clark"]. *The Spiritualist* (London) 11(23): 272 (7 Dec. 1877).
—reprint: *Banner of Light* (Boston) 42(14): 1d (29 Dec. 1877).

283. Psychological Curiosities of Scepticism. A Reply to Dr. Carpenter. *Fraser's Mag.* 16 (n.s.): 694–706 (Dec. 1877: no. 96).
—reprints: in Supplement IX of *Popular Science Monthly Supplement VII–XII*: 230–239 (1878); in *The Psycho-Physiological Sciences, and their Assailants* (Colby & Rich, Boston, 1878): 37–55.

284. Northern Affinities of Chilian Insects [lttE]. *Nature* 17: 182–183 (3 Jan. 1878: no. 427).

285. The Curiosities of Credulity. *Athenaeum* no. 2620: 54–55 (12 Jan. 1878).
—reprint: as 'Letter from Mr. Wallace,' first half of 'Carpenter, Wallace, and Spiritualism,' in Supplement XI of *Popular Science Monthly Supplement VII–XII*: 463–464 (followed by 'Dr. Carpenter's Rejoinder' on pp. 464–465) (1878).

286.* Distribution [by Alfred Russel Wallace and William T. Thiselton Dyer]. in *Encyclopaedia Britannica* (9th ed., 25 vols., Adam & Charles Black, Edinburgh, 1875–1889), Vol. 7: 267–290 (pp. 267–286 by Wallace; pp. 286–290 by Thiselton Dyer) (Jan. 1878).
—reprint: as pamphlet entitled *The Distribution of Life, Animal and Vegetable, in Space and Time* (Humboldt Library of Popular Science Literature, Vol. 6, No. 64: J. Fitzgerald, New York, Jan. 1885; pp. (1)–48; 2 tables).

287.* Psychological Curiosities of Credulity [lttE concerning remarks by William B. Carpenter]. *Athenaeum* no. 2623: 157 (2 Feb. 1878).

288.* The Climate and Physical Aspects of the Equatorial Zone. Essay I of *TNOE*: 1–26 (April 1878).
—reprint: in *NSTN*: 217–237 (May 1891).

289.* Equatorial Vegetation. Essay II of *TNOE*: 27–68 (April 1878).
—reprint: in *NSTN*: 238–269 (May 1891).

290.* Animal Life in the Tropical Forests. Essay III of *TNOE*: 69–123 (April 1878).
—reprint: in *NSTN*: 270–311 (May 1891).

291. A Twenty Years' Error in the Geography of Australia [lttE]. *Nature* 18: 193–194 (20 June 1878: no. 451).

292. Epping Forest. *Fortnightly Rev.* 24 (n.s.; 30, o.s.): 628–645 (1 Nov. 1878: no. 143, n.s.).
—reprints: *Littell's Living Age* 139 (5th s., vol. 24): 707–717 (21 Dec. 1878: no. 1801); with a few additions as 'Epping Forest, and the Temperate Forest Regions' in *SSS*, Vol. 2: 74–98 (Nov. 1900).

293. Remarkable Local Colour-variation in Lizards [lttE introducing remarks by the Baron de Basterot on colour-variation in Capri lizards]. *Nature* 19: 4 (7 Nov. 1878: no. 472).

294.* *Scilla autumnalis* in Essex [note on distribution]. *J. of Botany* 7 (n.s.): 346 (Nov. 1878).

295. lttE [one of two printed as 'The Formation of Mountains']. *Nature* 19: 121 (121) (12 Dec. 1878: no. 476).

296.* discussion/N [of paper on the Australasian–Pacific races of man by S. J. Whitmee read at the Anthropological Institute meeting of 7 Jan. 1879]. *J. Anthro. Instit. Great Britain & Ireland* 8: 367–368 (360–369) (1879).

297.* discussion/N [of paper on New Guinea tribes by W. G. Lawes read at the Anthropological Institute meeting of 7 Jan. 1879]. *J. Anthro. Instit. Great Britain & Ireland* 8: 377 (369–377) (1879).

298. The Formation of Mountains [lttE]. *Nature* 19: 244 (16 Jan. 1879: no. 481).

299.* review [of *Leisure-time Studies, Chiefly Biological: A Series of Essays and Lectures* by Andrew Wilson, 1878]. *Nature* 19: 286–287 (30 Jan. 1879: no. 483).

300. lttE [one of two printed as 'The Formation of Mountains']. *Nature* 19: 289 (289) (30 Jan. 1879: no. 483).

301. New Guinea and its Inhabitants. *Contemporary Rev.* 34: 421–441 (Feb. 1879).
—reprints: *Littell's Living Age* 140 (5th s., vol. 25): 771–783 (29 March 1879: no. 1815); in two parts in *Popular Science Monthly* 14: 742–750 (April 1879) / 15: 57–68 (May 1879); expanded version printed in *SSS*, Vol. 1: 416–460 (Nov. 1900).

302.* Animals and their Native Countries. *Nineteenth Century* 5: 247–259 (Feb. 1879: no. 24).
—reprint: as 'Evolution and the Distribution of Animals' in *SSS*, Vol. 1: 267–284 (Nov. 1900).

303. Organisation and Intelligence [review of *Habit and Intelligence: Essays on the Laws of Life and Mind* (2nd ed.) by Joseph John Murphy, 1879; and *Life and Habit* by Samuel Butler, 1877]. *Nature* 19: 477–480 (27 March 1879: no. 491).

304. Colour in Nature [review of *The Colour Sense, its Origin and Development; An Essay in Comparative Psychology* by Grant Allen, 1879]. *Nature* 19: 501–505 (3 April 1879: no. 492).

305. review [of *The Evolution of Man: A Popular Exposition of the Principal Points of Human Ontogeny and Phylogeny* by Ernst Haeckel, 1879]. *Academy* 15: 326–327 (12 April 1879: no. 362, n.s.) / 15: 351–352 (19 April 1879: no. 363, n.s.).

306.* Reciprocity the True Free Trade. *Nineteenth Century* 5: 638–649 (April 1879: no. 26).
—reprint: as 'Reciprocity the Essence of Free Trade' in *SSS*, Vol. 2: 167–187 (includes ref 312 as pp. 183–187) (Nov. 1900).

307.* Waterton's Life and Travels [review of *Wanderings in South America, the North-west of the United States, and the Antilles, in the Years 1812, 1816, 1820, and 1824* (new ed., ed. by Rev. John G. Wood) by Charles Waterton, 1879]. *Nature* 19: 576–578 (24 April 1879: no. 495).

308. discussion [of comments by Grant Allen on ref 304; printed with Allen's letter as 'Colour in Nature']. *Nature* 19: 581 (580–581) (24 April 1879: no. 495).

309. Did Flowers Exist during the Carboniferous Epoch? [lttE]. *Nature* 19: 582 (24 April 1879: no. 495).

310.* Bounties and Countervailing Duties [lttE followed by reply by the Editor]. *Spectator* 52: 531 (531) (26 April 1879: no. 2652).

311. review [of *Evolution, Old and New; or, The Theories of Buffon, Dr. Erasmus Darwin, and Lamarck, as Compared with that of Mr. Charles Darwin* by Samuel Butler, 1879]. *Nature* 20: 141–144 (12 June 1879: no. 502).

312. A Few Words in Reply to Mr. Lowe. *Nineteenth Century* 6: 179–181 (July 1879: no. 29).
—reprint: in *SSS*, Vol. 2: 183–187 (Nov. 1900).

313. Glacial Epochs and Warm Polar Climates [running title for notice of books on glacial theory by James Croll, James Geikie, George S. Nares, Oswald Heer & Julius Payer; anonymous, but referred to in a letter from Wallace to Darwin dated 9 Jan. 1880 (Marchant, 1916, p. 250)]. *Quarterly Rev.* 148: 119–135 (July 1879: no. 295).
—reprint: portions incorporated into Chapters 7–9 of *Island Life* (Oct. 1880).

314.* review [of *Scientific Lectures* by Sir John Lubbock, 1879]. *Nature* 20: 335–336 (7 Aug. 1879: no. 510).

315. review [of *The Natural History of the Agricultural Ant of Texas* by Henry Christopher McCook, 1879]. *Nature* 20: 501 (25 Sept. 1879: no. 517).

316.* Islands, as Illustrating the Laws of the Geographical Distribution of Animals [from a lecture delivered in the Hulme Town Hall, Manchester, 15 Oct. 1879]. in *Science Lectures for the People.* Eleventh Series (Manchester & London, April 1880): No. 1 (Jan. 1880; pp. 1–18).

317. Wallace's *Australasia* [lttE]. *Nature* 20: 625–626 (30 Oct. 1879: no. 522).

318.* The Protective Colours of Animals. in *Science for All* ed. by Robert Brown (5 vols., Cassell, Petter, Galpin & Co., London, Paris & New York, 1877–1882), Vol. 2: 128–137 (Oct. 1879).

319.* Protective Mimicry in Animals. in *Science for All* ed. by Robert Brown (5 vols., Cassell, Petter, Galpin & Co., London, Paris & New York, 1877–1882), Vol. 2: 284–296 (Oct. 1879).

320.* discussion/N [of paper on a Dutch expedition to Sumatra by P. J. Veth read at the RGS meeting of 10 Nov. 1879]. *Proc. RGS & Monthly Record of Geography* 1(12) (new monthly s.): 775–777 (759–777) (Dec. 1879).

321.* Popular Natural History [review of *Animal Life; Being a Series of Descriptions of the Various Sub-kingdoms of the Animal Kingdom* by E. Perceval Wright, 1879]. *Nature* 21: 232–235 (8 Jan. 1880: no. 532).

322. The Origin of Species and Genera. *Nineteenth Century* 7: 93–106 (Jan. 1880: no. 35).
—reprints: *Eclectic Mag.* 31(3) (n.s.; 94(3), o.s.): 317–326 (March 1880); in French as 'L'Origine des Espèces et des Genres' in *La Revue Scientifique* 18 (2e s.): 672–678 (17 Janvier 1880); revised version printed in *SSS*, Vol. 1: 285–304 (Nov. 1900).

323. Wallace's *Australasia* [lttE responding to published criticism]. *Nature* 21: 562 (15 April 1880: no. 546).

324. Two Darwinian Essays [review of *Studies in the Theory of Descent Pt. I* by August Weismann, 1880; and *Degeneration: A Chapter in Darwinism* by E. Ray Lankester, 1880]. *Nature* 22: 141–142 (17 June 1880: no. 555).

325.* Dr. Croll's Excentricity Theory [lttE]. *Geological Mag.* 7 (Decade II, n.s.), no. 6: 284–285 (June 1880).

326.* European Caddis-flies [review of *A Monographic Revision and Synopsis of the Trichoptera of the European Fauna* by Robert McLachlan, 1874–1880]. *Nature* 22: 314–315 (5 Aug. 1880: no. 562).

327.* English and American Bee-keeping [review of books on bee-keeping by Henry Taylor (1880), James Robinson (1880), and Albert J. Cook (1879)]. *Nature* 22: 433–434 (9 Sept. 1880: no. 567).

328. The Emotions and Senses of Insects [review of *Insect Variety, its Propagation and Distribution* by Archibald H. Swinton, 1880]. *Academy* 18: 294–296 (23 Oct. 1880: no. 442, n.s.).

329. How to Nationalize the Land: A Radical Solution of the Irish Land Problem. *Contemporary Rev.* 38: 716–736 (Nov. 1880).
—reprinted with note in *SSS*, Vol. 2: 265–295 (Nov. 1900).

330. review [of *Siberia in Europe: A Visit to the Valley of the Petchora, in North-east Russia. With Descriptions of the Natural History, Migrations of Birds, etc.* by Henry Seebohm, 1880]. *Academy* 18: 408–409 (4 Dec. 1880: no. 448, n.s.).

331. Geological Climates [lttE]. *Nature* 23: 124 (9 Dec. 1880: no. 580).

332. review [of *New Guinea: What I Did and What I Saw* by Luigi M. D'Albertis, 1880]. *Nature* 23: 152–155 (16 Dec. 1880: no. 581) / 23: 175–178 (23 Dec. 1880: no. 582).
—reprint: *Littell's Living Age* 148 (5th s., vol. 33): 186–189 (15 Jan. 1881: no. 1909).

333. Climates of Vancouver Island and Bournemouth [lttE]. *Nature* 23: 169 (23 Dec. 1880: no. 582).

334. Correction of an Error in *Island Life* [lttE]. *Nature* 23: 195 (30 Dec. 1880: no. 583).

335. lttE [one of two printed as 'Geological Climates']. *Nature* 23: 217 (217) (6 Jan. 1881: no. 584).

336. lttE [one of two printed as 'Geological Climates']. *Nature* 23: 266–267 (266–267) (20 Jan. 1881: no. 586).

337. review [of *Anthropology; An Introduction to the Study of Man and Civilization* by Edward B. Tylor, 1881]. *Nature* 24: 242–245 (14 July 1881: no. 611).

338. review [of *Studies in the Theory of Descent Pt. II* by August Weismann, 1881]. *Nature* 24: 457–458 (15 Sept. 1881: no. 620).

339.* lttE [written at the request of the Editor; one of two letters printed as 'Nationalisation of the Land']. *Mark Lane Express* 51: 1351a–b (1351a–c) (3 Oct. 1881).

340. Abstract of Four Lectures on the Natural History of Islands [from a series of lectures given at Rugby School on 6, 13, 20, & 27(?) Oct. 1881; consists of four summaries/U composed of excerpts]. *Report Rugby School Nat. Hist. Soc., 1881*: 1–17 ('Lecture I. Sea and Land' (pp. 1–4); 'Lecture II. Recent Continental Islands' (pp. 4–9); 'Lecture III. Ancient Continental Islands' (pp. 9–12); 'Lecture IV. Oceanic Islands' (pp. 12–17)) (1882).
—different but equally thorough summaries/P printed in Rugby School's magazine *The Meteor* of 3 Nov. 1881 (1st and 2nd lectures, pp. 127–129), 29 Nov. 1881 (3rd, pp. 141–142), and 21 Dec. 1881 (4th, pp. 151–152).

341.* Nationalisation of the Land [letter to, and written at the request of, the Editor]. *Mark Lane Express* 51: 1383a-c (10 Oct. 1881).

342. review [of *The Head-hunters of Borneo: A Narrative of Travel up the Mahakkam and down the Barito; also Journeyings in Sumatra* by Carl Bock, 1881]. *Nature* 25: 3–4 (3 Nov. 1881: no. 627).

343.* lttE [one of several printed as 'Nationalisation of the Land']. *Mark Lane Express* 51: 1544a–b (1544a–b) (14 Nov. 1881).

344.* The Land Question [lttE]. *The Times* (London) no. 30362: 11f (26 Nov. 1881).

345.* The Land Question [lttE]. *The Times* (London) no. 30371: 10d (7 Dec. 1881).

346.* The Land Question [lttE]. *The Times* (London) no. 30394: 3c (3 Jan. 1882).

347.* Land Nationalisation Society. Conference this Day. [note/N summarizing the meeting and Wallace's address to it]. *The Echo* (London) no. 4075: 3c (16 Jan. 1882).

348. review [of *Vignettes from Nature* by Grant Allen, 1881]. *Nature* 25: 381–382 (23 Feb. 1882: no. 643).

349. review [of *The Story of our Museum; Showing how We Formed It, and what It Taught Us* by Rev. Henry Housman, 1881]. *Nature* 25: 407–408 (2 March 1882: no. 644).

350. Monkeys. *Contemporary Rev.* 41: 417–430 (March 1882).
—reprints: *Popular Science Monthly* 21: 20–34 (May 1882); *Littell's Living Age* 152 (5th s., vol. 37): 771–780 (31 March 1882: no. 1971); *Eclectic Mag.* 35(5) (n.s.; 98(5), o.s.): 664–674 (May 1882); revised version printed as 'Monkeys: Their Affinities and Distribution' in *SSS*, Vol. 1: 146–184 (Nov. 1900).

351.* review [of *Rhopalocera Malayana: A Description of the Butterflies of the Malay Peninsula* by William L. Distant, 1882]. *Nature* 26: 6–7 (4 May 1882: no. 653).

352. review [of *Studies in the Theory of Descent Pt. III* by August Weismann, 1882]. *Nature* 26: 52–53 (18 May 1882: no. 655).

353. Dr. Fritz Müller on some Difficult Cases of Mimicry. *Nature* 26: 86–87 (25 May 1882: no. 656).

354.* review [of *Worked Examination Questions in Plane Geometrical Drawing* by Frederick E. Hulme, 1882]. *Nature* 26: 103 (1 June 1882: no. 657).

355.* review [of *Wanderings South and East* by Walter Coote, 1882; and *Pioneering in the Far East, and Journeyings to California in 1849, and the White Sea in 1878* by Ludwig Verner Helms, 1882; anonymous, but attributed to Wallace in the Index for Vol. 26]. *Nature* 26: 476 (14 Sept. 1882: no. 672).

356.* lttE [one of several printed as 'Materialisation and Exposures']. *Light* (London) 2: 447–448 (447–448) (7 Oct. 1882: no. 92).

357.* review [of *The Sportman's Handbook to Practical Collecting, Preserving, and Artistic Setting up of Trophies and Specimens* (2nd ed.) by Rowland Ward, 1882]. *Nature* 27: 146 (14 Dec. 1882: no. 685).

358.* The Debt of Science to Darwin. *Century Mag.* 25(3): 420–432 (Jan. 1883).
—reprint: in *NSTN*: 450–475 (May 1891).

359. lttE [one of two printed as 'Difficult Cases of Mimicry']. *Nature* 27: 481–482 (481–482) (22 March 1883: no. 699).

360. On the Value of the "Neoarctic" as One of the Primary Zoological Regions [lttE]. *Nature* 27: 482–483 (22 March 1883: no. 699).
—reprint: in 'On the Value of the "Nearctic" as One of the Primary Zoological Regions. Replies to Criticisms by Mr. Alfred Russel Wallace and Prof. Theodore Gill.' by Angelo Heilprin, in *Proc. Acad. Nat. Sci., Philadelphia* 35(3): 266–267 (266–275) (Nov. & Dec. 1883).

361.* President's Address, Summarised [summary/U of Wallace's address to the second annual meeting of the LNS, 27 June 1883]. in *Report of the LNS. 1881–3.* (LNS, London, 1883): (5)–6.
—note/N entitled 'Land Nationalization' describing the meeting and Wallace's address printed in *The Times* (London) no. 30859: 4f (29 June 1883).

362. review [of *Ants and their Ways* by Rev. W. Farren White, 1883]. *Nature* 28: 293–294 (26 July 1883: no. 717).

363.* review [of *Manual of Taxidermy; A Complete Guide in Collecting and Preserving Birds and Mammals* by Charles J. Maynard, 1883]. *Nature* 28: 317–318 (2 Aug. 1883: no. 718).

364.* *How Land Nationalisation will Benefit Householders, Labourers, and Mechanics.* LNS Tract No. 3: LNS, London, n.d. (probably published in 1882 or 1883); pp. (1)–7.

365. The "Why" and the "How" of Land Nationalisation. I. *Macmillan's Mag.* 48: 357–368 (Sept. 1883: no. 287) / II. 48: 485–493 (Oct. 1883: no. 288).
—reprints: as pamphlet (Macmillan & Co., London, 1883; pp. (1)–(24)); with some revision as LNS Tract No. 6 (LNS, London, n.d.; pp. (1)–32); with slight revision in *SSS*, Vol. 2: 296–332 (Nov. 1900); *SSS* version reprinted incompletely in *Land and Labour* 13(11): 98–100 (Nov. 1902) / 13(12): 107–109 (Dec. 1902) / 14(1): 2–4 (Jan. 1903) / 14(2): 14–15 (Feb. 1903).

366.* Lord Jersey on Land Nationalization [lttE]. *The Times* (London) no. 30964: 12a (30 Oct. 1883).

367.* Mr. Wallace's Reply to Mr. T. Mellard Reade on the Age of the Earth [lttE]. *Geological Mag.* 10 (Decade II, n.s.), no. 10: 478–480 (Oct. 1883).

368.* letter [of support to the Berne International Anti-Vaccination Congress held on 28 Sept. 1883 "and following days"]. *The Vaccination Inquirer and Health Review* 5: 160 (Nov. 1883: no. 56).

369.* lttE [one of several printed as 'Mr. [Henry] George's Theories']. *The Times* (London) no. 31042: 3a–b (3a–b) (29 Jan. 1884).

370. The Morality of Interest—The Tyranny of Capital. *The Christian Socialist* no. 10: 150–151 (March 1884).
—reprint: in *My Life*, Vol. 2: 244–249 (Oct. 1905).

371.* *How to Experiment in Land Nationalisation.* LNS Tract No. 8: LNS, London, n.d. (probably published in late 1883 or 1884); pp. (1)–3.

372.* Two Bee Books [review of *A Collection of Papers on Bee-keeping in India* published by the Dept. of Revenue and Agriculture, Government of India, 1883; and *The Honey-bee; Its Nature, Homes, and Products* by William H. Harris, 1884]. *Nature* 31: 1–2 (6 Nov. 1884: no. 784).

373.* review [of *Éléments de Mécanique, avec de Nombreux Exercises* by "F. I. C."]. *Nature* 31: 78 (27 Nov. 1884: no. 787).

374.* *(To Members of Parliament and Others.) Forty-five Years of Registration Statistics, Proving Vaccination to be both Useless and Dangerous.* E. W. Allen, London, Dec. 1884 or Jan. 1885; pp. (1)–38 (in two parts: 'Part I. Small-pox Mortality and Vaccination.' (pp. 3–24) and 'Part II. Comparative Mortality of the Vaccinated and the Unvaccinated.' (pp. 25–38)); fold-out.
—Second Edition ["with corrections, notes, and an Appendix by Alexander Wheeler"]. E. W. Allen, London, 1888; pp. (1)–45 (Wheeler's Appendix on pp. 41–45); fold-out.

375.* How to Cause Wealth to be more Equally Distributed [from a paper read in Prince's Hall, Piccadilly, on 30 Jan. 1885]. in *Industrial Remuneration Conference; The Report of the Proceedings and Papers* (Cassell & Co., Ltd., London, Paris, New York, & Melbourne, June 1885): 368–392.
—abstract/Y entitled 'How to Cause Wealth to be more Equably Distributed' [prepared "for the newspapers"] reprinted in *My Life*, Vol. 2: 251–253 (Oct. 1905).

376.* The Industrial Remuneration Conference [lttE "correcting misapprehension" concerning the aims and methods of the LNS]. *The Times* (London) no. 31360: 6d (3 Feb. 1885); also printed as 'The Industrial Remuneration Conference and Land Nationalisation' in *The Daily News* (London) no. 12110: 3a (3 Feb. 1885).

377.* review [of *Differential Calculus for Beginners, with a Selection of Easy Examples* by Alexander Knox, 1885]. *Nature* 31: 527 (9 April 1885: no. 806).

378. The Colours of Arctic Animals [lttE]. *Nature* 31: 552 (16 April 1885: no. 807).

379.* Modern Spiritualism. Are its Phenomena in Harmony with Science? *The Sunday Herald* (Boston) 26 April 1885: 9c–d.
—reprints: as 'Are the Phenomena of Spiritualism in Harmony with Science?' in *Light* (London) 5: 255–256 (30 May 1885: no. 230); again in *Light* (London) 33: 31–32 (18 Jan. 1913: no. 1671); under same title as a pamphlet (London Spiritualist Alliance, 1913; pp. 4); under same title but "revised and corrected by the author" in *The Medium and Daybreak* (London) 16: 809b–810b (18 Dec. 1885: no. 820); as 'Science and Spiritualism—Their Conclusions not necessarily Antagonistic' in *Banner of Light* (Boston) 57(11): 8a–b (30 May 1885); as 'Science and Spiritualism; Are the Phenomena of Spiritualism in Harmony with Science?' in *The Christian Register* (Boston) 65(9): 132a–d (4 March 1886); as part of a pamphlet entitled *Harmony of Spiritualism with Science* (Southern Science Record Office, Collingwood, Victoria, Australia, n.d.): (1)–(4). [excerpts, summaries, etc., under various titles: *J. of Science* 7 (3rd s.; 22, o.s.): 400–407 (in article by "R. M. N.") (July 1885); *Banner of Light* (Boston) 57(7): 4c (2 May 1885); *Religio-Philosophical J.* (Chicago) 38(11): 4a–b (9 May 1885)]

380.* President's Address [to the fourth annual meeting of the LNS, 13 May 1885]. in *Report of the LNS. 1884–5.* (LNS, London, 1885): (5)–15.

381. review [of *A Naturalist's Wanderings in the Eastern Archipelago, a Narrative of Travel and Exploration from 1878 to 1883* by Henry O. Forbes, 1885]. *Nature* 32: 218–220 (9 July 1885: no. 819).

382.* The *Journal of Science* on Spiritualism [a reply to criticism of ref 379 by "R. M. N."]. *Light* (London) 5: 327–328 (11 July 1885: no. 236).
—excerpt reprinted as part of a pamphlet entitled *Harmony of Spiritualism with*

Science (Southern Science Record Office, Collingwood, Victoria, Australia, n.d.): (4).

383.* Harmony of Spiritualism and Science [a reply to criticism of ref 379 by Frederick F. Cook]. *Light* (London) 5: 352 (25 July 1885: no. 238).

384. review [of *The Wanderings of Plants and Animals from their First Home* by Victor Hehn, 1885]. *Nature* 33: 170–171 (24 Dec. 1885: no. 843).

385.* *State-tenants versus Freeholders*. LNS Tract No. 15: The LNS, London, n.d. (probably published in the second half of 1885 or the first half of 1886); pp. (1)–7.
—reprints: as Appendix II in *Land Nationalisation* (new ed.): 239–245 (July 1892); as part of ref 585 in *SSS*, Vol. 2: 345–351 (Nov. 1900); reprinted from *SSS* in *Land and Labour* 12(10): 115–116 (Oct. 1901).

386.* President's Address [summary/U of Wallace's address to the fifth annual meeting of the LNS, 29 June 1886]. in *Report of the LNS. 1885–6.* (LNS, London, 1886): (3)–7.

387. The Depression of Trade, its Causes and its Remedies. in *The Claims of Labour; A Course of Lectures Delivered in Scotland in the Summer of 1886, on Various Aspects of the Labour Problem* by John Burnett and others (Co-operative Printing Co., Ltd., Edinburgh, Dec. 1886): 112–154.
—reprints: as pamphlet (Claims of Labour Lecture No. 4: Co-operative Printing Co., Ltd., Edinburgh, 1886; pp. (1)–40); in *SSS*, Vol. 2: 188–222 (Nov. 1900).

388. Central American Entomology [review of *Biologia Centrali-Americana. Insecta: Coleoptera Vol. III, Part 2, Malacodermata* by Rev. Henry Stephen Gorham, 1880–86; and *Vol. V, Longicornia* by Henry W. Bates and *Bruchides* by David Sharp, 1879–86]. *Nature* 34: 333–334 (12 Aug. 1886: no. 876).

389.* Romanes *versus* Darwin. An Episode in the History of the Evolution Theory. *Fortnightly Rev.* 40 (n.s.; 46, o.s.): 300–316 (1 Sept. 1886: no. 237, n.s.).

390. lttE [one of several printed as 'Physiological Selection and the Origin of Species']. *Nature* 34: 467–468 (467–468) (16 Sept. 1886: no. 881).

391.* *Note on Compensation to Landlords*. LNS Tract No. 16: LNS, London, n.d. (probably published in late summer or early autumn 1886, per information provided in *Report of the LNS. 1886–7.*); pp. (1)–4.

392.* review [of *Lessons in Elementary Dynamics* arranged by Henry G. Madan, 1886]. *Nature* 35: 51 (18 Nov. 1886: no. 890).

393.* Oceanic Islands: Their Physical and Biological Relations [synopsis/P of a lecture delivered at the American Geographical Society meeting of 11 Jan. 1887]. *Bull. Amer. Geogr. Soc.* 19(1): 1–17 (followed on pp. 17–21 by related material by D. Morris originally printed in *Nature*) (1887).

394. The American School of Evolutionists [review of *The Origin of the Fittest: Essays on Evolution* by Edward D. Cope, 1887]. *The Nation* (New York) 44: 121–123 (10 Feb. 1887: no. 1128).

395. Mr. Romanes on Physiological Selection [letter dated 30 Jan., Washington, U. S. A.]. *Nature* 35: 366 (17 Feb. 1887: no. 903).

396. Letter from Dr. Alfred R. Wallace, in re Mrs. Ross [dated 23 Feb., Washington, D. C.; concerning the medium Mrs. Ross]. *Banner of Light* (Boston) 60(25): 4d–e (5 March 1887).

397. review [of *The Origin of the Fittest: Essays on Evolution* by Edward D. Cope, 1887; signed 'A. H.' but referred to in *My Life* (Vol. 2, p. 133)]. *The Independent* (New York) 39: 335–336 (17 March 1887: no. 1998).

398.* If a Man Die, Shall He Live Again? [a lecture given at Metropolitan Temple, San Francisco, on 5 June 1887]. *Golden Gate* (San Francisco) (exact date of publication unknown, but undoubtedly in early June 1887).
—reprints: as pamphlet (Banner of Light Publishing Co., Boston, 1887 / Colby & Rich, Boston, 1887; pp. (1)–24); another pamphlet (E. W. Wallis, Manchester, England, 1887; pp. (1)–16); another pamphlet (Albert Morton, San Francisco, April? 1888; pp. 1–22; portrait front.); *Banner of Light* (Boston) 61(15): 1a–e, 2a–b (25 June 1887); *Harbinger of Light* (Melbourne) no. 211: 3529–3534 (1 Sept. 1887); *Light* (London) 7: 295–297 (2 July 1887: no. 339) / 7: 308–309 (9 July 1887: no. 340); *The Carrier Dove* (San Francisco) 4(7): 237b–243a (July 1887). [excerpts, summaries, etc., under various titles: *Banner of Light* (Boston) 61(14): 5a (18 June 1887); *Religio-Philosophical J.* (Chicago) 42(18): 2a–c (25 June 1887); *The Daily Alta California* no. 13793: 1f (6 June 1887); *The Daily Examiner* (San Francisco) 44(157): 2f–g (6 June 1887); *San Francisco Chronicle* 45(142): 6d (6 June 1887); *The Morning Call* (San Francisco) 6 June 1887: 8a–b]

399.* letter [extract from letter on American land taxes read at the sixth annual meeting of the LNS, 8 June 1887; dated 8 Feb., Washington, D.C.]. in *Report of the LNS. 1886–7.* (LNS, London, 1887): 7.

400.* account of slate-portrait drawing [referred to in *My Life* (Vol. 2, p. 349) and *The Medium and Daybreak* (24 June 1887)]. *Golden Gate* (San Francisco) (exact date of publication unknown, but probably in mid-June 1887).

401. American Museums. The Museum of Comparative Zoology, Harvard University. *Fortnightly Rev.* 42 (n.s.; 48, o.s.): 347–359 (1 Sept. 1887: no. 249, n.s.).
—reprint: in *SSS*, Vol. 2: 16–37 (Nov. 1900).

402. The British Museum and American Museums [lttE]. *Nature* 36: 530–531 (6 Oct. 1887: no. 936).

403.* *Land Lessons from America* [from an address delivered at Essex Hall, Essex Street, Strand, on 1 Nov. 1887]. Land Nationalisation Tract No. 18: LNS, London, n.d.; pp. (1)–15.

404.* American Museums. Museums of American Pre-historic Archaeology. *Fortnightly Rev.* 42 (n.s.; 48, o.s.): 665–675 (1 Nov. 1887: no. 251, n.s.).
—reprint: in *SSS*, Vol. 2: 37–58 (Nov. 1900).

405.* Note on American Museums and the British Museum [lttE]. *Fortnightly Rev.* 42 (n.s.; 48, o.s.): 740 (1 Nov. 1887: no. 251, n.s.).

406.* The Antiquity of Man in North America. *Nineteenth Century* 22: 667–679 (Nov. 1887: no. 129).
—reprints: *Littell's Living Age* 175 (5th s., vol. 60): 472–480 (26 Nov. 1887: no. 2265); *Eclectic Mag.* 47(1) (n.s.; 110(1), o.s.): 44–53 (Jan. 1888); in *NSTN*: 433–449 (May 1891).

407.* address [Presidential Address to the seventh annual meeting of the LNS, 8 May 1888]. in *Report of the LNS, 1887–8* (LNS, London, 1888): 10–16.

408.* replies [to questions regarding land purchase and rental posed during the seventh annual meeting of the LNS, 8 May 1888]. in *Report of the LNS, 1887–8* (LNS, London, 1888): 23–24.

409.* Account of a Spiritualistic Test. Nellie Morris. [a letter by Wallace introducing a series of letters by others concerning this subject]. *J. SPR* 3 (1887–1888): 273–274 (273–288) (June 1888: no. 51).

410.* Mr. Gulick on Divergent Evolution [lttE]. *Nature* 38: 490–491 (20 Sept. 1888: no. 986).

411.* Nellie Morris [edited discussion/N concerning the "spirit" Nellie Morris consisting of three letters by Wallace and several by others]. *J. SPR* 3 (1887–1888): 313–314, 315, 316–317 (312–318) (Oct. 1888: no. 55).
—most of the three letters reprinted in discussion entitled 'Materialisation: Evidence of Mr. Alfred Russel Wallace' in *Light* (London) 8: 530–531 (27 Oct. 1888).

412.* Introductory Note by Alfred R. Wallace. to *A Colonist's Plea for Land Nationalisation* by Arthur J. Ogilvy (Land Nationalisation Tract No. 23: LNS, London, n.d. (probably published in 1888)): (2).

413. Which are the Highest Butterflies? [lttE introducing remarks by W. H. Edwards on Colorado butterflies]. *Nature* 39: 611–612 (25 April 1889: no. 1017).

414. Address [Presidential Address to the eighth annual meeting of the LNS, 20 June 1889; includes notice of *Poverty and the State* by Herbert V. Mills (1889)]. in *Report of the LNS, 1888–9* (LNS, London, 1889): 15–23.
—excerpt reprinted in *Land and Labor* no. 4: 2 (Feb. 1890).

415. Lamarck *versus* Weismann [lttE]. *Nature* 40: 619–620 (24 Oct. 1889: no. 1043).

416.* lttE [one of two printed as 'Mr. Angelo Lewis and Dr. Monck' concerning the medium Francis W. Monck]. *J. SPR* 4 (1889–90): 143–144 (143–146) (Oct. 1889: no. 63).

417. Protective Coloration of Eggs [lttE introducing remarks by Rev. Fred. F. Grensted]. *Nature* 41: 53 (21 Nov. 1889: no. 1047).

418.* letter [responding to an inquiry regarding "*The Star*'s" (London?) comments on Wallace's support of land nationalisation and socialism]. *Land and Labor* no. 1: 7–8 (Nov. 1889).

419. The Action of Natural Selection in Producing Old Age, Decay, and Death [added as footnote to essay/N entitled 'The Duration of Life;' written "between 1865 and 1870," according to source]. in *Essays upon Heredity and Kindred Biological Problems* by August Weismann, authorised translation ed. by Edward B. Poulton, Selmar Schönland, and Arthur E. Shipley (The Clarendon Press, Oxford, Dec. 1889): 23 (5–35).

420.* testimony [the text of testimony Wallace presented before a Royal Commission on vaccination on 26 Feb., 5 March, 12 March & 21 May 1890]. in *Third Report of the Royal Commission Appointed to Inquire into the Subject of Vaccination; with Minutes of Evidence and Appendices* (printed for Her Majesty's Stationery Office by Eyre & Spottiswoode, London, 24 Dec. 1890): 6–35, 121–131 (plus Appendix 2: 'Papers Handed in by Mr. Alfred Russel Wallace 26 February and 21 May,' consisting of Diagrams and Tables A through P).

—excerpts, summaries, etc., under various titles: *The Vaccination Inquirer and Health Review* 12: 164–168 (163–176) (1 Jan. 1891: no. 142); *The Times* (London) no. 33204: 4e (4e–f) (25 Dec. 1890); as pamphlet entitled *Mr. Alfred Russel Wallace, LL.D., D.C.L., on Vaccination* (London Society for the Abolition of Compulsory Vaccination, London, n.d.; pp. (1)–4).

421.* review [of *A Naturalist among the Head-hunters. Being an Account of Three Visits to the Solomon Islands in the Years 1886, 1887, and 1888.* by Charles Morris Woodford, 1890]. *Nature* 41: 582–583 (24 April 1890: no. 1069).

422.* The Instability of Peasant-proprietorship—The Necessity of Rent [first part of an article/N entitled 'The New Round Table: Land Nationalisation' with contributions from 10 writers]. *Westminster Rev.* 133(5): 541–543 (541–559) (May 1890).

423.* *Presidential Address Delivered by Dr. Alfred R. Wallace, at the Ninth Annual Meeting, June 19th, 1890* (LNS Tract No. 33: LNS, London, June 1890; pp. (1)–12); also printed untitled in *Land and Labor* no. 9: 2–5 (July 1890).
—reprint: a portion of address reprinted as part of ref 585 in *SSS*, Vol. 2: 351–354 (Nov. 1900).

424. review [of *The Colours of Animals: Their Meaning and Use especially Considered in the Case of Insects* by Edward B. Poulton, 1890]. *Nature* 42: 289–291 (24 July 1890: no. 1082).

425. Birds and Flowers [lttE]. *Nature* 42: 295 (24 July 1890: no. 1082).

426.* Taxation or Compensation [lttE sent in response to queries regarding Wallace's LNS annual meeting address]. *The Democrat* (London) 6(8): 192 (1 Aug. 1890).
—reprint: *Land and Labor* no. 10: 3–4 (Aug. 1890).

427. Human Selection. *Fortnightly Rev.* 48 (n.s.; 54, o.s.): 325–337 (1 Sept. 1890: no. 285, n.s.).
—reprints: *Popular Science Monthly* 38: 93–106 (Nov. 1890); *Eclectic Mag.* 52(4) (n.s.; 115(4), o.s.): 547–555 (Oct. 1890); in German as 'Menschliche Auslese' in *Die Zukunft* (Berlin) 8: 10–24 (7 Juli 1894); in *SSS*, Vol. 1: 509–526 (Nov. 1900).

428. Dr. Romanes on Physiological Selection [lttE]. *Nature* 43: 79 (27 Nov. 1890: no. 1100).

429. Dr. Romanes on Physiological Selection [lttE]. *Nature* 43: 150 (18 Dec. 1890: no. 1103).

430.* Are there Objective Apparitions? *Arena* 3: 129–146 (Jan. 1891: no. 14).
—reprints: in French as 'Étude sur les Apparitions' in *Annales des Sciences Psychiques* 1: 131–147 (Mai 1891); in *Miracles and Modern Spiritualism* (3rd ed.): 231–254 (Jan. 1896); excerpts reprinted in *Light* (London) 11: 327–329 (11 July 1891: no. 549).

431.* An English Nationalist. Alfred Russell [*sic*] Wallace Converted by Bellamy's Book. [note/N containing extracts from two Wallace letters concerning Richard T. Ely's book *An Introduction to Political Economy* (1889)]. *New York Times* no. 12303: 9d (1 Feb. 1891).

432. Modern Biology and Psychology [review of *Animal Life and Intelligence* by C. Lloyd Morgan, 1890]. *Nature* 43: 337–341 (12 Feb. 1891: no. 1111).

433. Remarkable Ancient Sculptures from North-west America [notice of *Sculptured Anthropoid Ape Heads Found in or near the Valley of the John Day River, a Tributary of the Columbia River, Oregon* by James Terry, 1891]. *Nature* 43: 396 (26 Feb. 1891: no. 1113).

434.* What are Phantasms, and Why do They Appear? *Arena* 3: 257–274 (Feb. 1891: no. 15).
—reprint: in *Miracles and Modern Spiritualism* (3rd ed.): 255–278 (Jan. 1896).

435.* lttE. *Light* (London) 11: 133–134 (21 March 1891: no. 533).
—reprint: as 'A Notable Example of the Natural Organic Power through which Mediums are Influenced' in *The Two Worlds* (Manchester) 4: 435–436 (24 July 1891: no. 193).

436.* Mr. S. J. Davey's Experiments [lttE concerning the medium S. J. Davey; followed by comment by the Editor]. *J. SPR* 5 (1891–1892): 43 (43–45) (March 1891: no. 78).

437. Another Darwinian Critic [review of *On the Modification of Organisms* by David Syme, 1890]. *Nature* 43: 529–530 (9 April 1891: no. 1119).

438.* Presidential Address of Dr. Alfred Russel Wallace [to the tenth annual meeting of the LNS, 18 June 1891]. in *Report of the LNS. 1890–91.* (LNS, London, 1891): 16–23; also printed in *Land and Labor* no. 21: 1–4 (July 1891).

439.* Introductory Note. to *The History of Human Marriage* by Edward Westermarck (Macmillan & Co., London & New York, July 1891): (v)–vi.

440. Variation and Natural Selection [lttE commenting on an address given by C. Lloyd Morgan to the Bristol Naturalists' Society]. *Nature* 44: 518–519 (1 Oct. 1891: no. 1144).

441. English and American Flowers. I. *Fortnightly Rev.* 50 (n.s.; 56, o.s.): 525–534 (1 Oct. 1891: no. 298, n.s.) / English and American Flowers. II. Flowers and Forests of the Far West. 50 (n.s.; 56, o.s.): 796–810 (1 Dec. 1891: no. 300, n.s.).
—reprints: *Littell's Living Age* 191 (5th s., vol. 76): 742–747 (19 Dec. 1891: no. 2477) / 192 (5th s., vol. 77): 467–476 (20 Feb. 1892: no. 2486); with slight changes in *SSS*, Vol. 1: 199–212, 213–234 (Nov. 1900); summary/N entitled 'English and American Flowers' printed in *New York Times* no. 12585: 4d–e (27 Dec. 1891).

442. discussion [of comments by David Syme on ref 437; printed with Syme's letter as 'Topical Selection and Mimicry']. *Nature* 45: 31 (30–31) (12 Nov. 1891: no. 1150).

443.* Commons [letter]. *Land and Labour* no. 25: 2 (Nov. 1891).
—reprint: as 'Commons Enclosures' in *Report of the LNS. 1891–92.* (LNS Tract No. 48: LNS, London, July 1892): 10–11.

444.* Popular Zoology [review of *Animal Sketches* by C. Lloyd Morgan, 1891]. *Nature* 45: 291–292 (28 Jan. 1892: no. 1161).

445. Human Progress: Past and Future. *Arena* 5: 145–159 (Jan. 1892: no. 26).
—reprint: in *SSS*, Vol. 2: 493–509 (Nov. 1900).

446.* H. W. Bates, the Naturalist of the Amazons [obituary]. *Nature* 45: 398–399 (25 Feb. 1892: no. 1165).

447. A Remarkable Book on the Habits of Animals [review of *The Naturalist in La*

Plata by William H. Hudson, 1892]. *Nature* 45: 553–556 (14 April 1892: no. 1172).

448. Correction in *Island Life* [lttE]. *Nature* 46: 56 (19 May 1892: no. 1177).

449.* remarks [on receiving the Founder's Medal during the RGS meeting of 23 May 1892]. *Proc. RGS & Monthly Record of Geography* 14(7) (new monthly s.): 485–486 (July 1892).

450. Presidential Address [to the eleventh annual meeting of the LNS, 23 June 1892; consists in large part of a discussion of *Justice* by Herbert Spencer (1891) and *Freeland; A Social Anticipation* by Theodor Hertzka (1891)]. in *Report of the LNS. 1891–92.* (LNS Tract No. 48: LNS, London, July 1892): 15–26; also printed in *Land and Labor* no. 33: 1–4 (July 1892).
—a portion reprinted as 'Herbert Spencer on the Land Question: A Criticism' in *SSS*, Vol. 2: 333–344 (Nov. 1900).

451.* Spiritualism. in *Chambers's Encyclopaedia* (new ed., 10 vols., William & Robert Chambers, Ltd., London & Edinburgh, 1888–1892), Vol. 9: 645–649 (July 1892).
—reprint: as pamphlet (J. B. Lippincott Co., Philadelphia, 1892; pp. (1)–14).

452.* Psychography in the Presence of Mr. Keeler [description of experiences at three séances]. *Psychical Rev.* 1(1): 16–18 (July 1892).

453. The Permanence of the Great Oceanic Basins. *Natural Science* 1: 418–426 (Aug. 1892: no. 6).
—revised version printed in *SSS*, Vol. 1: 28–39 (Nov. 1900).

454. Why I Voted for Mr. Gladstone. IV. [one of eight invited replies to an inquiry]. *Nineteenth Century* 32: 182–185 (177–193) (Aug. 1892: no. 186).

455. Our Molten Globe [an analysis of *Physics of the Earth's Crust* (2nd ed.) by Rev. Osmond Fisher, 1891]. *Fortnightly Rev.* 52 (n.s.; 58, o.s.): 572–584 (1 Nov. 1892: no. 311, n.s.).
—reprints: *Littell's Living Age* 195 (5th s., vol. 80): 771–779 (24 Dec. 1892: no. 2530); *Eclectic Mag.* 56(6) (n.s.; 119(6), o.s.): 836–843 (Dec. 1892); slightly revised version printed in *SSS*, Vol. 1: 40–58 (Nov. 1900); summary/N entitled 'Is our Globe Molten?' printed in *New York Times* no. 12909: 4e–f (8 Jan. 1893).

456. An Ancient Glacial Epoch in Australia [notice of *Notes on the Glacial Conglomerate, Wild Duck Creek* by Edward J. Dunn, 1892]. *Nature* 47: 55–56 (17 Nov. 1892: no. 1203).
—summary/N printed in *The Glacialists' Mag.* 1(6): 142–143 (Jan. 1894).

457. The Permanence of Ocean Basins [lttE]. *Natural Science* 1: 717–718 (Nov. 1892: no. 9).

458. lttE [one of two printed as 'The Earth's Age']. *Nature* 47: 175 (175) (22 Dec. 1892: no. 1208).

459. Note on Sexual Selection. *Natural Science* 1: 749–750 (Dec. 1892: no. 10).

460. lttE [one of two printed as 'The Earth's Age']. *Nature* 47: 227 (226–227) (5 Jan. 1893: no. 1210).

461.* Man and Evolution [review of *Evolution and Man's Place in Nature* by Henry Calderwood, 1893]. *Nature* 47: 385–386 (23 Feb. 1893: no. 1217).

462. The Glacier Theory of Alpine Lakes [lttE]. *Nature* 47: 437–438 (9 March 1893: no. 1219).
—synopsis/N printed in *The Glacialists' Mag.* 1(6): 144–145 (Jan. 1894).

463. Reveries of a Naturalist [review of *Idle Days in Patagonia* by William H. Hudson, 1893]. *Nature* 47: 483–484 (23 March 1893: no. 1221).

464. Inaccessible Valleys; A Study in Physical Geography. *Nineteenth Century* 33: 391–404 (March 1893: no. 193).
—reprints: *Littell's Living Age* 197 (5th s., vol. 82): 498–508 (20 May 1893: no. 2551); slightly revised version printed in *SSS*, Vol. 1: 1–27 (Nov. 1900).

465.* Note on Mr. Jukes-Browne's Paper [concerning the origin and classification of islands]. *Natural Science* 2: 193–194 (March 1893: no. 13).

466. The Social Quagmire and the Way out of It. I. The Farmers. *Arena* 7: 395–410 (March 1893: no. 40) / The Social Quagmire and the Way out of It. II. Wage-workers. 7: 525–542 (April 1893: no. 41).
—reprint: in *SSS*, Vol. 2: 394–431 (Nov. 1900).

467.* The Late Mr. S. J. Davey's Experiments [lttE followed by a lengthy reply from Richard Hodgson]. *J. SPR* 6 (1893–94): 33–36 (33–47) (March 1893: no. 98).

468. Are Individually Acquired Characters Inherited? I. *Fortnightly Rev.* 53 (n.s.; 59, o.s.): 490–498 (1 April 1893: no. 316, n.s.) / II. 53 (n.s.; 59, o.s.): 655–668 (1 May 1893: no. 317, n.s.).
—reprinted with slight revisions in *SSS*, Vol. 1: 315–344 (Nov. 1900).

469. Mr. H. O. Forbes's Discoveries in the Chatham Islands [lttE]. *Nature* 48: 27–28 (11 May 1893: no. 1228).

470. Reason *versus* Instinct [review of *The Intelligence of Animals* by Charles William Purnell, 1893]. *Nature* 48: 73–74 (25 May 1893: no. 1230).

471.* President's Address, 1893. The Conditions Essential to the Success of Small Holdings. [read by William Volckman, Chairman, at the twelfth annual meeting of the LNS, 15 June 1893]. in *Report of the LNS, 1892–3* (LNS Tract No. 50: LNS, London, June 1893): 15–23; also printed in *Land and Labor* no. 45: 3–4 (July 1893); also printed as LNS Tract No. 51 (LNS, London, 1893).
—reprinted with slight modification as part of ref 585 in *SSS*, Vol. 2: 354–363 (Nov. 1900).

472. discussion [of letter by Graham Officer; printed with Officer's letter as 'The Glacier Theory of Alpine Lakes']. *Nature* 48: 198 (198) (29 June 1893: no. 1235).

473. The Non-inheritance of Acquired Characters [lttE]. *Nature* 48: 267 (20 July 1893: no. 1238).

474.* The Response to the Appeal. From Prelates, Pundits and Persons of Distinction. [article/N posting results of a survey of opinion concerning the stated intended purpose of the journal; Wallace's one of many responses printed]. *Borderland* 1: 17 (10–23) (July 1893: no. 1).

475. The Bacon-Shakespeare Case. Verdict No. I. [one of many solicited responses concerning the authorship of Shakespeare's works]. *Arena* 8: 222–225 (July 1893: no. 44).

476. Prenatal Influences on Character [lttE]. *Nature* 48: 389–390 (24 Aug. 1893: no. 1243).

477. Habits of South African Animals [lttE introducing comments by R. R. Mortimer on some South African birds]. *Nature* 48: 390–391 (24 Aug. 1893: no. 1243).

478.* Notes on the Growth of Opinion as to Obscure Psychical Phenomena during the Last Fifty Years [a paper communicated to the Psychical Congress in Chicago, held 21–25 Aug. 1893; sent from Parkstone, Dorset, England]. *Religio-Philosophical J.* (Chicago) 4 (n.s.), no. 15: 229a–230a (2 Sept. 1893).
—reprints: *The Two Worlds* (Manchester) 6: 440–441 (15 Sept. 1893: no. 305); as 'Science and Psychics. Fifty Years' Progress.' in *Borderland* 1: 149–150 (Oct. 1893: no. 2); reprinted with slight revision as 'Preface to the Third Edition' in *Miracles and Modern Spiritualism* (3rd ed.): ix–xvii (Jan. 1896); reprinted from the latter as 'The Evidence of Facts' as part of an article/N entitled 'Shall We Live after Death; and If so, How? Speculations.' in *Borderland* 3(3): 269–270 (258–272) (July 1896); extracts from paper entitled 'Some Notes by Alfred R. Wallace' reprinted in *Light* (London) 13: 439 (16 Sept. 1893: no. 662).

479.* On Malformation from Pre-natal Influence on the Mother [a letter communicated to the 15 Sept. 1893 (Nottingham) meeting of Section D, Biology, of the BAAS]. in *Rept. BAAS* 63 (1893) (John Murray, London, 1894): 798–799.

480. The Supposed Glaciation of Brazil [lttE]. *Nature* 48: 589–590 (19 Oct. 1893: no. 1251).
—summary/N printed in *The Glacialists' Mag.* 1(6): 147 (Jan. 1894).

481. The Ice Age and its Work. I. Erratic Blocks and Ice-sheets. *Fortnightly Rev.* 54 (n.s.; 60, o.s.): 616–633 (1 Nov. 1893: no. 323, n.s.) / The Ice Age and its Work. II. Erosion of Lake Basins. 54 (n.s.; 60, o.s.): 750–774 (1 Dec. 1893: no. 324, n.s.).
—reprints: in four parts in *Popular Science Monthly* 44: 681–689 (March 1894) & 781–791 (April 1894) / 45: 40–50 (May 1894) & 244–258 (June 1894); slightly revised versions printed in *SSS*, Vol. 1: 59–93, 94–128 (Nov. 1900); selections from each part reprinted in *Annual Report of the Board of Regents of the Smithsonian Institution (to July 1893)* (Government Printing Office, Washington, D. C., 1894): 277–286, 286–300; this subsequently issued as Octavo pamphlet No. 945 (pp. 24; plate); summary/N entitled 'The Origin of Lake Basins' printed in *The Glacialists' Mag.* 1(5): 113–116 (Dec. 1893).

482. The Recent Glaciation of Tasmania [lttE]. *Nature* 49: 3–4 (2 Nov. 1893: no. 1253).

483.* The Programme of Land Nationalisers [a paper read at a general meeting of the LNS held 15 Nov. 1893]. *Land and Labor* no. 50: 1–2 (Dec. 1893).

484. Sir Henry H. Howorth on "Geology in Nubibus" [lttE concerning glacial movement]. *Nature* 49: 52 (16 Nov. 1893: no. 1255).

485. Recognition Marks [letter responding to inquiry regarding rabbits' use of their tails as danger signals]. *Nature* 49: 53 (16 Nov. 1893: no. 1255).

486. "Geology in Nubibus" [letter responding to remarks by Henry H. Howorth concerning glacial movement]. *Nature* 49: 101 (30 Nov. 1893: no. 1257).

487. lttE [one of two printed as 'The Origin of Lake Basins']. *Nature* 49: 197 (197) (28 Dec. 1893: no. 1261).

488.* Preface. to *The Dispersal of Shells; An Inquiry into the Means of Dispersal Possessed by Fresh-water and Land Mollusca* by Harry Wallis Kew (International Scientific Series, Vol. LXXV: Kegan Paul, Trench, Trübner & Co., Ltd., London, Dec. 1893): v–viii.

489. lttE [one of two printed as 'The Origin of Lake Basins']. *Nature* 49: 220–221 (220–221) (4 Jan. 1894: no. 1262).

490.* *Why does Man Exist?* [letter to Arthur J. Bell, author of *Why does Man Exist?* (1890); treated here as a notice of that book]. *Borderland* 1: 272 (Jan. 1894: no. 3).

491. How to Preserve the House of Lords. *Contemporary Rev.* 65: 114–122 (Jan. 1894).
—reprint: as 'A Representative House of Lords' in *SSS*, Vol. 2: 223–234 (Nov. 1900).

492.* Richard Spruce, Ph.D., F.R.G.S. [obituary]. *Nature* 49: 317–319 (1 Feb. 1894: no. 1266).

493. A Critic Criticised [review of *Darwinianism: Workmen and Work* by James Hutchison Stirling, 1893]. *Nature* 49: 333–336 (8 Feb. 1894: no. 1267).

494. What are Zoological Regions? [a paper read at the 500th meeting of the Cambridge Natural Science Club, 12 March 1894]. *Nature* 49: 610–613 (26 April 1894: no. 1278).

495. President's Address [presented at the thirteenth annual meeting of the LNS, 9 April 1894; includes a notice of *Social Evolution* by Benjamin Kidd (1894)]. in *13th Report of the LNS, 1893–4* (LNS Tract No. 56: LNS, London, April 1894): 15–24; also printed in *Land and Labor* no. 54: 34–37 (May 1894); also printed as pamphlet entitled *The Local Government (England and Wales) Act, 1894, Relating to Parish Councils, etc.* (Land Nationalisation Tract No. 57: LNS, London, May 1894; pp. (1)–11).

496. The Future of Civilisation [review of *Social Evolution* by Benjamin Kidd, 1894]. *Nature* 49: 549–551 (12 April 1894: no. 1276).

497.* lttE [one of two printed as 'Woman and Natural Selection']. *Humanitarian* 4 (n.s.), no. 4: 315 (315–319) (April 1894).

498. Economic and Social Justice. in *Vox Clamantium; The Gospel of the People* by "writers, preachers & workers brought together by Andrew Reid" (A. D. Innes & Co., London, April 1894): 166–197.
—reprint: in *SSS*, Vol. 2: 432–454 (Nov. 1900).

499.* Panmixia and Natural Selection [lttE]. *Nature* 50: 196–197 (28 June 1894: no. 1287).

500. The Palaearctic and Nearctic Regions Compared as regards the Families and Genera of their Mammalia and Birds. *Natural Science* 4: 433–445 (June 1894: no. 28).
—reprint: in *SSS*, Vol. 1: 235–249 (Nov. 1900).

501. A New Book on Socialism [review of *The Great Revolution of 1905; or The Story of the Phalanx* by Frederick W. Hayes, 1893]. *Land and Labor* no. 57: 52–54 (July 1894).

502.* The Influence of Previous Fertilisation of the Female on her Subsequent Offspring, and the Effect of Maternal Impressions during Pregnancy on the Offspring [interim report to the BAAS drawn up by James Clark, member of Wallace-chaired committee studying this subject; meeting held in Oxford in Aug. 1894]. in *Report of the Sixty-fourth Meeting of the BAAS* (1894) (John Murray, London, 1894): 346.

503. Rev. George Henslow on Natural Selection. *Natural Science* 5: 177–183 (Sept. 1894: no. 31).
—reprint: as 'A Critic of Natural Selection Answered' in *SSS*, Vol. 1: 305–314 (Nov. 1900).

504. Another Substitute for Darwinism [review of *Nature's Method in the Evolution of Life* by "Anonymous" [James W. Barclay], 1894]. *Nature* 50: 541–542 (4 Oct. 1894: no. 1301).

505. A Suggestion to Sabbath-keepers. *Nineteenth Century* 36: 604–611 (Oct. 1894: no. 212).
—reprint: as 'A Counsel of Perfection for Sabbatarians' in *SSS*, Vol. 2: 364–374 (Nov. 1900).

506. Why Live a Moral Life? The Answer of Rationalism. [one of several invited replies to an inquiry; in part based on ref 243]. in *The Agnostic Annual 1895* ed. by Charles A. Watts (London, W. Stewart & Co., Dec.? 1894): 6–12.
—reprints: in *SSS*, Vol. 2: 375–383 (Nov. 1900); edited and reprinted as 'Immortality and Morality' in *Borderland* 2: 10–11 (Jan. 1895: no. 7); excerpts printed as 'Why Live a Moral Life?' in *The Two Worlds* (Manchester) 7: 505–506 (26 Oct. 1894: no. 363).

507.* The Social Economy of the Future. in *The New Party Described by Some of its Members* (new ed.) ed. by Andrew Reid (Hodder Brothers, London, Dec. 1894): 177–211.

508.* Note on Compensation to Landlords [not the same article as ref 391]. *Land and Labor* no. 63: 5 (Jan. 1895).

509.* Forty-five Years' Registration Statistics. A Correction. [lttE]. *The Vaccination Inquirer and Health Review* 16: 159–160 (1 Feb. 1895: no. 191).

510. The Method of Organic Evolution. I. *Fortnightly Rev.* 57 (n.s.; 63, o.s.): 211–224 (1 Feb. 1895: no. 338, n.s.) / II. 57 (n.s.; 63, o.s.): 435–445 (1 March 1895: no. 339, n.s.).
—reprints: in *Annual Report of the Board of Regents of the Smithsonian Institution (to July 1894)* (Government Printing Office, Washington, D.C., 1896): 413–425, 426–435; this subsequently issued as Octavo pamphlet No. 1013 (pp. 22); reprinted with minor revisions in *SSS*, Vol. 1: 345–377 (Nov. 1900).

511. Tan-spots over Dogs' Eyes [lttE responding to an inquiry). *Nature* 51: 533 (4 April 1895: no. 1327).

512. *Suggestions for Solving the Problem of the Unemployed, etc., etc.* [Presidential Address read at the fourteenth annual meeting of the LNS, 8 April 1895]. Land Nationalisation Tract No. 64 (LNS, London, April? 1895; pp. (1)–20); also printed as 'President's Address' in *Land and Labor* no. 67: 40–45 (May 1895).
—reprints: with changes and additions as 'Re-occupation of the Land: The Only Solution to the Problem of the Unemployed' in *Forecasts of the Coming*

Century by a Decade of Writers ed. by Edward Carpenter (The Labour Press Ltd., Manchester, Aug. 1897 / W. Scott, Ltd., London, 1897): 9–26; this then reprinted as 'Re-occupation of the Land' in *Hand and Brain: A Symposium of Essays on Socialism* by William Morris and others (Roycroft Printing Shop, East Aurora, New York, 1898): 25–43; revised version revised further and printed as 'Re-occupation of the Land: The Only Immediate Solution to the Problem of the Unemployed' in *SSS*, Vol. 2: 478–492 (Nov. 1900); excerpt reprinted as 'Security of the Home' in *My Life*, Vol. 2: 264–265 (Oct. 1905).

513. lttE [one of two printed as 'The Age of the Earth']. *Nature* 51: 607 (607–608) (25 April 1895: no. 1330).

514. Uniformitarianism in Geology [lttE]. *Nature* 52: 4 (2 May 1895: no. 1331).

515. Another Book on Social Evolution [review of *The Evolution of Industry* by Henry Dyer, 1895]. *Nature* 52: 386–387 (22 Aug. 1895: no. 1347).

516. discussion [concerning a letter Wallace sent to Adolf B. Meyer in 1869; Meyer's letter (containing an excerpt from the 1869 letter) and Wallace's comments thereon printed as 'How was Wallace Led to the Discovery of Natural Selection?']. *Nature* 52: 415 (415) (29 Aug. 1895: no. 1348).

517.* Our Native Birds [review of *British Birds* by William H. Hudson, 1895]. *Saturday Rev.* 80: 342–343 (14 Sept. 1895: no. 2081).

518. The Expressiveness of Speech, or Mouth-gesture as a Factor in the Origin of Language. *Fortnightly Rev.* 58 (n.s.; 64, o.s.): 528–543 (1 Oct. 1895: no. 346, n.s.).
—reprints: *Littell's Living Age* 207 (6th s., vol. 8): 369–380 (9 Nov. 1895: no. 2679); *Eclectic Mag.* 62(5) (n.s.; 125(5), o.s.): 649–660 (Nov. 1895); in *SSS*, Vol. 2: 115–137 (Nov. 1900).

519.* Introductory Note. to *Psychic Philosophy as the Foundation of a Religion of Natural Law* by V. C. Desertis [pseudonym of Stanley DeBrath] (George Redway, London, Dec. 1895 / Bellairs & Co., London, Nov. 1895): v–vi.

520. lttE [one of two printed as 'The Cause of an Ice Age']. *Nature* 53: 220–221 (220–221) (9 Jan. 1896: no. 1367).

521. The Astronomical Theory of a Glacial Period [lttE]. *Nature* 53: 317 (6 Feb. 1896: no. 1371).

522.* The Theory of the Double. *Light* (London) 16: 87–88 (22 Feb. 1896: no. 789).
—reprint: as 'On the Theory of the Double. A Suggestion.' in *Borderland* 3(2): 173–175 (April 1896).

523.* Philosophy and Evolution [review of *Evolution and Man's Place in Nature* (2nd ed.) by Henry Calderwood, 1896). *Nature* 53: 435 (12 March 1896: no. 1376).

524. Old and New Theories of Evolution [review of *The Primary Factors of Organic Evolution* by Edward D. Cope, 1896; and *The Present Evolution of Man* by G. Archdall Reid, 1896]. *Nature* 53: 553–555 (16 April 1896: no. 1381).

525.* Letter from the President [to Joseph Hyder, read at the fifteenth annual meeting of the LNS, 20 April 1896]. *Land and Labour* no. 79: 33–34 (May 1896).

526. The Proposed Gigantic Model of the Earth. *Contemporary Rev.* 69: 730–740 (May 1896).
—reprints: *Littell's Living Age* 209 (6th s., vol. 10): 692–700 (13 June 1896: no. 2710); *Eclectic Mag.* 63(6) (n.s.; 126(6), o.s.): 791–798 (June 1896); as 'How Best to Model the Earth' in *SSS*, Vol. 2: 59–73 (Nov. 1900).

527. The Problem of Utility: Are Specific Characters always or generally Useful? [a paper read at the LSL meeting of 18 June 1896]. *J. Linn. Soc. Zool.* 25: 481–496 (1896).
—reprint: in *SSS*, Vol. 1: 378–398 (Nov. 1900).

528. lttE [concerning labour and militarism to Keir Hardie, on the occasion of the International Labour Congress]. *Labour Leader* 8: 251 (25 July 1896: no. 121, n.s.).

529. The Gorge of the Aar and its Teachings. *Fortnightly Rev.* 60 (n.s.; 66, o.s.,): 175–182 (1 Aug. 1896: no. 356, n.s.).
—reprints: *The Glacialists' Mag.* 4(2): 61–71 (Sept. 1896); slightly revised version printed in *SSS*, Vol. 1: 129–145 (Nov. 1900).

530.* Spiritualism [lttE]. *The Echo* (London) no. 8637: 4a (12 Sept. 1896).
—extract included in note/N entitled 'Mr. Alfred R. Wallace, F.R.S., in the *Echo*' in *Light* (London) 16: 449 (19 Sept. 1896: no. 819).

531.* Methods of Land Nationalisation. *Land and Labour* no. 85: 82–83 (Nov. 1896).

532. Darwin and Darwinism [review of *Charles Darwin and the Theory of Natural Selection* by Edward B. Poulton, 1896]. *Nature* 55: 289–290 (28 Jan. 1897: no. 1422).

533.* Lord Penrhyn and the Quarrymen [full version of a lttE first printed in edited form in *The Daily Chronicle* (London), date unknown]. *Land and Labour* no. 88: 9–10 (Feb. 1897).

534. The Problem of *Instinct* [review of *Habit and Instinct* by C. Lloyd Morgan, 1896]. *Natural Science* 10: 161–168 (March 1897: no. 61).
—reprints: *Scientific American Supplement* 43: 17724–17725 (3 April 1897: no. 1109); in *SSS*, Vol. 1: 497–508 (Nov. 1900).

535.* On the Colour and Colour-patterns of Moths and Butterflies [comments on a paper on this subject by Alfred Goldsborough Mayer]. *Nature* 55: 618–619 (29 April 1897: no. 1435).

536. *Vaccination a Delusion; Its Penal Enforcement a Crime: Proved by the Official Evidence in the Reports of the Royal Commission.* Swan Sonnenschein & Co., Ltd., London, March 1898 (according to *My Life* (Vol. 2, p. 353), "brought out" as a booklet for distribution in Jan. 1898); pp. (1)–96; 12 fold-out diagrams.
—reprints: as Chapter 18 of *The Wonderful Century*: 213–323 (includes the Appendix 'The Causes of the Improvement in the Health of London toward the End of the Eighteenth and Beginning of the Nineteenth Centuries,' not present in the 1898 pamphlet edition; American eds. pagination: 213–324) (June 1898); this then reprinted as pamphlet ("Second Edition": The National A-VL & Swan Sonnenschein & Co., Ltd., London, 1901 & 1904; pp. (i)–(iv), 213–323 (plus two pages of Index, unnumbered); 12 fold-out diagrams).

537.* *The Eagle and the Serpent* [letter commending the aims of *The Eagle and the Serpent*]. *The Clarion* (London) no. 328: 95b (19 March 1898).

538. Mr. A. R. Wallace and Vaccination [lttE]. *Lancet*, Vol. 1 for 1898: 894 (26 March 1898: no. 3891).
—reprint: as 'Dr. Alfred Russel Wallace and the *Lancet*' in *The Vaccination Inquirer and Health Rev.* 20: 12 (1 April 1898: no. 229).

539.* lttE [commending the aims of the journal]. *The Eagle and the Serpent* 1(2): 21 (15 April 1898).
—reprint: *The Eagle and the Serpent* 1(3): 46 (15 June 1898).

540.* Nietzsche as a Social Reformer, or, The Joys of Fleecing and being Fleeced [article/N posting results of a general inquiry; Wallace's one of several responses printed]. *The Eagle and the Serpent* 1(2): 26–27 (25–28) (15 April 1898).

541.* letter [to Joseph Hyder, read at the seventeenth annual meeting of the LNS, 22(?) April 1898]. *Land and Labour* no. 103: 35 (May 1898).

542. Dr. Bond v. Mr. A. R. Wallace [lttE concerning vaccination]. *Shrewsbury Chronicle* 20 May 1898: 3b.

543.* letter [included in note/N entitled 'Dr. Alfred R. Wallace on Altruism: Mr. Platt's Repudiation']. *The Eagle and the Serpent* 1(3): 36 (15 June 1898).

544. Dr. Bond and Mr. A. R. Wallace. Mr. Wallace Replies. [lttE concerning vaccination]. *The Echo* (London) no. 9183: 1c (16 June 1898).

545. Spiritualism and Social Duty [from an address delivered at the International Congress of Spiritualists on 23 June 1898, at St. James Hall, London]. *Light* (London) 18: 334–336 (selection followed by an account of related discussion on pp. 336–337) (9 July 1898: no. 913).
—reprints: main portion of address printed privately as pamphlet as *Justice, not Charity, as the Fundamental Principle of Social Reform. An Appeal to my Readers.* (summer 1898; pp. 14); reprinted under this title "with verbal modifications" in *SSS*, Vol. 2: 521–528 (Nov. 1900).

546.* discussion/N [of paper on Brazilian spiritism by A. Alexander read at the 23 June 1898 session of the International Congress of Spiritualists]. *Light* (London) 18: 337 (9 July 1898: no. 913).

547.* The Importance of Dust: A Source of Beauty and Essential to Life. Chapter 9 of *The Wonderful Century*: 69–85 (68–84, American edition; pp. 170–191 of the "New Edition" of Sept. 1903) (June 1898).
—reprint: in *English Prose* selected and ed. by Frederick William Roe & George Roy Elliott (Longmans, Green, & Co., New York, Nov. 1913): 278–291.

548. Dr. Bond and Mr. A. R. Wallace. Mr. Wallace's Final Reply. [lttE]. *The Echo* (London) no. 9234: 1c (15 Aug. 1898).

549. Darwinism in Sociology: Dr. Alfred Russell [sic] Wallace Replies to Mr. Thomas Common. *The Eagle and the Serpent* 1(4): 57–59 (1 Sept. 1898).

550.* letter [referring to the intent of the journal and to a book by Ragnar Redbeard]. *The Eagle and the Serpent* 1(4): 62 (1 Sept. 1898).

551.* The Vaccination Question [lttE]. *The Times* (London) no. 35610: 10e (1 Sept. 1898).

552.* letter [to columnist "Dangle" concerning socialist reforms and the monopoly of money]. *The Clarion* (London) no. 357: 325c–d (8 Oct. 1898).

553.* Is there Scarcity or Monopoly of Money? [letter to columnist "Dangle"]. *The Clarion* (London) no. 360: 348c-e (29 Oct. 1898).

554.* *The Wonderful Century.*—A Correction. *Land and Labour* no. 108: 82 (Oct. 1898).

555.* Introductory Note. to *The Third Factor of Production and other Essays* by Arthur J. Ogilvy (Swan Sonnenschein & Co., Ltd., London, Nov. 1898): vii–viii.

556.* A Complete System of Paper Money [letter to columnist "Dangle"]. *The Clarion* (London) no. 365: 389e–f (3 Dec. 1898).

557. Paper Money as a Standard of Value. *Academy* 55: 549–550 (31 Dec. 1898: no. 1391, n.s.).
—reprint: in *SSS*, Vol. 2: 145–149 (Nov. 1900).

558. The Utility of Specific Characters [lttE concerning recognition marks]. *Nature* 59: 246 (12 Jan. 1899: no. 1524).

559.* America, Cuba, and the Philippines [lttE]. *The Daily Chronicle* (London) 19 Jan. 1899: 3g.

560. The Inefficiency of Strikes: Is there not a Better Way? in *The Labour Annual: 1899* ed. and published by Joseph Edwards (Wallasey, Cheshire, Jan.? 1899): 105.

561. Mr. Podmore on Clairvoyance and Poltergeists [lttE]. *J. SPR* 9 (1899–1900): 22–30 (Feb. 1899: no. 156).

562. White Men in the Tropics. *The Independent* (New York) 51: 667–670 (9 March 1899: no. 2623).
—reprints: *New-York Daily Tribune* no. 19113: 5d–e (15 March 1899); in *SSS*, Vol. 2: 99–106 (Nov. 1900).

563.* Letter from the President [to Joseph Hyder, read at the eighteenth annual meeting of the LNS, 26 April 1899]. *Land and Labour* no. 115: 46 (May 1899).

564.* letter [to columnist Julia Dawson, accompanying a donation to the *Clarion* Van] *The Clarion* (London) no. 386: 130e (29 April 1899).

565. Clairvoyance and Poltergeists [lttE]. *J. SPR* 9 (1899–1900): 56–57 (April 1899: no. 158).

566.* Garden City [letter to Ebenezer Howard concerning planned communities]. *Land and Labour* no. 114: 38 (April 1899).

567. Les Causes de la Guerre—Comment y Remédier [abridged translation of one of many essays included in a special supplementary publication entitled 'Enquête sur la Guerre et le Militarisme']. *L'Humanité Nouvelle* Mai 1899: 245–250.
—reprints: entire (unabridged) essay printed in English as 'The Causes of War, and the Remedies' in *The Clarion* (London) no. 396: 213a-d (8 July 1899); this then reprinted in *SSS*, Vol. 2: 384–393 (Nov. 1900).

568.* Introductory Note. to 'Extract from Js-E de Mirville's *Des Esprits et de leurs Manifestations Fluidiques*' [read as 'Clairvoyance of Alexis Didier' at the SPR meeting of 23 June 1899]. *Proc. SPR* 14 (1898–9): 373–374 (373–381) (Supplement 5 to Part XXXV, July? 1899).

569.* lttE [one of many printed as 'Protests against War']. *Manchester Guardian* no. 16557: 7f (7f–h) (2 Sept. 1899).

570.* letter [a reply to remarks by Thomas Common concerning the "might vs. right" issue]. *The Eagle and the Serpent* 1(9): 136 (15 Oct. 1899).

571. The Transvaal War. Wanted Facts. [lttE]. *The Clarion* (London) no. 415: 365a–b (18 Nov. 1899).

572.* Facts from the Transvaal [lttE introducing remarks by an observer back from the Boer War]. *The Clarion* (London) no. 417: 380f (2 Dec. 1899).

573.* Mottoes for the New Year. Wise Words from Famous People. [article/N posting results of a general inquiry; Wallace's one of many responses printed]. *The Daily News Weekly* (London) no. 27: 11b–c (11a–d) (30 Dec. 1899).

—reprint: as 'Alfred Russel Wallace on Equality of Opportunity' in *Land and Labour* 11(2): 21 (Feb. 1900).

574.* Introductory Note. to *The Ascent of Man* by Mathilde Blind (T. Fisher Unwin, London, Dec. 1899): v–xii.

575. Is New Zealand a Zoological Region? [lttE]. *Nature* 61: 273 (18 Jan. 1900: no. 1577).

576.* Labour and the next General Election [article/N posting results of a general inquiry; Wallace's one of several responses printed]. in *The Labour Annual: The Reformers' Year-book for 1900* ed. and published by Joseph Edwards (Wallasey, Cheshire, Jan. 1900): 29 (26–30).

577.* letter [to columnist "Nunquam" concerning the death of *Clarion* columnist "The Bounder"]. *The Clarion* (London) no. 447: 205b (30 June 1900).

578.* lttE [concerning Edward Jenner's observations on the cuckoo; part of a discussion including another letter and comments by the Editor entitled 'Jenner and the Cuckoo']. *The Vaccination Inquirer and Health Rev.* 22: 58 (58–61) (2 July 1900: no. 256).

579.* Imperial Might and Human Right [comments on an article by George Bernard Shaw concerning imperialism]. *The Clarion* (London) no. 450: 230b (21 July 1900).

580.* letter [a reply to remarks by Ragnar Redbeard concerning the "might vs. right" issue]. *The Eagle and the Serpent* 1(11): 164 (July 1900).

581.* Letter from Dr. A. R. Wallace [concerning social problems]. in *The Anatomy of Misery: Plain Lectures on Economics* (2nd ed.) by John Coleman Kenworthy (Simpkin, Marshall, Hamilton, Kent, & Co., Ltd., London, Sept. 1900 / Small, Maynard & Co., Boston, 6 April 1901): 98–100 (as part of the Appendix).

582.* letter extracts [in Part XI and Appendix G of 'On the so-called Divining Rod. A Psycho-Physical Research on a Peculiar Faculty Alleged to Exist in Certain Persons Locally Known as Dowsers. Book II.' by William F. Barrett]. *Proc. SPR* 15 (1900–1901): 277, 374–375 (Part XXXVIII, Oct.? 1900).

583.* Affinities and Origin of the Australian and Polynesian Races [the first part of this article (pp. 461–473) forms the concluding section of Vol. 1, Chapter 5 of *Australasia* (ed. of Nov. 1893)]. in *SSS*, Vol. 1: 461–496 (Nov. 1900).

584.* Interest-bearing Funds Injurious and Unjust. in *SSS*, Vol. 2: 254–264 (Nov. 1900).

585.* Some Objections to Land Nationalization Answered [derived "from Tracts issued by the LNS": see refs 385, 423 & 471]. in *SSS*, Vol. 2: 345–363 (Nov. 1900).

586. Ralahine and its Teachings [comments on *The Irish Land & Labour Question, Illustrated in the History of Ralahine and Co-operative Farming* by Edward T. Craig, 1882]. in *SSS*, Vol. 2: 455–477 (Nov. 1900).

587. True Individualism—The Essential Preliminary of a Real Social Advance. in *SSS*, Vol. 2: 510–520 (Nov. 1900).

588.* Professor Alfred Russel Wallace [letter of support sent to the Eastbourne Conference of the National A-VL, held 5 Dec. 1900]. *The Vaccination Inquirer and Health Rev.* 22: 156 (1 Jan. 1901: no. 262).

589. Evolution [the first in a series of articles published in *The Sun* under the overall title 'The Passing Century']. *The Sun* (New York) 68(114): 4a–g, 5a (23 Dec. 1900).
—reprint: nearly verbatim in *The Progress of the Century* (Harper & Brothers, New York & London, 20 April 1901): 3–29.

590. article concerning social evolution in the twentieth century [not seen; part of a special feature entitled 'Dawn of the Century']. *The [Sunday] Journal* (New York) no. 6611: ? (23 Dec. 1900).

591.* A Message to my Fellow Spiritualists for the New Century. *The Two Worlds* (Manchester) 13: 867 (28 Dec. 1900: no. 685).

592.* letter [dated Nov. 1900; written in response to an inquiry]. "*The Sermon*" [title of serial not traced] Dec. 1900.
—reprint: as 'Dr. A. R. Wallace and the "Subliminal Self" ' in *Light* (London) 21: 29 (19 Jan. 1901: no. 1045).

593. Words of Counsel [article/N posting results of a general inquiry; Wallace's one of several responses printed]. *The Morning Leader* (London) no. 2689: 3a (3a) (2 Jan. 1901).

594.* letter [to columnist "Nunquam," concerning his novel *Julie*]. *The Clarion* (London) no. 485: 92e (23 March 1901).

595.* letter [to columnist Julia Dawson, concerning the subjects of peace and the *Clarion* Van]. *The Clarion* (London) no. 493: 160d (18 May 1901).

596.* Are Plant Diseases Hereditary? *The Garden* 61: 317 (17 May 1902: no. 1591).

597.* *Eucalyptus gunnii*. *The Garden* 62: 47 (19 July 1902: no. 1600).

598.* Dr. Russel Wallace's Advice [letter read at an A-VL meeting held 22 Oct. 1902]. *Vaccination Inquirer* 24: 158–159 (1 Nov. 1902: no. 284).

599. The Dawn of a Great Discovery (My Relations with Darwin in Reference to the Theory of Natural Selection). *Black and White* 25: 78–79 (17 Jan. 1903: no. 624).

600. Genius and the Struggle for Existence [lttE]. *Nature* 67: 296 (29 Jan. 1903: no. 1735).

601.* Our Sphinx's Fatal Question: Why do the Ungodly Prosper? Is Might Right? Can the Poor be Saved through the Pity of the Rich? [article/N posting results of a general inquiry; Wallace's one of several responses printed]. *The Eagle and the Serpent* no. 18: 69 (68–73) (Feb.? 1903).

602. Man's Place in the Universe. *The Independent* (New York) 55: 473–483 (26 Feb. 1903: no. 2830); also printed as 'Man's Place in the Universe: As Indicated by the New Astronomy' in *Fortnightly Rev.* 73 (n.s.; 79, o.s.): 395–411 (1 March 1903: no. 435, n.s.).
—reprints: *The Living Age* 237 (7th s., vol. 19): 1–13 (4 April 1903: no. 3065); *Eclectic Mag.* 9(5) (3rd s.; 140(5), o.s.): 561–573 (May 1903).

603.* Dr. Russel Wallace on the Use of the Vote [letter read at the annual meeting of the A-VL held 18 March 1903]. *Vaccination Inquirer* 25: 14 (1 April 1903: no. 289).

604.* lttE [one of several printed as 'Man's Place in the Universe']. *Knowledge* 26: 107–108 (107–110) (May 1903: no. 211).

605.* Dr. A. R. Wallace on the Strenuous Policy [lttE]. *Vaccination Inquirer* 25: 49 (comment by the Editor follows on pp. 49–50) (1 June 1903: no. 291).

606. Man's Place in the Universe: A Reply to Criticisms. *The Independent* (New York) 55: 2024–2031 (27 Aug. 1903: no. 2856); also printed in *Fortnightly Rev.* 74 (n.s.; 80, o.s.): 380–390 (1 Sept. 1903: no. 441, n.s.).
—reprints: *The Living Age* 239 (7th s., vol. 21): 193–201 (24 Oct. 1903: no. 3094); *Eclectic Mag.* 10(6) (3rd s.; 141(6), o.s.): 773–781 (Dec. 1903).

607.* letter [of support read at a meeting of the Gloucester League held 22 Oct. 1903]. *Vaccination Inquirer* 25: 153 (2 Nov. 1903: no. 296).

608. *The Wonderful Century* [letter responding to comments by a reviewer on the new edition of *The Wonderful Century* (Sept. 1903)]. *Academy* 65: 453 (24 Oct. 1903: no. 1642, n.s.).

609. Does Man Exist in other Worlds? "A Reply to my Critics." *The Daily Mail* (London) no. 2362: 4d (12 Nov. 1903).

610. Anticipations and Hopes for the Immediate Future [invited for publication by the *Berliner-Lokal-Anzeiger* but rejected by it]. *The Clarion* (London) no. 630: 1c–d (1 Jan. 1904).
—reprint: in *My Life*, Vol. 2: 221–223 (Oct. 1905).

611.* A Letter from Wallace [excerpt from letter concerning Wallace's continued support of spiritualism]. *The Sentinel* (Wood Green, etc., England) no. 451: 3d (15 Jan. 1904).
—entire letter printed in article/N entitled 'Dr. Alfred Russel Wallace and Spiritualism' in *Light* (London) 24: 26 (16 Jan. 1904: no. 1201).

612. An Unpublished Poem by Edgar Allan Poe. Leonaine [sic]. *Fortnightly Rev.* 75 (n.s.; 81, o.s.): 329–332 (1 Feb. 1904: no. 446, n.s.).

613.* From Dr. Russel Wallace [letter of support sent to the annual meeting of the A-VL held 15–16 March 1904]. *Vaccination Inquirer* 26: 18 (1 April 1904: no. 301).

614. The "Leonainie" Problem [lttE]. *Fortnightly Rev.* 75 (n.s.; 81, o.s.): 706–711 (1 April 1904: no. 448, n.s.).

615. The Birds of Paradise in the Arabian Nights. I. *Independent Rev.* 2(7): 379–391 (April 1904) / II. 2(8): 561–571 (May 1904).

616. *A Summary of the Proofs that Vaccination does not Prevent Small-pox but Really Increases It*. National A-VL, London, 1904 (probably April, May or June); pp. (1)–24 (incl. diagram).
—Second Edition [reprint of First Edition]. National A-VL, London, 1910.

617. Practical Politics [a reply to Editor Robert Blatchford's comments on military spending]. *The Clarion* (London) no. 669: 1b–c (30 Sept. 1904).
—reprint: as 'A Substitute for Militarism' in *My Life*, Vol. 2: 223–226 (Oct. 1905).

618.* letter [excerpt from letter concerning John F. Burton's *Story of the Vaccination Crusade in Hackney & Stoke Newington, 1902–1904, and What Came of It*]. *Vaccination Inquirer* 26: 162 (1 Nov. 1904: no. 308).

619.* letter [of support read at the annual meeting of the A-VL held 15–16 March 1905]. *Vaccination Inquirer* 27: 6 (1 April 1905: no. 313).

620.* In Memoriam A. C. Swinton [obituary]. *Land and Labour* 16(4): 33–34 (April 1905).

621. A Message from Dr. A. R. Wallace [concerning the inconsistency of the government's position on vaccination]. *Vaccination Inquirer* 27: 49 (1 June 1905: no. 315).

622.* If there were a Socialist Government—How should It Begin? *The Clarion* (London) no. 715: 5a–f (18 Aug. 1905).

623.* The South-Wales Farmer [written for publication about 1843 but not appearing at that time]. in *My Life*, Vol. 1: 206–222 (Oct. 1905).

624.* Alfred R. Wallace on *Isis Unveiled* [letter to Helena P. Blavatsky dated 1 Jan. 1878, commending her book]. *The Theosophist* (Adyar, India) 27(7): 559 (April 1906).

625.* letter [of support to Joseph Hyder, read at the 25th annual meeting of the LNS, 9 May 1906]. *Land and Labour* 17(6): 61–62 (June 1906).

626.* *A Statement of the Reasons for Opposing the Death Penalty*. Leaflet No. 2: Society for the Abolition of Capital Punishment, London, n.d.; pp. 2.
—reprint: excerpt reprinted untitled in *Light* (London) 26: 225 (12 May 1906: no. 1322).

627.* The Nativity of Dr. Alfred Russel Wallace [article by the Editor containing short excerpt from Wallace letter commenting on the results of an astrological charting]. *Modern Astrology* 3 (n.s.; o.s., 17), no. 5: 206–207 (May 1906).

628.* How to Nationalise Railroads. *The Daily News* (London) no. 18883: 6e (24 Sept. 1906).
—reprint: as 'Dr. Alfred Russel Wallace and Compensation by Terminable Annuities' in *Land and Labour* 17(10): 113–114 (Oct. 1906).

629.* Our Black Brother [lttE referring to ref 630]. *The Clarion* (London) no. 781: 5c (23 Nov. 1906).

630. The Native Problem in South Africa and Elsewhere. *Independent Rev.* 11: 174–182 (Nov. 1906: no. 38).

631.* Britain's Greatest Benefactor [article/N posting results of an opinion survey concerning the most important figures in British history; Wallace's one of many responses printed; his choice: Shakespeare]. *The Clarion* (London) no. 786: 3a (3a–c) (28 Dec. 1906).

632. Personal Suffrage. A Rational System of Representation and Election. *Fortnightly Rev.* 81 (n.s.; 87, o.s.): 3–9 (1 Jan. 1907: no. 481, n.s.).

633. Fertilisation of Flowers by Insects [lttE sent in response to an inquiry]. *Nature* 75: 320 (31 Jan. 1907: no. 1944).

634.* The Railways for the Nation. *Arena* 37: 1–6 (Jan. 1907: no. 206).

635. A New House of Lords: Representative of the Best Intellect and Character of the Nation. *Fortnightly Rev.* 81 (n.s.; 87, o.s.): 205–214 (1 Feb. 1907: no. 482, n.s.).

636.* Dr. Alfred Russel Wallace [excerpt from letter of support read at the annual meeting of the A-VL]. *Vaccination Inquirer* 29: 4 (1 April 1907: no. 337).

637.* Archdeacon Colley and Mr. Maskelyne [article/N including part of the transcript of Wallace's testimony at the libel trial of J. N. Maskelyne in late April 1907]. *Light* (London) 27: 208–209 (207–209, 213) (4 May 1907: no. 1373).

—reprints: testimony summarized as part of note/N entitled 'Colley v. Maskelyne' in *The Times* (London) no. 38318: 4c (4b-d) (27 April 1907); Wallace's own "proof" of what he intended to say printed in note/N entitled 'Dr. A. R. Wallace's Testimony to Materialisations' in *Light* (London) 33: 557 (22 Nov. 1913: no. 1715).

638.* Dr. Wallace's Letter [of support to Joseph Hyder, read at the 26th annual meeting of the LNS, 14 May 1907]. *Land and Labour* 18(6): 65 (June 1907).

—reprint: included in note/N entitled 'Land Nationalization Society' printed in *The Times* (London) no. 38334: 3f (16 May 1907).

639.* "Economic Chivalry" [article/N posting survey of opinions regarding Bishop Gore's report 'The Moral Witness of the Church on Economic Subjects;' Wallace's one of several criticisms printed]. *Public Opinion* (London) 91: 639 (639–640) (24 May 1907: no. 2383).

640.* The New Vaccination Bill [lttE]. *The Clarion* (London) no. 808: 2e–f (31 May 1907); highly edited version printed as 'A Simple Declaration' in *The Daily News* (London) no. 19093: 4e (27 May 1907); excerpt from letter entitled 'Dr. A. R. Wallace' included in article/N entitled 'The Government Vaccination Bill. Text of the Measure and some Opinions upon It.' printed in *Vaccination Inquirer* 29: 47 (44–48) (1 June 1907: no. 339).

641.* *Britain's Hope*: An Appreciation [notice of *Britain's Hope; An Open Letter concerning the Pressing Social Problems to the Rt. Hon. John Burns, M. P.* by Julie Sutter, 1907]. *Land and Labour* 18(6): 73 (June 1907).

642. lttE [one of two printed as 'The "Double Drift" Theory of Star Motions']. *Nature* 76: 293 (293–294) (25 July 1907: no. 1969).

643.* letter [to Editor Robert Blatchford; included in a Blatchford article entitled 'The Socialist Ideal']. *The Clarion* (London) no. 819: 1f (1f) (16 Aug. 1907).

644.* Dr. Alfred Russel Wallace on Socialist Poets [letter to columnist A. E. Fletcher]. *The Clarion* (London) no. 823: 2a (13 Sept. 1907).

645. Dr. A. R. Wallace & Sir W. M. Ramsay's Theory; Did Man Reach his Highest

Development in the Past? [letter]. *Public Opinion* (London) 92: 336 (13 Sept. 1907: no. 2399).

646. How Life became Possible on the Earth. in *Harmsworth History of the World* ed. by Arthur Mee, J. A. Hammerton, & A. D. Innes (8 vols., Carmelite House, London, 1907–1909), Vol. 1: 91–98 (11 Oct. 1907).

647.* letter [to columnist Julia Dawson, giving advice to Socialist women of Great Britain]. *The Clarion* (London) no. 838: 7c (27 Dec. 1907).

648.* Dr. Wallace Pleased with Advance in Natural History Exploration [remarks contributed to a feature article/N entitled 'Leaders of Thought Tell *The World* of Achievements of 1907;' Wallace's one of many invited responses printed]. *The World* (New York) no. 16931: 2a–b (1–3) (29 Dec. 1907).

649. Evolution and Character. *Fortnightly Rev.* 83 (n.s.; 89, o.s.): 1–24 (1 Jan. 1908: no. 493, n.s.).
—reprint: in *Character and Life; A Symposium* ed. and arranged by Percy L. Parker (Williams & Norgate, London, Nov. 1912): 3–50.

650. lttE [one of several printed as 'Dr. Russel Wallace and Woman']. *The Outlook* (London) 21: 89 (89–90) (18 Jan. 1908: no. 520).

651.* What to Eat, Drink, and Avoid [article/N posting results of a general inquiry; Wallace's one of many responses printed]. *The Review of Reviews* (London) 37: 137–138 (136–146) (Feb. 1908: no. 218).
—excerpts reprinted in *Land and Labour* 19(5): 52 (May 1908).

652.* Dr. A. R. Wallace's Message [letter of support read at the annual meeting of the A-VL held 11 March 1908]. *Vaccination Inquirer* 30: 8 (1 April 1908: no. 349).

653.* Letter from Dr. A. R. Wallace [to Joseph Hyder, read at the 27th annual meeting of the LNS, 29 April 1908]. *Land and Labour* 19(6): 65 (June 1908).

654.* letter [to columnist Julia Dawson, concerning the book that converted Wallace to Socialism, *Looking Backward*]. *The Clarion* (London) no. 861: 9c–d (5 June 1908).

655. The Remedy for Unemployment [in large part an account of *Poverty and the State* by Herbert V. Mills, 1889] I. *Socialist Rev.* 1: 310–320 (June 1908) / II. 1: 390–400 (July 1908).
—revised version printed as pamphlet (Pass On Pamphlets, No. 8: The Clarion Press, London, Jan. 1909; pp. 1–(24)).

656.* address [acceptance speech on receiving the Darwin-Wallace Medal on 1 July 1908]. in *The Darwin-Wallace Celebration Held on Thursday, 1st July 1908, by the Linnean Society of London* (printed for the Linnean Society by Burlington House, Longmans, Green & Co., London, Feb. 1909): 5–11.
—reprint: as 'The Origin of the Theory of Natural Selection' in *Popular Science Monthly* 74: 396–400 (April 1909).

657. The First Paper on Natural Selection [lttE]. *The Times* (London) no. 38689: 12b (3 July 1908).

658.* Dr. A. R. Wallace and Honours [lttE concerning an erroneous report that Wallace was soon to be knighted]. *Public Opinion* (London) 94: 78a (17 July 1908: no. 2443).

659. Is It Peace or War?; A Reply by Dr. Alfred R. Wallace [letter responding to article printed the week before]. *Public Opinion* (London) 94: 202–203 (14 Aug. 1908: no. 2447).

660. The Present Position of Darwinism. *Contemporary Rev.* 94: 129–141 (Aug. 1908).
—reprint: *The Living Age* 258 (7th s., vol. 40): 707–716 (19 Sept. 1908: no. 3350).

661. Nationalisation, *not* Purchase, of Railways [lttE]. *The New Age* 3: 417–418 (19 Sept. 1908: no. 21, n.s.; 732, o.s.).

662. The Facts Beat Me. *Delineator* 72: 542 (Oct. 1908) / The Dead have never really Died. 72: 852–853 (Nov. 1908). [included as short sections of a several part article/N entitled 'Are the Dead Alive?'; in large part consisting of excerpts from ref 398 and the Preface to the 3rd ed. of *Miracles and Modern Spiritualism* (Jan. 1896)]

663.* letter [concerning Wallace's notification of winning the Copley Medal; read at the Royal Society's annual meeting of 30 Nov. 1908]. *The Times* (London) no. 38818: 9e (1 Dec. 1908).

664.* A Veteran Scientist's Testimony [note responding to an inquiry regarding Wallace's thoughts on Jesus Christ and his teachings]. *The Christian Commonwealth* 29: 166d (9 Dec. 1908: no. 1417).
—reprinted untitled in *Light* (London) 28: 610 (19 Dec. 1908: no. 1458).

665. 1909. For What should We Strive? [article/N posting results of an opinion survey; Wallace's one of many responses printed]. *The Christian Commonwealth* 29: 231a (231a–b) (30 Dec. 1908: no. 1420).

666. Darwinism *versus* Wallaceism. *Contemporary Rev.* 94: 716–717 (Dec. 1908).

667.* Preface. to *Notes of a Botanist on the Amazon and Andes* by Richard Spruce (ed. and condensed by Alfred Russel Wallace, Dec. 1908), Vol. 1: v–ix.

668.* Biography. in *Notes of a Botanist on the Amazon and Andes* by Richard Spruce (ed. and condensed by Alfred Russel Wallace, Dec. 1908), Vol. 1: xxi–xlvii.

669. The World of Life: As Visualised and Interpreted by Darwinism [revision of a lecture a portion of which was delivered at the Royal Institution on 22 Jan. 1909]. *Fortnightly Rev.* 85 (n.s.; 91, o.s.): 411–434 (1 March 1909: no. 507, n.s.); also printed in *The Independent* (New York) 66: 398–404 (25 Feb. 1909: no. 3143) / 66: 449–457 (4 March 1909: no. 3144).
—excerpts, summaries, etc., under various titles: *Nature* 79: 384–386 (28 Jan. 1909: no. 2048); *The Times* (London) no. 38864: 9e (23 Jan. 1909); *Popular Science Monthly* 75: 452–457 (Nov. 1909); *Scientific American Supplement* 68: 350–351 (27 Nov. 1909: no. 1769).

670. Flying Machines in War. Dr. A. R. Wallace Calls to Action. [lttE]. *The Daily News* (London) no. 19626: 4f (6 Feb. 1909).

671.* Dr. A. R. Wallace and Woman Suffrage [note/N including letter from Wallace read at a meeting to support woman suffrage held 10 Feb. 1909 at Godalming]. *The Times* (London) no. 38880: 10d (11 Feb. 1909).

672. To-day's Centenaries. Charles Darwin. *The Daily Mail* (London) no. 4007: 4d (12 Feb. 1909).

673. The Centenary of Darwin. *The Clarion* (London) no. 897: 5c–d (12 Feb. 1909).

674.* Dr. A. R. Wallace [letter of support read at the annual meeting of the A-VL held 24 Feb. 1909]. *Vaccination Inquirer* 30: 202 (1 March 1909: no. 360).

675.* Note to the Present Edition. to *Psychic Philosophy as the Foundation of a Religion of Natural Law* (new ed.) by V. C. Desertis [pseudonym of Stanley DeBrath] (William Rider & Son, Ltd., London, Feb. 1909): vi.

676.* Note on the Passages of Malthus's *Principles of Population* which Suggested the Idea of Natural Selection to Darwin and Myself. in *The Darwin–Wallace Celebration Held on Thursday, 1st July 1908, by the Linnean Society of London* (printed for the Linnean Society by Burlington House, Longmans, Green & Co., London, Feb. 1909): 111–118.

677. Aerial Fleets. Dr. Russel Wallace on Idle Panics. [lttE]. *The Daily News* (London) no. 19678: 4e (8 April 1909).
—reprint: as 'Aerial Fleets and Idle Panics' in *Land and Labour* 20(5): 55 (May 1909).

678.* letter [of support to Joseph Hyder, read at the 28th annual meeting of the LNS, 13 May 1909]. *Land and Labour* 20(6): 64 (June 1909).

679.* The Development Fund. Letter from Dr. A. R. Wallace. *Land and Labour* 20(10): 112–113 (Oct. 1909).

680.* letter [of support (as Honorary Vice-President) read at the second annual conference of the A-VL of America, held 29 Dec. 1909 in Philadelphia]. *Vaccination Inquirer* 31: 253 (1 March 1910: no. 372).

681.* Dr. Wallace's High Praise [note/N including letter to Ernest McCormick regarding latter's article 'Is Vaccination a Disastrous Delusion?']. *Vaccination Inquirer* 31: 228 (1 Feb. 1910: no. 371).

682.* Alfred Russel Wallace, LL.D., O.M., etc. [letter of support read at the annual meeting of the A-VL held 15 March 1910]. *Vaccination Inquirer* 32: 5–6 (1 April 1910: no. 373).

683.* telegram [of support read at the 29th annual meeting of the LNS, 18 April 1910]. *Land and Labour* 21(5): 51 (May 1910).

684. A New Era in Public Opinion; Some Remarkable Changes in the Last Half-century. *Public Opinion* (London) 98: 377 (14 Oct. 1910: no. 2555).

685.* Life after Life. "Star-shine and Immortal Tears." [article/N including a letter by Wallace concerning spiritualism]. *The Clarion* (London) no. 998: 5c (5c–e) (20 Jan. 1911).
—excerpts printed as 'Life after Life' in *Light* (London) 31: 43 (28 Jan. 1911: no. 1568).

686.* letter [concerning the still-holding arguments posed in *Vaccination a Delusion*; presented at interview of Deputation of the A-VL with the Edinburgh Public Health Committee on 24 Jan. 1911]. *Vaccination Inquirer* 32: 293 (1 March 1911: no. 384).
—reprint: as 'Dr. A. R. Wallace and Vaccination' in *Public Opinion* (London) 99: 206c (3 March 1911: no. 2575).

687.* Alfred Russel Wallace, LL.D, O.M., etc. [letter of support read at the annual meeting of the A-VL held 28 March 1911]. *Vaccination Inquirer* 33: 42 (1 May 1911: no. 386).

688.* letter [of support to Joseph Hyder, read at the 30th annual meeting of the LNS, 15 May 1911]. *Land and Labour* 22(6): 61 (June 1911).

689.* letter extracts [passages from a letter to Clement Reid concerning the relation of the present flora of the British Isles to the Glacial Period; communicated to the 4 Sept. 1911 (Portsmouth) meeting of Section K, Botany, of the BAAS]. in *Report of the Eightieth Meeting of the BAAS* (John Murray, London, 1912): 577–578.

—letter first printed in note/N entitled 'A Letter from Dr. A. R. Wallace' in *The Times* (London) no. 39682: 10d–e (5 Sept. 1911); also described and quoted from as part of 'Botany at the British Association' in *Nature* 88: 57 (57–58) (9 Nov. 1911: no. 2193).

690.* Mr. A. R. Wallace and the Insurance Act [lttE]. *The Times* (London) no. 39781: 6c (29 Dec. 1911).

691. Dr. Russel Wallace on Insurance Act [letter]. *The Daily Chronicle* (London) no. 15581: 4d (25 Jan. 1912).

692.* Naturalist Answers Birthday Greeting. A. R. Wallace Sends Letter. *Silver and Gold* [student newspaper at the University of Colorado, Boulder] 20(52): 4c (2 Feb. 1912).

—excerpt reprinted untitled in *Nature* 89: 12–13 (7 March 1912: no. 2210).

693.* Dinner to Mr. Hyndman. Mr. Bernard Shaw on the Strike. [note/N containing summary of letter Wallace sent on the occasion of Henry H. Hyndman's 70th birthday]. *The Times* (London) no. 39841: 7c (8 March 1912).

694.* Alfred Russel Wallace, LL.D. O.M., etc. [letter of support read at the annual meeting of the A-VL held 19 March 1912]. *Vaccination Inquirer* 34: 11 (1 April 1912: no. 397).

695.* letter [to Isaac Bickerstaffe, printed in Part I of his article 'Some Principles of Growth and Beauty']. *The Field, The Country Gentleman's Newspaper* 119: 946b (946–948) (11 May 1912: no. 3098).

—reprint: in *The Curves of Life; Being an Account of Spiral Formations and their Application to Growth in Nature, to Science and to Art* by Theodore Andrea Cook (Constable & Co., London / H. Holt & Co., New York, Oct. 1914): 3–4.

696.* letter [of support to Joseph Hyder, read at the 31st annual meeting of the LNS, 16 May 1912]. *Land and Labour* 23(6): 63–64 (June 1912).

697.* Dr. Alfred Russel Wallace's Letter [letter to Bombay medical authorities concerning the accuracy of statistics Wallace presented before the Royal Commission on Vaccination]. *Vaccination Inquirer* 34: 121 (1 Aug. 1912: no. 401).

698. A Policy of Defence [lttE]. *The Daily News & Leader* (London & Manchester) no. 20723: 4f (9 Aug. 1912).

699.* Mr. Blatchford's Dogmatism [letter commenting on discussion on spiritualism between *The Clarion* Editor Robert Blatchford and *The Christian Commonwealth*]. *The Christian Commonwealth* 32: 815b (11 Sept. 1912: no. 1613).

—excerpt reprinted untitled in *Light* (London) 32: 436 (14 Sept. 1912: no. 1653).

700. The Origin of Life. A Reply to Dr. Schäfer. *Everyman* 1(1): 5–6 (18 Oct. 1912).

—excerpts reprinted in article/N entitled 'Doctor Alfred Russel Wallace's Challenge to the Chemical Theory of Life' in *Current Literature* 53(6): 657–659 (Dec. 1912).

701.* Dr. A. R. Wallace's Birthday [note/N including one-line response by Wallace to letter sent by the London Spiritualist Alliance on his 90th birthday]. *Light* (London) 33: 28 (18 Jan. 1913: no. 1671).

702.* letter [of support read at the annual meeting of the A-VL held 13 March 1913]. *Vaccination Inquirer* 35: 9 (1 April 1913: no. 409).
—reprint: in note/N entitled 'Vaccination a Crime' describing the annual meeting printed in *The Times* (London) no. 40159: 4d (14 March 1913).

703.* letter [of support to Joseph Hyder, read at the 32nd annual meeting of the LNS, 28 May 1913]. *The Land Nationaliser* 24(6): 88 (June 1913).

704.* A Scientist's Sleepless Hours [note/N printing a letter to columnist Solomon Eagle concerning the authorship of two poems; first referred to in Eagle's column in *New Statesman* (London) 1: 790 (27 Sept. 1913: no. 25)]. *Public Opinion* (London) 104: 568 (21 Nov. 1913: no. 2717).

705.* Dr. A. R. Wallace's Faith [note/N containing one-sentence reply to an inquiry]. *The Christian Commonwealth* 34: 120b (12 Nov. 1913: no. 1674).
—reply printed with others' in an article/N entitled 'The Faith I Live By' in *The Christian Commonwealth* 34: 193c (190–193) (10 Dec. 1913: no. 1678).

706.* Dr. Wallace on the Genesis of the Soul [letter dated 5 April 1903 to an unnamed correspondent; submitted by S. H. Leonard]. *Spectator* 111: 863 (22 Nov. 1913: no. 4456).

707.* Preface. to *The Case for Land Nationalisation* by Joseph Hyder (Simpkin, Marshall, Hamilton, Kent & Co., Ltd., London, Nov. 1913): v–viii.

708.* *Edgar Allan Poe; A Series of Seventeen Letters concerning Poe's Scientific Erudition in Eureka and his Authorship of "Leonainie"* [letters to Ernest Marriott assembled by an unknown compiler and privately printed as a pamphlet]. New York, n.d. (possibly, and no later than, 1930, its cataloguing date); pp. (1)–18.

709.* Letters concerning Wallace, Newton and Tristram [note by A. L. Pickens reprinting 13 Jan. 1874 letter concerning bird families classification from Wallace to Canon Tristram; second letter is from Alfred Newton to Tristram]. *Auk* 51 (n.s.; 59, o.s.): 404–406 (July 1934).

710.* Letters of Alfred Russel Wallace to Lester F. Ward [article by Bernhard J. Stern centring on, and reprinting, Wallace's letters]. *The Scientific Monthly* 40: 375–379 (April 1935).

711.* A Letter from A. R. Wallace to F. P. Pascoe, Written from Ternate, 20 December 1860 [note by Hale Carpenter]. *Proc. Roy. ESL Series A* 14: 77–78 (1939).

712.* Letters from Edward B. Tylor and Alfred Russel Wallace to Edward Westermarck; ed. with Introductory Remarks concerning the Publication of *The History of Human Marriage* [article by K. Rob. V. Wikman]. *Acta Academiae Aboensis Humaniora* XIII.7; pp. (1)–22 (Wallace's letters presented on pp. 13–22) (1940).

Two unverified works: (a) letters concerning the storage of explosives [according to *My Life* (Vol. 2, p. 396), appearing in one or more newspapers in late June or July 1882]; (b) 'Why not British Guiana?' [according to Marchant (p. 482), appearing in *The Daily News* of 27 Oct. 1906, but apparently not so].

Part II: Books and excerpts from books

713. *Palm Trees of the Amazon and their Uses.* John Van Voorst, London, Oct. 1853; pp. (i)–viii, (1)–129; 48 plates (incl. map front.).

714. *A Narrative of Travels on the Amazon and Rio Negro, with an Account of the Native Tribes, and Observations on the Climate, Geology, and Natural History of the Amazon Valley.* Reeve & Co., London, Dec. 1853; pp. (i)–(x), (1)–541; front., map, illustrations, fold-out. [reprint: Macmillan & Co., London (Dec. 1870)]

—Second Edition [moderately revised; deletes Appendix on Indian languages]. Minerva Library of Famous Books, Vol. 8: Ward, Lock & Co., London, New York, & Melbourne, Nov. 1889; pp. (i)–(xviii), 1–363; 16 illus. (incl. portrait front. & map). [reprints: 3rd ed. (1890); 4th ed. (1892) / New (Fifth) Edition. Ward, Lock & Bowden, Ltd., London (1895 & May 1900) / *Travels on the Amazon.* World Library: Harper & Brothers, New York, & Ward, Lock & Co., Ltd., London (n.d. (1889); pp. (i)–(x), (1)–363; map, illustrations) / *Travels on the Amazon and Rio Negro.* New Amsterdam Book Co., New York (1900; pp. (i)–(xvi), (1)–363; 16 illus. (incl. portrait front. & map)) / *Travels on the Amazon.* World Library: Ward, Lock & Co., Ltd., London, Melbourne & Toronto (May 1911; pp. (i)–(x), (1)–363; map, illustrations).]

715. *The Malay Archipelago; The Land of the Orang-utan and the Bird of Paradise; A Narrative of Travel with Studies of Man and Nature.* 2 volumes. Macmillan & Co., London, Feb. 1869. Vol. 1: pp. (i)–(xxiv), (1)–478; 2 fold-out & 3 other maps, 27 illus. (incl. front.). Vol. 2: pp. (i–vi), (1)–524; 4 maps, 24 illus. (incl. front.). [reprint: 2nd ed. (Sept. 1869; illustrations, etc., as above; Vol. 1: pp. (i)–(xx), (1)–312; Vol. 2: pp. (i)–(iv), (1)–341)] / *The Malay Archipelago: The Land of the Orang-utan, and the Bird of Paradise. A Narrative of Travel, with Studies of Man and Nature.* Harper & Brothers, New York, 1 May 1869; pp. (i)–xii, (13)–638; 2 fold-out & 8 other maps, 51 illus. (incl. 2 fronts.) [reprint: 1885]

—Third Edition [lightly revised]. Macmillan & Co., London & New York, Feb. 1872; pp. (i)–xvi, (1)–653; 2 fold-out & 8 other maps, 51 illus. (incl. front.). [reprints: 4th ed. (Oct. 1872); 5th ed. (1874); 6th ed. (1877); 7th ed. (1880); 8th ed. (1883); 9th ed. (1886)]

—Tenth Edition [lightly revised]. Macmillan & Co., London & New York, Jan. 1891 (London) & Feb. 1891 (New York); pp. (i)–(xx), (1)–515; 2 fold-out & 8 other maps, 51 illus. (incl. front.). [reprints: 1894, 1898, 1902, 1906, 1913, 1922]

—[reprinted excerpts: 'The Habits of the Orang-utan' in *Scientific Opinion* 2: 110–111 (7 July 1869: no. 36); 'The Birds of Paradise' in *Harper's New Monthly Mag.* 39: 183–195 (July 1869: no. 230); 'How the Rajah Took the Census' & 'Life in the Malay Archipelago' in *Library of the World's Best Literature; Ancient and Modern* ed. by Charles Dudley Warner (45 vols., International Society, New York, 1896), Vol. 38: 15519–15530.]

716. *Contributions to the Theory of Natural Selection. A Series of Essays.* Macmillan & Co., London & New York; April 1870; pp. (i)–xvi, (1)–384.

—Second Edition ["with corrections and additions"]. Macmillan & Co., London & New York, March 1871; pp. (i)–xvi, (1)–384 (plus additional lettered pages). [reprints: Oct. 1872, Feb. 1875]

—[reprinted excerpts: 'Ueber Schützende Aehnlichkeiten bei Thieren,' selections from *Beiträge zur Theorie der Natürlichen Zuchtwahl. Eine Reihe von Essais* (Adolf Bernhard Meyer's German translation of *CTNS*: Eduard Besold, Erlangen, Juni 1870), in *Der Zoologische Garten* 12(9): 263–269 (Sept. 1871) / 12(10): 300–303 (Okt. 1871) / 12(11): 334–339 (Nov. 1871).]

717. *On Miracles and Modern Spiritualism. Three Essays.* [each essay "carefully revised"]. James Burns, London, March 1875; pp. (i)–viii, (1)–236. [reprint: Second Edition. Trübner & Co., London (Oct. 1881; pp. (iii)–(x), (1)–236).]

—*Miracles and Modern Spiritualism.* Revised [Third] Edition ["with chapters on apparitions and phantasms," and a new Preface]. George Redway, London, Jan. 1896; pp. (i)–(xx), (1)–292. [reprint: Nichols & Co., London (1901)]

718. *The Geographical Distribution of Animals; with A Study of the Relations of Living and Extinct Faunas as Elucidating the Past Changes of the Earth's Surface.* 2 volumes. Macmillan & Co., London, May 1876 / Harper & Brothers, New York, 9 Dec. 1876. Vol. 1: pp. (i)–(xxiv), (1)–503; 5 maps (incl. fold-out front.), 13 plates. Vol. 2: pp. (i)–(xii), (1)–607; 2 maps, 7 plates.

—[reprinted excerpts: 'The Journeyings and Dispersal of Animals' in *Popular Science Monthly* 10: 576–582 (March 1877); 'The Geographical Distribution of Animals: General Conclusions' (Chapter 16) in *Amer. Naturalist* 11(3): 157–165 (March 1877).]

719. *Tropical Nature, and Other Essays.* Macmillan & Co., London & New York, April 1878; pp. i–(xvi), 1–356; illustrations.

—[reprinted excerpts: 'La Nature Tropicale,' selection translated into French in *La Revue Scientifique* 16 (2[e] s.): 837–843 (8 Mars 1879); 'Tropical Vegetation' & 'Orchids' in 'Alfred Russel Wallace,' 4-page selection in *The University of Literature*, Editor-in-Chief, William H. De Puy (J. S. Barcus & Co., New York, 1895), Vol. XIX (not paginated); 'A Profitable Sugar Palm' (pp. 43–45) in *Timehri* 1 (n.s.): 134–136 (June 1887).]

720. *Australasia* ["edited and extended by Alfred R. Wallace, with Ethnological Appendix by A. H. Keane"]. Stanford's Compendium of Geography and Travel: Edward Stanford, London, May 1879; pp. (i)–(xx), (1)–672; 20 maps, 56 illus. (incl. front.). [reprint: 2nd ed. (1880)]

—Third Edition [moderately revised; pagination, etc., as above]. June 1883. [reprints: 4th ed. (1884); 5th ed. (Feb. 1888)]

—*Australasia. Vol. 1. Australia and New Zealand.* [complete revision of one-half of 3rd ed. (other half revised by another author)]. Stanford's Compendium of Geography and Travel (New Issue): Edward Stanford, London, Nov. 1893; pp. (i)–(xvi), (1)–505; 14 maps, 69 illus. (incl. front.).

721. *Island Life: or, The Phenomena and Causes of Insular Faunas and Floras, Including a Revision and Attempted Solution of the Problem of Geological Climates.* Macmillan & Co., London, Oct. 1880; pp. (iii)–(xx), (1)–526; 26 maps & illus. (incl. map front.). / Harper & Brothers, New York, 1881; pp. (i)–xvi, (1)–522; 26 maps & illus. (incl. map front.).

—Second and [heavily] Revised Edition. Macmillan & Co., London & New York, April 1892 (London) & 14 May 1892 (New York); pp. (i)–xx, (1)–563; 26 maps & illus. (incl. map front.).

—"Second Edition" [but with additional corrections thereto; pagination, etc., as above]. 1895.

—Third and [heavily] Revised Edition. Macmillan & Co., Ltd., London / The Macmillan Co., New York, 1902; pp. (i)–xx, (1)–563 (plus additional lettered pages); 26 maps & illus. (incl. map front.). [reprint: Macmillan & Co., Ltd., London (1911)]

722. *Land Nationalisation; Its Necessity and its Aims; Being a Comparison of the System of Landlord and Tenant with that of Occupying Ownership in their Influence on the Well-being of the People.* Trübner & Co., London, May 1882; pp. (i)–xiv, (1)–240 (plus 3 unnumbered pages entitled "Works to be Consulted on the Land Question" appended). [reprint: 2nd ed. (Oct. 1882)]

—Third Edition [lightly revised]. W. Reeves, London, April 1883; pp. (i)–xiv, (1)–240 (plus 3 unnumbered pages entitled "Works to be Consulted on the Land Question" appended).

—"Third Edition" ["with appendix on the nationalisation of house property"]. W. Reeves, London, Sept. 1883; pp. (i)–xiv, (1)–247 (plus 3 unnumbered pages entitled "Works to be Consulted on the Land Question" appended).

—[new edition; lightly revised]. Social Science Series, No. 57: Swan Sonnenschein & Co., London, & Charles Scribner's Sons, New York, July 1892; pp. (i)–xiv, (1)–252 (plus 4 unnumbered pages entitled "A Short Bibliography of the Land Question" appended). [reprints: 2nd ed. (Sept. 1896); 3rd ed. (Oct. 1902); 4th ed. (Nov. 1906); 5th ed. (June or July 1909)]

723. *Bad Times: An Essay on the Present Depression of Trade, Tracing It to its Sources in Enormous Foreign Loans, Excessive War Expenditure, the Increase of Speculation and of Millionaires, and the Depopulation of the Rural Districts; with Suggested Remedies.* Macmillan & Co., London & New York, Nov. 1885 (London) & Jan. 1886 (New York); pp. (iii)–(xvi), (1)–118; 3 diagrams.

—[reprinted excerpt: 'The Increasing Curse of European Militancy' (Chapter 5) in *Popular Science Monthly* 28: 521–525 (Feb. 1886).]

724. *Darwinism; An Exposition of the Theory of Natural Selection with Some of its Applications.* Macmillan & Co., London & New York, May 1889 (London) & June 1889 (New York); pp. (i)–xvi, (1)–494; portrait front., fold-out map, 37 figs.

—Second Edition [lightly revised]. Macmillan & Co., London & New York, Aug. 1889; pp. (i)–xvi, (1)–494; portrait front., fold-out map, 37 figs. [reprints: Oct. 1889, March 1890, 1891, 1896, 1897 / Humboldt Library of Popular Science, Nos. 115–116: Humboldt Publishing Co., New York (Dec. 1889; pp. (i)–x, (1)–332; portrait front., fold-out map, 37 figs.); also issued as individual pamphlets Nos. 115 (Oct. 1889; pp. (i)–x, (1)–160) & 116 (Nov. 1889; pp. 161–332)]

—Third Edition [heavily revised]. Macmillan & Co., Ltd., London / The Macmillan Co., New York, 1901; pp. (i)–xx, (1)–494 (plus additional lettered pages); portrait front., fold-out map, 37 figs.

—"Third Edition" [but with additional corrections thereto; pagination, etc., as above]. 1905. [reprints: 1912, 1923]

—[reprinted excerpts: 'Darwinism as Applied to Man' (Chapter 15) in *Representative Essays in Modern Thought; A Basis for Composition* ed. by Harrison Ross Steeves & Frank Humphrey Ristine (American Book Co., New York, etc., Feb. 1913): 238–271; Chapter 15 translated into Italian and reprinted as pamphlet entitled *Il Darwinismo Applicato all' Uomo* (Napoli, 1912; pp. 94).]

725. *Natural Selection and Tropical Nature; Essays on Descriptive and Theoretical Biology* ["new edition with corrections and additions;" this collection combines the majorities of *CTNS* and *TNOE*, with some newly reprinted material added]. Macmillan & Co., London & New York, May 1891 (London) & June 1891 (New York); pp. (i)–xii, (1)–492; illustrations. [reprint: 1895]

726. *The Wonderful Century; Its Successes and its Failures*. Swan Sonnenschein & Co., London, June 1898 / Dodd, Mead & Co., New York, Sept. 1898 / George N. Morang, Toronto, 1898; pp. (i)–xii, (1)–400 (internal pagination of English & American editions slightly different); portrait front., 12 fold-out diagrams. [reprints: Swan Sonnenschein & Co., London, 2nd ed. (Dec. 1898); 3rd ed. (May 1899); 4th ed. (Feb. 1901) / Dodd, Mead & Co., New York (1899 & 1909)]

—*The Wonderful Century Reader* [abridgment of the 4th ed.]. Swan Sonnenschein & Co., Ltd., London, Nov. 1901; pp. (i)–(viii), (1)–234; 127 illus.

—*The Wonderful Century; The Age of New Ideas in Science and Invention*. New (Fifth) Edition ["revised and largely re-written"]. Swan Sonnenschein & Co., Ltd., London, Sept. 1903; pp. (i)–xii, (1)–527; portrait front., 107 illus. [reprints: 6th ed. (April 1905); 7th ed. (Jan. 1908) / E. W. Cole, Melbourne (1906) / George Allen & Unwin, Ltd., London (1925)]

—[reprinted excerpts: 'Our Wonderful Century' in *New York Times Saturday Rev. of Books and Art* 10 Sept. 1898: 605; 'The Discoveries of the Century' in *The Living Age* 219 (7th s., vol. 1), *Supplement* to no. 2830 (*Readings from New Books*): 63–65 (1 Oct. 1898); nine selections incorporated into *Achievements in Science* (Modern Achievement, Vol. 7: P. F. Collier & Son, New York, 1902) and *Men of Achievement: Inventors and Scientists* (Modern Achievement, Vol. 7: The University Society, New York, 1902), both ed. by Edward Everett Hale; same nine selections incorporated into *Wonders of Science and Invention* ed. by Daniel Edwin Wheeler (The After School Library, Vol. 8: The After School Club, Philadelphia, 1909 / Young Folks' Treasury, Vol. 8: The University Society Inc., New York, 1909).]

727. *Studies Scientific and Social* [a collection of essays]. 2 volumes. Macmillan & Co., Ltd., London, Nov. 1900 / The Macmillan Co., New York, Dec. 1900. Vol. 1: pp. (i)–(xvi), (1)–532; map, 89 illus. Vol. 2: pp. (i)–(viii), (1)–535; 25 illus.

728. *Man's Place in the Universe; A Study of the Results of Scientific Research in relation to the Unity or Plurality of Worlds*. Chapman & Hall, Ltd., London, Oct. 1903; pp. (i)–(xii), (1)–330; 8 diagrams, 2 fold-out star-charts. [reprints: 2nd ed. (190?); 3rd ed. (1904)] / McClure, Phillips & Co., New York, Nov. 1903; pp. (i)–(xii), (1)–326; 8 diagrams, 2 fold-out star charts. [reprint: 2nd ed. (1904)]

—Fourth Edition [lightly revised, "with new chapter entitled 'An Additional Argument Dependent on the Theory of Evolution'"]. Chapman & Hall, Ltd., London, Sept. 1904; pp. (i)–(viii), (1)–341; 8 diagrams, 2 fold-out star-charts. [reprints: 5th ed. (1905); 6th ed. (190?); 7th ed. (1908) / New and Cheaper Edition (Sept. 1912 & 1914; pp. (i)–(vi), 1–283; 8 diagrams)] / Third Edition [lightly revised, "with new chapter entitled 'An Additional Argument Dependent on the Theory of Evolution'"]. McClure, Phillips & Co., New York, 1905; pp. (i)–(xii), (1)–338; 8 diagrams, 2 fold-out star-charts.

729. *My Life; A Record of Events and Opinions.* 2 volumes. Chapman & Hall, Ltd., London, Oct. 1905. Vol. 1: pp. (i)–(xii), (1)–435; portrait front., 2 fold-out maps, illustrations. Vol. 2: pp. (i)–viii, (1)–459; front., illustrations. [reprint: Bell's Indian and Colonial Library: George Bell & Sons, London & Bombay (1905)] / Dodd, Mead & Co., New York, Nov. 1905. Vol. 1: pp. (i)–xii, (1)–435; 2 fold-out maps, portrait front., illustrations. Vol. 2: pp. (i)–(viii), (1)–464; front., illustrations. [reprint: 1906]

—New Edition ["condensed and revised"]. Chapman & Hall, Ltd., London, autumn 1908; pp. (i)–xii, (1)–408; portrait front., illustrations.

—[reprinted excerpts: 'The Inclosure of Commons' in *Land and Labour* 16(12): 134–136 (Dec. 1905); *Spiritualistic Experiences* [pamphlet reprinting Wallace's discussion of spiritualism] (Propaganda Publications, No. V: Spiritualists' National Union, Ltd., Halifax, 1911; pp. 71).]

730. *Is Mars Habitable? A Critical Examination of Professor Percival Lowell's Book "Mars and its Canals," with an Alternative Explanation.* Macmillan & Co., Ltd., London, Dec. 1907 / The Macmillan Co., New York, 25 Jan. 1908; pp. (i)–xii, (1)–110; map front., diagram.

731. *Notes of a Botanist on the Amazon and Andes* [by Richard Spruce, "edited and condensed by Alfred Russel Wallace"]. 2 volumes. Macmillan & Co., Ltd., London, etc., Dec. 1908. Vol. 1: pp. (i)–lii, 1–518; portrait front., 3 maps, 49 illus. Vol. 2: pp. (i)–xii, 1–542; 4 maps, 22 illus.

732. *The World of Life; A Manifestation of Creative Power, Directive Mind and Ultimate Purpose.* Chapman & Hall, Ltd., London, Dec. 1910; pp. (i)–xvi, 1–408; 110 illus. [reprints: 2nd ed. (1911); 3rd ed. (July 1911); 4th ed. (1911); 5th ed. (1911) / New and Cheaper Edition (Oct. 1914) / G. Bell & Sons, Ltd., London (1911)] / Moffat, Yard & Co., New York, Jan. 1911; pp. (i)–xvi, 1–441; 110 illus. [reprint: 1916 (with portrait front.)]

733. *Social Environment and Moral Progress.* Cassell & Co., Ltd., London, New York, Toronto & Melbourne, March 1913; pp. (i)–viii, 1–(164); portrait front. / Cassell & Co., New York, 1913; pp. (i)–(vi), 7–181. [reprints: Funk & Wagnalls Co., New York & London (1913 & 1914; pp. (i)–(vi), 7–181)]

—[reprinted excerpt: 'Love, Marriage, and Population' in *Hearst's Mag.* 24: 309–311 (Aug. 1913).]

734. *The Revolt of Democracy.* Cassell & Co., Ltd., London, New York, Toronto & Melbourne, Oct. 1913; pp. (i)–(xlvi), 1–82; portrait front. [reprint: Funk & Wagnalls Co., Edinburgh, New York & London (1914)]

Part III: Interviews

735.* The English Naturalist [anonymous interview]. *The Sunday Herald* (Boston) 31 Oct. 1886: 13a–c.

—excerpts reprinted as 'Dr. Alfred Russel Wallace, the Great English Naturalist, on Darwin, Agassiz, the Coming Democracy of England, "Home Rule," the Land Question, etc.' in *Banner of Light* (Boston) 60(9): 2a–c (13 Nov. 1886).

736.* interview [in large part based on refs 427 & 445]. *The Daily Chronicle* (London). [not traced; referred to in *My Life* (Vol. 2, p. 210)]

—reprint: as pamphlet entitled *Woman and Natural Selection* (Central National Society for Women's Suffrage, London, n.d.; pp. 4) [a copy of this is item no. 3031 of the Gerritsen Collection of Women's History].

737.* Heredity and Pre-natal Influences. An Interview with Dr. Alfred Russel Wallace. [by Sarah A. Tooley; in part based on ref 445]. *Humanitarian* 4 (n.s.), no. 2: 80–88 (Feb. 1894).
—excerpts reprinted as 'Dr. A. R. Wallace on Heredity' in *Light* (London) 14: 151–152 (31 March 1894: no. 690).

738.* A Visit to Dr. Alfred Russel Wallace, F.R.S. [interview by "A. D."]. *The Bookman* (London) 13: 121–124 (Jan. 1898: no. 76).
—excerpts reprinted as 'A Veteran Spiritualist. Mr. Alfred Russel Wallace, F.R.S.' in *The Two Worlds* (Manchester) 11: 25–26 (14 Jan. 1898: no. 531) and untitled in *The Bookman* (New York) 8(2): 100–102 (Oct. 1898).

739.* Spiritualism—and Things [interview by columnist "The Whatnot"]. *The Clarion* (London) no. 343: 213b–e (2 July 1898).

740.* interview. *The Christian Commonwealth* (1899?). [not traced; referred to in *The Christian Commonwealth* 10 Dec. 1903, p. 176a, & 12 Nov. 1913, p. 112c]

741.* A Visit to Dr. Alfred Russel Wallace [by Albert Dawson]. *The Christian Commonwealth* 23: 176a–177d (10 Dec. 1903: no. 1156).
—excerpts included in article entitled 'Interview with Dr. Alfred R. Wallace' in *Light* (London) 23: 602–603 (19 Dec. 1903: no. 1197) and in article entitled 'Dr. Wallace and the Spirit World' in *The Two Worlds* (Manchester) 16: 659 (25 Dec. 1903: no. 841).

742.* Master Workers. XVII.—Dr. Alfred Russel Wallace. [by Harold Begbie]. *The Pall Mall Mag.* 34: 73–79 (Sept. 1904: no. 137).
—extensive excerpts reprinted as 'Interview with Dr. Alfred R. Wallace' in *Light* (London) 24: 410–411 (27 Aug. 1904: no. 1233).

743.* A Man of the Time; Dr. Alfred Russel Wallace and his Coming Autobiography [interview by James Marchant]. *The Book Monthly* 2(8): 545–549 (May 1905).

744.* conversations [portions of conversations with William Allingham on the dates 2 August 1884 and 6 Nov. 1884, noted down in Allingham's diary (the second conversation including Lord Tennyson)]. in *William Allingham; A Diary* ed. by Helen P. Allingham & Dollie Radford (Macmillan & Co., Ltd., London, Oct. 1907): 329–330, 332–335
—quoted in an article by James Robertson entitled 'Tennyson and Dr. Alfred Russel Wallace' in *Light* (London) 31: 425 (9 Sept. 1911: no. 1599).

745.* Dr. Alfred Russel Wallace at Home [interview by Ernest H. Rann]. *The Pall Mall Mag.* 43: 274–284 (March 1909: no. 191).
—excerpts reprinted as 'A Glimpse of Dr. A. R. Wallace' in *Light* (London) 29: 135 (20 March 1909: no. 1471) and as 'Dr. Alfred Russel Wallace at Home' in *Public Opinion* (London) 95: 180 (19 Feb. 1909: no. 2474).

746.* New Thoughts on Evolution [interview by Harold Begbie]. *The Daily Chronicle* (London) no. 15197: 4d (3 Nov. 1910) / no. 15198: 4d–e (4 Nov. 1910).
—reprinted as pamphlet (1911; pp. 14). [not seen]

747.* A Scientist on Politics [interview by Harold Begbie]. *The Daily Chronicle* (London) no. 15206: 5a–c (14 Nov. 1910).

748.* Scientist's 88th Birthday. Interview with Dr. Russel Wallace. Social Problems. [anonymous interview]. *The Daily News* (London) no. 20227: 1e–f (9 Jan. 1911).

—excerpts reprinted as 'Dr. Alfred Russel Wallace. Activity at 88.' in *Land and Labour* 22(2): 16 (Feb. 1911) and as 'Russel Wallace and Socialism' in *The Clarion* (London) no. 997: 1f (13 Jan. 1911).

749.* The Great Strike—and After. Hopes of a National Peace. [interview by Harold Begbie]. *The Daily Chronicle* (London) no. 15622: 4d–e (13 March 1912).

—excerpts reprinted as 'Dr. Wallace and the Coal Strike' in *Land and Labour* 23(4): 39–40 (April 1912).

750.* The Last of the Great Victorians. Special Interview with Dr. Alfred Russel Wallace. [by Frederick Rockell]. *The Millgate Monthly* 7, part 2: 657–663 (Aug. 1912: no. 83).

—excerpts reprinted as 'Dr. Alfred Russel Wallace and Eugenics' in *Light* (London) 32: 392 (17 Aug. 1912: no. 1649).

751.* The Problem of Life [anonymous interview regarding a paper on the origin of life given by Prof. E. A. Schäfer in Dundee at the 82nd annual BAAS meeting]. *The Daily News & Leader* (London & Manchester) no. 20748: 1a–b (7 Sept. 1912).

—excerpts reprinted as part of 'Life Involved Before It is Evolved' in *Light* (London) 32: 440–441 (plus untitled excerpt on p. 436) (14 Sept. 1912: no. 1653) and untitled in *Supplement to The Christian Commonwealth* no. 93: 12 (18 Sept. 1912).

752.* The Spectre of Poverty [anonymous interview]. *The Daily News & Leader* (London & Manchester) no. 20850: 1a–b (6 Jan. 1913).

753.* Alfred Russel Wallace [interview by W. B. Northrop]. *The Outlook* (New York) 105: 618–622 (22 Nov. 1913).

Appendix: A chronology of important events in Wallace's life

8 January 1823: born at Usk, Monmouthshire, Wales.

early 1837: leaves school and is sent to London to be apprenticed to a builder; is first exposed to the ideas of Robert Owen.

summer 1837 to December 1843: apprenticed as a surveyor to his brother William in southern England and Wales.

early 1844 to early 1845: employed as a master at a private school in Leicester; meets Henry Walter Bates; reads *The vestiges of the natural history of creation*, by Robert Chambers.

summer 1845 to summer 1847: employed as surveyor and builder.

March 1848: leaves for South America with Bates.

May 1848 to July 1852: collecting expeditions in Amazonian South America (for a detailed chronology, see George 1964, p. 287).

6 August 1852: on return trip to London, Wallace's ship catches fire and sinks; Wallace rescued at sea.

October 1852 to March 1854: writes *Palm trees of the Amazon* (ref 713) and *Narrative of travels on the Amazon and Rio Negro* (ref 714); prepares for next collecting expedition.

March 1854: leaves England for Singapore.

April 1854 to February 1862: collecting expeditions throughout Indonesia (for a detailed chronology, see George 1964, pp. 288–90).

September 1855: 'On the law . . .' (ref 20) published.

February 1858: 'On the tendency . . .' (ref 43) composed and sent to Darwin.

1 July 1858: 'On the tendency . . .' and Darwin's abstracts read to the Linnean Society of London.

1 April 1862: returns to London.

1 March 1864: 'The origin of human races . . .' (ref 93) read to the Anthropological Society of London.

April 1866: marries Annie Mitten, daughter of botanist William Mitten.

August to September 1866: 'The scientific aspect of the supernatural' (ref 118) published.

August 1867, summer 1869, and 1875: botanizing excursions to North Wales.

February 1869: *The Malay Archipelago* (ref 715) published.

early 1870 to early 1872: President of the Entomological Society of London.

May and June 1874: 'A defence of modern spiritualism' (ref 243) published.

May 1876: *The geographical distribution of animals* (ref 718) published.

September 1876: President, Section D, Biology, of the British Association for the Advancement of Science annual meetings.

spring 1877: botanizing excursion to Belgium.

October 1880: *Island life* (ref 721) published.

1881: Land Nationalisation Society formed and Wallace made its first President; moves to Frith Hill, Godalming.

May 1882: *Land nationalisation* (ref 722) published.

September and October 1883: 'The "why" and the "how" of land nationalisation' (ref 365) published.

October 1886 to August 1887: lecture tour in the United States and Canada.

May 1889: *Darwinism* (ref 724) published.

June 1889: moves to Parkstone, Dorset.

February to May 1890: presents testimony to the Royal Commission on Vaccination.

September 1890: 'Human selection' (ref 427) published.

November and December 1893: 'The Ice Age and its work' (ref 481) published.

July 1895: botanizing excursion to Switzerland.

September 1896: lectures on scientific progress in Davos, Switzerland.

March 1898: *Vaccination a delusion* (ref 536) published.

June 1898: attends International Congress of Spiritualists.

December 1902: moves into Old Orchard, Broadstone, Wimborne.

October 1903: *Man's place in the universe* (ref 728) published.

October 1905: *My life* (ref 729) published.

July 1908: receives Darwin–Wallace Medal from the Linnean Society of London.

December 1908: receives the Copley Medal from the Royal Society and the Order of Merit from the Crown.

January 1909: gives lecture at the Royal Institution.

December 1910: *The world of life* (ref 732) published.

7 November 1913: dies at Old Orchard, Broadstone, Dorset.

Selected secondary literature

Beddall, Barbara G. (1968). Wallace, Darwin, and the theory of natural selection. *J. Hist. Biol.* 1, 261–323.

Beddall, Barbara G. (ed.) (1969). *Wallace and Bates in the tropics.* Macmillan, London.

Beddall, Barbara G. (1972). Wallace, Darwin, and Edward Blyth: further notes on the development of evolutionary theory. *J. Hist. Biol.* 5, 153–8.

Brackman, Arnold C. (1980). *A delicate arrangement; the strange case of Charles Darwin and Alfred Russel Wallace.* Times Books, New York.

Brooks, John L. (1972). Extinction and the origin of organic diversity. *Trans. Conn. Acad. Arts & Sciences* 44, 17–56.

Brooks, John L. (1984). *Just before the origin: Alfred Russel Wallace's theory of evolution.* Columbia University Press, New York.

Browne, Janet (1983). *The secular ark; studies in the history of biogeography.* Yale University Press, New Haven & London.

Clements, Harry (1983). *Alfred Russel Wallace; biologist and social reformer.* Hutchison & Co., London, etc.

Cooter, Roger (1984). *The cultural meaning of popular science; phrenology and the organization of consent in nineteenth-century Britain.* Cambridge University Press, Cambridge, etc.

Durant, John R. (1979). Scientific naturalism and social reform in the thought of Alfred Russel Wallace. *Brit. J. Hist. Science* 12, 31–58.

Eiseley, Loren (1958). *Darwin's century; evolution and the men who discovered it.* Doubleday, New York.

Fichman, Martin (1977). Wallace: zoogeography and the problem of land bridges. *J. Hist. Biol.* 10, 45–63.

Fichman, Martin (1981). *Alfred Russel Wallace.* Twayne Publishers, Boston.

George, Wilma (1964). *Biologist philosopher; a study of the life and writings of Alfred Russel Wallace.* Abelard-Schuman, London.

George, Wilma (1981). Wallace and his line. In Whitmore, T. C. (ed.), *Wallace's Line and plate tectonics* (Oxford University Press, Oxford, etc.), pp. 3–8.

Gould, Stephen Jay (1980). Wallace's fatal flaw. *Natural History* 89, 26–40.

Henderson, Gerald M. (1958). Alfred Russel Wallace: his role and influence in nineteenth century evolutionary thought. University of Pennsylvania (Ph.D. thesis).

Hogben, Lancelot (1918). *Alfred Russel Wallace, the story of a great discoverer.* Society for Promoting Christian Knowledge, London.

Kleiner, Scott A. (1985). Darwin's and Wallace's revolutionary research programme. *Brit. J. Phil. Sci.* 36, 367–92.

Kottler, Malcolm J. (1974). Alfred Russel Wallace, the origins of man, and spiritualism. *Isis* 65, 145–92.

Loewenberg, Bert J. (1959). *Darwin, Wallace, and the theory of natural selection.* Arlington Books, Cambridge, Mass.

McKinney, H. Lewis (1966). Alfred Russel Wallace and the discovery of natural selection. *J. Hist. Med. & Allied Sciences* 21, 333–57.

McKinney, H. Lewis (1969). Wallace's earliest observations on evolution: 28 December 1845. *Isis* 60, 370–3.

McKinney, H. Lewis (1972). *Wallace and natural selection*. Yale University Press, New Haven & London.

McKinney, H. Lewis (1976). Wallace, Alfred Russel. *Dictionary of scientific biography* (Charles Scribner's Sons, New York), Vol. 14, pp. 133–140.

Malinchak, Michele (1987). Spiritualism and the philosophy of Alfred Russel Wallace. Drew University (Ph.D. thesis).

Marchant, James (ed.) (1916, 1975). *Alfred Russel Wallace; letters and reminiscences*. Arno Press, New York (reprint of the 1916 Harper & Brothers edn.).

Nelson, Gareth (1978). From Candolle to Croizat: comments on the history of biogeography. *J. Hist. Biol.* 11, 269–305.

Oppenheim, Janet (1985). *The other world; spiritualism and psychical research in England, 1850–1914*. Cambridge University Press, Cambridge, etc.

Osborn, Henry Fairfield (1924). *Impressions of great naturalists*. Charles Scribner's Sons, New York & London.

Peirce, Charles S. (1906). *My life* [book review]. *The Nation* 82, 160–1.

Remington, Charles (1963). Historical backgrounds to mimicry. *Proc. XVI Int. Congress of Zoology* 4, 145–9.

Schadewald, R. (1978). He knew earth is round, but his proof fell flat. *Smithsonian* 9, 101–13.

Schwartz, Joel S. (1984*a*). Darwin, Wallace, and *The descent of man*. *J. Hist. Biol.* 17, 271–89.

Schwartz, Joel S. (1984*b*). Alfred Russel Wallace and "Leonainie": a hoax that would not die. *Victorian Periodicals Rev.* 17, 3–15.

Smith, Charles H. (1984). Alfred Russel Wallace. *Geographers Biobibliographical Studies* (Mansell, London), Vol. 8, 125–33.

Smith, Charles H. (1989). Historical biogeography: geography as evolution, evolution as geography. *New Zealand J. Zool.* 16, 773–785.

Smith, Roger (1972). Alfred Russel Wallace: philosophy of nature and man. *Brit. J. Hist. Science* 6, 177–99.

Tuan, Yi-Fu (1963). Latitude and Alfred Russel Wallace. *J. Geography* 62, 258–61.

Turner, Frank M. (1974). *Between science and religion; the reaction to scientific naturalism in late Victorian Britain*. Yale University Press, New Haven & London.

Vorzimmer, Peter (1970). *The years of controversy:* The origin of species *and its critics*. Temple University Press, Philadelphia.

Williams-Ellis, Anabel (1966). *Darwin's moon; a biography of Alfred Russel Wallace*. Blackie, London.

Young, Robert M. (1970). *Mind, brain and adaptation in the nineteenth century: cerebral localization and its biological context from Gall to Ferrier*. Clarendon Press, Oxford.

Index

Entries in the Index are followed by either, or both, Bibliography reference numbers (underlined) or page listings. Note that general subject entries are followed by references to related material appearing in the Selections portion of the book; the subject indices presented at the end of each Section should be consulted for referrals to the content of Bibliography entries.